LIBRARY

ASTRONOMY AND
ASTROPHYSICS LIBRARY

Springer
Berlin
Heidelberg
New York
Barcelona
Budapest
Hong Kong
London
Milan
Paris
Singapore
Tokyo

ASTRONOMY AND
ASTROPHYSICS LIBRARY

Series Editors: I. Appenzeller · G. Börner · M. Harwit · R. Kippenhahn
P. A. Strittmatter · V. Trimble

Theory of Orbits (2 volumes)
Volume 1: Integrable Systems and Non-perturbative Methods
Volume 2: Perturbative and Geometrical Methods
By D. Boccaletti and G. Pucacco

Galaxies and Cosmology
By F. Combes, P. Boissé, A. Mazure and A. Blanchard

The Solar System 2nd Edition   By T. Encrenaz and J.-P. Bibring

Nuclear and Particle Physics of Compact Stars
By N. K. Glendenning

The Physics and Dynamics of Planetary Nebulae   By G. A. Gurzadyan

Astrophysical Concepts 2nd Edition   By M. Harwit

Stellar Structure and Evolution   By R. Kippenhahn and A. Weigert

Modern Astrometry   By J. Kovalevsky

Supernovae   Editor: A. Petschek

General Relativity, Astrophysics, and Cosmology
By A. K. Raychaudhuri, S. Banerji and A. Banerjee

Tools of Radio Astronomy   2nd Edition
By K. Rohlfs and T. L. Wilson

Atoms in Strong Magnetic Fields
Quantum Mechanical Treatment and Applications
in Astrophysics and Quantum Chaos
By H. Ruder, G. Wunner, H. Herold and F. Geyer

The Stars   By E. L. Schatzman and F. Praderie

Gravitational Lenses   By P. Schneider, J. Ehlers and E. E. Falco

**Relativity in Astrometry, Celestial Mechanics
and Geodesy**   By M. H. Soffel

The Sun An Introduction   By M. Stix

Galactic and Extragalactic Radio Astronomy   2nd Edition
Editors: G. L. Verschuur and K. I. Kellermann

Reflecting Telescope Optics (2 volumes)
Volume I: Basic Design Theory and its Historical Development
Volume II: Manufacture, Testing, Alignment, Modern Techniques
By R. N. Wilson

Galaxy Formation   By M. S. Longair

Astrophysical Formulae   3rd Edition
By K. R. Lang

Malcolm S. Longair

Galaxy Formation

With 141 Figures and 12 Tables

 Springer

Prof. Malcolm S. Longair
University of Cambridge, Department of Physics
Cavendish Laboratory, Mendengley Road
Cambridge CB3 OHE
United Kingdom

Cover picture: A Hubble Space Telescope image of the cluster of galaxies CL1358 + 62 at a redshift z = 0.33. Towards the bottom of the picture, there is a red arc, which is the gravitationally lensed image of a pair of very distant galaxies that have been lensed by the cluster of galaxies. The redshift of this pair of galaxies is z = 4.92; they are the most distant galaxies known. (Courtesy of M. Franx, G. Illingworth, the Space Telescope Science Institute and NASA.)

ISSN 0941-7834
ISBN 3-540-63785-0 Springer-Verlag Berlin Heidelberg New York

Library of Congress Cataloging-in-Publication Data.
Longair, M. S., 1941– Galaxy formation / Malcolm S. Longair. p. cm. – (Astronomy and astrophysics Library) Includes bibliographical references and index. ISBN 3-540-63785-0 (hc.) 1. Cosmology. 2. Galaxies–Formation. 3. Astrophysics. I. Title. II. Series. QB981.L846 1998 523.1–dc21 98-29744 CIP

Typesetting: Camera ready by authors
Cover design: *design & production* GmbH, Heidelberg
SPIN: 10022216   55/3142 – 5 4 3 2 1 0 – Printed on acid-free paper

Foreword

For Deborah, Mark and Sarah

'Not another book on cosmology!', I hear the reader exclaim. 'Surely there are quite enough books on cosmology to satisfy everyone's needs?'

I was asked by Springer-Verlag to expand into a full-length book the set of lecture notes that I prepared in 1988 for the First Astrophysics School organised by the European Astrophysics Doctoral Network. The set of notes was entitled *Galaxy Formation* and was published as a chapter of the volume *Evolution of Galaxies: Astronomical Observations* (eds. I. Appenzeller, H.J. Habing and P. Lena, pages 1 to 93, Springer-Verlag Berlin, Heidelberg, 1989). In that chapter, I attempted to bridge the gap between elementary cosmology and the technical papers appearing in the literature, which can seem quite daunting on first encounter. The objective was to present the physical concepts and key results as clearly as possible as an introduction and guide to the technical literature.

The revision of these lecture notes into a full-length book was delayed by other projects. Specifically, I am completing a three-volume work for Cambridge University Press, entitled *High Energy Astrophysics*, (Volume 1, 1992; Volume 2, 1994; Volume 3, Cambridge University Press, Cambridge 1998). In addition, a further series of lecture notes on *The Physics of Background Radiation* was prepared for the 1993 23rd Advanced Course of the Swiss Society of Astrophysics and Astronomy, the topic of which was *The Deep Universe* (A.R. Sandage, R.G. Kron and M.S. Longair, Springer-Verlag Berlin, Heidelberg, 1995). Finally, I have completed a history of twentieth century astrophysics and cosmology, which has been published as Chap. 23 of a three-volume work entitled *Twentieth Century Physics* (eds. L.M. Brown, A. Pais and A.B. Pippard, IOP Publications, AIP Press Bristol, and New York 1995). It will be published in enlarged form as a full-length book by Cambridge University Press in 1999.

All these works contain material central to the problems of galaxy formation and it therefore seemed a good idea to abstract from them an improved version of the original plan for the book on galaxy formation. A revised plan was all the more necessary because of the quite remarkable developments in observational and theoretical cosmology which have taken place in the decade since the original set of notes was written. Thus, the present volume is much more than a recycled and concatenated version of published works. I have rewritten and rethought the original versions, expanded some parts, brought everything up-to-date and included new material.

I often find that I understand things best, and present them most clearly, when I have to prepare them for students, at either the undergraduate or the post-graduate level, and so I have adopted the same form of presentation here. I have intentionally presented the material in an informal, pedagogical manner, and tried to avoid getting bogged down in formalities and technicalities. If the material becomes too difficult, I simply summarise the key points, give some appropriate references and pass on. My approach is to reduce the problems to their simplest form and rationalise from these examples the results of more complete analyses. Wherever it is feasible without excessive effort, we will attempt to derive exact results. The level of presentation is intended to be appropriate for a final-year undergraduate or first-year post-graduate course of lectures. In other words, it is assumed that the reader has a good grasp of basic physics but does not necessarily have the appropriate background in astronomy, astrophysics or cosmology. My aim has been to write a user-friendly book, taking particular care to expound carefully areas where I have found students have difficulty.

When I wrote the original set of lecture notes on galaxy formation, my objective was to tell the story of modern astrophysical cosmology from the perspective of one of its most important and fundamental problems of cosmology — how did the galaxies come about? I enjoy this approach to the exposition of modern cosmology because, to do the problem justice, it is essential to introduce the whole of what I call *classical cosmology*, as the framework for the discussion. This approach has, for me, the great advantage of concentrating upon a crucial problem of astrophysical cosmology rather than regarding the objective of cosmology as being simply the delineation of a preferred cosmological model, however interesting that is in its own right. As we will show, the origin of galaxies is one of the great cosmological problems and it can potentially provide us with unique and direct information about the physics of the very early Universe.

One final warning is in order. I make no claim that this presentation is complete, unbiased or objective. You should regard the book as my own impressions and opinions of what I consider to be the important issues of modern astrophysical cosmology. Others would tell the story in a completely different way and put emphasis upon different parts of what is unquestionably a multi-dimensional story. I will endeavour to include as wide a spectrum of ideas and opinions as possible but the text will inevitably be incomplete. I do not worry about this – it should encourage you to read as widely as possible in order to neutralise my prejudices and biases.

Good Luck!

Acknowledgements

Many people have contributed directly, or indirectly, to my understanding of the contents of this book. Perhaps the most important influence has been Peter Scheuer, who first introduced me to the physics of astrophysical cosmology. His approach and methods have very strongly influenced the way I have understood and taught this material over the years. I am very grateful to Immo Appenzeller, Harm Habing and Pierre Léna for the opportunity to give the lecture course in Les Houches in 1988. In preparing that set of lecture notes, I greatly benefitted from the advice of John Peacock and Alan Heavens who read parts of the typescript and offered very helpful comments. John Peacock very kindly allowed me to use part of his lecture notes in preparing some of the material for the original chapter. John has now written up his own notes in book form and it will soon be published by Cambridge University Press under the title *Cosmological Physics*. In my view, John's book is a brilliant achievement and I urge all interested readers to become familiar with his deep insights. Some of the text of the present book is based upon joint work with Rashid Sunyaev which dates from the period 1968 to 1980. I fully acknowledge Rashid's contributions to clarifying my own understanding.

The invitation to deliver the 1993 course of lectures on the background radiation, as part of the 23rd Course of the Swiss Society of Astrophysics and Astronomy, came from Gustav Tammann, Bruno Binggeli and Hermann Buser and I am grateful for their kindness and hospitality at Les Diablerets. The history of 20th Century Astrophysics and Cosmology was commissioned by Brian Pippard on behalf of the editors of *20th Century Physics* and I am grateful to him for his perceptive comments on that article.

I am particularly grateful to Bob Williams and his colleagues at the Space Telescope Science Institute at Baltimore, where I was a Visiting Fellow from September to December 1997. Without this sabbatical term at the Institute, the completion of this text would have been impossible. Special thanks are due to the staff, graduate students and research fellows at the Institute and at Johns Hopkins University, who kindly acted as guinea pigs on whom I 'battle-tested' portions of this book. Many of the research workers at the Institute gave generously of their time in discussing many of the topics discussed in the text; the discussions with Ron Allen, Michael Fall, Harry Fergusson, Mario Livio, Duccio Macchetto and Piero Madau were especially helpful. The writing was enormously aided by access to the excellent library facilities at the Institute and I am most grateful to Sarah Stevens-Rayburn and her colleagues, especially 'Chopin', for their kind assistance. Martin Harwit kindly read the book very carefully and made a number of helpful suggestions, which I have gratefully incorporated into the final text.

Finally, the book is dedicated to my family, Deborah, Mark and Sarah, whose constant love, support and patience have made it possible.

Summer 1998   *Malcolm Longair*, Cambridge and Baltimore.

Contents

X   Contents

Part IV The Post-Recombination Universe

Part I

Preliminaries

1 Introduction, History and Outline

We begin with a brief historical overview of the development of ideas and concepts concerning galaxies, cosmology and galaxy formation. In doing so, this chapter summarises qualitatively many of the topics dealt with in this book and we will return to all of them in the succeeding chapters. If you do not need this gentle introduction, or misguidedly think that history is boring, you may pass straightaway to Chap. 2[1].

1.1 Prehistory

It always comes as a surprise to me to realise how recent the history of galaxies, cosmology and galaxy formation is. The Sun, Moon, planets and the stars have been known from ancient times, but the scientific study of their motions in the modern sense only began in the 16th century. The precise determination of the motions of the planets about the Sun and the resulting discovery of Newton's Law of Gravity is one of the first great achievements of what can be generally recognised as modern science.

The origin of these developments can be traced to the technological and observational achievements of *Tycho Brahe* who, in the period 1576 to about 1596, measured the positions of the Sun, Moon, planets and the stars with unprecented accuracy. I have told this remarkable story elsewhere (Longair 1991). To summarise his great achievement, Tycho Brahe measured the positions of the Sun, Moon, planets and 777 stars over a period of 20 years, resulting in an order of magnitude improvement in accuracy over all previous measurements. In his last years, he employed *Johannes Kepler* as his assistant and his task was to work out the orbits of the planets from Tycho's data. In the period 1601 to 1619, Kepler interpreted the mass of Tycho's data on the motions of the planets in terms of elliptical orbits about the Sun which was located in one of the foci of each ellipse. Kepler's discovery of the three laws of planetary motion is a miracle of geometrical analysis. The three laws embody not only the elliptical orbits of the planets (the first law), but also

---

[1] Comprehensive references to the original papers discussed in this chapter can be found in my chapter *Astrophysics and Cosmology* in the three-volume work *Twentieth Century Physics* (Longair 1995)

the areal law – that equal areas are swept out by the radius vector from the Sun to the planet in equal times (the second law) – and the dependence of the period T of the planet's orbit about the Sun upon the three-halves power of its mean distance r from the Sun, $T \propto r^{3/2}$, (the third law).

In 1664 at the age of only 21, *Isaac Newton* first derived from Kepler's third law his law of gravity which we will use repeatedly throughout this volume

$$\mathbf{f} = -G\frac{M_1 M_2}{r^2}\,\mathbf{i}_r, \tag{1.1}$$

where $\mathbf{f}$ is the gravitational force acting between two point masses M_1 and M_2 separated by a distance r and $\mathbf{i}_r$ is the unit vector in the direction from one mass to the other. It is no exaggeration to say that astronomy, astrophysics and cosmology are the sciences of gravity – all the systems we study in astronomy and cosmology are attempting to counteract the omnipresent attractive force of gravity by one means or another.

In 1692 *Richard Bentley* gave the first series of Boyle Lectures which Robert Boyle had founded 'to combat atheism'. Bentley took as his theme Newton's 'sublime discoveries' and entered into a short but profound correspondence with Newton about the nature of our physical Universe. The question which they discussed was the stability under gravity of a finite or infinite Universe filled with stars. The conclusion of the correspondence was that the Universe must be infinite because, if it were not, it would collapse to the centre under the influence of the attractive force of gravity. They recognised, however, with remarkable physical insight, that an infinite Universe filled with stars is gravitationally unstable. To quote Harrison (1987), '(Newton) agreed with Bentley that providence had designed a universe of infinite extent in which uniformly distributed stars stand poised in unstable equilibrium like needles on their points'. It was only in the 20th century that the nature of this instability was fully appreciated. For a static medium, the stability criteria and the growth rate of the instability were derived by James Jeans in 1902 and for an expanding medium by Evgenii Lifshitz in 1946. These results will dominate much of the discussion of this book. As part of the dialogue with Bentley, Newton proposed that the stars are objects like the Sun and he made star counts in an attempt to show that the stars are uniformly distributed in space. He had, however, no way of measuring their distances. This had to await the measurement of the parallaxes of nearby stars, which was first announced in 1838 by the German astronomer *Friedrich Bessel* for the star 61 Cygni, although the first parallax measurements had been made by Thomas Henderson in 1832 for the star Alpha Centauri at the Cape Observatory.

In *Galileo Galilei's* extraordinary text of 1609, *Sidereus Nuncius* or *The Siderial Messenger*, he demonstrated that the Milky Way can be resolved into stars when observed through the telescope. The first proposal that the Milky Way is a flattened disc of stars was made by *Thomas Wright* of Durham and was reinforced by the work of *William Herschel* (Fig.1.1). The definition of

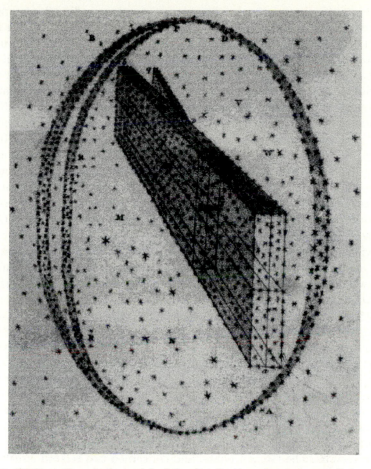

Fig. 1.1. William Herschel's model of the distribution of stars in our Galaxy (Herschel 1785). The Sun is located towards the centre of the flattened rectangular disc.

its structure was based upon star counts, assuming that all the stars were similar to our Sun, and so was hampered by the lack of distances for any of the stars, as well as by the fact that the effects of the scattering and absorption of light by interstellar dust was not appreciated – the full significance of dust extinction was only appreciated in the 1930s.

Even before the discovery of the telescope, it had been realised that there exist 'nebulous' objects which differ from the stars in having a diffuse or fuzzy appearance. During the eighteenth century, there had been various arguments of a philosophical nature due to Kant, Lambert, Swedenborg and Wright that these objects were 'island Universes' similar to the Milky Way, but too distant to be resolved into stars. There was, however, no observational basis for this hypothesis.

The cataloguing of the bright nebulae was begun by *Charles Messier* whose catalogue of 109 objects was compiled during the years 1771 to 1784. Messier's interest was primarily in comets and his objective in compiling the catalogue was to enable him to distinguish between diffuse nebulae and comets. The catalogue contains a mixture of what we now know to be among the brighter Galactic and extragalactic nebulae and these objects are still often referred to by their Messier numbers. For example, the Orion Nebula, the region of massive star formation nearest to the Earth is M42, the Crab Nebula, which we now know was a supernova which exploded in 1054, is M1, and the Andromeda Nebula, the nearest giant spiral galaxy to our own Galaxy, is catalogued as M31. The systematic cataloguing of the nebulae was begun by William Herschel and his sister Caroline and continued into the nineteenth century by his son *John Herschel*. The results of these huge endeavours was the publication by John Herschel in 1864 of the *General Catalogue of Galaxies* containing 5079 objects. These catalogues were based upon visual observations made before photography became a standard tool of the astronomer. In 1888, *John Dreyer* published an expanded catalogue which was known as the *New General Catalogue of Nebulae and Clusters of Stars* which, together with the two supplementary *Index Catalogues* contain some 15,000 objects. These objects are still commonly referred to by their NGC or IC numbers.

While the cataloguing of the nebulae proceeded apace, their nature remained a mystery. Undoubtedly, some of them were gas clouds, as demonstrated by William Huggins' pioneering spectroscopic observations of diffuse nebulae. The big question was whether or not the 'spiral nebulae' were objects within our own Galaxy or more distant systems. The problem was that these nebulae were beyond the distances at which conventional techniques of distance measurement could be used. This problem culminated in what became known as 'The Great Debate' and concerned two related issues. First, what is the size of our own Galaxy and, second, are the spiral nebulae members of our Galaxy or are they separate 'island universes', well beyond the confines of our Galaxy? This key episode in the history of modern astronomy should be required reading for all observers and theorists (see Sandage 1961, Hoskin 1976, Smith 1982).

I will not review that fascinating story here but simply note that the issue was resolved finally and conclusively in 1925 by *Edwin Hubble's* observations of Cepheid variables in the Andromeda Nebula. Using the very tight correlation between the periods and luminosities of Cepheid variables discovered by *Henrietta Leavitt* in 1912, he established to everyone's satisfaction that the spiral nebulae are distant extragalactic systems. Within a year, Hubble had published the first major survey of the properties of galaxies as extragalactic systems. In his remarkable paper of 1926, Hubble used number counts of galaxies to show that they are uniformly distributed in space and estimated the mean density of matter in the Universe in the form of galaxies. Adopting

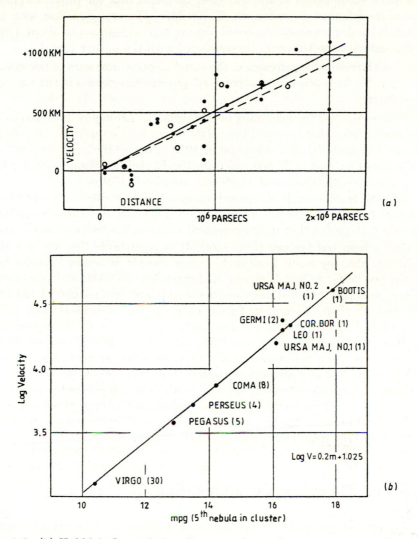

Fig. 1.2. (a) Hubble's first velocity–distance relation for nearby galaxies (Hubble 1929). The filled circles and the full line represent a solution for the solar motion using the nebulae individually; the open circles and the dashed line is a solution combining the nebulae into groups. **(b)** The velocity–apparent magnitude relation for the fifth brightest members of clusters of galaxies, corrected for galactic obscuration (Hubble and Humanson 1934). Each cluster velocity is the mean of the radial velocities of galaxies in the cluster, the number of galaxies being indicated by the figure in brackets. (See Sect. 2.3 for a discussion of the use of apparent magnitudes to measure distances in this relation.)

Einstein's static model of the Universe, he found that the radius of curvature of its spherical geometry was 27,000 Mpc. He estimated that with the 100-inch Hooker telescope, he could observe typical galaxies to about 1/600 of the radius of the Universe. He concluded with the remark that

> ... with reasonable increases in the speed of plates and sizes of telescopes, it may become possible to observe an appreciable fraction of the Einstein universe.

This paper marked the beginning of extragalactic astronomy. It comes as no surprise to learn that George Ellery Hale began his campaign to raise funds for the Palomar 200-inch telescope in 1928 – before the year was out, he had secured a grant of $6,000,000 from the Rockefeller Foundation for the telescope, the construction of which was completed in 1949.

In 1929, Hubble made his second fundamental contribution to cosmology in which he showed that the extragalactic nebulae are all moving away from our own Galaxy and that their recessional velocities are proportional to their distances from our Galaxy (Fig. 1.2a). It is remarkable that he was able to deduce this key result from such a small sample of nearby galaxies but, within five years, he and Humason had extended the relation to very much greater velocities and distances using the apparent magnitudes of the fifth brightest members of clusters of galaxies as distance indicators (Fig. 1.2b). The velocity–distance relation is commonly referred to as *Hubble's Law*. The significance of this discovery was that, combined with the isotropy of the Universe, Hubble's law demonstrates that the whole system of galaxies is dispersing as part of a uniform expansion. This was the first great discovery of modern observational cosmology and we will have a great deal to say about it in due course. This feature, the expansion of the Universe, turns out to be the cause of many of the problems in our attempts to understand how the galaxies and the large-scale structure of the Universe formed.

1.2 The Theory of the Expanding Universe

As we have noted, in the 1690s Newton and Bentley conducted a thought-provoking correspondence about the stability of the Universe under gravity on the large scales. Another aspect of Newton's views was to lead to a long-standing debate which has not yet been fully resolved. In all his writings, Newton emphatically took the position that all motion takes place with respect to a system of absolute space and time. The idea that the motion of a body could only be described relative to other bodies was fiercely rejected by him. His position was challenged by Bishop Berkeley, Christiaan Huygens and others but, at least until the late nineteenth century, Newton's view prevailed. The issue was revived by *Ernst Mach* who argued that motion can only be defined relative to other bodies. Specifically, he took the view that the local inertial frame of reference is determined by the frame of the distant stars, or galaxies, as we would now say. Thus, a freely swinging Foucault pendulum swings in a reference frame which is fixed relative to the distant galaxies. Einstein gave the name *Mach's Principle* to this idea.

Newton and his sucessors did not have the tools with which to address this proposition but, by 1915, *Albert Einstein* did. In that year, after a long struggle, he discovered the definitive version of his General Theory of Relativity which describes how space-time is distorted by the presence of matter and how, in turn, matter moves along trajectories in bent space-time. This theory enabled self-consistent models of the Universe as a whole to be constructed for the first time and, characteristically, Einstein did not hesitate to do so. In seeking a solution to his field equations for the Universe, he had explicitly in mind that it should be possible to incorporate Mach's principle into a model of the large-scale structure of the Universe. He had, however, a major problem. Without modification, his field equations predicted that the Universe was unstable. He could only find static solutions by introducing what is now known as the *cosmical* or *cosmological constant* which appears as a constant in Einstein's field equations – in classical General Relativity, it is set equal to zero. In his great paper of 1917, Einstein believed that the introduction of the cosmological constant, not only provided static solutions for our Universe, but also enabled Mach's Principle to be incorporated into General Relativity, in the sense that no solution of the equations would exist if there were no matter present. This was, however, shown to be incorrect by *Willem de Sitter*, who found solutions of the equations even if there were no matter present in the Universe.

The irony of the situation is that this debate took place *before* it was realised that the Universe is in fact non-stationary. In 1922, *Aleksander Friedman*[2] published the first of two classic papers in which he discovered both static and expanding solutions of Einstein's field equations. In the first paper, Friedman found solutions for expanding universes with closed spatial geometries, including those which expand to a maximum radius and eventually collapse to a singularity. In the second paper of 1924, he showed that there exist expanding solutions which are unbounded and which have hyperbolic geometry. These solutions correspond exactly to the standard world models of general relativity and are known as the *Friedman world models*.

In 1925, Friedman died of typhoid in Leningrad before the fundamental significance of his work was appreciated. The neglect of Friedman's work in these early days is somewhat surprising since Einstein had commented, incorrectly as he admitted, on the first of the two papers in 1923. It was not until *Georges Lemaître* independently rediscovered the same solutions in 1927, and then became aware of Friedman's papers, that the pioneering nature of Friedman's contributions was appreciated.

---

[2] My preferred spelling of Friedman's name is with one 'n'. The transliteration of his name from his native Russian is Fridman; the German transliteration, which was used in his papers of 1922 and 1924 in *Zeitschrift für Physik*, was Friedmann.

Einstein's field equations without the cosmological constant contain perfectly satisfactory solutions in which the Universe is uniformly expanding. The story goes that, when the expansion of the Universe was discovered, Einstein regarded the introduction of the cosmological constant as 'the biggest blunder of my life'. The cosmological constant was not consigned to oblivion for long however. As Zeldovich remarked, 'The genie is out of the bottle and, once he is out, he is very difficult to put back in again.' The cosmological constant immediately found a rôle in reconciling the age of the Earth with the expansion age of the Universe as given by the inverse of Hubble's constant H_0^{-1}. If the cosmological constant is zero, all Friedman models of the Universe have ages less than H_0^{-1}. At that time, Hubble's estimate of H_0 was about 500 km s^{-1} Mpc^{-1}, corresponding to $H_0^{-1} = 2 \times 10^9$ years. This time-scale was less than the age of the Earth as determined by nucleocosmochronology, that is, from dating using long-lived radioactive isotopes. A positive value of the cosmological constant can resolve this discrepancy since its effect is to stretch out the expansion time-scale of the Universe, a picture advocated by Eddington and Lemaître. It turned out that Hubble's estimate of H_0 was seriously overestimated and, following revisions in the 1950s by Baade and Sandage, this conflict was eliminated. This is, however, an ongoing saga and the same problem may have recurred in reconciling the present best estimates of the expansion age of the Universe with the ages of the oldest stars. We will investigate some of the other reincarnations of the cosmological constant, perhaps the most intriguing being its role in the inflationary model of the early Universe.

As the standard models of General Relativity became better understood, a major thrust of cosmological research became the determination of the large-scale dynamical and geometrical properties of the Universe – its rate of expansion, its deceleration, its mean density, its geometry and its age. These remain among the most difficult programmes of modern observational cosmology.

1.3 The Big Bang

The next major advance occurred soon after the Second World War when *George Gamow* realised that, in an expanding Universe, the early stages must have been very hot indeed – the temperature became so high that the dynamics of the expansion were dominated by the energy density of thermal radiation rather than by its matter content – the universe was radiation-dominated. In his work of the late 1940s, he attempted to explain the origin of the chemical elements by primordial nucleosynthesis, that is, by nuclear fusion processes as the Universe cooled down from its very hot initial stages. This programme was not successful because of the problem of synthesising elements heavier than helium – there are no stable isotopes with atomic mass numbers 5 and 8. His coworkers *Ralph Alpher* and *Robert Herman*, however, made the prescient prediction that there should be present in the Universe now a background of black-body radiation with temperature about 5 K, the cooled remnant of the hot early phases. The detection of this background radiation was far beyond the capabilities of the technology of the 1940s and the lack of success of Gamow's programme of primordial nucleosynthesis resulted in the neglect of this key prediction for a number of years. Furthermore, in the 1950s, Fred Hoyle discovered the triple-α resonance, which leads to the formation of carbon from three helium nuclei. Soon after, he and his colleagues, Burbidge, Burbidge and Fowler, showed how the heavy elements could be accounted for by nucleosynthesis in stars.

Interest in what is now referred to as the *Big Bang* model of the Universe, however, grew steadily through the 1950s and early 1960s as evidence was found for cosmological evolutionary effects in the distribution of faint radio sources. On the theoretical side, interest was rekindled in the question of the synthesis of elements in the early Universe, not now with the objective of creating all the elements, but rather to account for the cosmic abundance of helium. By 1964, it was appreciated that, wherever helium could be observed in the Universe, it is present in remarkably high proportions, about 24% by mass. This figure far exceeded what could be produced by stellar nucleosynthesis. I remember very vividly attending a course of lectures given by Fred Hoyle in Cambridge in 1964 on Problems of Extragalactic Astrophysics in which this problem was discussed. During the lecture course, Fred Hoyle, Roger Tayler, and John Faulkner carried out detailed calculations of the expected abundance of helium produced by primordial nucleosynthesis. Within three weeks of the topic being raised, they had shown that about 23 to 25% of helium by mass is created by this process and that the percentage is remarkably independent of the precise initial conditions. The paper by *Fred Hoyle* and *Roger Tayler* was published in *Nature* in 1964. Subsequent more detailed calculations by Wagoner, Fowler and Hoyle confirmed these conclusions and suggested that other elements which are difficult to account for by stellar nucleosynthesis, specifically, the abundances of the light isotope of helium, ^{3}He, deuterium, D, and lithium, ^{7}Li, could also be accounted for in this way.

By the early 1960s, it became feasible to search for the cool background radiation left over from the early stages of the Big Bang as the sensitivity of receivers at centimetre wavelengths improved. The predicted remnant of the Big Bang was discovered, more or less by accident, by *Arno Penzias* and *Robert Wilson* in 1965. The *Cosmic Microwave Background Radiation* was the second great discovery of 20th century observational cosmology. Since its discovery, it has been shown that the radiation is quite remarkably uniform over the sky and that its spectrum is of black-body form with a quite remarkable precision. These observations provide very convincing evidence that our Universe went through a very hot, dense phase when the matter and radiation were in thermal equilibrium in its early stages.

The upshot of these discoveries was that there are now four independent pieces of evidence for Big Bang picture of the origin and evolution of our Universe. First, the expansion of the distribution of galaxies discovered by Hubble; second, the black-body spectrum and isotropy of the Cosmic Microwave Background Radiation; third, the formation of the light elements by primordial nucleosynthesis; and fourth, the age of the oldest stars and nucleochronology ages are of the same order as the expansion age of the Universe. Thus, the Big Bang provides a natural framework within which to tackle the problems of galaxy formation.

1.4 Galaxy Formation

The Friedman world models are isotropic and homogeneous and so the enormous diversity of structure we observe in the Universe is absent. The next step in developing more realistic models of the Universe is to include small density perturbations into the homogeneous models and study their development under gravity. This problem was solved by *James Jeans* in 1902 for the case of a stationary medium. The criterion for collapse is that the size of the perturbation should exceed the *Jeans' length* $\lambda_J = c_s/(G\varrho_0/\pi)^{1/2}$, where c_s is the speed of sound in the medium and ϱ_0 its density. The significance of the instability criterion is that, on large enough scales, the gravitational force of attraction by the matter of the perturbation exceeds the pressure gradients which resist collapse.

The analysis was repeated for the case of an expanding medium in the 1930s by Lemaître and by Tolman for the case of spherically symmetric perturbations and the solution for the general case was found by *Evgenii Lifshitz* in 1946. Lifshitz found that the condition for gravitational collapse is exactly the same as the Jeans' criterion at any epoch but, crucially, the growth-rate of the density perturbation is no longer exponential but only algebraic. In the case in which the Universe has the critical density, $\Omega_0 = 1$, the density contrast $\Delta = \delta\varrho/\varrho$ grows with time as $\Delta \propto t^{2/3}$. The implication of this result is that the fluctuations from which the large-scale structure of the Universe formed cannot have grown from infinitesimal perturbations. For this reason, Lemaître, Tolman and Lifshitz inferred that galaxies could not have formed by gravitational collapse.

Other authors took the point of view that the solution to the problem was to include finite perturbations into the model of the early Universe and then to follow in detail how their spectrum would evolve with time. In the 1960s, the Moscow school led by *Yakov Zeldovich, Igor Novikov* and their colleagues and *James Peebles* at Princeton pioneered the study of the development of structure in the Universe. If perturbations on a particular physical scale are tracked backwards into the past, it is found that, at some large redshift, the scale of the perturbation is equal to the horizon scale, that is $r \approx ct$, where t is the age of the Universe. In 1964, Novikov showed that, to form structures on the scales of galaxies and clusters of galaxies, the density perturbations on the scale of the horizon had to have amplitude $\Delta = \delta\varrho/\varrho \sim 10^{-4}$ in order to guarantee the formation of galaxies by the present epoch. These were certainly *not* infinitesimal perturbations and their origin had to be ascribed to processes occurring in the very early Universe.

The discovery of the *Cosmic Microwave Background Radiation* in 1965 had an immediate impact upon these studies since the thermal history of the pregalactic gas could be worked out in detail and this was essential in order to determine how the speed of sound, and hence the Jeans' length, varied with cosmic epoch. If there is no energy input into the background radiation, the temperature of the thermal background changes with scale factor as $T = T_0/R = T_0(1 + z)$, exactly as in the adiabatic expansion of a photon gas. Therefore, at redshifts $z \sim 1500$, the temperature of the radiation was about 4,000 K, at which there were sufficient photons in the tail of the Planck distribution to ionise all the intergalactic hydrogen. This epoch is referred to as the *epoch of recombination* and at earlier epochs the hydrogen was fully ionised; at correspondingly earlier epochs, the primordial helium was ionised as well. At a still earlier epoch, the inertial mass density of the radiation was equal to the mass density of the matter, $\varrho c^2 = aT^4$, and before this time the dynamics of the Universe were radiation-dominated.

The coupling of matter and radiation by electron scattering was worked out by *Ray Weymann* in 1966 and in much more detail by Zeldovich and *Rashid Sunyaev* in 1969. The pioneering papers by Zeldovich and Sunyaev were based upon the theory of induced Compton scattering developed by Kompaneets which had been published in 1956, long after this remarkable classified work had been completed. What these papers showed was that, during the radiation-dominated epochs, the matter and radiation were maintained in very close thermal contact by Compton scattering as long as the intergalactic gas remained ionised. This enabled the speed of sound to be determined at all epochs before the epoch of recombination and hence the evolution of the Jeans' length and the mass of baryonic matter within this length, what is known as the *Jeans mass*, could be determined.

In 1968, *Joseph Silk* showed that, during the pre-recombination epochs, sound waves in the radiation-dominated plasma are damped by repeated electron scatterings. The effect of this damping is to dissipate fluctuations with masses less than about $10^{12} M_\odot$ by the epoch of recombination. In this picture, all fine-scale structure is wiped out and only large-scale structures on the scale of large galaxies and clusters of galaxies could form after recombination. In the early 1970s, Zeldovich and *Edward Harrison* independently put together information about the spectrum of the initial fluctuations on different physical scales and showed that the observed structures in the Universe could be accounted for if the mass fluctuation spectrum had the form $\Delta(M) \propto M^{-2/3}$ in the very early Universe, corresponding to a power spectrum of initial fluctuations of the form $|\Delta_k|^2 \propto k^n$ with $n = 1$. The amplitude of this scale-free power spectrum was inferred to be $\sim 10^{-4}$ and is known as the Harrison–Zeldovich spectrum of initial perturbations.

A key test of these models is provided by the fact that the presence of density fluctuations at the epoch of recombination should leave some imprint upon the intensity distribution of the Cosmic Microwave Background Radiation on the sky. In the simplest picture, if the process of recombination were instantaneous, the adiabatic perturbations would be expected to result in temperature fluctuations $\Delta T/T = \frac{1}{3}\Delta\varrho/\varrho$. In fact, the problem is much more complicated than this because the process of recombination is not instantaneous. The principal source of temperature fluctuations on the scale of clusters and superclusters of galaxies was expected to be associated with first order Doppler scattering due to the collapse of the primordial perturbations on these scales. These predictions provided a challenge for the observers since the predicted amplitudes of the fluctuations in these early theories were in the range $\Delta T/T \sim 10^{-3} - 10^{-4}$ and this was well within the capability of isotropy measurements of the Microwave Background Radiation.

In the 1970s, these concepts gave rise to two principal scenarios for the origin of structure in the Universe. The first, known as the *adiabatic* model, was based upon the picture outlined above in which the perturbations were adiabatic sound waves before recombination and the structure in the Universe formed by the fragmentation of large-scale structures which reached amplitude $\delta\varrho/\varrho \sim 1$ at relatively late epochs. A realisation of this scenario was described by Doroshkevich, Sunyaev and Zeldovich in 1974. An alternative picture was one in which the perturbations were not sound waves but simply *isothermal* perturbations in the pre-recombination plasma which were in pressure balance with the background radiation. Small mass perturbations would not be damped in this picture and so perturbations on all scales would survive to the recombination epoch. Galaxies and clusters of galaxies would then form by a process of hierarchical clustering. Both models predicted similar amplitudes of perturbations at the epoch of recombination as the perturbations began to collapse to form bound objects, but their subsequent behaviour was entirely different. The adiabatic picture could be thought of as a 'top-down' process of galaxy formation in which the largest scale structures formed first, whereas the isothermal picture corresponded to a 'bottom-up' process in which small-scale objects came together to form larger structures. In the adiabatic picture, the galaxies, the stars and the chemical elements all formed at late epochs whereas in the isothermal picture, the galaxies, stars and heavy elements began to form at large redshifts.

Upper limits to the spatial intensity fluctuations in the Cosmic Microwave Background Radiation continued to improve and it became apparent that models with low density parameters were in serious conflict with the upper limits to the background fluctuations because, in these, there is very little growth of the perturbations after the epoch of recombination. The limits to the density parameter in the form of baryons from cosmological nucleosynthesis arguments showed that, if the density parameter were to be 1, most of the matter in the Universe would have to be in some non-baryonic form.

A solution to these problems appeared in 1980 when Lubimov and his collaborators reported experiments which suggested that the electron neutrino had a finite rest-mass of about 30 eV. In 1966, Gershtein and Zeldovich had noted that relic neutrinos of finite rest mass could make an appreciable contribution to the mass density of the Universe and, in the 1970s, Marx and Szalay had considered the role of neutrinos of finite rest-mass as candidates for the dark matter, as well as studying their role in galaxy formation. The intriguing aspect of Lubimov's result was that, if the relic neutrinos had this rest-mass, the Universe would just be closed, $\Omega_0 = 1$. Zeldovich and his colleagues developed a new version of the adiabatic model in which the Universe was dominated by neutrinos with finite rest-mass.

Neutrino fluctuations would begin to grow as soon as they became non-relativistic but, since the neutrinos are weakly interacting, they would stream freely out of the perturbations and so small-scale density perturbations would be quickly damped out. The matter and radiation fluctuations remained at a low level but, after recombination, the baryonic matter would fall into the larger amplitude neutrino fluctuations and then evolve more or less as in the standard adiabatic scenario. Because of the free streaming of the neutrinos, only the very largest scale perturbations with masses $\sim 10^{16} M_\odot$ would survive to the epoch of recombination and so, just as in the adiabatic model, the largest scale perturbations would form first and then the smaller scale structures form by the process of fragmentation. This model had the great advantage of reducing very significantly the expected amplitude of the fluctuations in the Microwave Background Radiation since the fluctuations in the baryonic matter would be of low amplitude during the critical phases when the background photons were last scattered.

In 1970, Zeldovich had discovered a solution for the non-linear development of a collapsing cloud and used it to show that the large-scale perturbations would form sheets and pancakes which he believed would resemble the large-scale filamentary structure seen in the distribution of galaxies. This scenario for galaxy formation became known as the *Hot Dark Matter* picture of galaxy formation since the neutrinos were relativistic when they decoupled from the primordial plasma.

There were, however, concerns about this picture. First of all, there were reservations about the experiments which claimed to have measured the rest mass of the electron neutrino and it is now believed that the result was erroneous – the present upper limit to the rest mass of the electron neutrino is a few eV, although the muon and tau neutrinos could be more massive. Second, constraints could be set to the masses of the neutrinos if they were to constitute the dark matter in galaxies, groups and clusters of galaxies. In 1979, *James Gunn* and *Scott Tremaine* showed how the phase space constraints associated with fermions such as neutrinos could be used to set lower limits to their masses. While 30 eV neutrinos could bind clusters and the haloes of giant galaxies, those needed to bind dwarf galaxies would have to have masses much greater than 30 eV. This was not necessarily a fatal flaw because it could be that some other form of dark matter was present in the haloes of the dwarf galaxies.

There was also the realisation about this time that there were several alternative possibilities for the dark matter which came from theories of elementary particles. Examples included the axions, supersymmetric particles such as the gravitino or photino and ultraweakly interacting neutrino-like particles, all of which would be relics of the very early Universe. The period 1980 to 1982 marked the period when the particle physicists began to take the early Universe very seriously as a laboratory for particle physics. In 1982, Peebles introduced the term *Cold Dark Matter* to encompass many of these exotic types of particle suggested by the particle physicists. Generically, they were particles created in the very early Universe which decoupled from the primaeval plasma when they were already non-relativistic. Consequently, they would be very cold by now.

The Cold Dark Matter scenario is similar in many ways to the isothermal model. Since the matter is very cold, perturbations are not destroyed by free streaming. Fluctuations on all scales can survive and so, when the pre-recombination Universe became matter dominated, these perturbations began to grow, completely decoupled from the matter and radiation. As in the Hot Dark Matter scenario, after the epoch of recombination, the baryonic matter collapsed into the growing potential wells in the dark matter. After recombination, galaxies, groups and clusters formed by a process of hierarchical clustering. A remarkably useful formalism for the process of hierarchical clustering was described by *William Press* and *Paul Schechter* in 1974 which gives a good description of how the mass function of objects of different masses evolves with time.

These alternative dark matter pictures of galaxy formation have been the subject of a great deal of analysis and computer simulation. The Cold Dark Matter scenario can account for many features of the distribution of galaxies in the Universe, and it has the great advantage that it can make testable predictions. One of the most important tests has been the prediction of fluctuations in the Cosmic Microwave Background Radiation and these were detected in 1992 by the COBE satellite. Fluctuations on an angular scale $\theta \geq 7°$ were discovered at an amplitude of $\Delta I / I \approx 10^{-5}$ by *George Smoot* and his colleagues. These fluctuations correspond to physical dimensions about ten times the size of the largest holes and voids observed in the distribution of galaxies. On these large angular scales, the source of intensity fluctuations is the gravitational redshift associated with photons originating from within the density fluctuations at the last scattering surface. This source of fluctuations was first described by Sachs and Wolfe in 1967 and it can be shown that the Harrison–Zeldovich spectrum results in temperature fluctuations which are independent of angular scale for all scales greater than a few degrees. A number of groups have now measured temperature fluctuations in the Cosmic Microwave Background Radiation at roughly the same amplitude on different angular scales. The amplitude of these fluctuations as a function of angular scale is a key diagnostic for many aspects of cosmology and several major ground-based and space experiments are planned to determine the dependence in detail. As we will show, there is the real possibility of determining the values of cosmological parameters rather precisely from observations of the angular spectrum of these fluctuations.

The Cold Dark Matter picture has become the preferred picture for galaxy formation, but it does need some patching up to achieve consistency with all the observations. Viable models have been constructed which include a tilted power-spectrum of its initial fluctuations as compared with the standard Harrison–Zeldovich spectrum, others include the cosmological constant, yet others consider a mixture of hot and cold dark matter and others consider that the Universe may be open. This list gives some flavour of the remarkable activity which has been stimulated by recent observational developments in cosmology.

1.5 The Very Early Universe

Despite the success of the standard Big Bang model, it is incomplete in the sense that the initial conditions have to be arranged so that we end up with the Universe as we observe it today. There are four pieces of information which have to be incorporated as initial conditions:

1. The Universe must be isotropic.
2. There must have been a very small baryon–antibaryon asymmetry in the very early Universe.
3. The Universe must have been set up remarkably close to the critical cosmological model, $\Omega_0 = 1$, in the first place.
4. An initial spectrum of fluctuations must have been present from which the present large-scale structure of the Universe formed.

The origin of the fourth set of initial data was described in the last section. Condition 1 arises from the fact that as we look further back in time, the horizon scale encompasses less and less mass and so it is a problem to understand why the Universe should be so isotropic on the largest scales we can observe today.

Condition 2 arises from the fact that the photon-to-baryon ratio today is $N_\gamma/N_B = 4 \times 10^7/\Omega_B h^2$ where Ω_B is the density parameter in baryons. If photons are neither created or destroyed, this ratio is conserved. At temperatures of about 10^{10} K, electron–positron pair production takes place from the photon field. At a correspondingly higher temperature, baryon–antibaryon pair production takes place with the result that there must be a slight asymmetry in the baryon–antibaryon ratio in the very early Universe if we are to end up with the correct photon-to-baryon ratio at the present day. In 1965, Zeldovich showed that if the Universe were completely symmetric with respect to matter and antimatter, the present day photon to baryon/antibaryon ratio would be about 10^{18}. Various baryon symmetric models of the Universe were proposed by Alfvén and Klein in 1962 and by Omnes in 1969 but none of these demonstrated convincingly how the matter and antimatter could be separated in the early Universe.

The third problem, pointed out by Dicke and Peebles in 1979, arises from the fact that, according to the standard world models, if the Universe were set up with a value of the density parameter differing from the critical value $\Omega_0 = 1$, then it would have departed very rapidly from $\Omega_0 = 1$ at later epochs. If the value of Ω_0 is within a factor of 10 of the value $\Omega_0 = 1$ now, it must have been extremely close to the critical value in the remote past. There is nothing in the standard world models which would lead us to prefer any particular value of Ω_0.

The first suggestion that some of these problems might be resolved by appeal to particle physics was made by Sakharov in 1967 who suggested that the baryon–antibaryon asymmetry might be associated with the type of symmetry breaking observed in the decays of the K mesons, in other words, that the asymmetry is associated with the type of symmetry-breaking which occurs in Grand Unified Theories of elementary particles in the early Universe.

The most important conceptual development came in 1981 with *Alan Guth's* proposal of the inflationary model for the very early Universe. There had been earlier suggestions foreshadowing his proposal. For example, Zeldovich had noted in 1968 that there is a physical interpretation of the cosmological constant Λ associated with the zero-point fluctuations of a vacuum. Linde in 1974 and Bludman and Ruderman in 1977 had shown that the scalar Higgs fields, which have been introduced to give the $W^{\pm}$ and Z^0 particles mass, have similar properties to those which would result in a positive cosmological constant.

In Guth's paper of 1981, he realised that if there were an early exponential expansion of the Universe, this would solve both the problem of the isotropy of the Universe on a large scale and would also drive the Universe towards a flat spatial geometry. The effects of the exponential expansion is to drive neighbouring particles apart at an exponentially increasing rate so that, although the particles were in causal contact in the very early Universe, the exponential inflation quickly moves them far beyond their local horizons and can account for the large-scale isotropy of the Universe by the end of the inflation epoch. At the end of this phase of exponential inflation, the Universe transforms into the standard Friedman world model, which, since it has very precisely flat geometry, must have $\Omega_0 = 1$. The model also had the advantage of eliminating a problem with Grand Unified Theories, on which the model was based, that there should be too many magnetic monopoles present in the Universe now. The model was revised in 1982 by Linde and by Albrecht and Steinhardt in which the transition to the Friedman solutions was continuous and avoided problems associated with Guth's original picture in which the change to the Friedman solution takes place in a first order phase transition.

Since 1982, the inflationary scenario for the early evolution of the Universe between the epochs when the Universe was 10^{-34} to 10^{-32} seconds old has been studied very intensively by many authors. Among the further successes claimed for the theory has been the realisation that quantum fluctuations in the Higgs fields which drive the inflation are also amplified in the process of inflation and the Harrison–Zeldovich spectrum comes naturally out the theory.

The methodological problem with these ideas is that they are based upon extrapolations from physics tested in the laboratory to energies vastly exceeding those which can possibly be tested in a terrestrial laboratory. Cosmology and particle physics come together in the early Universe and they bootstrap their way to a self-consistent solution. This may be the best that we can hope for but it would be preferable to have independent constraints upon the theories.

2 The Large Scale Structure of the Universe

Our current picture of how matter and radiation are distributed in the Universe on a large scale is derived from a wide variety of different types of observation. In this chapter, we concentrate upon the large-scale distribution of matter and radiation in the Universe and deal with galaxies and clusters of galaxies in Chaps. 3 and 4 respectively. The observations described in this chapter provide much of the essential underpinning of modern cosmological research.

2.1 The Spectrum and Isotropy of the Cosmic Microwave Background Radiation

On the very largest scales, the best evidence for the overall isotropy of the Universe comes from observations of the *Cosmic Microwave Background Radiation*. This intense diffuse background radiation was discovered in 1965 by Penzias and Wilson whilst commissioning a sensitive receiver system for centimetre wavelengths at the Bell Telephone Laboratories (Penzias and Wilson 1965). It was quickly established that this radiation is remarkably uniform over the sky and that, in the wavelength range 1 m $> \lambda > 1$ cm, the intensity spectrum had the form $I_\nu \propto \nu^2$, corresponding to the Rayleigh-Jeans region of a black-body spectrum at a radiation temperature of about 2.7 K. The maximum intensity of such a spectrum occurs at a wavelength of about 1 mm at which atmospheric emission makes precise absolute measurements of the background spectrum very difficult from the surface of the Earth. Several high-altitude balloon experiments carrying millimetre and sub-millimetre spectrometers were flown during the 1970s and 1980s and evidence was found for the expected turn-over in the Wien region of the spectrum but there were discrepancies between the experiments and the absolute accuracy was not great (see Weiss (1980) for a discussion of the early spectral measurements). It was realised in the 1970s that the only way of studying the detailed spectrum and isotropy of the background radiation over the whole sky was from above the Earth's atmosphere. After a long period of gestation, the Cosmic Background Explorer (COBE) of NASA was launched in November 1989 and was dedicated to studies of the background radiation, not only in the millimetre and submillimetre wavebands, but also throughout the infrared waveband from 2 to 1000 μm.

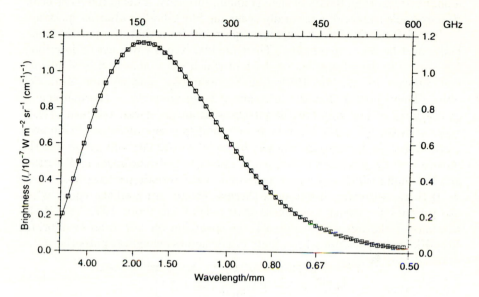

Fig. 2.1. The first published spectrum of the Cosmic Microwave Background Radiation as measured by the COBE satellite in the direction of the North Galactic Pole (Mather *et al.* 1990). Within the quoted errors, the spectrum is precisely that of a perfect black body at radiation temperature 2.735 ± 0.06 K. The more recent spectral measurements are discussed in the text. The units adopted for frequency on the ordinate are cm^{-1}. A useful conversion to more familiar units is 10^{-7} W m^{-2} sr^{-1} $(cm^{-1})^{-1} = 3.34 \times 10^{-18}$ W m^{-2} Hz^{-1} $sr^{-1} = 334$ MJ sr^{-1}.

2.1.1 The Spectrum of the Cosmic Microwave Background Radiation

The Far Infrared Absolute Spectrophotometer (FIRAS) measured the *spectrum* of the Cosmic Microwave Background Radiation in the wavelength range 0.5 to 2.5 mm with very high precision during the first year of the mission. The FIRAS detectors and a reference black-body source were cooled to liquid helium temperatures and there was only sufficient liquid cryogen for one year of observation. The first observations made by FIRAS revealed that the spectrum of the background is very precisely of black-body form. In the early spectrum shown in Fig. 2.1, it can be seen that the error boxes, which are shown as 1% of the peak intensity, are overestimates of the uncertainty because the black-body curve passes precisely through the centre of each error box. Much more data have become available since Fig. 2.1 was published and, in particular, the thermometry involved in making absolute measurements has been very carefully studied. More recent analyses reported by Fixsen *et al.* (1996) have shown that the spectrum is a perfect black body with a radiation temperature $T = 2.728 \pm 0.002$ K. Furthermore, the deviations from a perfect black-body spectrum in the wavelength interval $2.5 > \lambda > 0.5$ mm amount to less than 0.03% of the maximum intensity. This is the most beautiful example I know of a naturally occurring black-body radiation spectrum.

The deviations of the background spectrum from that of a perfect blackbody must be very small indeed. There are two convenient ways of describing how large the deviations can be, both of which were pioneered by Zeldovich and Sunyaev in the late 1960s (see, for example, Sunyaev and Zeldovich 1980). They showed that the injection of large amounts of thermal energy in the form of hot gas into the intergalactic medium can produce various types of distortion of the black-body radiation spectrum because of Compton scattering of the background photons by hot electrons. We will not go into the physics of these processes at this point, except to note the forms of distortion and the limits which can be set to certain characteristic parameters.

If there were early injection of thermal energy prior to the epoch when the primordial plasma recombines at a redshift of about 1000, and if the numbers of photons were conserved, the spectrum relaxes to an equilibrium Bose-Einstein spectrum with a finite dimensionless chemical potential μ.

$$I_\nu = \frac{2h\nu^3}{c^2} \left[\exp\left(\frac{h\nu}{k_\mathrm{B} T_\mathrm{r}} + \mu \right) - 1 \right]^{-1}. \tag{2.1}$$

The simplest way of understanding this result is to note that the Bose-Einstein distribution is the equilibrium distribution for photons when there is a mismatch between the total energy and the number of photons over which to distribute this energy. In the case of a black-body spectrum, both the energy density and number density of photons are determined solely by the temperature T_r. In contrast, the Bose-Einstein distribution is determined by two parameters, the temperature T_r and the dimensionless chemical potential μ.

In the case of Compton scattering by hot electrons at late epochs, the energies of the photons are redistributed about their initial values and, to second order, there is an increase in the mean energies of the photons (see, for example, Longair 1992) so that the spectrum is shifted to slightly greater frequencies. Zeldovich and Sunyaev (1969) showed that the distortion of the black-body spectrum takes the form

$$\frac{\Delta I_\nu}{I_\nu} = y \frac{x e^x}{(e^x - 1)} \left[x \left(\frac{e^x + 1}{e^x - 1} \right) - 4 \right], \tag{2.2}$$

where y is the Compton scattering optical depth $y = \int (k_\mathrm{B} T_e / m_e c^2) \sigma_\mathrm{T} N_e \, dl$, $x = h\nu / k_\mathrm{B} T_\mathrm{r}$ and σ_T is the Thomson scattering cross-section. In the limit of small distortions, $y \ll 1$, the intensity in the Rayleigh-Jeans region decreases as $\Delta I_\nu / I_\nu = -2y$ and the total energy under the spectrum increases as $\varepsilon = \varepsilon_0 e^{4y}$.

Limits to the parameters y and μ have been derived from the very precise spectral measurements made by the FIRAS instrument. The results quoted by Page (1997) are as follows:

$$|y| \leq 1.5 \times 10^{-5} \qquad |\mu| \leq 10^{-4}. \tag{2.3}$$

These are very powerful limits indeed and will prove to be of great astrophysical importance in studying the physics of the intergalactic gas, as well as constraining the amount of star and metal formation which could have taken place in young galaxies.

2.1.2 The Isotropy of the Cosmic Microwave Background Radiation

Equally remarkable have been the COBE observations of the *isotropy* of the Cosmic Microwave Background Radiation over the sky. The prime instruments for these studies are the Differential Microwave Radiometers which operate at 31.5, 53 and 90 GHz, thus sampling the Rayleigh-Jeans region of the spectrum. The angular resolution of the radiometers is 7°. The choice of observing frequency is crucial in these observations. At the higher frequencies, the millimetre and submillimetre emission of diffuse dust at high galactic latitude, known as 'cirrus', confuses the picture, whilst at the lower frequencies the synchrotron radiation of our Galaxy becomes important. In increasing levels of sensitivity, the results are as follows.

At a sensitivity of about one part in 1000 of the total intensity, there is a large scale anisotropy over the whole sky associated with the motion of the Earth through the frame of reference in which the radiation would be the same in all directions (Fig. 2.2a). This global anisotropy is due to aberration effects (sometimes loosely, but somewhat inaccurately, referred to as Doppler boosting) associated with the Earth's motion. As a result, the radiation is about one part in a thousand more intense in one direction and exactly the same amount less intense in the opposite direction. The temperature distribution is found to have precisely the expected dipole distribution over the sky, $T = T_0[1 + (v/c)\cos\theta]$, where θ is measured with respect to the direction of motion of the Earth. The amplitude of the cosmic microwave dipole is 3.353 ± 0.024 mK towards galactic coordinates $l = 264.25° \pm 0.33°; b = 48.22° \pm 0.13°$ (Bennett *et al.* 1996). It is inferred that the Earth is moving at about 350 km s^{-1} with respect to the frame of reference in which the radiation would be 100% isotropic. It is intriguing that, although not designed to undertake this task, exactly the same form of large scale anisotropy has been observed by the FIRAS instrument. At about the same level of intensity, the plane of our Galaxy can be observed as a faint band of emission over the sky (Fig. 2.2a).

The velocity of the Sun relative to the Cosmic Microwave Background Radiation is in itself an important result for understanding the large-scale

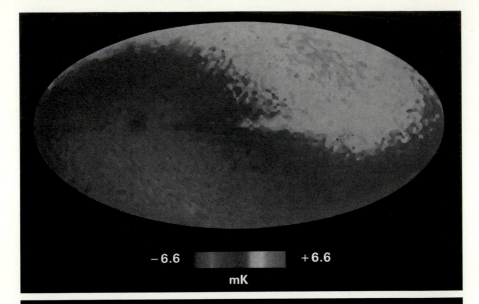

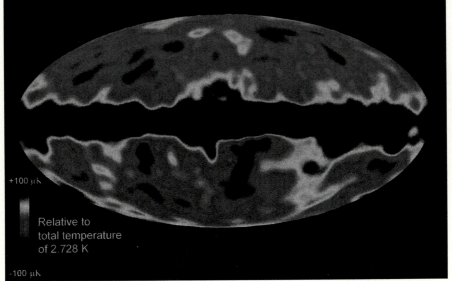

Fig. 2.2a-b. Maps of the whole sky in galactic coordinates as observed in the millimetre waveband at a wavelength of 5.7 mm (53 GHz) by the COBE satellite. The centre of our Galaxy is in the centre of the diagram and the plane of the Galaxy lies along the central horizontal axis. (**a**) The distribution of the background radiation at a sensitivity level about one part in 1000 of the total intensity. The dipole distribution of intensity over the sky can be seen, as well as evidence for emission from the Galactic plane. The latter is seen more clearly in the lower panel. (**b**) The distribution of the background radiation once the dipole component associated with the motion of the Earth has been removed. The residual radiation from the plane of the Galaxy can be seen as a bright band across the centre of the picture. The intensity fluctuations seen at high galactic latitudes are real fluctuations in the brightness distribution of the background intensity, the amplitude of the variations corresponding to an average root mean square signal of 35 ± 2 μK on an angular scale of $7°$, or to 29 ± 1 μK when smoothed to $10°$ angular scale (Bennett *et al.* 1996).

distribution of mass in the Universe. Once allowance is made for the motion of the Sun about the centre of our Galaxy, an estimate of the peculiar velocity of our Galaxy and the local group of galaxies relative to the frame of reference in which the background radiation would be perfectly isotropic can be found. This motion is assumed to be caused by irregularities in the distribution of mass on very large scales.

On angular scales of 7° and greater, Bennett *et al.* (1996) have achieved sensitivity levels of better than one part in 100,000 of the total intensity from analyses of the complete microwave data-set obtained over the four years of the COBE mission (Fig. 2.2b). At this sensitivity level, the radiation from the plane of the Galaxy is intense, but is confined to a broad strip lying along the galactic equator. Away from this region, the sky appears quite smooth on a large scale, but there are significant fluctuations in the intensity from beamwidth to beamwidth over the sky. These fluctuations are present at the level of only about 1 part in 100,000 of the total intensity and, when averaged over the clear region of sky at $|b| > 20°$ amount to a root-mean-square amplitude of 35 ± 2 μK on an angular scale of 7°, or to 29 ± 1 μK when smoothed to 10° angular scale. These values are found to be frequency independent for the three COBE frequency channels at 31.5, 53 and 90 GHz. The detection of these fluctuations is a crucial result for understanding the origin of the large scale structure of the Universe. The COBE observations allow information to be obtained about the angular spectrum of the intensity fluctuations on all scales $\theta \geq 7°$. Similar angular fluctuations have now been detected on smaller angular scales and, as we will show, the variation of the amplitude of the fluctuations with angular scale is of central importance for understanding the origin of the large-scale distribution of galaxies.

The COBE observations are crucial results for cosmology. From the point of view of the structure of the Universe on the largest angular scales, they show that the Cosmic Microwave Background Radiation is isotropic to one part in 100,000 on the large scale. Whatever its origin, this observation in itself shows that the Universe must be extraordinarily isotropic on the large scale. As we will show, it is wholly convincing that this radiation is the cooled remnant of the very hot early phases of the Big Bang.

How is the distribution of this radiation related to the distribution of ordinary matter in the Universe? We will take up this topic in much more detail in Chap. 9, but it is useful to outline here how they are related. In the standard Big Bang picture, when the Universe was squashed to only about one thousandth of its present size, the temperature of the Cosmic Background Radiation must have been about one thousand times greater than it is now. The temperature of the background radiation varies with redshift z as $T_r = 2.728(1 + z)$ K and so, at a redshift $z = 1500$, the temperature of the radiation field was about 4000 K. At that temperature, there were sufficient Lyman continuum photons in the Wien region of the background spectrum to photoionise all the neutral hydrogen in the Universe. At that stage, the galaxies could not have formed and so all the ordinary matter, which was eventually to become galaxies as we know them, was still in the form of re-markably smooth, ionised pre-galactic gas. At earlier epochs, the pre-galactic gas was fully ionised and was very strongly coupled to the background ra-diation by Thomson scattering. When we look back to these epochs, it is as if we were looking at the surface of a star surrounding us in all direc-tions, but the temperature of the radiation we observe has been cooled by a cosmological redshift factor of 1500, so that what we observe is redshifted into the millimetre waveband. This analogy makes it clear that, because of Thomson scattering of the radiation, we can only observe the very surface layers of our 'star'. Therefore, we cannot obtain any direct information about what was going on at earlier epochs. This 'surface' at which the Universe be-comes opaque to radiation is known as the *last scattering surface* and the fluctuations observed by COBE are interpreted as the very low intensity rip-ples present on that surface on angular scales of 7° and greater. As we will show, the COBE results provide information about the isotropy of the diffuse ionised pregalactic gas when the Universe was only about one thousandth of its present size.

In the standard interpretation of the COBE observations described in the last paragraph, it is assumed that the intergalactic gas was not reionised and heated at some later epoch. If that were to occur, the last scattering surface could occur at a significantly smaller redshift. This picture is not particularly plausible because the energy demands for reheating the neutral intergalactic gas are very severe and the heating process itself would probably cause much greater fluctuations in the background intensity than those observed.

One important aspect of these studies is that the energy density of the Cosmic Microwave Background Radiation amounts to $\varepsilon_{\mathrm{rad}} = aT_{\mathrm{r}}^4 = 4.2 \times 10^{-14}$ J m$^{-3} = 2.64 \times 10^5$ eV m^{-3}. This energy density of radiation pervades the whole Universe at the present epoch and provides by far the greatest contribution to the average energy density of the universal back-ground radiation.

2.2 The Large-scale Distribution of Galaxies

The visible Universe is highly inhomogeneous on a small scale, the matter being condensed into stars, which are themselves congregated into galaxies. The galaxies can be thought of as *the building blocks of the Universe*. They are themselves clustered, the associations ranging from small groups to giant regular clusters of galaxies, and even larger scale structures. If we take aver-ages over larger and larger scales, however, the inhomogeneities becomes less and less. This statement can be formalised using 2-point correlation func-tions. These can be described either in terms of the distribution of galaxies on the sky or in terms of spatial two-point correlation functions. On the sky, we can define the *angular two-point correlation function, $w(\theta)$,* by

$$N(\theta)\,\mathrm{d}\Omega = n_{\mathrm{g}}[1 + w(\theta)]\,\mathrm{d}\Omega, \tag{2.4}$$

where $w(\theta)$ describes the excess probability of finding a galaxy at an angular distance θ from any given galaxy; $\mathrm{d}\Omega$ is the element of solid angle and n_{g} is a suitable average surface density of galaxies. $w(\theta)$ contains information about the clustering properties of galaxies *as observed on the sky* to a given limiting apparent magnitude and can be measured with some precision from large statistical surveys of galaxies such as the Cambridge APM surveys.

It is more meaningful physically to work in terms of the *spatial two-point correlation function* $\xi(r)$ which describes the clustering properties of galaxies in three-dimensions about any galaxy.

$$N(r)\,\mathrm{d}V = N_0[1 + \xi(r)]\,\mathrm{d}V, \tag{2.5}$$

where $N(r)\,\mathrm{d}V$ is the number of galaxies in the volume element $\mathrm{d}V$ at distance r from any galaxy and N_0 is a suitable average space density. $\xi(r)$ describes the excess number of galaxies at distance r from any given galaxy.

In order to derive $\xi(r)$, we need to know the distribution of galaxies in space and this is a huge job. If a number of reasonable assumptions are made, it is possible to derive $\xi(r)$ from $w(\theta)$. It is found that the function $\xi(r)$ can be well represented by a power-law of the form

$$\xi(r) = \left(\frac{r}{r_0}\right)^{-\gamma}, \tag{2.6}$$

on physical scales from about $100\ h^{-1}$ kpc to $10\ h^{-1}$ Mpc in which the scale $r_0 = 5h^{-1}$ Mpc and the exponent $\gamma = 1.8$.[3] On scales greater than about $10h^{-1}$ Mpc the two-point correlation function decreases more rapidly than the power-law (2.5). The form of the two point correlation function is illustrated in Fig. 2.3b. Note the key point that clustering occurs on a *very wide range of scales* in the Universe and that there are *no preferred scales* in the two-point correlation functions. Notice also that, on large enough scales, the clustering smoothes out and the Universe becomes isotropic if we consider large enough physical scales.

The homogeneity of the distribution of galaxies with increasing distance can be studied by measuring the two-point correlation function as a function of increasing apparent magnitude. This has been carried out by comparing the two-point correlation function derived from the Lick survey $w_{\mathrm{L}}(\theta)$ with those derived from the Cambridge APM survey at different faint apparent magnitudes. If the galaxies are sampled from a homogeneous, but clustered,

---

[3] The use of $h = H_0/(100\ \mathrm{km\ s^{-1}\ Mpc^{-1}})$ is a convenient device for adjusting the dimensions and luminosities of extragalactic objects to the reader's preferred value of Hubble's constant. If a value of $H_0 = 100\ \mathrm{km\ s^{-1}\ Mpc^{-1}}$ is preferred, $h = 1$; if the value $H_0 = 50\ \mathrm{km\ s^{-1}\ Mpc^{-1}}$ is adopted, $h = 0.5$ and so on.

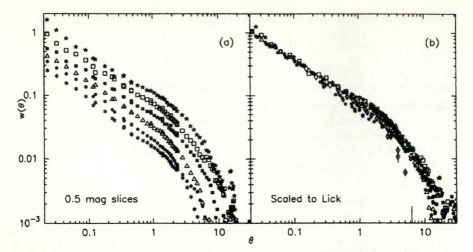

Fig. 2.3a-b. The two-point correlation function for galaxies over a wide range of angular scales. (a) The scaling test for the homogeneity of the distribution of galaxies can be performed using the correlation functions for galaxies derived from the APM surveys at increasing limiting apparent magnitudes in the range $17.5 < m < 20.5$. The correlation function are displayed at intervals of 0.5 magnitudes. (b) The two-point correlation functions scaled to the correlation function derived from the Lick counts of galaxies (Maddox *et al.* 1990).

distribution it is straightforward to show that the angular two-point correlation function should scale with increasing limiting distance D of the survey as

$$w(\theta, D) = \frac{D_0}{D} w_{\mathrm{L}} \left(\theta \frac{D}{D_0} \right), \tag{2.7}$$

where the function $w_{\mathrm{L}}(\theta)$ has been determined to distance D_0. Such analyses were first carried out by Groth and Peebles (1977, 1986) who showed that the two-point correlation functions determined from a bright sample of Zwicky galaxies, from the Lick counts of galaxies and from a deep sky survey plate in an area known as the Jagellonian field scaled exactly as expected if the distribution of the galaxies displayed the same degree of spatial correlation throughout the local Universe out to $z \sim 0.1$. A similar result has been found comparing the two-point correlation functions found at increasing apparent magnitude limits in the machine-scanned surveys carried out by the APM group at Cambridge (Maddox *et al.* 1990). Fig. 2.3a shows the angular two-point correlation functions $w(\theta)$ measured at increasing apparent magnitude limits in the magnitude range $17.5 < m < 20.5$. In Fig. 2.3b, these functions are scaled to the angular correlation function found from the Lick survey. As Peebles (1993) expresses it,

Fig. 2.4. The distribution of galaxies with $17 \leq b_j \leq 20.5$ shown in an equal area projection centred on the Southern Galactic Pole. This image was reconstructed from machine scans of 185 UK Schmidt plates by the Cambridge APM measuring machine. There are over two million galaxies in this image. The small empty patches in the map are regions that have been excluded about bright stars, nearby dwarf galaxies, globular clusters and step wedges (Maddox *et al.* 1990).

'... the correlation function analyses have yielded a new and positive test of the assumption that the galaxy space distribution is a stationary (statistically homogeneous) random process'.

The two-point correlation function for galaxies is a circularly symmetric average about each galaxy and so is only a crude measure of the overall clustering of galaxies. It does, however, provide the important result that on physical scales $r > 5h^{-1}$ Mpc, the mean amplitude of the density perturbations is less than one and that the distribution of galaxies becomes more and more uniform on the very largest scales. Notice that this means that the density perturbations on the largest scales are still in the linear regime, $\delta \varrho / \varrho \ll 1$.

A better representation of the overall distribution of galaxies is given in Fig. 2.4 which shows the distribution of galaxies centred on the direction of the South Galactic Pole. This remarkable picture was created from scans of 185 contiguous UK Schmidt plates, each of which covers an area of $6° \times 6°$ on the sky, made by the Cambridge APM high-speed measuring machine (Maddox *et al.* 1990). The image is centred on the South Galactic Pole and so the effects of Galactic obscuration by dust is negligible. Each plate was carefully calibrated and then stars distinguished from galaxies by their different image profiles. Fig. 2.4 includes over two million galaxies with apparent magnitudes in the range $17 \leq b_j \leq 20.5$ and so represents the distribution of galaxies in the Universe on the grandest scale.

It is apparent that the distribution of galaxies is far from uniform and, at best, the assumption of circular symmetry about each galaxy is only a rough approximation. The distribution of galaxies appears to include clumps of galaxies, elongated features and holes, but, of course, the eye is expert at finding structures among random data. In fact, much of the obvious clumping, the holes and the stringy structures are real features of the distribution of galaxies. To derive the three-dimensional distribution of galaxies, their distances need to be known and this is a very major undertaking for large sample of galaxies widely separated on the sky.

Fig. 2.5 shows the local three-dimensional distribution of galaxies derived from the Harvard–Smithsonian Astrophysical Observatory survey of over 14,000 bright galaxies (Geller and Huchra 1989). Our own Galaxy is located at the centre of the diagram and, if the galaxies were uniformly distributed in the local Universe, the points would be uniformly distributed over the diagram, which is certainly very far from the case. There are gross inhomogeneities and irregularities in the local Universe including large 'holes' or 'voids' in which the local number density of galaxies is significantly lower than the mean, and long 'filaments' of galaxies, including the feature known as the 'Great Wall', which extends from right ascensions 9^h to 17^h about half-way to the limit of the survey. There are a number of 'streaks' pointing toward our own Galaxy and these correspond to clusters of galaxies, the lengths of the 'streaks' corresponding to the components of the velocity dispersion of the galaxies in the clusters along the line of sight.

Further evidence for the overall homogeneity of the Universe on a large scale comes from the Las Campanas Redshift Survey which sampled the distribution of galaxies to a distance about four times that of the Harvard–Smithsonian survey. The redshifts of 26,418 galaxies have been measured in six narrow strips, three each in the Northern and Southern galactic caps (Lin et al. 1996). The resulting map of the distribution of galaxies is shown in Fig. 2.6, which shows the same large voids and walls of galaxies seen in Fig. 2.5. The sizes of the voids in the galaxy distribution are on roughly the same physical scale as those in Fig. 2.5, indicating that the Universe is homogeneous on a large enough scale and consistent with the scaling arguments from the angular 2-point correlation functions (Fig. 2.3b).

In both Figs. 2.5 and 2.6, the scale of the large holes is about $30 - 50$ times the scale of a cluster of galaxies. These are the largest known structures in the Universe and we note that one of the great cosmological challenges is to reconcile the gross irregularity in the large-scale spatial distribution of galaxies with the remarkable smoothness of the Cosmic Microwave Background Radiation seen in Fig. 2.2b. Despite the presence of the huge voids, the am-right ascension

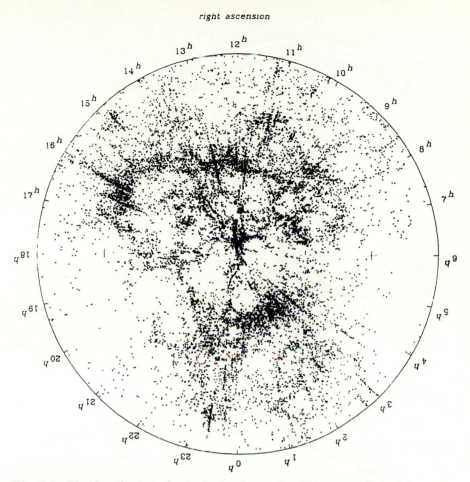

Fig. 2.5. The distribution of galaxies in the nearby Universe as derived from the Harvard–Smithsonian Center for Astrophysics survey of galaxies. The map contains over 14,000 galaxies which form a complete statistical sample around the sky between declinations $\delta = 8.5°$ and $44.5°$. All the galaxies have recession velocities less than 15,000 km s^{-1}. Our Galaxy is located at the centre of the map and the radius of the bounding circle is $150h^{-1}$ Mpc. The galaxies within this slice have been projected onto a plane to show the large scale features in the distribution of galaxies. Rich clusters of galaxies which are gravitationally bound systems with internal velocity dispersions of about 10^3 km s^{-1} appear as 'fingers' pointing radially towards our Galaxy at the centre of the diagram. The distribution of galaxies is highly irregular with huge holes, filaments and clusters of galaxies throughout the local Universe (Geller and Huchra 1989).

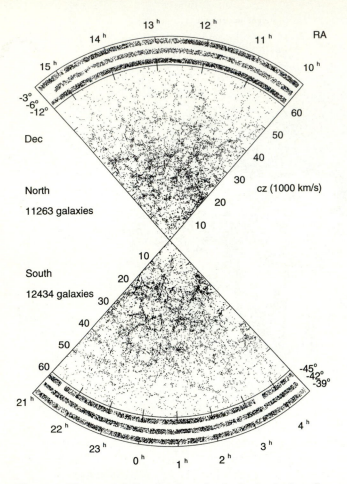

Fig. 2.6. The spatial distribution of galaxies in the Las Campanas Redshift Survey plotted in the same format as in Fig. 2.5. The survey covered 700 degree^2 in 6 strips, each $1.5° \times 80°$, three each in the Northern and Southern galactic caps. The limiting redshift of the survey is 60,000 km s^{-1}, the typical value being about 30,000 km s^{-1} (Lin *et al.* 1996).

plitude of these irregularities decreases with increasing scale so that on the very largest scales, one bit of Universe looks very much like another.

It is important to have a quantitative description of the large-scale topology of the galaxy distribution. Gott and his colleagues have developed techniques for evaluating the topology of the distribution of voids and galaxies from large redshift surveys of galaxies (Gott *et al.* 1986, Melott *et al.* 1988). They find that the distribution of the galaxies on the large scale is 'sponge-like' – the material of the sponge represents the location of the galaxies and the holes in the sponge correspond to the large voids seen in Figs. 2.5 and

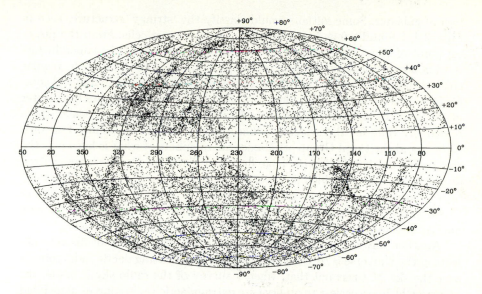

Fig. 2.7. An equal area projection drawn in galactic coordinates of the distribution of bright galaxies over the whole sky. In the north, the galaxies are from the UGC catalogue and, in the south, from the ESO, UGC and MCG catalogues. There is an absence of galaxies at low galactic latitudes because of extinction by interstellar dust. The prominent band of galaxies intersecting the Galactic Plane at right angles at $l \sim 320°$ is the Local Supercluster of Galaxies (Kolatt, Dekel and Lahav 1995).

2.6. Thus, both the holes and the distribution of galaxies can be thought of as being continuously connected throughout the local Universe. This topology is possible in three dimensions but not in two. The homogeneity of the clustering properties as indicated by the scaling laws in Fig. 2.3b and the comparison of Figs. 2.5 and 2.6 suggest that the sponge-like structure in the distribution of galaxies extends to large distances. Just like a sponge, overall the distribution of material of the sponge and the holes is homogeneous, but, on a small scale, it is highly inhomogeneous. Some of the largest holes in the distribution of galaxies seen in Figs. 2.5 and 2.6 have diameters about $50h^{-1}$ Mpc.

On fine scales, the clustering of galaxies takes place on a very wide variety of scales from pairs and small groups of galaxies, like the Local Group of galaxies, to giant *clusters* of galaxies such as the Coma and Pavo clusters which can contain thousands of members and we will discuss some of their properties in Chap. 4. The rich regular clusters are self-gravitating bound systems, but there are also irregular clusters which have an irregular, extended appearance and it is not so clear that these are bound systems.

The term *supercluster* is used to describe structures on scales larger than those of clusters of galaxies. They may consist of associations of clusters of galaxies, or a rich cluster with associated groups and an extended distribu-

tion of galaxies. Some authors would classify the 'stringy' structures seen in
Figs. 2.4, 2.5 and 2.6 as superclusters, or supercluster cells. From the physical point of view, the distinction between the clusters and the superclusters
is whether or not they are gravitationally bound. Even in the rich, regular
clusters of galaxies, which have had time to relax to a state of dynamical equilibrium, there has only been time for individual galaxies to cross the cluster
up to about 10 times in the age of the Universe and so, on larger scales, there
is scarcely time for the systems to become gravitationally bound. Our own
Galaxy and the Local Group of galaxies are members of what is known as the
Local Supercluster. This is the huge flattened distribution of galaxies centred
on the Virgo cluster, which lies at a distance of about $15 - 20$ Mpc from our
own Galaxy. It can be seen very prominently in maps of the distribution of
bright galaxies running more or less perpendicular to the plane of the Galaxy
and is the feature located at $l \sim 320°$ in Fig. 2.7 (Kolatt *et al.* 1995).

Another way of investigating the large-scale distribution of discrete objects in the Universe is to study the distribution of extragalactic radio sources
over the sky. It turns out that, when a survey of the radio sky is made, the
objects which are easiest to observe are extragalactic radio sources associated
with certain rare classes of galaxy, the radio quasars and radio galaxies, at
very great distances. Because they are rare objects, they sample the isotropy
of the Universe on a very large scale. Fig. 2.8 shows the distribution of the
brightest 31,000 extragalactic radio sources at a wavelength of 6 cm in the
northern hemisphere – the sources are listed in the Greenbank Catalogue of
radio sources (Gregory and Condon 1991). There is a hole in the centre of
the distribution corresponding to an area of sky about the North Celestial
Pole which was not observed as part of the survey and two other holes in the
vicinity of the intense radio sources Cygnus A and Cassiopeia A. There is also
a small excess of sources lying along the Galactic plane but otherwise the distribution is entirely consistent with the sources being distributed uniformly
at random over the sky. The radio sources are ideal for probing the large scale
distribution of discrete objects since they are so readily observed at large distances. It is probable that the bulk of the radio sources plotted in Fig. 2.8 lie
at redshifts $z \geq 1$ and so they sample the distribution of discrete sources on
the largest physical scales accessible to us at the present epoch. Notice that
the extragalactic radio sources provide complementary information to that
provided by the Cosmic Microwave Background Radiation, in that they refer
to the large scale distribution of discrete objects, such as galaxies, once they
have formed.

We conclude that, on the very largest scale, the distribution of matter and
radiation is remarkably isotropic and homogeneous. This greatly simplifies
the construction of cosmological models.

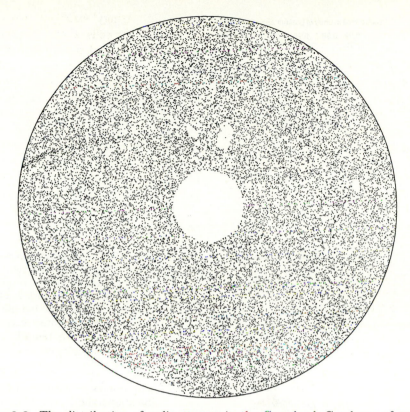

Fig. 2.8. The distribution of radio sources in the Greenbank Catalogue of radio sources at 6 cm (Gregory and Condon 1991). The picture includes 31,000 radio sources. In this equal area projection, the north celestial pole is in the centre of the diagram and the celestial equator around the solid circle. The area about the north celestial pole was not surveyed. There are 'holes' in the distribution about the bright sources Cygnus A and Cassiopeia A and a small excess of sources associated with the Galactic plane. Otherwise, the distribution does not display any significant departure from a random distribution (from Peebles 1993).

2.3 Hubble's Law and the Expansion of the Universe

Hubble made his great discovery of the velocity–distance relation for galaxies in 1929. A modern version of Hubble's Law, in the form of an apparent magnitude–redshift relation or Hubble diagram, is shown in Fig. 2.9a for the brightest galaxies in clusters. It is found empirically that the brightest galaxies in nearby clusters all have more or less the same intrinsic luminosity and so the apparent magnitudes of these galaxies can be used to estimate relative distances by application of the inverse square law. For a class of galaxy of fixed intrinsic luminosity L, the observed flux density S is given by the inverse square law, $S = L/4\pi r^2$ and so, converting this relation into astronomical apparent magnitudes m using the standard relation $m = \text{constant} - 2.5\log_{10} S$,

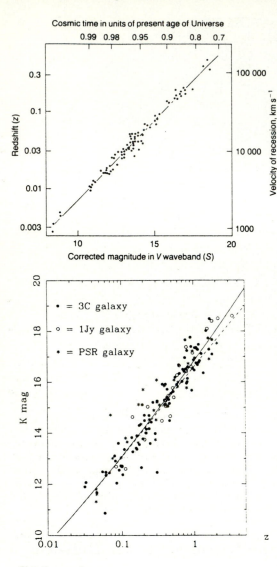

Fig. 2.9. (a) A modern version of the velocity–distance relation for galaxies for the brightest galaxies in rich clusters of galaxies. This correlation indicates that the brightest galaxies in clusters have remarkably standard properties and that their velocities of recession from our own Galaxy are proportional to their distances. (Sandage 1968).
(b) The redshift–K-magnitude relation for radio galaxies associated with strong radio sources. The radio galaxies are selected from the 3CR, 1 Jy and PKS samples of extragalactic radio sources (Dunlop and Peacock 1990).

it follows that

$$m = 5 \log_{10} r + \text{constant}. \tag{2.8}$$

This was the approach adopted by Hubble and Humason in their pioneering analysis of 1934 – they assumed that the 5th brightest galaxy in a cluster would have more or less the same intrinsic luminosity (Fig. 1.2b). In Fig. 2.9a, the corrected apparent magnitude in the V waveband is plotted against the logarithm of the redshift of the brightest galaxies in a number of rich clusters of galaxies which span a wide range of redshifts. The redshift is defined by the usual formula

$$z = \frac{\lambda_{\text{obs}} - \lambda_{\text{em}}}{\lambda_{\text{em}}}, \tag{2.9}$$

where λ_{em} is the emitted wavelength of some spectral feature and λ_{obs} is the wavelength at which is it observed. For velocities much less than the speed of light, the relativistic Doppler shift formula for motion along the line of sight,

$$1 + z = \left(\frac{1 + v/c}{1 - v/c} \right)^{1/2}, \tag{2.10}$$

reduces to $z = v/c$. It is an unfortunate tradition in optical astronomy to convert the splendidly dimensionless quantity, the redshift z, into a velocity by multiplying by the speed of light c, $v = cz$. The solid line shown in Fig. 2.9a is $m = 5\log_{10} z + \text{constant}$, which corresponds to $v \propto r$, and it runs precisely through the observed points – correlations do not come any better than this in cosmology. The velocity–distance relation is normally written $v = H_0 r$, where H_0 is known as Hubble's constant.

The velocity–distance relation appears to hold good for all classes of extragalactic system, including the active galaxies and quasars. My own personal favourite is the relation for the galaxies associated with strong radio sources, a recent example being shown in Fig. 2.9b. The narrow dispersion in absolute magnitude for the radio galaxies extends to redshifts of 2 and greater.

We discussed in detail recent evidence for the homogeneity of the distribution of galaxies in space in Sect. 2.2. In fact, Hubble realised in the 1930s that a simple test of the homogeneity of the Universe is provided by the number counts of galaxies. As we will show in Sect. 17.1.1, in a homogeneous Universe it is expected that the number counts of galaxies follow the law $N(\geq S) \propto S^{-3/2}$, where S is the observed flux density of the galaxy. This result is independent of the luminosity function of the sources so long as the counts do not extend to such large distances that the effects of the cosmological redshift have to be taken into account. In terms of apparent magnitudes, this relation becomes $N(\leq m) \propto 10^{0.6m}$. Hubble found that the counts of galaxies to about 20th magnitude more or less followed this relation, although they showed some convergence at the faintest apparent magnitudes, which Hubble interpreted as evidence for the effects of space curvature at large distances, showing directly that on average they are homogeneously distributed in space. More recent counts of galaxies are shown in Fig. 2.10 which extend to the faintest apparent magnitudes observed by the Hubble Space Telescope (Metcalfe et al. 1996). The results are similar to those of Hubble at $m \leq 20$ – the counts are slightly flatter than the Euclidean predictions, but are entirely consistent with the expectations of uniform world models once the effects of observing the populations at significant cosmological distances are taken into account. Divergences from the expectations of the uniform models do occur at much fainter apparent magnitudes ($B \geq 22$), in the sense that there is an excess of faint blue galaxies – we will take up the origin of this excess in detail in Chap. 17.

The combination of the observed large-scale isotropy and homogeneity of the Universe with Hubble's law shows that the Universe as a whole is expanding uniformly at the present time. Let me show this formally by the following

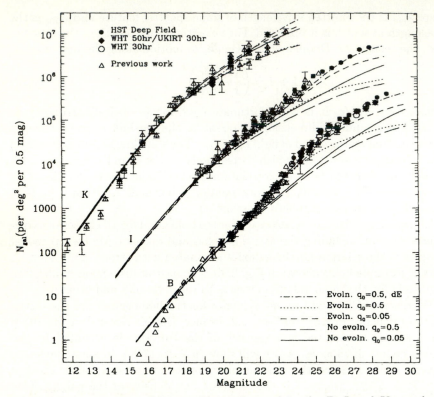

Fig. 2.10. The counts of faint galaxies observed in the B, I and K wavebands compared with the expectations of various uniform world models, as well as other models in which various forms of the evolution of the luminosity function of galaxies with redshift is assumed (Metcalfe *et al.* 1996). It can be seen that the counts follow closely the expectations of uniform world models at magnitudes less than about 21, but there is a excess of galaxies in the B and I wavebands at fainter magnitudes.

simple sum. Consider a uniformly expanding system of points (Fig. 2.11). Then, the definition of a uniform expansion is that the distances to any two points should increase by the same factor in a given time interval, that is, we require

$$\frac{r_1(t_2)}{r_1(t_1)} = \frac{r_2(t_2)}{r_2(t_1)} = \ldots = \frac{r_n(t_2)}{r_n(t_1)} = \ldots = \alpha = \text{constant}, \qquad (2.11)$$

for any set of points. The recession velocity of galaxy 1 from the origin is therefore

$$v_1 = \frac{r_1(t_2) - r_1(t_1)}{t_2 - t_1} = \frac{r_1(t_1)}{t_2 - t_1}\left[\frac{r_1(t_2)}{r_1(t_1)} - 1\right]$$

$$= \frac{r_1(t_1)}{t_2 - t_1}(\alpha - 1) = H_0 r_1(t_1). \qquad (2.12)$$

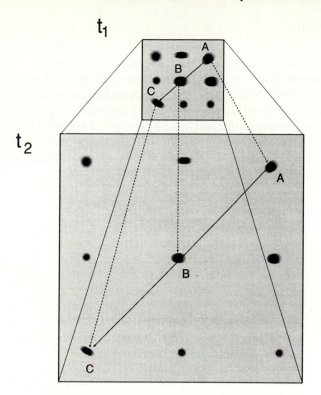

Fig. 2.11. Illustrating the origin of the velocity–distance relation for an isotropically expanding distribution of galaxies. The distribution of galaxies expands uniformly between the epochs t_1 and t_2. If, for example, we consider the motions of
the galaxies relative to the galaxy A, it can be seen that galaxy C travels twice as
far as galaxy B between the epochs t_1 and t_2 and so has twice the recession velocity
of galaxy B relative to A. Since C is always twice the distance of B from A, it
can be seen that the velocity–distance relation is a general property of isotropically
expanding Universes.

Similarly,

$$v_n = \frac{r_n(t_1)}{t_2 - t_1}(\alpha - 1) = H_0 r_n(t_1).\qquad (2.13)$$

Thus, a uniformly expanding distribution of galaxies automatically results in
a velocity–distance relation of the form $v \propto r$.

2.4 Conclusion

The upshot of the considerations of this chapter is that the correct starting point for the construction of models for the large-scale dynamics of the Universe is that they should be isotropic and homogenous on the large scale and that they should be uniformly expanding. These are enormous simplifications and, taken in conjunction with the General Theory of Relativity, provide a set of simple world models which provide the framework within which we can study the problems of the origin of galaxies and the other large-scale structures we observe in the Universe today.

3 Galaxies

Galaxies are the building blocks of the Universe. Since this book is about their formation, we should summarise what is known about their properties. Unfortunately, the astrophysics of normal galaxies is only partially understood and many important questions remain unresolved. Part of the problem arises from the fact that galaxies are complex, many-body systems. Typically, a galaxy consists of hundreds of millions or billions of stars, it can contain considerable quantities of interstellar gas and dust and can be subject to environmental influences through interactions with other galaxies and with the intergalactic gas. Furthermore, it is certain that dark matter is present in galaxies and in clusters of galaxies and their dynamics are largely dominated by this invisible dark component. Its nature is, however, unknown. Thus, the physics of galaxies is much more poorly understood as compared with, say, the physics of the stars.

This is an enormous subject and we will only touch on some of the more important aspects of galaxies, as well as elucidating some of the essential physics. In the next chapter, we will perform a similar exercise for clusters of galaxies.

3.1 The Revised Hubble Sequence for Galaxies

Galaxies come in a bewildering variety of shapes and forms. In order to put some order into this diversity, classification schemes have been devised on the basis of the appearances, or morphologies, of galaxies on photographic plates. The basis of most schemes is the famous *Hubble Sequence of Galaxies*, described in Hubble's monograph *The Realm of the Nebulae* (1936). The Hubble sequence, often referred to as a 'tuning-fork' diagram, arranged galaxies into a continuous sequence of types with elliptical galaxies at left-hand end and spirals at the right-hand end (Fig. 3.1). The spiral galaxies were ordered into two branches named 'normal' and 'barred' spirals.

Such classification schemes become an integral part of astrophysics when independent properties of galaxies are found to correlate with their morphological classes. This has been found to be the case for the overall properties of galaxies such as their integrated colours, the fraction of the mass of the

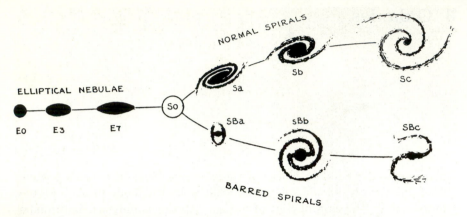

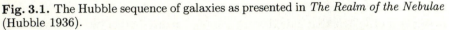

Fig. 3.1. The Hubble sequence of galaxies as presented in *The Realm of the Nebulae* (Hubble 1936).

galaxy in the form of neutral and molecular gas, and so on (Sect. 3.7). The vast majority of galaxies can be accommodated within the *revised Hubble Sequence of Galaxies*, which is described in detail by de Vaucouleurs (1974), Sandage (1975) and Kormendy (1982). It has been pointed out by van den Bergh that the classical Hubble types refer primarily to intrinsically luminous galaxies and that, in addition, there exists a large population of *dwarf galaxies*. The latter are intrinsically low luminosity galaxies and so can only be observed relatively nearby. There are also various other categories of galaxies with special characteristics, for example, the Seyfert galaxies, cD galaxies, N galaxies, radio galaxies and so on. Many of these types of galaxies contain active galactic nuclei.

Elliptical Galaxies E. These galaxies show no structural features in their brightness distributions but have an elliptical appearance, as if they were spheroids or ellipsoids of revolution (Fig. 3.2a). In absolute magnitude, elliptical galaxies range from among the most luminous galaxies known, having $M_B \approx -24$, to dwarf ellipticals (dE), which are found in the Local Group of galaxies. In the original Hubble scheme, the observed ellipticity of the galaxy was included in the morphological designation according to the rule that the number $10 \times (a - b)/a$ was written after the letter E, where a and b are the observed major and minor axes of the ellipse. Thus E0 galaxies are circular and E7 galaxies, the most extreme ellipticities found in elliptical galaxies, have $b/a = 0.3$. Galaxies flatter than E7 all show a distinct disc and bulge structure and hence are classified as lenticular (S0) rather than E galaxies.

Spiral Galaxies S, SA, SB. The characteristic feature of spiral galaxies is that they have a disk-like appearance with well-defined spiral arms emanating from their central regions (Fig. 3.2b and c). Very often the spiral pattern is double with a remarkable degree of symmetry with respect to the centre of the galaxy, but many more complicated configurations of spiral structure

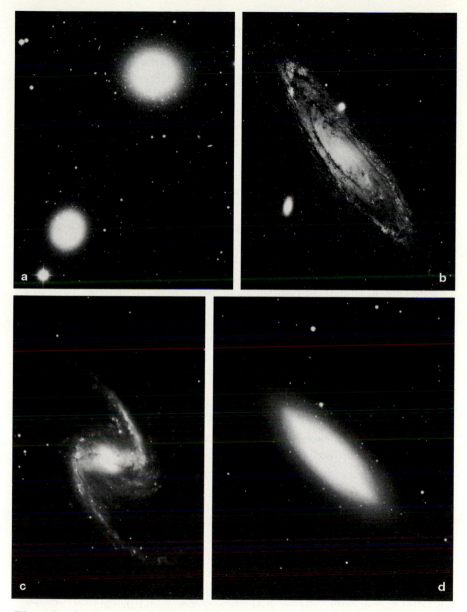

Fig. 3.2a-d. Examples of different types of normal galaxy. (**a**) The elliptical galaxies NGC 1399 and 1404, the brightest members of the Fornax cluster of galaxies. (**b**) The normal spiral galaxy M31, or the Andromeda Nebula (Courtesy of the Mount Palomar Observatory). (**c**) The barred spiral galaxy NGC 1365. (**d**) The S0 or lenticular galaxy NGC 3115. (Figs. (**a**), (**c**) and (**d**) courtesy of D. Malin and the Anglo-Australian Observatory.)

are known. The light distributions of what Hubble termed 'normal' spiral galaxies (or SA galaxies) can be decomposed into a *central bulge* or *spheroidal component*, similar in character to an elliptical galaxy, and a *disk component*, within which the spiral arms lie. In the case of the *barred spirals* (or SB galaxies), the central bulge has an elongated or ellipsoidal appearance, the spiral arms originating from the ends of the bar (Fig. 3.2c). There are as many 'barred' spiral galaxies as 'normal' spirals and, furthermore, there are just as many spirals intermediate between these two classes.

Spiral galaxies are classified as Sa, Sb, Sc according to the following criteria, in decreasing order of importance: (i) the openness of the winding of the spiral arms, (ii) the degree of resolution of the arms into stars and (iii) the size of the spheroidal component or central bar relative to the disk component. Thus,

- *Sa galaxies* have tightly wound spiral arms which are smooth showing no resolution into stars. The central bulge or bar is dominant, shows no structure and is unresolved into star clusters.
- *Sb galaxies* have more open spiral arms, which show resolution into stars. The central spheroidal component or bar is generally smaller than in Sa galaxies.
- *Sc galaxies* have very open spiral arms which are patchy and are resolved into star clusters and regions of ionized hydrogen. The spheroidal component is very small. In barred spiral galaxies, the bar is resolved into clusters and HII regions and is not as prominent as in classes Sa or Sb.
- The revised Hubble scheme extends this classification beyond Sc to include 'nearly chaotic' structures which would have been classified as very late Sc spirals in the standard sequence but are now as classified Sd spirals.

These morphological classes are rather broad and intermediate stages along the sequence are defined as Sab, Sbc and Scd.

Lenticular Galaxies S0 *or* L. All galaxies with smooth light distributions and axial ratios $b/a < 0.3$ show evidence of a disk-like component and these are called *lenticular* (lens-like) or S0 galaxies (Fig. 3.2d). They are similar to spiral galaxies in that their light distributions can be decomposed into a central bulge, similar in properties to elliptical galaxies, and an extensive disk. The lenticular galaxies appear intermediate in morphological type between elliptical and spiral galaxies.

In many cases, the central bulges of the S0 galaxies have a bar-like appearance and hence, as in the case of spirals, they can be divided into 'ordinary' and 'barred' lenticulars as well as intermediate types. In a number of lenticular galaxies, there is evidence for obscuring matter, often in the form of rings. In the revised Hubble classification, the lenticular galaxies which are free of obscuring matter are termed 'early' S0$^-$ with stages S0^0 and S0$^+$ representing 'later' stages with increasing amounts of obscuring material. By the intermediate stage between lenticular and spiral galaxies, S0/a, the obscuring matter begins to show 'incipient spiral structure'.

Irregular Galaxies. In Hubble's original classification, irregular galaxies were systems 'lacking both dominating nuclei and rotational symmetry' and the class included everything which could not be readily incorporated into the standard Hubble sequence. Many of these irregulars were similar to the companion galaxies of our own Galaxy, the Magellanic Clouds, and these became known as Irr I or Magellanic irregulars. There remained a small class of irregulars consisting of galaxies such as M82, NGC 520 and NGC 3077, in which there was no evidence of resolution into stars; these galaxies were classified Irr II galaxies.

Evidence that the Irr I galaxies form a natural extension of the Hubble sequence was provided by de Vaucouleurs' discovery of weak but definite spiral structure in the Large Magellanic Cloud (LMC). Galaxies like the LMC can be considered to belong to stages later than Sd and are denoted Sm. Thus, the late stages of the sequence reads: Scd, Sd, Sdm, Sm, Im. The Irr II systems find no natural place in the revised sequence and are designated I0 by de Vaucouleurs. The characteristics of the I0 irregular galaxies are that they are very rich in interstellar matter and contain young stars and active regions of star formation.

Table 3.1. The revised Hubble sequence of galaxies according to de Vaucouleurs' classification. (de Vaucouleurs 1974)

-6	-5	-4	-3	-2	-1	0	1	2	3	4	5	6	7	8	9	10	11
E^-	E^0	E^+	$S0^-$	$S0^0$	$S0^+$	S0a	Sa	Sab	Sb	Sbc	Sc	Scd	Sd	Sdm	Sm	Im	I0

In the Revised Hubble Sequence, shown in tabular form in Table 3.1, the various stages along the sequence are assigned numbers ranging from -5 to 10. All transitions along the sequence are smooth and continuous. The frequencies with which different types of galaxy are found among catalogues of bright galaxies are shown in Tables 3.2a and b (de Vaucouleurs 1963). Striking features of this table are the large percentage of lenticular galaxies and the roughly equal proportions of normal, barred and intermediate spiral galaxies. The latter statistics indicate that, in well over half the known examples of spiral galaxies, there are bar-like structures in their central regions and this has important implications for the origin of spiral structure.

Table 3.2a. The frequencies with which galaxies of different morphological types are found among samples of bright galaxies (from de Vaucouleurs 1963).

Class of Galaxy	E	L	S	Im	I0	Pec	Total
Number	199	329	934	39	13	14	1528
Percentage	13.0	21.5	61.1	2.55	0.85	0.9	100

Table 3.2b. The frequencies of different sub-types among 994 spiral galaxies (from de Vaucouleurs 1963).

Class of Galaxy	0/a	a	ab	b	bc	c	cd	d	dm	m	?	Total	Percentage
SA	17	25	25	57	57	82	30	9	3	4	2	311	31.3
SAB	13	15	23	45	50	71	35	11	3	7	1	274	27.6
SB	26	43	33	83	27	55	27	28	9	30	10	366	36.8
S	4	1	0	6	1	13	1	10	0	0	7	43	4.3

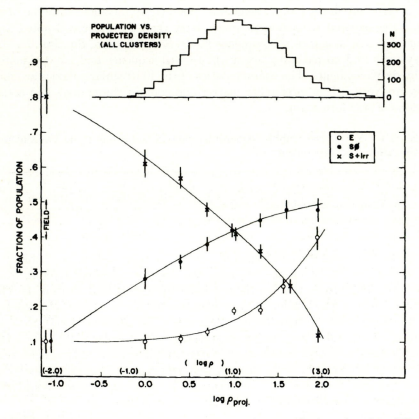

Fig. 3.3. The fractions of different morphological types of galaxy found in different galaxy environments. The local number density of galaxies is given as a projected surface density, n_{proj} of galaxies, that is, numbers Mpc^{-2} (Dressler 1980).

The figures given in Table 3.2 must be treated with considerable caution. First of all, the galaxies included in the above statistics are those present in bright galaxy catalogues. They therefore refer to objects of a very wide range of intrinsic luminosities, from among the most luminous galaxies known, such as M87, to nearby dwarf galaxies. Secondly, as mentioned above, the classical Hubble types refer almost exclusively to the most luminous galaxies. Third, the above statistics include galaxies belonging to the general field, to weak groups and to rich clusters of galaxies and it is known that the percentages of the different morphological types vary with the galaxy environment. Dressler (1980) has plotted the frequency of different galaxy types as a function of the number density of galaxies in which they are found (Fig. 3.3). Field galaxies, that is, galaxies which are not members of groups or clusters of galaxies, are located towards the left of the diagram, while rich clusters of galaxies are towards the right. It can be seen that, in the rich clusters, the elliptical and S0 galaxies are much more common than the spiral galaxies, whereas in the general field, most galaxies are spirals. Evidently, the environment in which a galaxy finds itself is correlated with its morphological characteristics. It is entirely plausible that there is a causal relation between morphology and galaxy environment.

3.2 Peculiar and Interacting Galaxies

The revised Hubble classification can encompass the forms of virtually all galaxies. There are, however, galaxies with very strange appearances and these are referred to collectively as *peculiar galaxies*. A few galaxies are known in which the stellar component is in the form of a ring rather than a disc or spheroid, the Cartwheel (Fig. 3.4) being a beautiful example of this type of galaxy – they are known as *ring galaxies*.

It is likely that many of these remarkable structures are due to gravitational interactions or collisions between galaxies. In 1973, Toomre and Toomre carried out pioneering computer simulations of close encounters between galaxies which showed how such events could give rise to remarkable asymmetric structures. In Fig. 3.5a, a deep image of the pair of interacting galaxies known as the Antennae is shown, revealing the extraordinary long 'tails' which seem to be emanating from the pair of closely interacting galaxies. The Toomres showed how even such a structure could be accounted for by a gravitational interaction between two spiral galaxies. In the simulation, the two spiral galaxies pass close to each other on prograde orbits, that is, the rotation axes of the two discs are parallel and also parallel to the rotational axis of the two galaxies about their common centre of mass. In this simulation, the spiral galaxies are represented by differentially rotating discs of stars and, while they are at their distance of closest approach, the stars in the outermost rings feel the same mutual force acting upon them for a much longer time than if the passage had been, say, retrograde. As a result, the outer rings of stars are stripped off and they form the types of extended structure observed in the Antennae (Fig. 3.5b). Similar structures are found

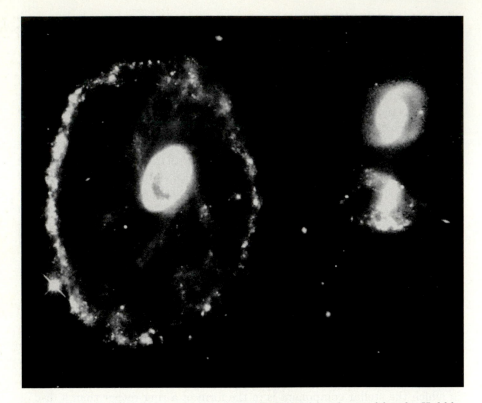

Fig. 3.4. The peculiar galaxy known as the Cartwheel as observed by the Hubble Space Telescope (Courtesy of NASA and the Space Telescope Science Institute). Its strange appearance is almost certainly due to a recent collision or strong interaction with one of its nearby companions. The simulations by Toomre and Toomre show that such a 'tidal wave' is expected if a compact mass had passed through a spiral galaxy close to its centre (Toomre 1974).

in recent supercomputer simulations of colliding and interacting galaxies, for example, in the splendid video by Barnes (1992).

Interactions between galaxies have assumed a central role in many aspects of galactic evolution. From the observational point of view, an important clue was provided by observations made by the IRAS satellite which showed that colliding galaxies are among the most luminous extragalactic far-infrared sources. The inference is that, when galaxies collide, the interstellar media in the galaxies are compressed to high densities and the rate of star formation is greatly enhanced, resulting in intense far-infrared emission.

Collisions between galaxies have also assumed a central role in models of galaxy formation. In perhaps the most popular scenario of galaxy formation, the *cold dark matter picture*, galaxies are built up by the process of hierarchical clustering in which larger galaxies are formed by the coalescence

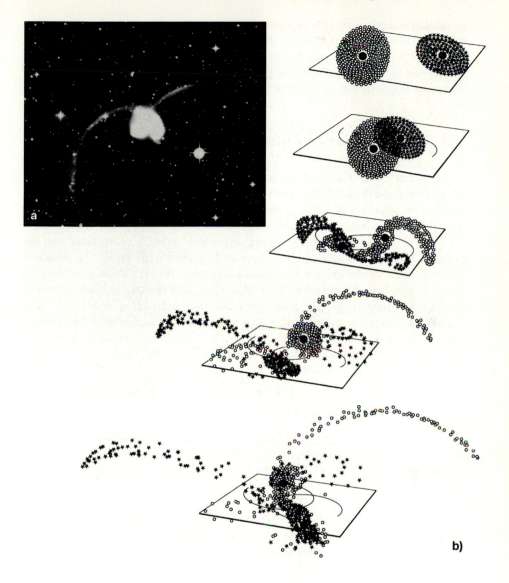

Fig. 3.5. (a) The strange pair of interacting galaxies known as the Antennæ, named after the elongated structures apparently torn off in a collision between the galaxies (Courtesy of the UK Schmidt Telescope and the ROE Photolabs). (b) A simulation of a close encounter between two disc galaxies which approach each other on prograde orbits. It can be seen that the outer rings of stars are torn off each galaxy, forming the remarkable 'Antenna' structures (Toomre and Toomre 1972).

of smaller galaxies. In this picture, strong gravitational encounters between galaxies are essential in forming the structures we observe today. An important point is to note the small percentage of peculiar and interacting systems among the present population of galaxies. We will find that this percentage increases dramatically as we look further and further back in time.

3.3 The Luminosity Function of Galaxies

The frequency with which galaxies of different intrinsic luminosities are found in space is described by the *luminosity function* of galaxies. The luminosity function of galaxies $n(L)\,\mathrm{d}L$ is defined to be the space density of galaxies with intrinsic luminosities in the range L to $L + \mathrm{d}L$. If S is the flux density (in W m^{-1} Hz^{-1}) of a nearby galaxy, for which redshift corrections can be neglected, the luminosity of the galaxy is $L = 4\pi r^2 S$ (in W Hz^{-1}), where r is the distance of the galaxy. In optical astronomy, it is traditional to work in terms of absolute magnitudes, M, rather than luminosities and so, in terms of absolute magnitudes, the luminosity function $n(L)\,\mathrm{d}L = n(M)\,\mathrm{d}M$. The important difference between these two forms of the luminosity function is that, in terms of magnitudes, the luminosity function is presented on a logarithmic scale of luminosity. The absolute magnitude M^* and the luminosity L^* are related by the expression

$$\log\left(\frac{L}{L^*}\right) = -0.4\,(M - M^*). \tag{3.1}$$

In 1977, Felten made a careful comparison of 9 different determinations of the luminosity function for bright galaxies, reducing them all to the same value of Hubble's constant, the same magnitude system and the same corrections for Galactic extinction. In this heroic analysis, he found that the independent determinations were in remarkably good agreement. Felten's analysis is summarized in Fig. 3.6, using reduced absolute magnitudes, $M_{B_T^0}$, in de Vaucouleurs' B_T^0 magnitude system and using a Galactic extinction law $A_B = 0.25\,\mathrm{cosec}\,|b|$. The solid line shows a best fit to the data of the form of luminosity function proposed by Schechter (1976)

$$n(x)\,\mathrm{d}x = \phi^* x^a \mathrm{e}^{-x}\,\mathrm{d}x, \tag{3.2}$$

where $x = L/L^*$ and L^* is the luminosity which characterises the 'break' in the luminosity function seen in Fig. 3.6. The form of the Schechter luminosity function is as simple as it could be – a power law with a high luminosity exponential cut-off. It is characterised by two parameters, the slope of the power-law a at low luminosities and the 'break' luminosity L^*.

It is traditional among optical astronomers to write the luminosity function in terms of astronomical magnitudes rather than luminosities and then the beautiful simplicity of the Schechter function is somewhat spoiled:

$$n(M)\,\mathrm{d}M = \tfrac{2}{5}\phi^* \ln 10 \left\{\mathrm{dex}[0.4(M^* - M)]\right\}^{a+1}$$
$$\times \exp\left\{-\mathrm{dex}[0.4(M^* - M)]\right\}\,\mathrm{d}M, \qquad (3.3)$$

where M^* is the absolute magnitude corresponding to the luminosity L^*. We have used the notation dex y to mean 10^y. In his reassessment of the luminosity function for galaxies, Felten (1985) preferred the following best-fit values: $a = -1.25$ and $M_{B_T^0} - 20.05 + 5\log_{10}h$. Binggeli $et\ al.$ (1988) reviewed a number of more recent determinations of the overall luminosity function for field galaxies and found roughly the same value of $M_{B_T^0}$, but the typical value of a was about -1.0.

The normalisation factor ϕ^* determines the space density of galaxies and allowance has to be made for the fact that the galaxies used in the determination mostly lie within the local supercluster. Hence, the value of ϕ^* is an overestimate as compared with what would be found for a sample of genuine field galaxies. Felten's preferred value of ϕ^* for the general field was $1.20 \times 10^{-2}h^3$ Mpc^{-3}. More recently, Loveday $et\ al.$ (1992) found the best fitting parameters for the field galaxy population to be $a = -1$, $M_{B_J} = -19.5 + 5\log h$ and $\phi^* = 1.40 \times 10^{-2}h^3$ Mpc^{-3}, where B$_J$ rather than B$_T^0$ magnitudes have been used.

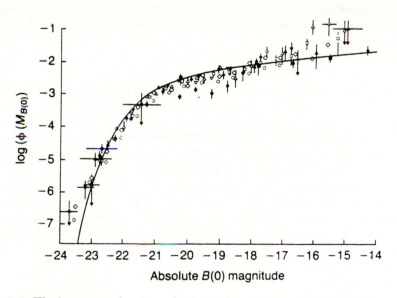

Fig. 3.6. The luminosity function of galaxies from 9 independent estimates considered by Felten (1977).

A number of features of the luminosity function of galaxies should be noted.

- Ideally, one would like to determine the luminosity function for galaxies of different morphological types in different environments, such as groups, clusters and the general field. The best-fits to the luminosity functions for E/S0 galaxies and Sp/Irr suggest that they are different. The detailed studies of Binggeli *et al.* (1988) of the luminosity functions of galaxies in the Virgo cluster and in the local general field indicate that there are indeed important differences between the functions for different classes of galaxy. This fact combined with the differences in the relative numbers of galaxies of different morphological types as a function of galaxy density indicate that the approximation of a universal luminosity function for all galaxies wherever they are found in the Universe is, at best, a crude approximation.

- Abell (1962) suggested that the luminosity of the break in the luminosity function of rich clusters L^*, corresponding to M^*, could be used as a 'standard candle' in the redshift–apparent magnitude relation. He found excellent agreement with the expected slope of the redshift–magnitude relation using this technique. Subsequent studies of the luminosity functions of individual clusters of galaxies have show that they are similar in form to the standard Schechter function with more or less the same parameters as those described above. Schechter (1976) found that if only those clusters for which good fits to his proposed function were included, the dispersion in absolute magnitude of M^* was only 0.25 magnitudes, which is as good a result as has been obtained from studies of the brightest galaxies in clusters.

- At the very highest luminosities, the brightest galaxies in clusters do not fit smoothly onto an extrapolation of the Schechter luminosity function. These massive galaxies, a splendid example of which can be seen in Fig. 4.1, are known as cD galaxies, their characteristic being that they similar to giant elliptical galaxies but in addition possess extensive stellar envelopes. They are the most luminous galaxies found in rich clusters and groups of galaxies. It appears that these are very special galaxies and not just the highest luminosity members of the luminosity function of galaxies, as shown statistically by Tremaine and Richstone (1977). Evidently, there must be some physical reason why the first ranked cluster galaxies have these unique properties – we will return to this issue in Chap. 4.

- The luminosity function is quite poorly known at low luminosities, because these galaxies can only be observed in nearby groups and clusters. According to Binggeli *et al.* (1988), the lowest luminosity regions of the luminosity function are exclusively associated with irregular and dwarf elliptical galaxies.

An important calculation is the integrated luminosity of all the galaxies within a given volume of space. For a cluster of galaxies, the result would be the integrated optical luminosity of the cluster; if this were a typical unit volume of space, the result would be the luminosity density of the radiation due to all the galaxies in the Universe. Although the number of galaxies in the luminosity function diverges at low luminosities, the total light remains finite. The luminosity density is

$$\varepsilon_{B(0)} = \int_0^\infty n(L)\,\mathrm{d}L = \phi^* L^* \int_0^\infty x^{a+1} e^{-x} dx$$
$$= \phi^* L^* \Gamma(a+2), \tag{3.4}$$

where Γ is the gamma-function. For a cluster of galaxies, ϕ^* is the normalization factor in the luminosity function. To evaluate the luminosity density of a typical volume of space, we adopt the values for the field luminosity function quoted above, $a = -1.25$, $\phi^* = 1.2 \times 10^{-3} h^3$ Mpc^{-3} and $M^* = -20.05 + 5\log_{10}h$, corresponding to $1.24 \times 10^{10} h^{-2}\, L_\odot$. Then,

$$\varepsilon_{B(0)} = 1.8 \times 10^8 h L_\odot\ \mathrm{Mpc}^{-3}. \tag{3.5}$$

A number of independent estimates are in agreement with this figure within about 50%.

A useful reference value for cosmological studies is the average mass-to-luminosity ratio for the Universe, if it is assumed to have the critical cosmological density, $\varrho_c = 3H_0^2/8\pi G = 2.0 \times 10^{-26} h^2$ kg m^{-3}, this density corresponding to that of the Einstein-de Sitter world model, the simplest of all the standard cosmological models (see Sects. 7.2.2 and 7.2.3). In terms of solar units, the mass-to-luminosity ratio would be

$$\frac{\varrho_c}{\varepsilon_B} = \left(\frac{M}{L}\right)_B = 1600\,h\left(\frac{M_\odot}{L_\odot}\right)_B. \tag{3.6}$$

Quite a range of values of M/L for the Universe as a whole is found in the literature. Felten quotes what he believes to be a lower limit to M/L of $920\,hM_\odot/L_\odot$ while Binggeli et al. prefer a value of $1670\,hM_\odot/L_\odot$. The importance of these values is that they are significantly greater than the typical mass-to-luminosity ratios of galaxies and clusters of galaxies, even when account is taken of the dark matter which must be present. This result has suggested that the mass present in galaxies and clusters of galaxies is not sufficient to close the Universe.

For many purposes it is useful to have figures for the mean space density and luminosity of galaxies. Using the mean luminosity of galaxies for Felten's best estimate of the luminosity function with $a = -1.25$, we find $\langle L \rangle = 1.25 L^* = 1.55 \times 10^{10} h^{-2}\, L_\odot$. Adopting the mean luminosity density of the Universe given by (3.5), the typical number density of galaxies $\bar{n} = \varepsilon_{B(0)}/\langle L \rangle = 10^{-2} h^3$ Mpc^{-3}. In other words, the typical galaxies which contribute most of the integrated light of galaxies are separated by a distance of about $5h^{-1}$ Mpc, if they were uniformly distributed in space, which we know to be very far from the truth. For reference, galaxies such as our own and M31 have luminosities $L_{\mathrm{Gal}}(B) = 10^{10} L_\odot$. Evidently, if the ratio of mass-to-luminosity were the same for all galaxies, the 'mean' galaxies would also contribute most of the visible mass in the Universe.

3.4 The Masses of Galaxies

All direct methods of measuring masses in astronomy are dynamical. For systems such as star clusters, galaxies and clusters of galaxies, it can generally be assumed that they have reached some form of dynamical equilibrium and then, by measuring the velocities of the stars or galaxies in the system and knowing its dimensions, mass estimates can be made. A key result for determining the masses of galaxies and clusters of galaxies is the *Virial Theorem*, first derived for star clusters by Eddington in 1915.

3.4.1 The Virial Theorem for Clusters of Stars, Galaxies and Clusters of Galaxies

Star clusters, galaxies and clusters of galaxies can be considered to be gravitationally bound configurations, meaning that the objects of which they are composed have come into dynamical equilibrium under gravity. This supposition is supported by comparison of the *crossing time* of an object within the system with its age. The crossing time is defined to be $t_{cr} = R/\langle v \rangle$ where R is the size of the system and $\langle v \rangle$ is the typical velocity, or velocity dispersion of the objects of which it is composed. For example, the orbital rotational velocity of the stars in our Galaxy at our distance from its centre, 8.5 kpc, is about 220 km s^{-1}. Therefore, the time it takes the stars to make one complete rotation about the centre of our Galaxy is $t = 2\pi R/v \approx 2.5 \times 10^8$ years. This is very much less than the age of the Galaxy, which is about 10^{10} years, and so the system must be gravitationally bound. Similarly, in systems like the Coma cluster of galaxies, the crossing-time is less than about one tenth the age of the Universe, indicating that the cluster must be a bound system, or else the galaxies would have dispersed long ago.

The *virial* was first introduced by Clausius in 1870 in connection with the thermal energy of gases. The virial was defined to be the quantity $\Xi_i = -\frac{1}{2}\langle \mathbf{r}_i \cdot \mathbf{F}_i \rangle$ where the force $\mathbf{F}_i$ acts on the particle i located at the position $\mathbf{r}_i$. The angle brackets represent the time average of the force acting on the particle and Clausius showed that Ξ_i is then its average kinetic energy. In the astronomical context, the theorem refers to the energy balance in systems in equilibrium under gravity and it is found in a variety of different guises. In the case of the internal properties of stars, the virial theorem describes the relation between the thermal energy of the gas, its gravitational potential energy, its magnetic energy, the energy in the form of turbulence, rotational energy and so on. In stellar dynamics, in which the 'gas' of stars may be considered collisionless, the *tensor virial theorem* relates the equilibrium state to the energies associated with the velocity distribution of the stars at each point, which will in general be anisotropic (Binney and Tremaine 1987). We will consider the simplest form of virial theorem for a self-gravitating system of point masses.

Suppose the particles (stars or galaxies), each of mass m_i, interact with each other only through their mutual forces of gravitational attraction. Then, the acceleration of the ith particle due to all other particles may be written vectorially

$$\ddot{\mathbf{r}}_i = \sum_{j \neq i} \frac{Gm_j(\mathbf{r}_j - \mathbf{r}_i)}{|\mathbf{r}_i - \mathbf{r}_j|^3}. \tag{3.7}$$

Now, take the scalar product of both sides with $m_i \mathbf{r}_i$.

$$m_i \mathbf{r}_i \cdot \ddot{\mathbf{r}}_i = \sum_{j \neq i} Gm_i m_j \frac{\mathbf{r}_i \cdot (\mathbf{r}_j - \mathbf{r}_i)}{|\mathbf{r}_i - \mathbf{r}_j|^3}. \tag{3.8}$$

Now

$$\frac{d}{dt}(\mathbf{r}_i \cdot \mathbf{r}_i) = 2\dot{\mathbf{r}}_i \cdot \mathbf{r}_i, \tag{3.9}$$

and hence

$$\frac{1}{2}\frac{d^2}{dt^2}(\mathbf{r}_i^2) = \frac{d}{dt}(\dot{\mathbf{r}}_i \cdot \mathbf{r}_i) = (\ddot{\mathbf{r}}_i \cdot \mathbf{r}_i + \dot{\mathbf{r}}_i \cdot \dot{\mathbf{r}}_i). \tag{3.10}$$

Therefore, (3.8) can be rewritten

$$\frac{1}{2}\frac{d^2}{dt^2}(m_i \mathbf{r}_i^2) - m_i \dot{\mathbf{r}}_i^2 = \sum_{j \neq i} Gm_i m_j \frac{\mathbf{r}_i \cdot (\mathbf{r}_j - \mathbf{r}_i)}{|\mathbf{r}_i - \mathbf{r}_j|^3}. \tag{3.11}$$

Now we sum over all the particles in the system,

$$\frac{1}{2}\frac{d^2}{dt^2}\sum_i m_i \mathbf{r}_i^2 - \sum_i m_i \dot{\mathbf{r}}_i^2 = \sum_i \sum_{j \neq i} Gm_i m_j \frac{\mathbf{r}_i \cdot (\mathbf{r}_j - \mathbf{r}_i)}{|\mathbf{r}_i - \mathbf{r}_j|^3}. \tag{3.12}$$

By inspection or otherwise, we can show that

$$\sum_i \sum_{j \neq i} \mathbf{r}_i \cdot (\mathbf{r}_j - \mathbf{r}_i) = -\frac{1}{2}\sum_{\substack{i,j \\ j \neq i}}(\mathbf{r}_i - \mathbf{r}_j)^2. \tag{3.13}$$

Therefore,

$$\frac{1}{2}\frac{d^2}{dt^2}\sum_i m_i \mathbf{r}_i^2 - \sum_i m_i \dot{\mathbf{r}}_i^2 = -\frac{1}{2}\sum_{\substack{i,j \\ j \neq i}} \frac{Gm_i m_j}{|\mathbf{r}_i - \mathbf{r}_j|}. \tag{3.14}$$

Now, $\sum_i m_i \dot{\mathbf{r}}_i^2$ is just twice the total kinetic energy, T, of all the particles in the system, that is,

$$T = \frac{1}{2}\sum_i m_i \dot{\mathbf{r}}_i^2. \tag{3.15}$$

The gravitational potential energy of the system is

$$U = -\frac{1}{2}\sum_{\substack{i,j \\ j \neq i}} \frac{Gm_i m_j}{|\mathbf{r}_i - \mathbf{r}_j|}. \tag{3.16}$$

Therefore,

$$\frac{1}{2}\frac{\mathrm{d}^2}{\mathrm{d}t^2}\sum_i m_i \mathbf{r}_i^2 = 2T - |U| \qquad (3.17)$$

If the system is in statistical equilibrium

$$\frac{\mathrm{d}^2}{\mathrm{d}t^2}\sum_i m_i \mathbf{r}_i^2 = 0, \qquad (3.18)$$

and therefore

$$T = \tfrac{1}{2}|U|. \qquad (3.19)$$

This is the equality known as the *virial theorem*.

Notice that, at no point, have we made any assumption about the orbits or velocity distributions of the particles. The velocities might be random, as can often be assumed to be the case for globular clusters or spherical elliptical galaxies. Alternatively, they could be highly ordered, as in the case of the discs of spiral galaxies. In all these cases, the virial theorem must hold, if the system is in statistical equilibrium. Thus, the theorem tells us nothing about the velocity distribution of particles within the system.

Despite the elegance of the theorem, its application to astronomical systems is not straightforward. In most cases, we can only measure directly radial velocities from the Doppler shifts of spectral lines, and positions on the sky. In some cases, independent distance measures of the galaxies are available, but generally, within star clusters and clusters of galaxies, it is not possible to distinguish whether the objects are on the near-side or far-side of the cluster. In a few cases, the proper motions of the objects can be measured and then the three-dimensional space motions can be found. Generally, for clusters of galaxies, we need to make assumptions about the spatial distribution of galaxies in the cluster. For example, if we assume that the velocity distribution of the galaxies is isotropic, the same velocity dispersion is expected in the two perpendicular directions as along the line of sight and so $\langle v^2 \rangle = 3\langle v_\parallel^2 \rangle$, where $v_\parallel$ is the velocity along the line of site. If the velocity dispersion is independent of the masses of the stars or galaxies, we can find the total kinetic energy $T = (1/2)\sum_i m_i \dot{\mathbf{r}}_i^2 = (3/2)M\langle v_\parallel^2 \rangle$, where M is the total mass of the system. If the velocity dispersion varies with mass, then $\langle v_\parallel^2 \rangle$ is a mass-weighted velocity dispersion. If it is assumed that the system is spherically symmetric, we can work out from the observed surface distribution of stars or galaxies a suitably weighted mean separation R_{cl}, so that the gravitational potential energy can be written $|U| = GM^2/R_{\mathrm{cl}}$. Thus, the mass of the system can be found from the virial theorem

$$T = \tfrac{1}{2}|U| \qquad M = 3\langle v_\parallel^2 \rangle R_{\mathrm{cl}}/G. \qquad (3.20)$$

Notice that, in general, we have to estimate some characteristic velocity, or velocity dispersion, and size of the system in order to find its mass. This general result is widely applicable in astrophysics.

Expression (3.20) can be used to estimate the masses of elliptical galaxies. Doppler broadening of the widths of stellar absorption lines in galaxies can be used to estimate the velocity dispersion $\langle \Delta v_\parallel^2 \rangle$ of stars along the line of sight through the galaxy. Typical mass-to-luminosity ratios for elliptical galaxies are about $10 - 20\ M_\odot / L_\odot$.

3.4.2 The Rotation Curves of Spiral Galaxies

In the case of spiral galaxies, masses can be estimated from their *rotation curves*, that is, the variation of the rotational velocity $v_{\rm rot}(r)$ about the centre of the galaxy with distance r from its centre. Examples of the rotation curves of spiral galaxies derived from optical and radio studies are shown Fig. 3.7. In a few galaxies, there is a well defined maximum in the rotation curve and the velocity of rotation decreases monotonically with increasing distance from the centre. If this decrease continues to infinite distance, the total mass of the galaxy converges and is similar to that derived from the rotation curve in the central regions. In many cases, however, the rotational velocities in the outer regions of galaxies are remarkably constant with increasing distance from the centre.

The significance of these flat rotation curves can be appreciated from a simple application of Gauss' theorem to Newton's law of gravity. For simplicity, let us assume that the distribution of mass in the galaxy is spherically symmetric, so that we can write the mass within radius r as $M(\leq r)$. According to Gauss's law for gravity, for any spherically symmetric variation of mass with radius, we can find the radial acceleration at radius r by placing the mass within radius r, $M(\leq r)$, at the centre of the galaxy. Then, equating the centripetal acceleration at radius r to the gravitational acceleration, we find

$$\frac{GM(\leq r)}{r^2} = \frac{v_{\rm rot}^2(r)}{r} \qquad M(\leq r) = \frac{v_{\rm rot}^2(r)r}{G}. \tag{3.21}$$

For a point mass, say the Sun, $M(\leq r) = M_\odot$, and we recover Kepler's third law of planetary motion, the orbital period T being equal to $2\pi r/v_{\rm rot} \propto r^{3/2}$. This result can also be written $v_{\rm rot} \propto r^{-1/2}$ and is the variation of the circular rotational velocity expected in the outer regions of a galaxy if most of the mass is concentrated within the central regions.

If the rotation curve of the spiral galaxy is flat, $v_{\rm rot} = $ constant, $M(\leq r) \propto r$ and so the mass within radius r increases linearly with distance from the centre. This contrasts dramatically with the distribution of light in the discs, bulges and haloes of spiral galaxies which decrease much more rapidly with increasing distance from the centre. Consequently, the local mass-to-luminosity ratio must increase in the outer regions of spiral galaxies. If averages are taken over the visible regions of galaxies, global mass-to-luminosity ratios $M/L \leq 10\ M_\odot / L_\odot$ are found. This value must increase to much larger values at large values of r. These data provide crucial information about the presence of dark matter in galaxies.

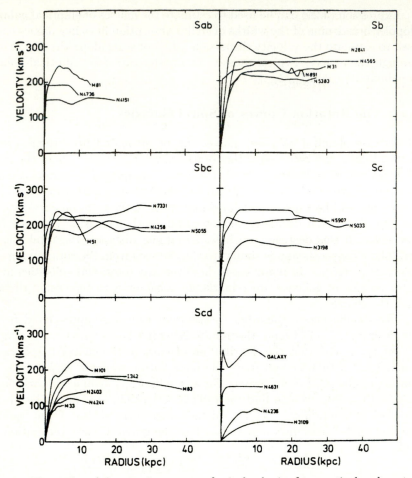

Fig. 3.7. Examples of the rotation curves of spiral galaxies from optical and neutral hydrogen observations (Kormendy 1982).

There are theoretical reasons why spiral galaxies should possess dark haloes. Ostriker and Peebles (1973) showed that, without a halo, a differentially rotating disc of stars is subject to a bar instability. Their argument has been confirmed by subsequent detailed computations and suggests that dark haloes can stabilise the discs of spiral galaxies. We will return to the thorny question of the nature of the dark matter in Chap. 4.

It is most convenient to quote the results in terms of mass-to-luminosity ratios relative to that of the Sun. For the inner parts of galaxies for which the rotation curves are well determined, mean mass-to-light ratios in the B waveband in the range $1 - 10$ are found. This is composed of two parts. The discs of spiral galaxies have M/L ratios $\approx 1 - 3 \ M_\odot/L_\odot$, whereas the bulge components have $M/L \approx 10 - 20\ M_\odot/L_\odot$, similar to the values found for elliptical galaxies.

3.5 The Properties of Elliptical Galaxies

At first glance, it would seem that the elliptical galaxies should be among the simpler stellar systems to interpret theoretically because they can be approximated as single spheroidal stellar distributions. This turns out to be an over-optimistic expectation.

3.5.1 The Light Distribution of Elliptical Galaxies

The earliest expression for the observed surface brightness distribution of elliptical galaxies as a function of radius is commonly referred to as *Hubble's law*

$$I(r) = I_0 \left(\frac{r}{r_c} + 1 \right)^{-2} , \qquad (3.22)$$

where r_c is a core radius for the galaxy. This expression provides a reasonable description of the intensity distribution in the central regions of elliptical galaxies but is a poor fit in the outer regions.

It turns out that a much better description of the surface brightness distribution of elliptical galaxies and the bulges of spiral galaxies is provided by the empirical law proposed by de Vaucouleurs (1948) which is usually referred to as the $r^{1/4}$ law

$$\log_{10} \left[\frac{I(r)}{I(r_e)} \right] = -3.3307 \left[\left(\frac{r}{r_e} \right)^{1/4} - 1 \right] . \qquad (3.23)$$

This expression provides a good representation of the luminosity profile over many decades of surface brightness. The expression has been normalised so that r_e is the radius within which half the total luminosity is emitted and $I(r_e)$ is the surface brightness at that radius. The corresponding total luminosity of the galaxy is

$$L = 7.215 \pi I_e r_e^2 \left(\frac{b}{a} \right) , \qquad (3.24)$$

where b/a is the apparent axis ratio of the elliptical galaxy (van der Kruit 1989).

The simplest physical models for the mass distribution in elliptical galaxies and stars clusters involve an analogy with isothermal gas spheres, in which a spherically-symmetric isothermal gas distribution is in hydrostatic equilibrium under gravity. We discuss these models in detail in the context of clusters of galaxies in Sect. 4.3.2. These models have a core radius $r_{1/2}$ at which the mass density falls to half the central density and they can provide a reasonable description of the central mass distribution in elliptical galaxies. The models also have the unfortunate feature of being unbounded since the density varies with radius at large values of r as $\varrho \propto r^{-2}$ and so the total mass of the galaxy is $M = \int 4\pi r^2 \varrho(r) \, dr \to \infty$ as $r \to \infty$. There are, however, good reasons why there should be a cut-off at large radii, as discussed in Sect. 4.3.2.

3.5.2 The Faber–Jackson Relation and the Fundamental Plane

Extensive studies have been made of correlations between various properties of elliptical galaxies, specifically, their luminosities, their sizes (as described by the de Vaucouleurs radius r_e), their central velocity dispersions, the abundance of heavy elements and so on. Of these, two studies are of particular importance. The first is the analysis of Faber and Jackson (1976) who found a strong correlation between luminosity L and central velocity dispersion σ of the form $L \propto \sigma^x$ where $x \approx 4$. This correlation has been studied by other authors who have found values of x ranging from about 3 to 5. The significance of this relation is that, if the velocity dispersion σ is measured for an elliptical galaxy, its intrinsic luminosity can be found from the Faber–Jackson relation and hence, by measuring its flux density, its distance can be found.

This procedure for measuring distances was refined by Dressler *et al.* (1987) and Djorgovski and Davis (1987) who introduced the concept of the *fundamental plane* for elliptical galaxies. The fundamental plane lies in a three-dimensional diagram in which luminosity L is plotted against central velocity dispersion σ and the surface brightness Σ_e within the half-light radius r_e, that is, $\Sigma_e = L(\leq r_e)/\pi r_e^2$. Dressler *et al.* found an even stronger correlation than the Faber–Jackson relation when the surface brightness was included.

$$L \propto \sigma^{8/3} \, \Sigma_e^{-3/5}. \tag{3.25}$$

They found just as good a correlation if they introduced a new diameter D_n, which was defined as the circular diameter within which the total mean surface brightness of the galaxy exceeded a particular value. The surface brightness was chosen to be 20.75 B magnitudes arcsec^{-2}. The correlation found was $\sigma \propto D_n^{3/4}$, thus incorporating the dependence of both L and Σ_e into the new variable D_n. Again, this empirical formula enables the distances of elliptical galaxies to be determined, independent of their redshifts. Dressler *et al.* estimate that, using these correlations, the distances of individual galaxies can be determined to about 25% and for clusters of galaxies to about 10%. The physical origin of these correlations are not yet understood.

3.5.3 Ellipticals Galaxies as Triaxial Systems

It might be thought that the internal dynamics of elliptical galaxies would be relatively straightforward. Their surface brightness distributions appear to be ellipsoidal, the ratio of the major to minor axes ranging from 1:1 to about 3:1. It is natural to attribute the flattening of the elliptical galaxies to the rotation of these stellar systems and this can be tested by observations of the mean velocities and velocity dispersions of the stars throughout the body of the galaxy. These measurements can be compared with the rotation and internal velocity dispersions expected if the flattening of the elliptical galaxies were wholly attributed to the rotation of an axisymmetric distribution of stars. In the simplest picture, it is assumed that the velocity distribution is isotropic at each point within the galaxy.

Bertola and Capaccioli (1975) and Illingworth (1977) first showed that elliptical galaxies rotate too slowly for centrifugal forces to be the cause of their observed flattening; in other words, the ratio of rotational to random kinetic energy is too small. This analysis has been repeated for a larger sample of elliptical galaxies and for the bulges of spiral galaxies by Davies *et al.* (1983) with the results shown in Fig. 3.8. The solid lines show the amount of rotation v_m necessary to account for the observed ellipticity of the elliptical galaxy relative to the velocity dispersion σ of the stars. It can be seen that, for low luminosity elliptical galaxies and for the bulges of spiral galaxies, the ellipticity of the stellar distribution can be attributed to rotation. The most luminous ellipticals with $M_B < -20.5$ generally do not possess enough rotation to account for the observed flattening of the galaxies. This means that the assumptions of an axisymmetric spatial distribution and/or an isotropic velocity distribution of stars at all points within the galaxy must be wrong. As a consequence, these massive elliptical galaxies must be *triaxial* systems, that is, systems with three unequal axes and consequently with anisotropic stellar velocity distributions. There is no reason why the velocity distribution should be isotropic because the time-scale for the exchange of energy between stars through gravitational encounters is generally greater than the age of the galaxy. Therefore, if the velocity distribution began by being anisotropic, it would not have been isotropised by now.

Further evidence for the triaxial nature of massive elliptical galaxies has come from studies of their light distributions. In many systems not only does the ellipticity of the isophotes of the surface brightness distribution vary with radius, but also the position angle of the major axis of the isophotes can change as well. All types of variation of ellipticity with radius are known. In some cases there is a monotonic change with radius but in others there can be maxima and minima in the radial variation of the ellipticity. The dynamics of such galaxies must be much more complicated than those of a rotating isothermal gas sphere. Another piece of evidence for the complexity of the shapes and velocity distributions within elliptical galaxies comes from the observation that, in some ellipticals, rotation takes place along the minor as

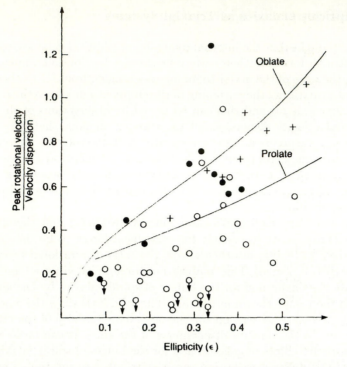

Fig. 3.8. A diagram showing the flattening of elliptical galaxies as a function of their rotational velocities. The open circles are luminous elliptical galaxies, the filled circles are lower luminosity ellipticals and the crosses are the bulges of spiral galaxies. If the ellipticity were entirely due to rotation with an isotropic stellar velocity distribution at each point, the galaxies would be expected to lie along the solid lines. This diagram shows that, at least for massive ellipticals, this simple picture of rotational flattening cannot be correct (Davies *et al.* 1983).

well as along the major axis. Thus, despite their simple appearances, some elliptical galaxies may be triaxial systems.

The theoretical position has been clarified by an elegant and original analysis by Schwarzschild (1979). By applying linear programming techniques to the determination of orbits in general self-gravitating systems, he showed that there exist stable triaxial configurations not dissimilar from those necessary to explain some of the internal dynamical properties of what appear on the surface to be simple ellipsoidal stellar distributions. His analysis showed that there exist stable orbits about the major and minor axes but not about the immediate axis of the triaxial figure. With this new understanding of the stellar motions in elliptical galaxies, galaxies can be characterised as oblate-axisymmetric, prolate-axisymmetric, oblate-triaxial, prolate-triaxial and so on.

3.6 The Properties of Spiral and Lenticular Galaxies

3.6.1 The Light Distribution in Spiral and Lenticular Galaxies

The light distributions in most spiral and lenticular galaxies can be decomposed into two components, a spheroidal component associated with the central bulge and a disc component. The luminosity profile of the spheroidal component is the same as that of an elliptical galaxy and may be described by the de Vaucouleurs $r^{1/4}$ law discussed above.

In almost all galaxies in which there is evidence of a disc component, including spirals, barred spirals and lenticulars, the luminosity profile of the disc may be represented by an exponential light distribution

$$I(r) = I_0 \exp(-r/h),\qquad(3.26)$$

where h is called the disc scale length; for our own Galaxy $h \approx 3$ kpc. The total luminosity of the disc is then $L = 4\pi h^2 I_0$. According to Freeman (1970), this luminosity profile is also found in very late type galaxies in the revised Hubble sequence, such as Magellanic Irregulars which show evidence of a disc in rotation. In 1970, Freeman discovered the remarkable result that, although the disc components of large disc galaxies have a wide range of luminosities, there is remarkably little scatter in the value of the central surface brightness I_0. A mean value of $I_0 = 21.67 \pm 0.3$ B magnitudes arcsec^{-2} was found for the galaxies in the sample studied by Freeman, the differences in total luminosity being due to variations in the scale length of the light distribution h from galaxy to galaxy. There has been some debate about the validity of this result because the samples of galaxies studied could be strongly influenced by selection effects. The key point is that the galaxies which are suitable for such detailed photometric studies have to be bright enough and large enough for precise surface photometry to be possible. Disney (1976) suggested that the constancy of the central surface brightness of the discs of spiral galaxies could be largely attributed to this selection effect. Van der Kruit (1989) surveyed a number of attempts to remove the effects of observational selection from the samples studied and concluded that 'non-dwarf galaxies do have a relatively narrow dispersion of central surface brightnesses, and this is not the result of selection effects.' It is certainly the case that, for low luminosity spiral galaxies, the values of central surface brightness are smaller Freeman's standard value.

3.6.2 The Tully–Fisher Relation

The masses of spiral galaxies can be estimated from their rotation curves as described in Sect. 3.4.2. In 1975, Tully and Fisher discovered that, for spiral galaxies, the widths of the profiles of the 21-cm line of neutral hydrogen, once corrected for the effects of inclination, are strongly correlated with their intrinsic luminosities. In their studies, they correlated the total B luminosities with the corrected velocity width ΔV of the 21-cm line and found the relation

$$L_B \propto \Delta V^\alpha, \tag{3.27}$$

where $\alpha = 2.5$. A much larger survey carried out by Aaronson and Mould (1983) found a somewhat steeper slope, $\alpha = 3.5$, for luminosities measured in the optical B waveband and an even steeper slope, $\alpha = 4.3$ in the near infrared H waveband at 1.65 μm. The correlation was found to be much tighter in the infrared as compared with the blue waveband, because the luminosities of spiral galaxies in the blue waveband are significantly influenced by interstellar extinction within the galaxies themselves, whereas, in the infrared waveband the dust becomes transparent. What has come to be called the *infrared Tully–Fisher relation* is very tight indeed. As a result, measurement of the 21-cm velocity width of a spiral galaxy can be used to infer its absolute H magnitude and hence, by measuring its flux density in the H waveband, its distance can be estimated. This procedure has resulted in some of the best distance estimates for spiral galaxies and has been used in programmes to measure the value of Hubble's constant.

There is an interesting interpretation of the Tully–Fisher relation for exponential discs. Suppose that the mass distribution follows the same distribution as the optical surface brightness with radius, $\sigma = \sigma_0 \exp(-r/h)$. Then, the total mass of the disc is

$$M = \int_0^\infty 2\pi r \sigma_0 e^{-r/h} \, dr = 2\pi\sigma_0 h^2 \int_0^\infty x e^{-x} \, dx = 2\pi\sigma_0 h^2. \tag{3.28}$$

Thus, most of the mass of the disc lies within radius $r \sim h$. The maximum of the rotation curve therefore corresponds roughly to the Keplerian velocity at distance h from the centre. Placing all the mass at the centre of the disc and equating the centripetal and gravitational accelerations, the maximum of the rotation curve is expected to correspond to V_{max} where

$$\frac{V_{max}^2}{h} \approx \frac{2\pi G\sigma h^2}{h^2} \quad ; \quad V_{max} \propto (\sigma h)^{1/2}. \tag{3.29}$$

Eliminating h from (3.28) and (3.29), we find that $M \propto V_{max}^4$. If we now adopt Freeman's result that the central surface brightnesses of bright spiral galaxies have a roughly constant value and assume that the mass-to-luminosity ratio is constant within the discs of spiral galaxies, we expect $L \propto V_{max}^4$, roughly the observed Tully–Fisher relation.

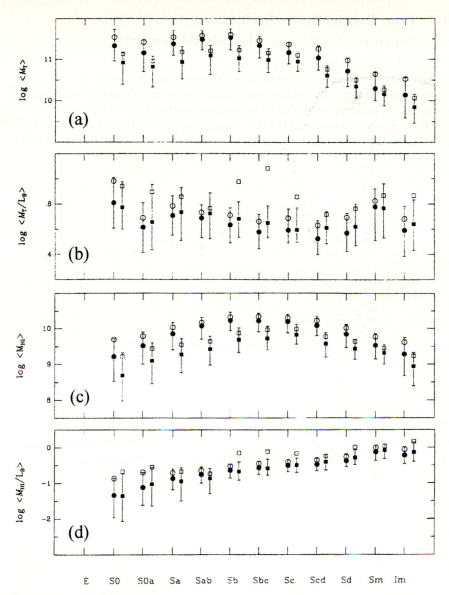

Fig. 3.9a-g. Global galaxy parameters as a function of stage along the Hubble sequence (After Roberts and Haynes 1994). The circles represent the galaxies in the RC3-UGC sample and the squares those within the local supercluster of galaxies. The filled circles are medians; the open symbols are mean values. The error bars represent the 25 and 75 percentiles of the distributions. (a) total masses, M_T; (b) total mass-to-luminosity ratio (M_T/L_B); (c) neutral hydrogen mass to total mass (M_{HI}/M_T); (d) neutral hydrogen mass to blue luminosity (M_{HI}/L_B); (e) total mass surface density (σ_T); (f) surface mass density of neutral hydrogen (σ_{HI}); (g) integrated (B − V) colour.

3.7 The Properties of Galaxies –
Correlations Along the Hubble Sequence

What gives the Hubble classification physical significance is the fact that certain independent physical properties are correlated with position along the sequence. A number of the more important correlations have been reviewed by Roberts and Haynes (1994). They emphasise that, although there are clear trends, there is also a wide dipersion about the correlations at any point along the sequence.

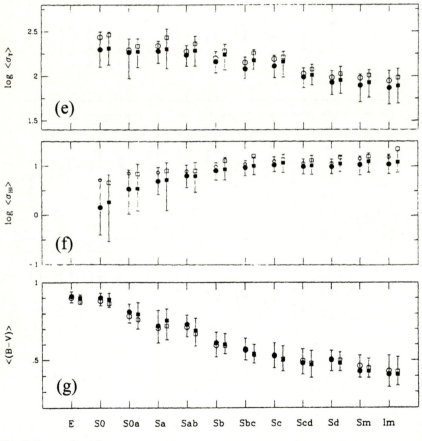

Fig. 3.9. continued.

The principal findings of Roberts and Haynes' survey are as follows:

- *Total Masses and Luminosities.* The average masses and range of masses are roughly constant for galaxies in classes S0 to Scd. At later stages beyond Scd, the masses of the galaxies decrease montonically (Fig. 3.9a).

The mass-to-luminosity ratios of the galaxies in the sample are roughly constant and so it is no surprise that the average luminosity for the S0 to Scd galaxies is roughly constant, whilst it decreases monotonically beyond Scd (Fig. 3.9b). These relations again quantify van den Bergh's remark that the classical Hubble types refer primarily to luminous, and consequently, massive galaxies.

- *Neutral Hydrogen.* There is a clear distinction between elliptical and spiral galaxies in that very rarely is any neutral hydrogen observed in ellipticals while all spiral and late-type galaxies have significant gaseous masses. The upper limit to the mass of neutral hydrogen in elliptical galaxies corresponds to $M_{\mathrm{HI}}/M_{\mathrm{tot}} \leq 10^{-4}$. For spiral galaxies, the fractional mass of the galaxy in the form of neutral hydrogen ranges from about 0.01 for Sa galaxies to about 0.15 at Sm, the increase being monotonic along the revised Hubble sequence (Fig. 3.9c). The fractional hydrogen mass is more or less independent of the mass of the galaxy at a particular point along the Hubble sequence. A consequence of the constancy of the $M_{\mathrm{tot}}/L_{\mathrm{B}}$ ratio for the galaxies in the sample is that there is also a significant trend for the ratio $M_{\mathrm{HI}}/L_{\mathrm{B}}$ to increase along the sequence (Fig. 3.9d).

- *Total Surface Density and Surface Density of Neutral Hydrogen.* These quantities change in opposite senses along the Hubble sequence. The total surface density, as determined by the total mass of the galaxy and its characteristic radius, decreases monotonically along the sequence (Fig. 3.9e), whereas the surface density of neutral hydrogen increases along the sequence (Fig. 3.9f).

- *Integrated Colour.* There is a strong correlation in the sense that elliptical galaxies are red whereas late-type galaxies are blue. This relation is shown quantitatively in Fig. 3.9g. Despite the systematic trend, there is a significant dispersion about the relation at each point in the sequence. For example, there are Sc galaxies which are red.

- *Luminosity Function of HII Regions.* In a pioneering study, Kennicutt *et al.* (1989) determined the luminosity function of HII regions in different galaxy types. Normalising to the same fiducial mass, it was found that there is a much greater frequency of HII regions in the late type galaxies as compared with early type galaxies and that the relation is monotonic along the sequence.

As Morton and Haynes point out, the obvious interpretation of these correlations is in terms of the different rates of star formation in different types of galaxy. In their interpretation, the various correlations provide information about the past, current and future star formation rates in galaxies. The correlation with colour along the sequence is related to the past star formation history of the galaxy; the changes in the luminosity function of HII regions refer to star formation rates at the present epoch; the large fraction of the mass of neutral hydrogen and its large surface density at late stages in the sequence show that these galaxies will continue to have high star formation rates in the future.

To put more flesh on this argument, the integrated colours of galaxies of different Hubble types can be plotted on a $(U - B, B - V)$ colour–colour diagram, the colours being corrected for internal and external reddening. Such a colour–colour diagram for a sample of galaxies selected from the Hubble Atlas of Galaxies is shown in Fig. 3.10a in which it can be seen that the colours of galaxies occupy a remarkably narrow region of the $(U - B, B - V)$ plane (Larson and Tinsley 1978). There is a monotonic variation of Hubble types along this locus, the bluest galaxies being the Sc and Sd galaxies and the reddest the elliptical galaxies (cf Fig. 3.9g). The colours of the galaxies cannot be represented by those of any single class of star which is hardly surprising since different classes of star make the dominant contribution at different wavelengths.

The integrated light of galaxies is principally the sum of the light of main sequence stars plus red giant stars, in particular, the K and M giants. To a rough approximation, the colours of galaxies can be represented by the sum of the numbers of luminous blue stars on the main sequence and of luminous giants on the giant branch. If all the stars in galaxies formed 10^{10} years ago, the main sequence termination point would now have reached roughly the mass of the Sun, $M \approx M_\odot$, and the brightest main-sequence stars would have spectral properties similar to that of the Sun, that is, a G2 star. There would therefore be no bright blue stars on the main sequence and the integrated light of the galaxy would be dominated by red giants. On the other hand, if star formation has continued over 10^{10} years, or if there were a burst of star formation in the recent past, there would be a significant population of hot blue stars on the main sequence giving the galaxy a significantly bluer colour.

To model the integrated colours of galaxies, evolutionary tracks for stars of different masses are required. In the simplest picture, it is sufficient to consider evolution on the main sequence and the giant branch because the subsequent phases of evolution, excursions along the horizontal branch and the planetary nebula stages, all proceed over very much shorter time scales. We write the evolution of the luminosity of a star of mass M in the i waveband as $L_i(M, t)$. As the star evolves through different stellar types and surface temperatures, the corresponding changes in its luminosity in the i band can be derived from the known spectral properties of stars of these surface properties. The birth rate of stars with masses in the range M to $M + \mathrm{d}M$ at epoch t is written $B(M, t) \, \mathrm{d}M$. Suppose a group of stars is born at epoch t_0 and there is no subsequent star formation. Then, the integrated luminosity of the group in the i waveband at epoch τ is

$$I_i(\tau) = \int L_i(M, \tau - t_0) \, B(M, t_0) \, \mathrm{d}M. \tag{3.30}$$

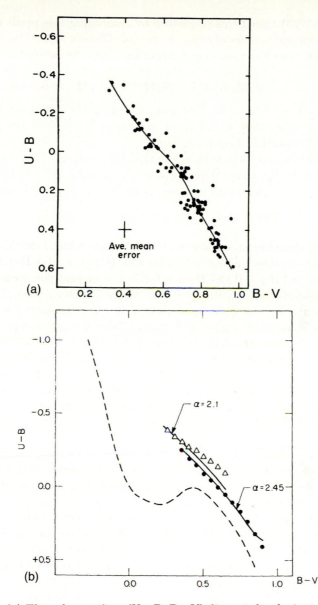

Fig. 3.10. (a) The colour–colour (U − B, B − V) diagram for the integrated colours of galaxies of different mophological types for galaxies selected from the Hubble Atlas of Galaxies (Larson and Tinsley 1978). (b) Illustrating the expected location of galaxies on the two-colour diagram for the simple models discussed in the text in which the star formation rate decreases exponentially with time. The location of main sequence stars on the colour–colour diagram is indicated by the dashed S-shaped curve. The solid lines show the predicted colours for values of τ_F which range from 10% to 100% of the age of the galaxy, the slopes of the initial mass function being assumed to be 2.45 and 2.1. The solid dots show the mean colours of Sc galaxies and the triangles the mean colours of IrrI galaxies (Searle, Sargent and Bagnuolo 1973).

Searle *et al.* (1973) tested this procedure by comparing this prediction with the luminosities and colours of open clusters of different ages. They assumed that the initial mass function was of power-law form, as derived by Salpeter (1955)

$$B(M, t_0) \, dM = B_0 M^{-\alpha} \delta(t_0) \, dM \qquad (3.31)$$

where $\delta(t_0)$ is the Dirac delta function. Adopting the exponent found by Salpeter, $\alpha = 2.35$, they found good agreement with both the observed colours and luminosities of open clusters.

To extend these results to normal galaxies, the birth rate function $B(M, t)$ must be specified. For illustrative purposes, a suitable analytic form which can accommodate a wide range of cases, is an exponential birth rate

$$B(M, t) = H(t - t_0) B_0 M^{-\alpha} \exp[-(t - t_0)/\tau_F] \qquad (3.32)$$

t_0 is the epoch when the galaxy started forming stars, which is normally taken to correspond to an epoch about 10^{10} years ago. $H(t-t_0)$ is the Heaviside step function, $H(x) = 1$ if $x \geq 0$ and $H(x) = 0$ if $x < 0$, and t is the present epoch. τ_F is the characteristic decay time over which star formation took place. If all star formation took place in a rapid burst soon after the galaxy formed, τ_F is small, that is, $\tau_F \ll (t - t_0) \sim 10^{10}$ years; if star formation has continued at a constant rate throughout the lifetime of the galaxy, $\tau_F = \infty$. The results of model calculations of this type are shown in Fig. 3.10b. It can be seen that they are in good general agreement with the observations. These results suggest that, in elliptical galaxies, most of the star formation took place in a relatively short period about 10^{10} years ago. On the other hand, in the bluest normal galaxies, Sc, Sd and Im, star formation is continuing at the present day.

Since these pioneering efforts, much more detailed models of galaxy evolution have become available. In particular, in modern studies, atlases of stellar spectra are used in order to synthesise the integrated photometric spectra of galaxies of different types, rather than just their colours. The models of Bruzual (1983), Guiderdoni and Rocca-Volmerange (1988) and Bruzual and Charlot (1993) have been widely used to interpret observations of galaxies at large redshifts, at which they are much younger than their counterparts at the present epoch. These studies have become a major industry as faint and distant samples of galaxies have become available in which the effects of evolution over cosmological timescales cannot be neglected. We will take up this story in Chap. 17.

4 Clusters of Galaxies

Associations of galaxies range from pairs and small groups of galaxies, through the giant clusters containing thousands of galaxies, to the vast structures on scales much greater than clusters, the vast 'walls' which surround the large voids seen in Fig. 2.5. Very few galaxies can be considered to be truly isolated. Rich clusters of galaxies are of particular interest because they are the largest gravitationally bound systems we know of in the Universe. In addition, hot intergalactic gas has been detected in rich clusters of galaxies, both through its bremsstrahlung X–ray emission and through the decrements which it causes in the Cosmic Microwave Background Radiation due to Compton scattering of the background photons by the electrons in the hot ionised gas. Clusters, therefore, provide laboratories for studying many aspects of galactic evolution within a well-defined astrophysical environment. Interactions of galaxies with each other and with the intergalactic medium in the cluster can be studied, as well as the distribution and nature of the dark matter, which dominates their dynamics. These are key topics for the physics of galaxy formation.

4.1 The Large-scale Distribution of Clusters of Galaxies

4.1.1 Catalogues of Rich Clusters of Galaxies

The 48-inch Schmidt telescope was constructed on Palomar mountain during the late 1940s to complement observations made with the Palomar 200-inch telescope, which was being commissioned at that time. The Palomar Sky Survey took seven years to complete and comprised 879 pairs of plates, having limiting magnitudes 21.1 in the blue and 20.0 in the red wavebands. George Abell was one of the principal observers for the survey and, while the plates were being taken, he systematically catalogued the rich clusters of galaxies appearing on the plates. The word 'rich' meant that there was no doubt as to the reality of the associations of galaxies appearing on the plates (Abell 1958). A typical example of a rich, regular cluster of galaxies is the Pavo cluster shown in Fig. 4.1. A corresponding catalogue for the southern hemisphere

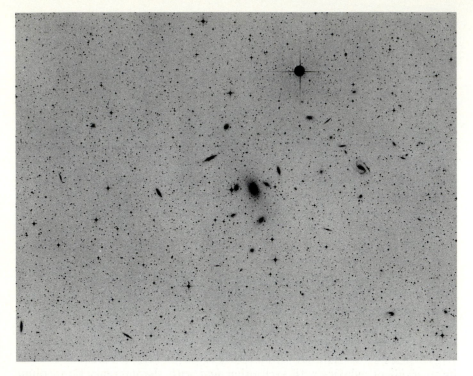

Fig. 4.1. The rich, regular cluster of galaxies in the constellation of Pavo in the southern hemisphere. The central galaxy is a supergiant or cD galaxy, which is very much brighter than all the other galaxies in the cluster. It is located close to the dynamical centre of the cluster (Courtesy of the Royal Observatory, Edinburgh).

became possible with the completion of the ESO-SERC Southern Sky Survey which was carried out by the 48-inch UK Schmidt Telescope at the Siding Spring Observatory in New South Wales (Abell, Corwin and Olowin 1989). In both cases, the clusters were found by visual inspection of the Sky Survey plates. Since that time, machine-scanned surveys of the plates have been completed using the high speed measuring machines at Edinburgh (COSMOS) and Cambridge (APM).

The 4073 clusters in the combined northern and southern catalogue of Abell, Corwin and Olowin fulfil the following selection criteria:

- *Richness Criterion.* The clusters must have 50 members brighter than 2 magnitudes $(m_3 + 2)$ fainter than the third brightest member (m_3). *Richness classes* are defined by the number of galaxies with magnitudes between m_3 and $m_3 + 2$, as described in Table 4.1. A richness class 0 is included for clusters which have between 30 and 50 members within the same magnitude interval. Studies of a number of nearby clusters of different richness classes have shown that richness is proportional to the total number of galaxies in the cluster.

- *Compactness Criterion.* Galaxies are only counted within a radius of $1.5\,h^{-1}$ Mpc of the cluster centre. The radius corresponds to an angular radius of $1.7/z$ arcmin, the redshift of the cluster being estimated from the apparent magnitude of its tenth brightest member, m_{10}.
- *Distance Criteria.* Abell clusters with redshifts less than 0.02 spanned more than one Sky Survey plate and hence this lower redshift limit is adopted. The upper redshift limit is set by the fact that galaxies could not be counted to magnitudes fainter than $m_r = 20$ and therefore the third brightest galaxy must be brighter than $m_3 = 17.5$. This redshift limit corresponds to $z \approx 0.2$. From redshift measurements of a number of clusters, it was found that m_{10}, the apparent magnitude of the tenth brightest cluster member, is a reliable distance indicator. The clusters were then divided into *distance classes* in such a way that there was a small probability of the class assigned to the cluster being more than one class wrong. The criteria are given in Table 4.2.

Table 4.1. Definitions of the richness classes of Abell clusters and the numbers of clusters within Abell's complete northern sample of 1682 clusters. N is the number of galaxies in the cluster between magnitudes m_3 and $m_3 + 2$ (Abell 1958, Bahcall 1988).

Richness Class R	N	Number of clusters in the complete northern sample
(0)[a]	$(30 - 49)$	$(\geq 10^3)$
1	$50 - 79$	1224
2	$80 - 129$	383
3	$130 - 199$	68
4	$200 - 299$	6
5	300 or more	1

[a] The sample is not complete for richness class zero.

Within the northern sample, Abell defined a *complete statistical sample* of 1682 clusters which fulfilled distance criteria 1 to 6 and richness criteria 1 to 5. The numbers of clusters in each distance and richness class in this sample are included in Tables 4.1 and 4.2. The counts of clusters in Table 4.2, $N(\leq m)$, follow closely the relation

$$\log_{10} N(\leq m) = 0.6m + \text{constant}, \tag{4.1}$$

expected for a uniform distribution of clusters, being the equivalent of a source count $N(\geq S) \propto S^{-3/2}$. The space density of Abell clusters with richness classes greater than or equal to 1 is

$$N_{\text{cl}}(R \geq 1) = 10^{-5} h^3 \, \text{Mpc}^{-3}, \tag{4.2}$$

so that the typical distance between cluster centres, if they were uniformly distributed in space, would be $\sim 50 h^{-1}$ Mpc. These figures can be compared

Table 4.2. The definitions of the distance classes of Abell clusters and the numbers of clusters in different classes within the complete sample of 1682 clusters in the northern hemisphere. The typical redshift of the distance classes is also listed (Abell 1958, Bahcall 1988).

Distance Class D	Range of m_{10}	Mean estimated redshift	Number of clusters in complete northern sample with $R \geq 1$
1	$13.3 - 14.0$	0.0283	9
2	$14.1 - 14.8$	0.0400	2
3	$14.9 - 15.6$	0.0577	33
4	$15.7 - 16.4$	0.0787	60
5	$16.5 - 17.2$	0.131	657
6	$17.3 - 18.0$	0.198	921

with the space density of 'mean galaxies' of $10^{-2}h^3$ Mpc^{-3} and their typical separations of $5h^{-1}$ Mpc (see Sect. 3.3).

4.1.2 Abell Clusters and the Large-scale Distribution of Galaxies

Abell clusters themselves are strongly correlated in space. Spatial two-point correlations functions $\xi_{cc}(r)$ have been evaluated for the complete sample of clusters in distance classes 1 to 4, for all of which redshifts have been measured, and the amplitude of the function is much greater than that of galaxies selected at random (Bahcall 1988). In her analysis, Bahcall finds that the function has the form

$$\xi_{cc}(r) = \left(\frac{r}{r_0}\right)^{-1.8} \qquad \text{where} \qquad r_0 = 26h^{-1}\,\text{Mpc}. \qquad (4.3)$$

In other words, the scale at which the cluster–cluster correlation function has a value of unity is about 5 times greater than that for the galaxy–galaxy correlation function, $r_0 = 5h^{-1}$ Mpc. It is found that the correlation function scales exactly as expected for clusters in distance classes 1-4, 5 and 6, indicating that the clustering is real and not the result of patchy Galactic or intergalactic obscuration.

Another important test is the cross-correlation of Abell clusters with the distribution of galaxies in general. Seldner and Peebles (1977) cross-correlated the counts of galaxies in the Shane–Wirtanen catalogue, which extend to apparent magnitude $m \approx 19$, with the positions of Abell clusters of distance class 5. The amplitude of this cross-correlation function is found to be intermediate between the galaxy–galaxy correlation function and the cluster–cluster correlation function

$$\xi_{gc} = \left(\frac{r}{r_1}\right)^{-2.5} + \left(\frac{r}{r_2}\right)^{-1.7}, \qquad (4.4)$$

with $r_1 = 7h^{-1}$ Mpc and $r_2 = 12.5h^{-1}$ Mpc. The enhancement of the Shane–Wirtanen counts extends to distances about $40h^{-1}$ Mpc from the cluster centres, before it disappears into the noise. The first term represents the correlation of galaxies within the Abell clusters themselves while the second represents the correlation of the Abell clusters with galaxies in the immediate neighbourhood of the cluster, but outside what would normally be considered to be the cluster boundary.

It is useful to have a simple model to interpret the amplitudes of the different correlation functions. Peebles (1980) gave the following schematic picture of the distribution of galaxies on the large scale. The basic structural unit is a *cloud of galaxies* which has scale $\sim 50h^{-1}$ Mpc. About 25% of all galaxies belong to these clouds. All Abell clusters are members of clouds and the mean number per cloud is two. The Abell clusters in the cloud contain about 25% of all galaxies in the cloud. The association of Abell clusters within clouds corresponds to Abell's superclusters. The remaining 75% of galaxies in the cloud are distributed according to the spatial correlation functions. The amplitude of the two-point correlation function for elliptical galaxies ξ_{EE} is greater than that for spiral galaxies ξ_{SS}, consistent with the morphology–galaxy density relation found by Dressler (1980) (Fig. 3.3). An equivalent way of envisaging this large-scale structure is in terms of the 'sponge' or 'shell' model for the distribution of galaxies (see Sect. 2.2 and Fig. 2.5). Most of the galaxies are located along the walls of the large voids, what are sometimes referred to as the 'cell-walls', and the rich clusters of galaxies are found at the intersection of the sheets or cell-walls. Bahcall (1988) has found that this 'shell' picture can account for the amplitudes of the galaxy–galaxy, galaxy–cluster and cluster–cluster correlation functions.

The important message of these considerations is that, although the rich clusters are certainly gravitationally bound systems, they are strongly correlated with galaxies on a much larger scale in the vicinity of the cluster.

4.2 The Distribution of Galaxies in Clusters of Galaxies

4.2.1 The Galaxy Content and Spatial Distribution of Galaxies in Clusters

A wide range of structural types is found among the clusters of galaxies. Abell (1958) classified clusters as *regular* if they are more or less circularly symmetrical with a central concentration, similar in structure to globular clusters. The member galaxies are predominantly elliptical and S0 galaxies. Some of these are among the richest of all clusters having more than 1000 members. Examples include the Coma and Corona Borealis clusters. All the other clusters in which there is much less well defined structure were classified as *Irregular* clusters. Examples of the latter include the Hercules and Virgo clusters.

Oemler (1974) studied systematically a representative sample of fifteen Abell clusters of different richness classes and morphological types and distinguished three principal types of cluster according to their *galaxy content*. These are:

- *cD clusters* have one (sometimes two) unique and dominant cD galaxy and have ratios of elliptical (E) to lenticular (S0) to spiral (S) galaxies roughly in the ratio 3 : 4 : 2, that is, only about 20% are spirals. They are very rich in E and S0 galaxies.
- *Spiral-rich clusters* have galaxy-type ratios E : S0 : S roughly 1 : 2 : 3, that is, about 50% spirals, a distribution similar to the proportions of types found in the general field.
- The remaining clusters are called *spiral-poor clusters*. They have no dominant cD galaxy and have galaxy type ratios E : S0 : S roughly 1 : 2 : 1.

Abell noted that there is a correlation between the structure of clusters and galaxy content and this result has been quantified by Oemler who established the following relations:

- In *cD clusters* or *regular clusters*, the spatial distribution of galaxies resembles the distribution of stars in a globular cluster (see Sect. 4.2.2). The space density of galaxies increases rapidly towards the centre of the cluster. The *spiral-rich clusters* and *irregular clusters* are not symmetric and there is little central concentration; the spatial density of galaxies is roughly uniform towards the central regions and is lower than that in cD clusters. The *spiral-poor clusters* are intermediate between these two extremes.
- In the case of *spiral-rich clusters*, the radial distribution of elliptical, lenticular and spiral galaxies is more or less the same. In *cD and spiral-poor clusters*, however, the relative space density of spiral galaxies decreases markedly towards the central regions, reflecting Dressler's correlation of galaxy type with galaxy number density. In these clusters, the spiral galaxies form a halo around a central core of elliptical and S0 galaxies. There is therefore *segregation by galaxy type* in cD and spiral-poor clusters.
- In addition to evidence for segregation by galaxy type, there is some evidence for *mass segregation* as a function of cluster type. Adopting apparent magnitude as a measure of mass, Oemler found that, *in cD and spiral-poor clusters*, the massive galaxies are located closer to the centre than less massive galaxies. However, this mass segregation is only important for the brightest members of clusters, $m \leq m_1 + 2$, where m_1 is the apparent magnitude of the brightest galaxy; galaxies fainter than $m_1 + 2$ appear to be uniformly distributed throughout the clusters. No such mass segregation is found in *spiral-rich clusters*.

The cD galaxies found in the centres of cD clusters have a number of quite distinct features which distinguish them from giant elliptical galaxies (Kormendy 1982). Their most distinctive feature is the presence of an extensive stellar envelope which can be as large as 100 kpc in size. In addition to being found at the centres of rich clusters, examples of cD galaxies are known in small groups of galaxies but, in all cases, they are found in regions of significantly enhanced galaxy density relative to the general field. Dressler (1984) states that cD galaxies are only found in regions of galaxy density which exceed 1 galaxy Mpc^{-3}, compared with an average galaxy density of $10^{-2}h^{-3}$ Mpc^{-3}. According to Dressler, the local galaxy density rather than the richness of the cluster determines whether or not a cD galaxy is present in a group or cluster. The central cD galaxies in rich clusters are also distinctive in having multiple nuclei which are found in about $25 - 50\%$ of all cD galaxies. In contrast, multiple nuclei are an order of magnitude less common in second and third ranked cluster members.

These results are important in understanding the dynamics and evolution of clusters. Regular, cD clusters are systems which have had time to relax to dynamical equilibrium, whereas the other systems are still in the process of relaxation.

4.2.2 Clusters of Galaxies and Isothermal Gas Spheres

In a regular cluster, the space density of galaxies increases towards the central regions, which is called the *core* of the cluster. Outside the core, the space density of galaxies decreases steadily until it becomes impossible to distinguish cluster galaxies from the background of unrelated objects. The regular structures of these clusters suggest that they have relaxed to a stationary dynamical state which should be similar to those which describe the distribution of stars in globular clusters. The spatial distribution of galaxies in a regular cluster can be represented by the distribution of mass in an *isothermal gas sphere*. These distributions are important in the discussion which follows and so let us derive the relevant expressions for the density distribution of an isothermal gas sphere.

The term *isothermal* means that the temperature, or mean kinetic energy of the particles, is constant throughout the cluster. In physical terms, this means that the velocity distribution of the galaxies is Maxwellian with the same velocity dispersion (or temperature) throughout the cluster. If all the galaxies had the same mass, the velocity dispersion would be the same at all locations within the cluster.

To derive the structure of the cluster, we use the *Lane–Emden equation*, which describes the structure of a spherically symmetric object, normally a star, in hydrostatic equilibrium. The requirement of hydrostatic equilibrium is that, at all points in the system, the attractive gravitational force acting on a mass element $\varrho\,dV$ at radial distance r from the centre of the system is balanced by the pressure gradient at that point.

$$\nabla p = \frac{\mathrm{d}p}{\mathrm{d}r} = -\frac{GM\varrho}{r^2}, \tag{4.5}$$

where M is the mass contained within radius r,

$$M = \int_0^r 4\pi r^2 \varrho(r)\,\mathrm{d}r \qquad \mathrm{d}M = 4\pi r^2 \varrho(r)\,\mathrm{d}r. \tag{4.6}$$

Reordering (4.5) and differentiating, we find

$$\frac{r^2}{\varrho}\frac{\mathrm{d}p}{\mathrm{d}r} = -GM$$

$$\frac{\mathrm{d}}{\mathrm{d}r}\left(\frac{r^2}{\varrho}\frac{\mathrm{d}p}{\mathrm{d}r}\right) = -G\frac{\mathrm{d}M}{\mathrm{d}r}$$

$$\frac{\mathrm{d}}{\mathrm{d}r}\left(\frac{r^2}{\varrho}\frac{\mathrm{d}p}{\mathrm{d}r}\right) + 4\pi G r^2 \varrho = 0. \tag{4.7}$$

This is the Lane–Emden equation. We are interested in the simple case in which the pressure p and the density ϱ are related by the perfect gas law at all radii r, $p = \varrho k_{\mathrm{B}} T/\mu$, where μ is the mass of an atom, molecule or galaxy. In thermal equilibrium, $\frac{3}{2} k_{\mathrm{B}} T = \frac{1}{2}\mu\langle v^2 \rangle$, where $\langle v^2 \rangle$ is the mean square velocity of the atoms, molecules or galaxies. Therefore,

$$\frac{\mathrm{d}}{\mathrm{d}r}\left(\frac{r^2}{\varrho}\frac{\mathrm{d}\varrho}{\mathrm{d}r}\right) + \frac{4\pi G\mu}{k_{\mathrm{B}}T}r^2 \varrho = 0. \tag{4.8}$$

Equation (4.8) is a non-linear differential equation and, in general, must be solved numerically. There is, however, an analytic solution for large values of r. If $\varrho(r)$ is expressed as a power series in r, $\varrho(r) = \sum A_n r^{-n}$, there is a solution for large r with $n = 2$. In this case,

$$\varrho(r) = \frac{2}{Ar^2} \qquad \text{where} \qquad A = \left(\frac{4\pi G\mu}{k_{\mathrm{B}}T}\right). \tag{4.9}$$

This mass distribution has the unfortunate properties that the density diverges at the origin and that the total mass of the cluster diverges at large values of r

$$\int_0^\infty 4\pi r^2 \varrho(r)\,\mathrm{d}r = \int_0^\infty \frac{8\pi}{A}\,\mathrm{d}r \to \infty. \tag{4.10}$$

There are, however, at least two reasons why there should be a cut-off at large radii. First, at very large distances, the particle densities become so low that the mean free path between collisions is very long. The thermalisation timescales are consequently very long indeed, longer than the time-scale of the system. The radius at which this occurs is known as *Smoluchowski's envelope*. Second, in astrophysical systems, the outermost stars or galaxies are stripped from the system by tidal interactions with neighbouring systems. This process defines a *tidal radius* R_t for the cluster. Therefore, if clusters are modelled by isothermal gas spheres, it is perfectly permissible to introduce a cut-off at some suitably large radius, resulting in a finite total mass.

It is convenient to rewrite (4.8) in dimensionless form by writing $\varrho = \varrho_0 y$, where ϱ_0 is the central mass density, and introducing a *structural index* or *structural length* α, where α is defined by the relation

$$\alpha = \frac{1}{(A\varrho_0)^{1/2}}. \tag{4.11}$$

Distances from the centre can then be measured in units of the structural length α by introducing a dimensionless distance $x = r/\alpha$. Then, (4.8) becomes

$$\frac{\mathrm{d}}{\mathrm{d}x}\left[x^2 \frac{\mathrm{d}(\log y)}{\mathrm{d}x}\right] + x^2 y = 0. \tag{4.12}$$

Two versions of the solution of (4.12) are listed in Table 4.3 and illustrated in Fig. 4.2. In column 2, the solution of y as a function of distance x is given; in the third column, the projected distribution onto a plane is given, this being the observed distribution of a cluster of stars or galaxies projected onto the sky. It is a simple calculation to show that, if q is the projected distance from the centre of the cluster, the surface density $N(q)$ is related to $y(x)$ by the integral

$$N(q) = 2 \int_q^\infty \frac{y(x)x}{(x^2 - q^2)^{1/2}} \, \mathrm{d}x. \tag{4.13}$$

Inspection of Table 4.3 and Fig. 4.2, shows that α is a measure of the size of the *core* of the cluster. It is convenient to fit the projected distribution $N(q)$ to the distribution of stars or galaxies in a cluster and then a *core radius* for the cluster can be defined. It can be seen that the projected density falls to the value $N(q) = 1/2$ at $q = 3$, that is, at a core radius $R_{1/2} = 3\alpha$. $R_{1/2}$ is a convenient measure of the core radius of the cluster.

Having measured $R_{1/2}$, the central mass density of the cluster can be found if the velocity dispersion of the galaxies in this region is also known. From Maxwell's equipartition theorem, we know that $\frac{1}{2}\mu\langle v^2 \rangle = \frac{3}{2}k_\mathrm{B}T$ and therefore, from the definition of α,

$$\alpha^2 = \frac{1}{A\varrho_0} = \frac{k_\mathrm{B}T}{4\pi G\mu\varrho_0} = \frac{\langle v^2 \rangle}{12\pi G\varrho_0}. \tag{4.14}$$

Observationally, we can only measure the radial component of the galaxies' velocities $v_\parallel$. Assuming the velocity distribution of the galaxies in the cluster is isotropic,

$$\langle v^2 \rangle = \langle v_x^2 \rangle + \langle v_y^2 \rangle + \langle v_z^2 \rangle = 3\langle v_\parallel^2 \rangle. \tag{4.15}$$

Expressing the central density ϱ_0 in terms of $R_{1/2}$ and $\langle v_\parallel^2 \rangle$, we find

$$\varrho_0 = \frac{9\langle v_\parallel^2 \rangle}{4\pi G R_{1/2}^2}. \tag{4.16}$$

Table 4.3. The density distribution $y(x)$ and the projected density distribution $N(q)$ for an isothermal gas sphere

x, q	$y(x)$	$N(q)$	x, q	$y(x)$	$N(q)$
0	1.0	1.0	12	0.0151	0.0839
0.5	0.9597	0.9782	14	0.0104	0.0694
1.0	0.8529	0.9013	16	0.0075	0.0591
1.5	0.7129	0.8025	20	0.0045	0.0457
2	0.5714	0.6955	30	0.0019	0.0313
3	0.3454	0.5033	40	0.0010	0.0229
4	0.2079	0.3643	50	0.0007	0.0188
5	0.1297	0.2748	100	1.75×10^{-4}	0.0101
6	0.0849	0.2143	200	5.08×10^{-5}	0.0053
7	0.0583	0.1724	300	2.32×10^{-5}	0.0036
8	0.0418	0.1420	500	8.40×10^{-6}	0.0021
9	0.0311	0.1209	1000	2.0×10^{-6}	0.0010
10	0.0238	0.1050			

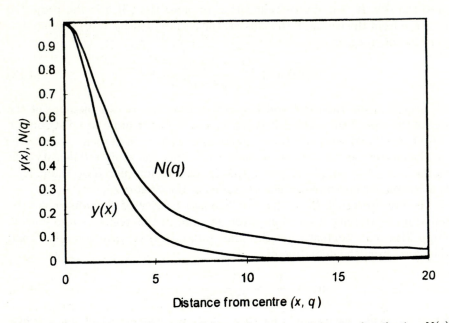

Fig. 4.2. The density distribution $y(x)$ and the projected density distribution $N(q)$ for an isothermal gas sphere.

Thus, assuming the central density of a cluster can be represented by an isothermal gas sphere, by measuring $\langle v_{\parallel}^2 \rangle$ and $R_{1/2}$, we can find the central mass density of the cluster.

An extension of the simple isothermal spheres which have proved to provide good fits to the light distribution in globular clusters and galaxies are the profiles evaluated by King (1966). The models are derived from solutions of the Fokker–Planck equation which describes the distribution function $f(v, r)$ for the stars in a cluster under the condition that there should be no particles present with velocities exceeding the escape velocity from the cluster. There is a corresponding tidal radius r_t in the numerical solutions. The forms of the luminosity profiles are shown in Fig. 4.3, in which the models are parameterised by the quantity $\log r_t/r_c$. In the limit $r_t/r_c \to \infty$, the models become isothermal gas spheres.

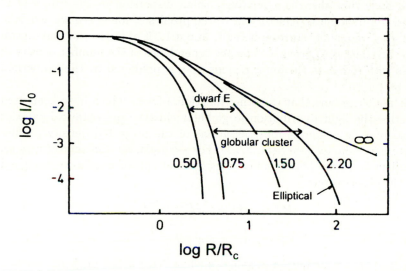

Fig. 4.3. King models for the distribution of stars in globular clusters, galaxies or clusters of galaxies (Freeman 1976, after King 1966). The curves are parameterised by the quantity r_t/r_c where r_t is the tidal radius and r_c the core radius.

The models which best represent elliptical galaxies are not too dissimilar from isothermal gas spheres. In terms of King's models, the giant elliptical galaxies have $\log r_t/r_c \approx 2.2$; for dwarf elliptical galaxies, the surface brightness decreases somewhat more rapidly in the outer regions and King models having small values of $\log r_t/r_c$ can give a good representation of the data. These profiles have also been found to provide a good description of the distribution of galaxies in clusters.

4.2.3 The Structures of Regular Clusters of Galaxies

According to Bahcall (1977), the observed distribution of galaxies in clusters can be described by truncated isothermal distributions $N(r)$ of the form

$$N(r) = N_0[f(r) - C], \qquad (4.17)$$

where $f(r)$ is the projected isothermal distribution normalised to $f(r) = 1$ at $r = 0$ and C is a constant which reduces the value of $N(r)$ to zero at some radius R_h such that $f(R_h) = C$. For regular clusters core radii lie in the range $R_{1/2} = 150 - 400$ kpc, the Coma cluster having $R_{1/2} = 220$ kpc. Bahcall finds that there is a relatively small dispersion in the values of C required to provide a satisfactory fit to the profiles of many regular clusters, typically the value of C corresponding to about 1.5% of the isothermal central density. Its effect is to provide a steeper dependence of the number density of galaxies with radius in the outer regions than is predicted by the isothermal model $\varrho(r) \propto r^{-2}$.

It is not surprising that the other functional forms which have been used to describe the light distribution in globular clusters and elliptical galaxies can also provide adequate representations of the projected density profiles in clusters of galaxies (see Sect. 3.5.1). For example, for the *central regions*, King's density profiles for globular clusters and elliptical galaxies are good fits to the observations

$$\varrho(r) = \varrho_0 \left(1 + r^2/R_c^2\right)^{-3/2}; \qquad (4.18)$$

$$N(r) = N_0 \left(1 + r^2/R_c^2\right)^{-1}. \qquad (4.19)$$

$\varrho(r)$ is the space distribution of galaxies while $N(r)$ is the projected surface distribution. For these distributions, there is a simple relation between the central values $N_0 = 2R_c\varrho_0$. De Vaucouleurs' law for elliptical galaxies (3.23) can also provide a good fit. The problem in distinguishing between these distributions is that the density profiles are only known with reasonable accuracy to projected densities of about 1% of the central values, and hence a wide range of functional forms is acceptable.

4.2.4 The Luminosity Function for Cluster Galaxies

Despite the fact that there are variations in the luminosity functions of different classes of galaxies in clusters as indicated in Sect. 3.3, the Schechter luminosity function is generally a good representation of the distribution of luminosities among cluster galaxies. Only at the very brightest end of the luminosity function is there evidence for variations in the function from cluster to cluster. In Sandage's analysis of the redshift–apparent magnitude relation, which is shown in Fig. 2.9a, he demonstrated that there is remarkably little scatter in the absolute magnitudes of the brightest members of rich clusters of galaxies, the standard deviation of the brightest cluster galaxies amounting to $\sigma_M = 0.28$ (Sandage 1988).

There has been some controversy about whether the constancy of the absolute magnitude of the first ranked cluster galaxy can be explained by randomly sampling the high luminosity end of the luminosity function, or whether there is some special property of the first ranked cluster member which is independent of the richness of the cluster. Tremaine and Richstone (1977) compared the dispersion in absolute magnitudes of the first ranked members $\sigma(M_1)$ with the mean value of the difference in magnitude between the first and second ranked members $\Delta M_{12} = \langle M_1 - M_2 \rangle$. They showed that for any statistical luminosity function $\sigma(M_1)/\Delta M_{12} = t_1 > 1$ must hold. For example, using Schechter's function, $t_1 = 1.20$. For Sandage's data on rich clusters, $t_1 = 0.48 \pm 0.10$, supporting his point of view that there is much less dispersion in the absolute magnitudes of first ranking cluster galaxies than would be expected if they were simply randomly sampled from the luminosity function. Geller and Postman (1983) failed to confirm this result. Nonetheless, as discussed by Kormendy (1982), the cD galaxies are quite distinct from normal giant elliptical galaxies.

A further classification scheme, due to Bautz and Morgan (1970) also bears upon the issue of the origins of the brightest galaxies in clusters. In the *Bautz–Morgan classification* scheme, clusters are classified according to the presence or absence of a dominant D or cD galaxy at the centre of the cluster. Bautz–Morgan class I clusters contain a dominant centrally located cD galaxy (for example Abell 2199); class II clusters have a central galaxy intermediate between a cD and giant elliptical galaxy (for example, the Coma cluster); class III clusters have no dominant central galaxy. Intermediate classes between classes I, II and III have been defined. A number of properties of clusters depend upon Bautz–Morgan class. For example, Sandage has found that there is a weak correlation between Bautz–Morgan class and the absolute magnitude of the brightest cluster member which cannot account for all the difference between the Bautz–Morgan classes. There must in addition be an inverse correlation between Bautz–Morgan class and the absolute magnitudes of the second and third brightest members in the sense that they are relatively brighter in Bautz–Morgan class III clusters. This phenomenon is illustrated by the data in Table 4.4 which lists the mean absolute magnitudes of the first, second and third brightest galaxies in a large sample of rich clusters studied by Sandage and Hardy (1973).

4.2.5 Summary of the Properties of Rich Clusters of Galaxies

Despite the wide range of types described above, there are some clear trends in the overall properties of clusters of galaxies. A sequence of types can be defined based on Abell's distinction between *regular* and *irregular* clusters with the addition of an *intermediate* class. Table 4.4 is adapted from Bahcall (1977) and summarises the properties described above as a function of

Table 4.4. The absolute magnitudes of the first, second and third ranked cluster galaxies as a function of Bautz–Morgan type (Sandage 1976). It is assumed that $h = 0.5$.

Bautz–Morgan Class	M_1	M_2	M_3
I	-23.09 ± 0.051	-22.72 ± 0.077	-22.47 ± 0.091
I-II	-23.30 ± 0.068	-22.83 ± 0.091	-22.56 ± 0.121
II	-23.37 ± 0.130	-22.60 ± 0.217	-22.34 ± 0.212
II-III	-23.46 ± 0.092	-22.18 ± 0.247	-21.96 ± 0.233
III	-23.68 ± 0.102	-22.22 ± 0.157	-21.82 ± 0.187

cluster type. Like the Hubble sequence for galaxies, these types are only part of a continuous sequence and there is considerable overlap in some of the properties.

Table 4.5. A summary of the typical properties of rich cluster of galaxies of different types (after Bahcall 1977).

Property/Class	Regular	Intermediate	Irregular
Bautz–Morgan type	I, I-II, II	(II), II-III	(II-III), III
Galaxy content	Elliptical/S0 rich	Spiral-poor	Spiral-rich
E : S0 : S ratio	3 : 4 : 2	1 : 4 : 2	1 : 2 : 3
Symmetry	Spherical	Intermediate	Irregular shape
Central concentration	High	Moderate	Very little
Central profile	Steep gradient	Intermediate	Flat gradient
Mass segregation	Marginal evidence for $m - m(1) < 2$	Marginal evidence for $m - m(1) < 2$	No segregation
Examples	Abell 2199, Coma	Abell 194, 539	Virgo, Abell 1228

4.3 Dark Matter in Clusters of Galaxies

Two of the key issues of modern astrophysics and cosmology are the nature of the dark matter, which we know must be present in galaxies and clusters of galaxies, and how much it contributes to the overall mass density of the Universe. We have already discussed the dark matter problem for galaxies in Sect. 3.4. Here we concentrate upon the problem in the context of clusters of galaxies.

4.3.1 Dynamical Estimates of the Masses of Clusters of Galaxies

It might seem that the problem of measuring the masses of clusters of galaxies is relatively straight-forward. The virial theorem (3.19) provides a simple relation between the velocity dispersion of the galaxies $\langle v_\parallel^2 \rangle$ in the bound cluster and the characteristic radius R_{cl} of the galaxy distribution. It turns out that there are remarkably few clusters for which a detailed analysis can be made. The reason is that it is essential to make a careful assessment of the galaxies which are truly cluster members and to measure accurate radial velocities for large enough samples of galaxies.

The Coma cluster is a good example of a regular rich cluster of galaxies which has been the subject of considerable study. The space density of galaxies increases smoothly towards the centre, resembling to a good approximation the spatial distribution expected of an isothermal gas sphere (Sect. 4.2.2). The inference is that the cluster has relaxed to a bound equilibrium configuration and this is confirmed by comparing the *crossing time* of a typical galaxy in the cluster with the age of the Universe. The crossing time is defined by $t_{cr} = R/\langle v \rangle$ where R is the size of the cluster and $\langle v \rangle$ is the typical random velocity for a galaxy. For the Coma cluster, taking $\langle v \rangle = 10^3$ km s^{-1} and $R = 2$ Mpc, the crossing time about 2×10^9 years, about a fifth to one tenth the age of the Universe. This is clear evidence that the cluster must be a bound system or else the galaxies would have dispersed long ago.

The surface distribution of galaxies in the Coma cluster and the variation of their velocity dispersion with radius have been determined by Kent and Gunn (1982) and are shown in Fig. 4.4. The profile fitted to the projected distribution of galaxies in Fig. 4.4a is one of King's models, with a cut-off at $R_t = 16h^{-1}$ Mpc, well outside the last data point. In Fig. 4.4b, the observed projected velocity dispersion as a function of angular radius is shown. The various lines in Fig. 4.4c illustrate possible variations of the mean velocity dispersion with distance from the cluster centre, model A corresponding to the assumption that the velocity distribution is isotropic at each point in the cluster.

These data have been analysed in detail by Merritt (1987) who considered a wide range of possible models for the mass distribution within the cluster. In the simplest reference model, with which the others can be compared, it is assumed that the mass distribution in the cluster follows the galaxy distribution, that is, that mass-to-luminosity ratio is a constant throughout the cluster, and that the velocity distribution is isotropic at each point in the cluster. With these assumptions, Merritt derived a mass for the Coma cluster of $1.79 \times 10^{15} h^{-1} M_\odot$, assuming that the cluster extends to $16h^{-1}$ Mpc. The mass within a radius of $1h^{-1}$ Mpc of the cluster centre is $6.1 \times 10^{14} M_\odot$. The corresponding value of the mass-to-blue luminosity ratio for the central regions of the Coma cluster is about $350h^{-1} M_\odot/L_\odot$.

This is the key result. The Coma cluster is a classic example of a rich regular cluster and the population is dominated in the central regions by

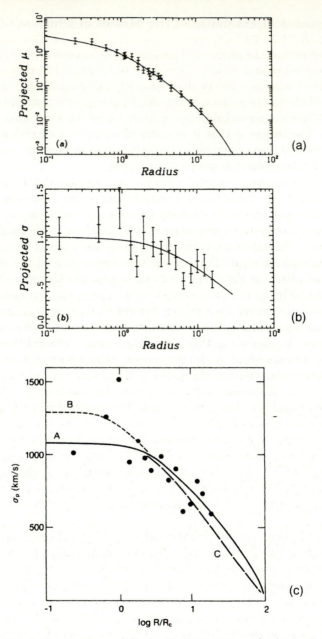

Fig. 4.4. (a) The surface density profile for the distribution of galaxies in the Coma cluster according to Kent and Gunn (1982). (b) The projected velocity dispersion as a function of angular radius for galaxies in the Coma cluster (Kent and Gunn 1982). (c) Three possible velocity dispersion profiles which are consistent with the data of Kent and Gunn, A corresponding to an isotropic velocity distribution, while B and C involve velocity anisotropies (Merritt 1987).

elliptical and S0 galaxies for which the typical mass-to-luminosity ratios are about $15\,M_\odot/L_\odot$. There is therefore a discrepancy of about a factor of 20 between the mass which can be attributed to galaxies and the total mass which must be present. This is perhaps the best defined case for the presence of dark matter in any system and was first noted by Zwicky in 1937. It is also where all the trouble begins. The dark matter dominates the mass of the cluster and there is no reason why it should have the same distribution as the visible matter. Likewise, there is no reason *a priori* why the velocity distribution of the galaxies should be isotropic. This is most simply expressed in terms of the velocity dispersion of the galaxies in the radial direction $\sigma_r^2(r)$ and the circumferential direction $\sigma_t^2(r)$ within the cluster. The assumption of isotropy is that $\sigma_r^2(r) = \sigma_t^2(r)$. If, however, the galaxies were on circular orbits about the centre, $\sigma_r^2(r) = 0$, or, if they were purely radial, $\sigma_t^2(r) = 0$. Merritt (1987) has provided a careful study of how the inferred mass-to-luminosity ratio would change for a wide range of different assumptions about the relative distributions of the visible and dark matter and the anisotropy of the velocity distribution. For the cluster as a whole, the mass-to-luminosity ratio varies from about 0.4 to at least three times the reference value, while the mass-to-luminosity ratio within $1\,h^{-1}$ Mpc is always very close to $350\,h^{-1}M_\odot/L_\odot$.

We have considered the case of the Coma cluster in some detail because for few clusters is such a detailed analysis feasible. The velocity dispersions of rich clusters are all about 10^3 km s^{-1} and they have similar length-scales. Therefore, dark matter must be present in all of them.

4.3.2 X-ray Observations of Hot Gas in Clusters of Galaxies

One of the most important discoveries of the UHURU X-ray Observatory was the detection of intense X-ray emission from rich clusters of galaxies. The nature of the emission was soon identified as the bremsstrahlung of hot intracluster gas, the key observations being the extended nature of the emission and the detection of the highly ionised iron line FeXXVI by the Ariel-V satellite. It was quickly appreciated that the X-ray emission of the gas provides a very powerful probe of the gravitational potential within the cluster. Let us repeat the simple calculation presented by Fabricant, Lecar and Gorenstein (1980) which shows how this can be done.

For simplicity, we assume that the cluster is spherically symmetric so that the total gravitating mass within radius r is $M(< r)$. The gas is assumed to be in hydrostatic equilibrium within the gravitational potential defined by the total mass distribution in the cluster, that is, by the sum of the visible and dark matter as well as the gaseous mass. If p is the pressure of the gas and ϱ its density, both of which vary with position within the cluster, the requirement of hydrostatic equilibrium is

$$\frac{\mathrm{d}p}{\mathrm{d}r} = -\frac{GM(< r)\varrho}{r^2}. \tag{4.20}$$

The pressure is related to the local gas density ϱ and temperature T by the perfect gas law

$$p = \frac{\varrho k_B T}{\mu m_H},$$ (4.21)

where m_H is the mass of the hydrogen atom and μ is the mean molecular weight of the gas. For a fully ionised gas with the standard cosmic abundance of the elements, a suitable value is $\mu = 0.6$. Differentiating (4.21) with respect to r and substituting into (4.20), we find

$$\frac{\varrho k_B T}{\mu m_H}\left(\frac{1}{\varrho}\frac{d\varrho}{dr} + \frac{1}{T}\frac{dT}{dr}\right) = -\frac{GM(<r)\varrho}{r^2}.$$ (4.22)

Reorganising (4.22),

$$M(<r) = -\frac{k_B T r^2}{G\mu m_H}\left[\frac{d(\log \varrho)}{dr} + \frac{d(\log T)}{dr}\right].$$ (4.23)

Thus, the mass distribution within the cluster can be determined if the variation of the gas density and temperature with radius are known. Assuming the cluster is spherically symmetric, these can be derived from high sensitivity X-ray intensity and spectral observations. A suitable form for the bremsstrahlung spectral emissivity of a plasma is

$$\kappa_\nu = \frac{1}{3\pi^2}\frac{Z^2 e^6}{\varepsilon_0^3 c^3 m_e^2}\left(\frac{m_e}{k_B T}\right)^{1/2} g(\nu, T) N N_e \exp\left(-\frac{h\nu}{k_B T}\right),$$ (4.24)

where N_e and N are the number densities of electrons and nuclei respectively, Z is the charge of the nuclei and $g(\nu, T)$ is the Gaunt factor, which can be approximated by

$$g(\nu, T) = \frac{\sqrt{3}}{\pi}\ln\left(\frac{k_B T}{h\nu}\right).$$ (4.25)

The spectrum of thermal bremsstrahlung is roughly flat up to X-ray energies $\varepsilon = h\nu \sim k_B T$, above which it cuts off exponentially (see, for example, Longair 1992). Thus, by making precise spectral measurements, it is possible to determine the temperature of the gas from the location of the spectral cut-off and the particle density along the line of sight from the emissivity of the gas. In practice, the spectral emissivity has to be integrated along the line of sight through the cluster. Performing this integration and converting it into an intensity, the observed surface brightness at projected radius a from the cluster centre is

$$I_\nu(a) = \frac{1}{2\pi}\int_a^\infty \frac{\kappa_\nu(r)r}{(r^2 - a^2)^{1/2}}\,dr.$$ (4.26)

Cavaliere (1980) noted that this is an Abel integral which can be inverted to find the emissivity of the gas as a function of radius

$$\kappa_\nu(r) = \frac{4}{r}\frac{d}{dr}\int_r^\infty \frac{I_\nu(a)a}{(a^2 - r^2)^{1/2}}\,da.$$ (4.27)

Ideally, one would like to measure precisely the spectrum of the X-ray emission along many lines of sight through the cluster. The problem is that, at present, X-ray telescopes either have high angular resolution and low spectral resolution, or good spectral resolution and low angular resolution. A further problem is that there is evidence that *cooling flows* are important in the central regions of a considerable fraction of all clusters which are strong X-ray emitters. In the dense central regions, the gas density is sufficiently high for the gas to cool over cosmological time-scales. The observation of peaks in the X-ray surface brightness distribution in a number of clusters and the fact that the X-ray temperature is lower in the centre than in the outer regions are the principal pieces of evidence for cooling flows.

X-rays maps of about 200 clusters were made by the Einstein X-ray Observatory (Forman and Jones 1982) and more recently the ROSAT observatory has provided high resolution maps of the X-ray emission of a number of nearby clusters (Böhringer 1995). A beautiful example of the quality of data now available is illustrated by the X-ray map of the central regions of the Virgo cluster obtained from the ROSAT All Sky Survey (Fig. 4.5a). A number of galaxies belonging to the Virgo cluster have been detected as X-ray sources, as well as a few background clusters and active galaxies. In addition, the X-ray emission of the diffuse intergalactic gas can be seen centred on the massive galaxy M87. Evidence that the intergalactic gas traces the mass distribution of the cluster is provided by comparison of the contours of the X-ray surface brightness distribution with the surface distribution of galaxies as determined by the photometric survey of the Virgo cluster by Binggeli, Tammann and Sandage (1987). The distribution of galaxies in the cluster and the diffuse X-ray emission are remarkably similar (Fig. 4.5b). This is an important result because it indicates that the dark matter, which defines the gravitational potential in the cluster and which is traced out by the distribution of hot gas, must be distributed like the galaxies. Within a radius of 1.8 Mpc, the total mass was found to lie in the range $1.5 - 5.5 \times 10^{14}$ $M_\odot$, the uncertainty in the mass resulting from uncertainties in the temperature profile. For comparison, the X-ray luminosity of the cluster is 8×10^{36} W in the energy band 0.1 to 2 keV, corresponding to a gaseous mass of $4 - 5.5 \times 10^{13}$ $M_\odot$.

Another beautiful example of the application of these procedures is to the Perseus cluster of galaxies (Böhringer 1995). In this case, the X-ray emission could be traced out to a radius $1.5h^{-1}$ Mpc. From the X-ray observations, it was possible to determine both the total gravitating mass, $M(< r)$, and the mass of gas within radius r, $M_{gas}(< r)$, and to compare these with the mass in the visible parts of galaxies. This comparison is shown in Fig. 4.6, in which it can be seen that the mass of hot intracluster gas is about five times greater than the mass in galaxies, but that it is insufficient to account for all the gravitating mass which must be present. Some form of dark matter must be present to bind the cluster gravitationally.

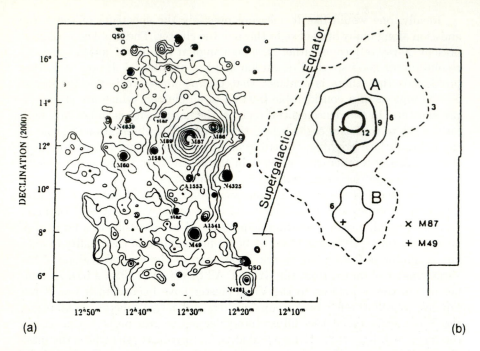

(a)   (b)

Fig. 4.5a-b. Comparison of the X-ray emission and the surface density of luminous matter in galaxies in the Virgo cluster (Böhringer 1995). (**a**) The X-ray image of the Virgo cluster from the ROSAT All Sky Survey in the X-ray energy band $0.4 - 2$ keV. The image has been smoothed with a variable Gaussian filter to enable low X-ray brightness regions to be detected. Some galaxies in the Virgo cluster have been detected as well as a few distant clusters and active galaxies. (**b**) The surface density of galaxies in the Virgo Cluster from the photometric survey of Binggeli, Tammann and Sandage (1987).

Böhringer (1995) has summarised the typical masses found in rich clusters of galaxies. The typical total masses lie in the range 5×10^{14} to 5×10^{15} $M_\odot$, of which about 5% is attributed to the mass contained in the visible parts of galaxies and about 10 to 30% to hot gas. The remaining 60 to 85% of the mass is in some form of dark matter. Typically, the iron abundance of the hot gas is between about 20 and 50% of the solar value, indicating that the intergalactic gas has been enriched by the products of stellar nucleosynthesis.

4.3.3 The Sunyaev–Zeldovich Effect in Hot Intracluster Gas

A completely independent method of studying the hot gas in clusters of galaxies is to search for decrements in the intensity of the Cosmic Microwave Background Radiation in the directions of clusters of galaxies. As the photons of the background radiation pass through the gas cloud, a few of them suffer Compton scattering by the hot electrons. Although, to first order, the photons

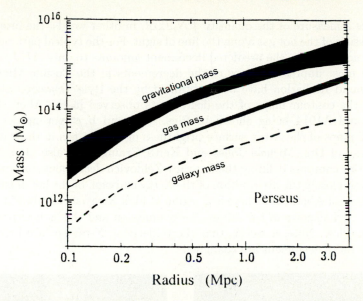

Fig. 4.6. Integrated radial profiles for the mass in the visible parts of galaxies, hot gas and total gravitating mass for the Perseus cluster of galaxies, as determined by observations with ROSAT. The upper band indicates the range of possible total masses and the central band the range of gaseous masses (Böhringer 1995).

are just as likely to gain as lose energy in these Compton scatterings, to second order there is a net statistical gain of energy. Thus, the spectrum of the Cosmic Microwave Background Radiation is shifted to slightly higher energies and so, in the Rayleigh–Jeans region of the spectrum, there is expected to be a decrement in the intensity of the background radiation in the direction of the cluster of galaxies, while in the Wien region there should be a slight excess. These predictions were made by Sunyaev and Zeldovich as long ago as 1969 but it was almost 20 years before what came to be known as the *Sunyaev–Zeldovich effect* was observed with confidence in the directions of clusters of galaxies, which were by then known to contain large masses of hot intracluster gas (Birkinshaw 1990).

The magnitude of the distortion is determined by the *Compton scattering optical depth* through the region of hot gas

$$y = \int \left(\frac{k_\mathrm{B}T_\mathrm{e}}{m_\mathrm{e}c^2}\right) \sigma_\mathrm{T}N_\mathrm{e}\,\mathrm{d}l. \qquad (4.28)$$

The resulting decrement in the Rayleigh–Jeans region of the spectrum is

$$\frac{\Delta I_\nu}{I_\nu} = -2y, \qquad (4.29)$$

(for a discussion of the physical process involved, see Longair (1992)). Thus, the magnitude of the decrement along any line of sight through the cluster provides a measure of the quantity $\int N_e T_e \, dl$, in other words, the integral of the pressure of the hot gas along the line of sight. For the typical parameters of hot intracluster gas, the predicted decrement amounts to only $\Delta I / I \approx 10^{-4}$.

The most impressive maps of the decrements in the Cosmic Microwave Background Radiation have been obtained by the Ryle Telescope at Cambridge, the contour maps of the decrements observed in the Abell clusters A1413 and A1914 being shown in Figs. 4.7a and b, superimposed upon ROSAT maps of the X-ray surface brightness of the hot gas in these clusters (Courtesy of Drs. Michael Jones and Keith Grainge; see also, Jones *et al.* 1993). The contours defining the Sunyaev-Zeldovich decrements in both clusters follow closely the distribution of the X-ray emission of the hot cluster gas. In the case of Abell 1914, there is a region of high X-ray surface brightness to the left of the centre of the diffuse X-ray emission and this is interpreted as a cooling flow. Note, however, that the underlying X-ray surface brightness and the radio decrement are coaligned.

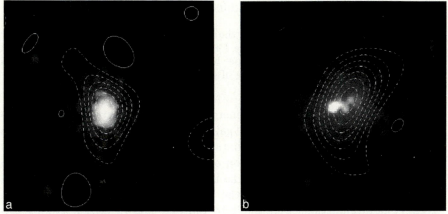

Abell 1413   Abell 1914

Fig. 4.7a-b. Comparison of the decrements in the Cosmic Microwave Background Radiation with the distribution of X-ray emission for the rich Abell clusters (**a**) Abell 1413 and (**b**) Abell 1914 (Courtesy of Drs. Michael Jones and Keith Grainge).

Maps of the Sunyaev–Zeldovich decrement have now been made for more than a dozen clusters which are known to be strong X-ray sources. In conjunction with the X-ray bremsstrahlung measurements, these observations enable the physical conditions in the intracluster gas to be determined in some detail. Indeed, with a number of reasonable assumptions, the physical properties of the hot gas are over-determined and so the physical dimensions of the gas clouds can be estimated. In conjunction with measurements of the angular sizes of the gas clouds, distances to the clusters can be mea-

sured which are independent of their redshifts, enabling direct estimates of
Hubble's constant to be made (see Sect. 8.3).

4.3.4 Gravitational Lensing by Clusters of Galaxies

A beautiful method for determining the mass distribution in clusters of galaxies has been provided by the discovery of gravitationally lensed arcs about
the central regions of rich clusters of galaxies. In his great paper of 1915 on
the General Theory of Relativity, Einstein showed that the deflection of light
by the Sun amounts to precisely twice that predicted by a simple Newtonian
calculation. According the General Relativity, the angular deflection of the
position of a background star due to the bending of space-time by a point
mass M is

$$\alpha = \frac{4GM}{pc^2},$$
(4.30)

where p is the 'collision parameter' (Fig. 4.8a). For the very small deflections
involved in the gravitational lens effect, p is almost exactly the distance of
closest approach of the light rays to the deflector.

Chwolson (1924) and Einstein (1936) realised that, if the background
star were precisely aligned with the deflecting point object, the gravitational
deflection of the light rays would result in a circular ring, centred upon the
deflector (Fig. 4.8b). It is a straightforward calculation to work out the radius
of what came to be known as an 'Einstein ring', although they should perhaps
be known as 'Chwolson rings'. In Fig. 4.8b, the distance of the background
source is D_s and that of the deflector, or lens, D_d, the distance between them
being D_{ds}. Suppose the observed angular radius of the Einstein ring is θ_E as
illustrated in Fig. 4.8b. Then, by simple geometry, since all the angles are
small,

$$\theta_E = \alpha \left(\frac{D_{ds}}{D_s} \right),$$
(4.31)

where α is the deflection given by (4.30). Therefore,

$$\theta_E = \alpha \left(\frac{D_{ds}}{D_s} \right) = \frac{4GM}{pc^2} \left(\frac{D_{ds}}{D_s} \right).$$
(4.32)

Since $p = \theta_E D_d$,

$$\theta_E^2 = \frac{4GM}{c^2} \left(\frac{D_{ds}}{D_s D_d} \right) = \frac{4GM}{c^2} \frac{1}{D},$$
(4.33)

where $D = (D_s D_d / D_{ds})$. Thus, the *Einstein angle* θ_E, the angle subtended
by the Einstein ring at the observer, is given by

$$\theta_E = \left(\frac{4GM}{c^2} \right)^{1/2} \frac{1}{D^{1/2}}.$$
(4.34)

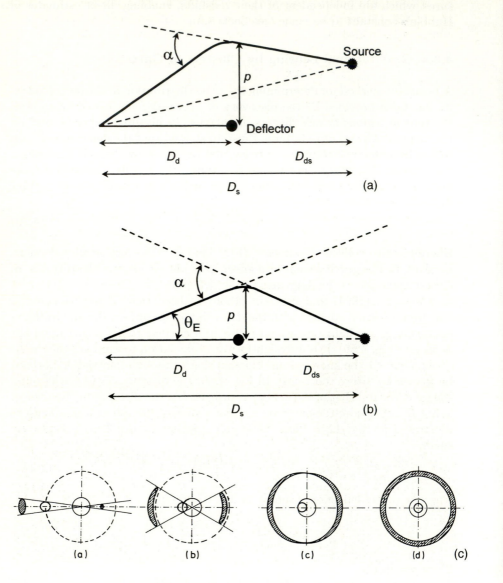

Fig. 4.8. (a) Illustrating the geometry of the deflection of light by a deflector, or lens, of mass M. (b) Illustrating the formation of an Einstein ring when the source and deflector are perfectly aligned. (c) Illustrating the changes of the appearance of a compact background source as it passes behind a point mass. The dashed circles correspond to the Einstein radius. When the lens and the background source are precisely aligned, an Einstein ring is formed with radius equal to the Einstein radius θ_E.

We have worked out this expression assuming the geometry of space is Euclidean. The above relation is also correct if the sources are at cosmological distances, provided the Ds are *angular diameter distances* (Blandford and Narayan 1992, see also Sect. 5.5.3).

Expressing the mass of the deflector in terms of solar masses $M_\odot$ and the distance D in Gpc ($= 10^9$ pc $= 3.056 \times 10^{25}$ m), we find

$$\theta_E = 3 \times 10^{-6} \left(\frac{M}{M_\odot} \right)^{1/2} \frac{1}{D_{Gpc}^{1/2}} \text{ arcsec.} \qquad (4.35)$$

Thus, clusters of galaxies with masses $M \sim 10^{15} M_\odot$ at cosmological distances can result in Einstein rings with angular radii tens of arcseconds. Such rings were first reported by Soucail *et al.* (1987) and Lynds and Petrosian (1986). A beautiful example of Einstein rings about the centre of the cluster Abell 2218 has been observed with the Hubble Space Telescope by Kneib, Ellis and their colleagues (Fig. 4.9). It can be seen that the rings are not complete and are elliptical rather than circular. The ellipticity and the incompleteness of the rings reflect the facts that the gravitational potential of the cluster is not precisely spherically symmetric and that the background galaxy and the cluster are slightly misaligned.

Fig. 4.9. An image of the central region of the rich cluster of galaxies Abell 2218 taken by the Hubble Space Telescope at a wavelength of 840 nm. Several arcs can be observed more or less centred on the Abell cluster. These are the gravitationally lensed images of background galaxies more or less perfectly aligned with the centre of the cluster (Courtesy of NASA, J.-P. Kneib, R. Ellis, and the Space Telescope Science Institute).

A comprehensive discussion of all aspects of gravitational lensing is presented in the monograph *Gravitational Lenses* by Schneider, Ehlers and Falco

(1992). The simplest generalisation of the above result is to consider the deflection due to a lens with an axially symmetric mass distribution. In that case, Schneider *et al.* show that the deflection is given by the expression

$$\alpha = \frac{4GM(<p)}{pc^2}, \tag{4.36}$$

where $M(<p)$ is the total projected mass within the radius p at the lensing galaxy or cluster, a result similar to Gauss's theorem for Newtonian gravity.

We can derive from this result the necessary condition for the formation of a gravitationally lensed image about an object of mass M and radius R. If the body is opaque, we require that the impact parameter p be greater than the R. Then, since

$$\alpha = \frac{p}{D} = \frac{4GM(<p)}{pc^2}, \tag{4.37}$$

we find that the requirement $p > R$ implies

$$\frac{4GM(<p)D}{c^2} > R^2 \quad \text{or} \quad \Sigma_{\text{lens}} = \frac{M(<p)}{\pi R^2} > \frac{c^2}{4\pi GD}, \tag{4.38}$$

where Σ_{lens} is the surface density of the material of the lens. If the source is at a cosmological distance, $D \approx c/H_0$, the critical surface density is

$$\Sigma_{\text{lens}} \approx \varrho_{\text{crit}} \left(\frac{c}{H_0}\right), \tag{4.39}$$

where $\varrho_{\text{crit}} = 3H_0^2/8\pi G = 2 \times 10^{-26}h^2$ kg m^{-3}. Taking $c/H_0 = 9.25 \times 10^{25} h^{-1}$ m, we find that the critical surface density is $2h$ kg m^{-2}. In fact, the Universe as a whole can be thought of as acting as a giant gravitational lens in understanding the effects of inhomogeneities upon the cosmological redshift–angular diameter relation (Zeldovich 1964, Zeldovich and Dashevskii 1964, Roeder and Dyer 1972, 1973. See also, Sect. 7.4).

Let us apply the result (4.36) to the case of an isothermal gas sphere, which provides a reasonable description of the mass distribution in clusters of galaxies. We consider the simple analytic solution (4.9), which has the unpleasant features of being singular at the origin and of having infinite mass when integrated to an infinite distance, but these are unimportant for our present analysis. Assuming that the velocity dispersion is isotropic and that $\langle v_{\parallel}^2 \rangle$ is the observed velocity dispersion along the line of sight,

$$\varrho(r) = \frac{2}{Ar^2} \quad \text{where} \quad A = \frac{4\pi G\mu}{k_B T} = \frac{4\pi G}{\langle v_{\parallel}^2 \rangle}. \tag{4.40}$$

We can now work out the projected mass density, or the surface density $\Sigma(p)$, at projected distance p by integrating along the line of sight, say, in the z-direction

$$\Sigma(p) = 2 \int_0^\infty \varrho(r)\,\mathrm{d}z = 2 \int_0^{\pi/2} \varrho(r)p\sec^2\theta\,\mathrm{d}\theta$$

$$= \frac{\langle v_\parallel^2 \rangle}{\pi G}\frac{1}{p} \int_0^{\pi/2} \mathrm{d}\theta = \frac{\langle v_\parallel^2 \rangle}{2G}\frac{1}{p}. \tag{4.41}$$

Therefore, the total mass within the distance p perpendicular to the line of sight at the deflector is

$$\int_0^p \Sigma(p)2\pi p\,\mathrm{d}p = \frac{\pi\langle v_\parallel^2 \rangle p}{G}. \tag{4.42}$$

The gravitational deflection of the light rays is therefore

$$\alpha = \frac{4GM(<p)}{pc^2} = \frac{4\pi\langle v_\parallel^2 \rangle}{c^2}. \tag{4.43}$$

This is the remarkable result we have been seeking. For an isothermal gas sphere, the gravitational deflection is *independent* of the distance at which the light rays pass by the lens. We can therefore find the Einstein radius θ_E directly from (4.32)

$$\theta_\mathrm{E} = \frac{4\pi\langle v_\parallel^2 \rangle}{c^2}\frac{D_\mathrm{ds}}{D_\mathrm{s}} = 28.8\,\langle v_{3\parallel}^2 \rangle\frac{D_\mathrm{ds}}{D_\mathrm{s}}\,\mathrm{arcsec}, \tag{4.44}$$

where $\langle v_{3\parallel}^2 \rangle$ means the observed velocity dispersion of the galaxies in the cluster measured in units of 10^3 km s^{-1}. As noted by Fort and Mellier (1994), this is a rather robust expression for comparing the masses of clusters of galaxies, as measured by the gravitational deflection of background galaxies, with their internal velocity dispersions. They find that these estimates are in good agreement to within about 10%.

In general, strong lensing of background sources only occurs if they lie within the Einstein angle θ_E of the axis of the lens. The changes in the appearance of the image of a compact source as it passes behind a point source deflector are illustrated in Fig. 4.8c. An excellent discussion of the shapes and intensities of the gravitationally distorted images of background sources is given by Fort and Mellier (1994). The gravitational lensing is not true lensing in the sense of geometric optics but rather the light rays come together to form caustics and cusps. Fig. 4.10 shows the types of images expected for gravitational lensing by an ellipsoidal gravitational potential. The background source is shown in panel (I) and, in the second panel labelled (S), different positions of the background source with respect to the critical inner and outer caustic lines associated with the gravitational lens are shown. These are lines along which the lensed intensity of the image is infinite. The images labelled (1) to (10) show the observed images of the background source when the source is located at the positions labelled on the second panel (S).

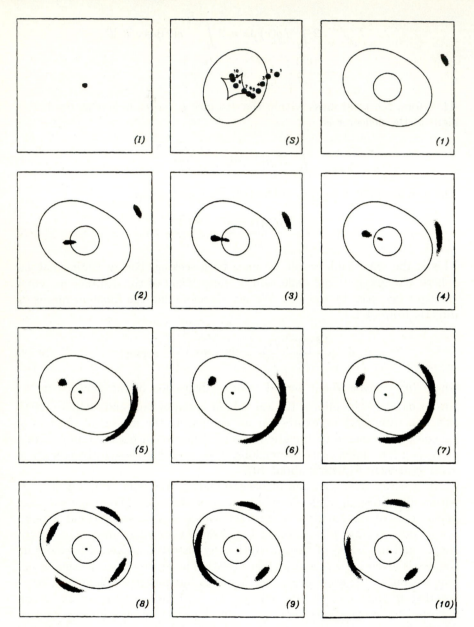

Fig. 4.10. The gravitational distortions of a background source (Panel I) when it is located at different positions with respect to the axis of the gravitational lens. In this example, the lens is an ellipsoidal non-singular squeezed isothermal sphere. The ten positions of the source with respect to the critical inner and outer caustics are shown in the panel (S). The panels labelled (1) to (10) show the shapes of the images of the lensed source (from J.-P. Kneib, Ph.D. Thesis (1993)). Note the shapes of the images when the source crosses the critical caustics. Positions (6) and (7) correspond to cusp catastrophes and position (9) to a fold catastrophe (Fort and Mellier 1994).

The numbers and shapes of the images depend upon the location of the source with respect to the caustic surfaces.

As mentioned above, for clusters of galaxies, the inferred masses are in good agreement with the values obtained by measuring the velocity dispersion of the cluster galaxies and with the X-ray methods of measuring total masses. An intriguing aspect of the gravitational lensing approach to the problem of determining cluster masses is that, just like the X-ray approach, it is possible to determine the details of the gravitational potential within the cluster. These studies show, for example, that the distribution of mass in Abell 2218 is more complex than a simple spherically symmetric distribution (Kneib *et al.* 1996). The extension of this approach to the weak gravitational lensing of distant galaxies in general by large scale structures has been developed by Kaiser (1992) who has shown how, in principle, the distorted images of large samples of galaxies can be used to determine the large-scale two-point correlation function for galaxies.

4.3.5 Summary

The upshot of all these considerations is that there is no doubt about the presence of dark matter in the regions within clusters of galaxies which are known to be in dynamical equilibrium. Typical values of the mass-to-blue luminosity ratios lie in the range $(100 \text{ to } 300)h^{-1} M_\odot/L_\odot$. This value is significantly less than the value needed to close the Universe. There may, however, be more dark matter with greater mass-to-luminosity ratio in the space between clusters of galaxies.

4.4 Forms of Dark Matter

One of the fundamental problems for the whole of cosmology is the nature of the dark matter which must be present in the outer regions of large galaxies, in clusters of galaxies and other large scale systems. This problem will haunt much of this book. Here, we discuss what can be said about the nature of the dark matter on the basis of observations of galaxies and clusters of galaxies – we will have much more to say about its central role in later chapters.

The problem is that we are certain that the dark matter is present on the basis of the dynamical arguments presented in Sect. 4.3 but we can only identify what it is if it emits radiation or absorbs the radiation of background sources. One amusing illustrative example is to suppose that the dark matter is in the form of standard bricks. There would have to be only one three kilogram brick per cube of side roughly a million kilometres to attain the critical cosmological density $\varrho_c = 3H_0^2/8\pi G \doteq 2.0 \times 10^{-26}h^2 \text{ kg m}^{-3} \Omega_0 = 1$ (see Sects. 7.2.2 and 7.2.3). If the bricks were uniformly distributed throughout the Universe, they would not obscure the most distant objects we can observe and they would be so cold that they would emit a negligible amounts of far infrared radiation. Thus, there could be many forms of ordinary baryonic matter present in the Universe which would be very difficult to detect, let alone the more exotic possibilities. Let us consider first the case of possible forms of baryonic matter. For convenience, we describe mean cosmological densities relative to the critical cosmological density ϱ_c through the *density parameter* $\Omega_0 = \bar{\varrho}/\varrho_c$, where the subscript 0 means that all gravitating contributions at the present epoch t_0 are included. Ω is given different subscripts to refer to different contributions to the overall density parameter Ω_0 (see Sect. 7.2.2).

4.4.1 Baryonic Dark Matter

By *baryonic matter*, we mean ordinary matter composed of protons, neutrons and electrons and for convenience we will include the black holes in this category. As illustrated by the example of the bricks, certain forms of baryonic matter are very difficult to detect because they are very weak emitters of electromagnetic radiation. Perhaps the most important examples of these are stars with masses $M \leq 0.05 M_\odot$, which have such low masses that their central temperatures are not hot enough to burn hydrogen into helium – they are referred to collectively as *brown dwarfs*. They have no internal energy source and so the source of their luminosity is the thermal energy with which they were endowed at birth. They are therefore very faint, cool objects and have proved very difficult to detect. By the same token, even lower mass solid bodies such as planets, asteroids and other small lumps of rock are extremely difficult to detect. It is quite conceivable that a significant fraction of the dark matter in the outer regions of galaxies and in clusters of galaxies is in one of these baryonic forms.

A strong limit to the total amount of baryonic matter in the Universe is provided by considerations of primordial nucleosynthesis. As will be shown in Sect. 10.4, the standard Big Bang model of the Universe is remarkably successful in accounting for the observed abundances of light elements such as helium-4, helium-3, deuterium and probably lithium-7 through the process of primordial nucleosynthesis. One of the important consequences of that success story is that, even adopting the most conservative numbers, the average baryonic mass density in the Universe Ω_B must be less than $0.036h^{-2}$ or else even the strong lower limit to the observed abundance of deuterium cannot be explained (Hogan 1997). If the mass density of the Universe corresponds to $\Omega_0 = 1$, the Universe must be dominated by non-baryonic matter. It can be seen from these figures that it would just be possible to bind clusters of galaxies with baryonic dark matter if $h \approx 0.5$. This value of h would be sufficient to increase Ω_B to about 0.15 and then the typical mass-to-luminosity ratio could be about 15 times greater than that of the visible parts of galaxies for which $\Omega_{\mathrm{vis}} \approx 0.01$ and $M/L \sim 10 M_\odot / L_\odot$. The same reasoning might apply to the outer regions of spiral galaxies.

If we were to adopt the somewhat bolder approach discussed in Sect. 10.4 of adopting a higher primordial deuterium abundance and accounting for the abundances of the other light elements, the upper limit to Ω_{B} would become $(6.2 \pm 1.1) \times 10^{-3} h^{-2}$ and then, even if $h = 0.5$, the average baryon density would be 0.025 and the dark matter in luminous galaxies and clusters of galaxies would have to be non-baryonic (Hogan 1997).

Another possible candidate for the dark matter is *black holes*. There are now useful limits to the number density of black holes in certain mass ranges. These are derived from studies of the numbers of gravitational lenses observed in large samples of extragalactic radio sources and from the absence of gravitational lensing effects by stellar mass black holes in the haloes of galaxies. In their survey of a very large sample of extragalactic radio sources, designed specifically to search for gravitationally lensed structures, Hewitt *et al.* (1987) were able to set limits to the number density of massive black holes with masses $M \sim 10^{10}$ to $10^{12} M_\odot$ and found that the limit corresponded to $\Omega_0 \ll 1$. It cannot be excluded that the dark matter might consist of a very large population of low mass black holes but these would have to be produced by a rather special initial perturbation spectrum in the very early Universe before the epoch of primordial nucleosynthesis. The fact that black holes of mass less than about 10^{12} kg evaporate by Hawking radiation on a cosmological timescale sets a firm lower limit to the possible masses of mini-black holes which could contribute to the dark matter at the present epoch (Hawking 1975)

An important programme to detect dark objects in the halo of our Galaxy has been described by Alcock *et al.* (1993a). The idea is to search for the characteristic signature of gravitational microlensing events when a 'massive compact halo object', or MACHO, passes in front of a background star. These are very rare events and so a very large number of background stars has to be monitored. In the MACHO project, the background objects are the stars in the Magellanic Clouds and over a million of them are observed regularly to search for the rare gravitational lensing events and to exclude known types of variable star. A similar project has been carried out by the European EROS collaboration. This technique is sensitive to discrete objects with masses in the range $10^{-7} < M < 100\ M_\odot$ and it does not matter whether they are brown dwarfs, planets, isolated neutron stars or black holes. When such dark objects pass in front of a background star, there is expected to be a characteristic brightening of the star which is independent of wavelength. The observations began in 1993 and the first candidate event was reported in October 1993 (Alcock 1993b). The observed brightening of the background star in blue and red wavebands is shown in Fig. 4.11, in which it is compared with the expected brightening due to the gravitational lens effect. The observed brightening is independent of wavelength. The mass of the invisible lensing object was estimated to lie in the range $0.03 < M < 0.5\ M_\odot$.

By mid-1996, many lensing events had been observed, including over 100 in the direction towards the Galactic bulge, which is about a factor of three more than expected. In addition, eight definite events were observed in the direction of the Large Magellanic Clouds (Alcock *et al.* 1997). The statistics are still small but they are consistent with MACHOs making up a significant fraction of the dark halo mass of our own Galaxy. Unfortunately, the technique does not provide distances and masses for individual objects, but the best statistical estimates suggest that the mean mass of the MACHOs is about $0.3 - 0.5\ M_\odot$, which is surprisingly large. Typically, these objects could account for about 50% of the dark halo mass and within the uncertainties could account for all of it. The most likely candidates would appear to be white dwarfs which would have to be produced in large numbers in the early evolution of the Galaxy, but other more exotic possibilities are certainly not excluded at this stage.

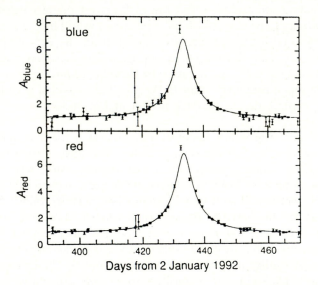

Fig. 4.11. A gravitational lensing event recorded by the MACHO project in February and March 1993. The horizontal axis shows the date in days as measured from day zero which was 2 January 1992. The vertical axes show the amplification of the brightness of the lensed star relative to the unlensed intensity in standard blue and red wavebands. The solid lines show the expected variation of the brightness of the lensed object with time. It has the same characteristic light-curve at blue and red wavelengths (Alcock *et al.* 1993b).

4.4.2 Non-baryonic Dark Matter

It is entirely plausible that the dark matter is in some non-baryonic form and this is of the greatest interest for particle physicists since the dark matter may consist of the types of particles predicted by theories of elementary particles but which have not yet been detected experimentally. Three of the most popular possibilities are as follows.

- The smallest mass candidates are the *axions* which were invented by the particle theorists in order to 'save quantum chromodynamics from strong CP violation'. If they exist, they must have been born when the thermal temperature of the Universe was about 10^{12} K but they were produced out of equilibrium and so never acquired thermal velocities – they remain 'cold'. Their rest mass energies are expected to lie in the range 10^{-2} to 10^{-5} eV. The role of such particles in cosmology and galaxy formation are discussed by Efstathiou (1990) and Kolb and Turner (1990).

- A second possibility is that the three known types of neutrino have finite rest masses. Laboratory experiments have provided an upper limit to the rest mass of the electron antineutrino of a few eV (Perkins 1987), but this does not exclude the possibility that the two other types of neutrino, the muon and τ neutrinos, have greater masses. The reason that this value is of interest is that a neutrino of rest mass about 10 – 20 eV is almost exactly the value needed to close the Universe, as may be seen from the following calculation. The number density of neutrinos of a single type in thermal equilibrium at temperature T is

$$N = \bar{N} = \frac{4\pi g}{h^3} \int_0^\infty \frac{p^2 \, \mathrm{d}p}{\mathrm{e}^{E/k_\mathrm{B}T} + 1} = 0.091 \left(\frac{2\pi k_\mathrm{B}T}{hc} \right)^3 \ \mathrm{m}^{-3}, \qquad (4.45)$$

where the statistical weight g for the neutrinos is $g = 1$. If there are N_ν neutrino types present, each with rest mass m_ν, the present mass density of neutrinos in the Universe would be

$$\varrho_\nu = N N_\nu m_\nu. \qquad (4.46)$$

The present temperature of the neutrino background radiation, which was in equilibrium with the matter prior to the epoch when the neutrinos decoupled, is only $(4/11)^{1/3}$ of the temperature of the Cosmic Microwave Background Radiation, that is, $T_\nu = 1.94$ K, and so the value of N is rather precisely known (see Sect. 10.5). Therefore, if ϱ_ν is to equal the critical density of the Universe $\varrho_\mathrm{c} = 1.88 \times 10^{-26} h^2$ kg m^{-3}, the necessary rest mass energy of the neutrino is $m_\nu = 184 h^2/N_\nu$ eV. Since we know that there are three neutrino species, each with its antiparticle, $N_\nu = 6$ and hence the necessary rest mass of the neutrino is 31 h^2 eV. Since h lies in the range 0.5 to 1, it follows that, if the neutrino rest mass were in the range 8 to 40 eV, the neutrinos could close the Universe. The lower end of this range of neutrino masses is marginally consistent with the upper limit to the electron neutrino mass measured in the laboratory. Another upper limit to the neutrino rest mass can be derived from the distribution of the energies and arrival times of the neutrinos associated with the explosion of the supernova SN1987A in the Large Magellanic Cloud (see Bahcall 1989). These all arrived within an interval of about 12 seconds. If they had finite rest mass, the more enegetic neutrinos would have arrived before the less energetic ones, since $E_\nu = \gamma_\nu m_\nu c^2$. Since the neutrinos travelled a distance of about 50 kpc, the dispersion in arrival times of at most 12 seconds for 35 and 7.5 MeV neutrinos enable an upper limit of $m_\nu \leq 20$ eV to be derived.

- A third possibility is that the dark matter is in some form of much more weakly interacting massive particle, or WIMP. This might be the gravitino, the supersymmetric partner of the graviton, or the photino, the supersymmetric partner of the photon, or some form of as yet unknown massive neutrino-like particle. The possible existence of these types of unknown particles represent theoretical extrapolations beyond the range of energies which have been explored experimentally, but these ideas are sufficiently compelling on theoretical grounds that many particle theorists take seriously the possibility that cosmological studies will prove to be important in constraining theories of elementary processes at ultra-high energies.

Useful astrophysical limits can be set to the number densities of different types of neutrino-like particles in the outer regions of giant galaxies and in clusters of galaxies. The WIMPs and massive neutrinos are collisionless fermions and therefore there are constraints on the phase space density of these particles, which translates into a lower limit to their masses since, for a given momentum, only a finite number of particles within a given volume is allowed. Let us give a simple derivation of the key result. More details of this calculation are given by Tremaine and Gunn (1979), who provide a slightly tighter constraint on the masses of the neutrinos.

Being fermions, neutrino-like particles are subject to the Pauli Exclusion Principle according to which there is a maximum number of particle states in phase space for a given momentum p_{max}. The elementary phase volume is h^3 and, recalling that there can be two particles of opposite spin per state, the maximum number of particles with momenta up to p_{max} is

$$N \leq 2\frac{g}{h^3}\frac{4\pi}{3}p_{max}^3, \qquad (4.47)$$

per unit volume, where g is the statistical weight of the neutrino species. If there is more than one neutrino species present, we multiply this number by N_ν. Bound gravitating systems such as galaxies and clusters of galaxies are subject to the *virial theorem* according to which the kinetic energy of the particles which make up the system must be equal to half of its gravitational potential energy (Sect. 3.4.1). If σ is the root-mean-square velocity dispersion of the objects which bind the system, $\sigma^2 = GM/R$, and the maximum velocity which particles within the system can have is the escape velocity from the cluster, $v_{\mathrm{max}} = (2GM/R)^{1/2} = \sqrt{2}\sigma$. Now, the neutrino-like particles bind the system and so its total mass is $M = NN_\nu m_\nu$ where m_ν is the rest mass of the neutrino. We therefore find the following lower limit to the rest mass of the neutrino from (4.47) in terms of observable quantities:

$$m_\nu^4 \geq \left(\frac{9\pi}{8\sqrt{2}g}\right) \frac{\hbar^3}{N_\nu G\sigma R^2} \qquad m_\nu \geq \frac{1.5}{(N_\nu \sigma_3 R_{\mathrm{Mpc}}^2)^{1/4}} \; \mathrm{eV}, \qquad (4.48)$$

where the velocity dispersion σ_3 is measured in units of 10^3 km s^{-1} and R is measured in Mpc. Let us insert different values for the velocity dispersions and radii of the systems in which there is known to be a dark matter problem. In clusters of galaxies, typical values are $\sigma = 1000$ km s^{-1} and $R = 1$ Mpc. In this case, if there is only one neutrino species, $N_\nu = 1$, we find $m_\nu \geq 1.5$ eV. If there were six neutrino species, namely, electron, muon, tau neutrinos and their antiparticles, $N_\nu = 6$ and then $m_\nu \geq 0.9$ eV. For giant galaxies, for which $\sigma = 300$ km s^{-1} and $R = 10$ kpc, $m_\nu \geq 20$ eV if $N_\nu = 1$ and $m_\nu \geq 13$ eV if $N_\nu = 6$. For small galaxies, for which $\sigma = 100$ km s^{-1} and $R = 1$ kpc, the corresponding figures are $m_\nu \geq 80$ eV and $m_\nu \geq 50$ eV respectively.

These are useful limits to the masses of neutrino-like particles which could form the dark haloes of massive galaxies and clusters of galaxies. Thus, neutrinos with rest masses $m_\nu \sim 10$ eV could bind clusters of galaxies but it is marginal whether or not they could bind the haloes of giant galaxies. It is clear that neutrinos with rest masses $m \sim 10$ eV cannot bind small galaxies. Some other form of dark matter would have to be present in the haloes of these galaxies.

There is a further constraint on the possible masses WIMPs. Studies of the decay of the $W^\pm$ and Z^0 bosons at CERN have shown that the width of the decay spectrum is consistent with there being only three neutrino species with rest mass energies less than about 40 GeV. Therefore, if the dark matter is in some form of ultraweakly interacting particle, its rest mass energy must be greater than 40 GeV.

Another important conclusion is that, if the masses of the particles were much greater than $10 - 20$ eV and they are as common as the neutrinos and photons, which is expected in a simplest picture of the Hot Big Bang model with massive ultraweakly interacting particles present, the present density of the Universe would far exceed the critical mass density $\Omega = 1$. Therefore there would have to be some suppression mechanism to ensure that, if $m \geq 40$ GeV, these particles are much less common than the photons and electrons neutrinos at the present day.

Part II

The Basic Framework

5 The Theoretical Framework

5.1 The Cosmological Principle

The observational evidence discussed in Chap. 2 indicates that the natural starting point for the construction of cosmological models is to assume that the Universe is isotropic, homogeneous and expanding uniformly at the present epoch. In other words, we smooth out all the fine-scale structure in the Universe. It is intriguing that this is precisely what Einstein assumed in his static model of 1917, the first fully self-consistent model of the Universe, derived long before these properties of our Universe had been established. Likewise, Friedman's discovery of what have become the standard models for the large-scale dynamics of the Universe predated the discovery of the expansion of the Universe, but were based upon expanding solutions of Einstein's equations, following clues provided by de Sitter and Lanczos. One of the problems facing the pioneers of relativistic cosmology was the interpretation of the space and time coordinates to be used in these calculations. For example, de Sitter's solution for an empty universe could be written in apparently stationary form, or as an exponentially expanding solution. By 1935, the problem was solved independently by Robertson and Walker. They derived the form of the metric of space-time for *all* isotropic, homogeneous, uniformly expanding models of the Universe. The form of the metric is independent of the assumption that the large-scale dynamics of the Universe are described by Einstein's General Theory of Relativity – whatever the physics of the expansion, the space-time metric must be of Robertson–Walker form, because of the assumptions of isotropy and homogeneity.

A key step in the development of these models was the introduction by Hermann Weyl in 1923 of what is known as *Weyl's Postulate*. To eliminate the arbitrariness in the choice of coordinate frames, Weyl introduced the idea that, in the words of Hermann Bondi (1960),

> The particles of the substratum (representing the nebulae) lie in space-time on a bundle of geodesics diverging from a point in the (finite or infinite) past.

The most important aspect of this statement is the postulate that the geodesics, which represent the world-lines of galaxies, do not intersect, except at a singular point in the finite, or infinite, past. Again, it is extraordinary that Weyl introduced this postulate *before* Hubble's discovery of the recession of the nebulae. By the term 'substratum', Bondi meant an imaginary medium which can be thought of as a fluid which defines the overall kinematics of the system of galaxies. A consequence of Weyl's postulate is that there is only one geodesic passing through each point in space-time, except at the origin. Once this postulate is adopted, it becomes possible to assign a notional observer to each world line and these are known as *fundamental observers*. Each fundamental observer carries a standard clock and time measured on that clock from the singular point is called *cosmic time*.

One further assumption is needed before we can derive the framework for the standard models. This is the assumption known as the *cosmological principle* and it can be stated:

We are not located at any special location in the Universe.

A corollary of this statement is that we are located at a *typical* position in the Universe and that any other fundamental observer located anywhere in the Universe at the same cosmic epoch would observe the same large-scale features which we observe. Thus, we assert that every fundamental observer at the same cosmic epoch observes the same Hubble expansion of the distribution of galaxies, the same isotropic Cosmic Microwave Background Radiation, the same large-scale spongy structure in the distribution of galaxies and voids, and so on. As we showed in Sect. 2.3, the combination of Hubble's law and the isotropy of the Universe shows that the whole system of galaxies is expanding uniformly and every observer on every galaxy partaking in the uniform expansion observes the same Hubble flow at the same epoch – all of them correctly believe that they are at the centre of a uniformly expanding Universe. The isotropy of the background radiation, evidence of the scaling of the two-point correlation function with apparent magnitude and the ubiquity of the sponge-like structure of the distribution of galaxies suggest that the cosmological principle is a sensible starting point for the construction of cosmological models.

The specific features of the observable Universe which we need in what follows are its overall isotropy and homogeneity, as well as Hubble's law. The combination of these features results in the *Robertson–Walker metric* for any isotropic, uniformly expanding world model.

5.2 Isotropic Curved Spaces

During the late 18th century, non-Euclidean spaces began to be taken seriously by mathematicians who realised that Euclid's fifth postulate, that parallel lines meet only at infinity, might not be essential for the construction of self-consistent geometries. The first suggestions that the global geometry of space might not be Euclidean were discussed by Lambert and Saccheri. In 1786, Lambert noted that, if space were hyperbolic rather than flat, the radius of curvature of space could be used as an absolute measure of distance. In 1816, Gauss repeated this proposal in a letter to Gerling and is reputed to have tested whether or not space is locally Euclidean by measuring the sum of the angles of a triangle between three peaks in the Harz mountains.

The fathers of non-Euclidean geometry were Nicolai Ivanovich Lobachevski in Russia and János Bolyai in Hungary. In his book, *On the Principles of Geometry* of 1829, Lobachevski at last solved the problem of the existence of non-Euclidean geometries and showed that Euclid's fifth postulate could not be deduced from the other postulates. Non-Euclidean geometry was placed on a firm theoretical basis by the studies of Bernhard Riemann and the English-speaking world was introduced to these ideas through the works of Clifford and Cayley.

Einstein's monumental achievement was to combine special relativity and the theory of gravity through the use of Riemannian geometry and tensor calculus to create the General Theory of Relativity. We will describe Einstein's achievement in Chap. 6. Within a couple of years of formulating the theory, Einstein realised that he now had the tools with which fully self-consistent solutions for the Universe as a whole could be found. In Einstein's model, which we discuss in Sect. 7.4, the Universe is static, closed and has isotropic, spherical geometry. The Friedman solutions, published in 1922 and 1924, were also isotropic models but they were expanding solutions and included geometries which were both spherical and hyperbolic.

It turns out that it is not necessary to become enmeshed in the details of Riemannian geometry to appreciate the geometrical properties of isotropic curved spaces. We can demonstrate simply why the only isotropic curved spaces are those in which the two-dimensional curvature of any space section κ is constant throughout the space and can only take positive, zero or negative values. The essence of the following argument was first shown to me by Dr. Peter Scheuer.

Let us consider first of all the simplest two-dimensional curved geometry, the surface of a sphere (Fig. 5.1). In the diagram, a triangle is shown consisting of two lines drawn from the north pole down to the equator, the triangle being completed by the line drawn along the equator. The three sides of the triangle are all segments of great circles on the sphere and so are the shortest distances between the three corners of the triangle. The three lines are *geodesics* in the curved geometry.

We need a procedure for working out how non-Euclidean the curved geometry is. The way this is done in general is by the procedure known as the *parallel displacement* or *parallel transport* of a vector on making a complete circuit around a closed figure such as the triangle in Fig. 5.1. Suppose we start with a little vector perpendicular to AC at the pole and lying in the surface of the sphere as shown. We then transport that vector from A to C, keeping it perpendicular to AC. At C, we rotate the vector through 90° so that it is now perpendicular to CB. We then transport the vector, keeping

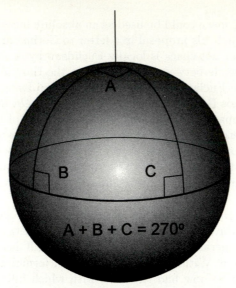

Fig. 5.1. Illustrating the sum of the angles of a triangle on the surface of a sphere.

it perpendicular to BC to the corner B. We make a further rotation through 90° to rotate the vector perpendicular to BA and then transport it back to A. At that point, we make a final rotation through 90° to bring the vector back to its original direction. Thus, the total rotation of the vector is 270°. Clearly, the surface of the sphere is a non-Euclidean space. This procedure illustrates how we can work out the geometrical properties of any two-space, entirely by making measurements within the two-space.

Another simple calculation illustrates an important feature of parallel transport on the surface of a sphere. Suppose the angle at A is not 90° but some arbitrary angle θ. Then, if the radius of the sphere is R_c, the surface area of the triangle ABC is $A = \theta R_c^2$. Thus, if $\theta = 90°$, the area is $\pi R_c^2/2$ and the sum of the angles of the triangle is 270°; if $\theta = 0°$, the area is zero and the sum of the angles of the triangle is 180°. Evidently, the difference of the sum of the angles of the triangle from 180° is proportional to the area of the triangle, that is

$$\text{(Sum of angles of triangle} - 180°) \propto \text{(Area of triangle)}. \qquad (5.1)$$

This result is a general property of isotropic curved spaces.

Let us now take a vector around a loop in some general two-space. We define a small area ABCD in the two-space by light rays which travel along geodesics in the two-space. We consider two light rays, OAD and OBC, originating from O and these are crossed by the light rays AB and DC as shown in Fig. 5.2a. The light rays AB and DC are chosen so that they cross the light ray OAD at right angles. We start at A with the vector to be transported round the loop parallel to AD as shown. We then parallelly transport

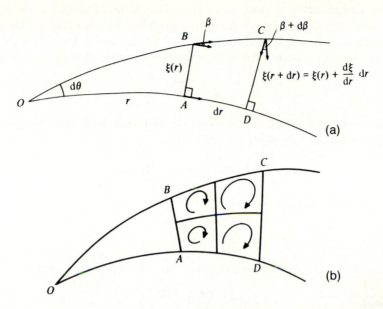

Fig. 5.2. (a) Illustrating the parallel transport of a vector around the small loop ABCD formed by the light rays OBC, OAD, AB and DC. **(b)** Illustrating how the sum of the rotations around the sub-loops must add up linearly to the total rotation $d\beta$.

the vector along AB until it reaches B. At this point, it has to be rotated through an angle of $90° - \beta$ in order to lie perpendicular to BC at B. It is then transported along BC to C, where it is rotated through an angle of $90° + (\beta + d\beta)$ as shown in the diagram in order to lie perpendicular to CD at C. It can be seen by inspection of the diagram that the extra rotations at B and C are in opposite directions. Since CD and AB are perpendicular to the light rays, the subsequent rotations of the vector on parallel transport amount to 180° and so the total rotation round the loop is $360° + d\beta$. Thus, $d\beta$ is a measure of the departure of the two-space from Euclidean space, for which $d\beta = 0$.

Now, the rotation of the vector $d\beta$ must depend upon the area of the loop. In the case of an isotropic space, we should obtain the same rotation wherever we place the loop in the two-space. Furthermore, if we were to split the loop up into a number of sub-loops, the rotations around the separate sub-loops must add up linearly to the total rotation $d\beta$ (Fig. 5.2b). Thus, in an isotropic two-space, the rotation $d\beta$ should be proportional to the area of the loop ABCD. We conclude that the constant relating the rotation $d\beta$ to the area of the loop must be a constant everywhere in the two-space, just as we found in the particular case of a spherical surface in Fig. 5.1.

The complication is that, since the space is non-Euclidean, we do not know how to relate the length AB to the geodesic distance OA $= r$ along the light path and the angle θ subtended by AB at the origin. Therefore, we have to write that the distance AB is an unknown function of r, $\xi(r)$, such that

$$\theta = \frac{\xi(r)}{r}. \tag{5.2}$$

We can now find the angle of rotation β by determining how ξ changes when we move a distance dr along the geodesic.

$$\xi(r + dr) = \xi(r) + \frac{d\xi}{dr}\, dr. \tag{5.3}$$

Thus, the rotation at B is given by

$$\beta = \frac{d\xi}{dr}. \tag{5.4}$$

At a distance Δr further along the geodesic, the rotation becomes

$$\beta + d\beta = \frac{d\xi}{dr} + \left(\frac{d^2\xi}{dr^2}\right)\Delta r, \tag{5.5}$$

and so the net rotation $d\beta$ is

$$d\beta = \left(\frac{d^2\xi}{dr^2}\right)\Delta r. \tag{5.6}$$

But, we have argued that the net rotation around the loop must be proportional to the area of the loop, $dA = \xi \Delta r$, in an isotropic two-space:

$$\left(\frac{d^2\xi}{dr^2}\right)\Delta r = -\kappa\,\Delta r\xi \quad \text{and so} \quad \left(\frac{d^2\xi}{dr^2}\right) = -\kappa\xi, \tag{5.7}$$

where κ is a constant, the minus sign being chosen for convenience. This is the equation of simple harmonic motion which has solution

$$\xi = \xi_0 \sin \kappa^{1/2} r. \tag{5.8}$$

We can find the value of ξ_0 from the value of ξ for very small values of r which must reduce to the Euclidean expression $\theta = \xi/r$. Therefore, $\xi_0 = \theta/\kappa^{1/2}$ and

$$\xi = \frac{\theta}{\kappa^{1/2}} \sin \kappa^{1/2} r. \tag{5.9}$$

κ is the *curvature* of the two-space and it can be either positive, negative or zero. If it is negative, we can write $\kappa = -\kappa'$, where κ' is positive and then the circular functions become hyperbolic functions

$$\xi = \frac{\theta}{\kappa'^{1/2}} \sinh \kappa'^{1/2} r. \tag{5.10}$$

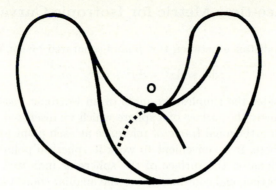

Fig. 5.3. Illustrating the geometry of an isotropic hyperbolic two-space. The principal radii of curvature of the surface are equal in magnitude but have opposite signs in orthogonal directions.

In the case $\kappa = 0$, we find the Euclidean result

$$\xi = \theta r. \tag{5.11}$$

The results we have derived include all possible isotropic curved two-spaces. The constant κ can be positive, negative or zero corresponding to spherical, hyperbolic and flat spaces respectively. In geometric terms, $R_c = \kappa^{-1/2}$ is the radius of curvature of a two dimensional section through the isotropic curved space and has the same value at all points and in all orientations within the plane. It is often convenient to write the expression for ξ in the form

$$\xi = \theta R_c \sin \frac{r}{R_c}, \tag{5.12}$$

where R_c is real for closed spherical geometries, imaginary for open hyperbolic geometries and infinite for the case of Euclidean geometry.

The simplest examples of such spaces are the spherical geometries in which R_c is just the radius of the sphere as illustrated in Fig. 5.1. The hyperbolic spaces are more difficult to envisage. The fact that R_c is imaginary can be interpreted in terms of the principal radii of curvature of the surface having opposite sign. The geometry of a hyperbolic two-sphere can be represented by a saddle-shaped figure (Fig. 5.3), just as a two-sphere provides an visualisation of the properties of a spherical two-space.

5.3 The Space-time Metric for Isotropic Curved Spaces

In flat space, the distance between two points separated by dx, dy, dz is

$$dl^2 = dx^2 + dy^2 + dz^2. \tag{5.13}$$

Now, let us consider the simplest example of an isotropic *two-dimensional* curved space, namely the surface of a sphere which we discussed in Sect. 5.2. We can set up an orthogonal frame of reference at each point locally on the surface of the sphere. It is convenient to work in spherical polar coordinates to describe positions on the surface of the sphere as indicated in Fig. 5.4. In this case, the orthogonal coordinates are the angular coordinates θ and ϕ, and the expression for the increment of distance dl between two neighbouring points on the surface can be written

$$dl^2 = R_c^2 \, d\theta^2 + R_c^2 \sin^2 \theta \, d\phi^2, \tag{5.14}$$

where R_c is the radius of curvature of the two-space, which in this case is just the radius of the sphere.

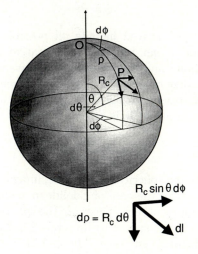

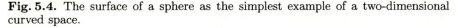

Fig. 5.4. The surface of a sphere as the simplest example of a two-dimensional curved space.

The expression (5.14) is known as the *metric* of the two-dimensional surface and can be written more generally in tensor form

$$dl^2 = g_{\mu\nu} \, dx^\mu dx^\nu. \tag{5.15}$$

It is a fundamental result of differential geometry that the *metric tensor* $g_{\mu\nu}$ contains all the information about the intrinsic geometry of the space. The problem is that we can set up a variety of different coordinate systems to define the coordinates of a point on any two dimensional surface. For example, in the case of a Euclidean plane, we could use rectangular *Cartesian coordinates* so that

$$dl^2 = dx^2 + dy^2, \tag{5.16}$$

or we could use *polar coordinates* in which

$$dl^2 = dr^2 + r^2 d\phi^2. \tag{5.17}$$

How can we determine the *intrinsic curvature* of the space simply in terms of the $g_{\mu\nu}$ of the metric tensor? Gauss first showed how it is possible to do this (see Weinberg 1972, Berry 1989). For the case of two-dimensional metric tensors which can be reduced to diagonal form, as in the cases of the metrics (5.14), (5.16) and (5.17), the intrinsic curvature of the space is given by the quantity

$$\kappa = \frac{1}{2g_{11}g_{22}} \left\{ -\frac{\partial^2 g_{11}}{\partial x_2^2} \right.$$
$$-\frac{\partial^2 g_{22}}{\partial x_1^2} + \frac{1}{2g_{11}} \left[\frac{\partial g_{11}}{\partial x_1} \frac{\partial g_{22}}{\partial x_1} + \left(\frac{\partial g_{11}}{\partial x_2}\right)^2 \right]$$
$$\left. + \frac{1}{2g_{22}} \left[\frac{\partial g_{11}}{\partial x_2} \frac{\partial g_{22}}{\partial x_2} + \left(\frac{\partial g_{22}}{\partial x_1}\right)^2 \right] \right\}. \tag{5.18}$$

It is a useful exercise to show that both metrics (5.16) and (5.17) have zero curvature and that, for the surface of a sphere, the curvature of space $\kappa = R_c^{-2}$ at all points on the sphere. κ is known as the *Gaussian curvature* of the two-space and is identical to the definition of the curvature introduced in Sect. 5.2. In general curved spaces, the curvature κ varies from point to point in the space. The extension to isotropic three-spaces is straightforward if we remember that any two-dimensional section through an isotropic three-space must be an isotropic two-space and we already know the metric tensor for this case.

We have already worked out the length of the distance increment dl (5.14). The natural system of coordinates for an isotropic two-space is a spherical polar system in which a radial distance ϱ round the sphere is measured from the pole and the angle ϕ measures angular displacements at the pole. From Fig. 5.4, the distance ϱ round the arc of a great circle from the point O to P is given by $\varrho = \theta R_c$ and so the metric can be written

$$dl^2 = d\varrho^2 + R_c^2 \sin^2\left(\frac{\varrho}{R_c}\right) d\phi^2. \tag{5.19}$$

The distance ϱ is the shortest distance between O and P on the surface of the sphere since it is part of a great circle and is therefore the *geodesic distance* between O and P in the isotropic curved space. Geodesics play the role of straight lines in curved space.

We can write the metric in an alternative form if we introduce a distance measure

$$x = R_{\mathrm{c}} \sin\left(\frac{\varrho}{R_{\mathrm{c}}}\right). \tag{5.20}$$

Differentiating and squaring, we find

$$\mathrm{d}x^2 = \left[1 - \sin^2\left(\frac{\varrho}{R_{\mathrm{c}}}\right)\right]\mathrm{d}\varrho^2 \qquad \mathrm{d}\varrho^2 = \frac{\mathrm{d}x^2}{1 - \kappa x^2}, \tag{5.21}$$

where $\kappa = 1/R_{\mathrm{c}}^2$ is the curvature of the two space. Therefore, we can rewrite the metric in the form

$$\mathrm{d}l^2 = \frac{\mathrm{d}x^2}{1 - \kappa x^2} + x^2\mathrm{d}\phi^2. \tag{5.22}$$

Notice the interpretation of the distance measure x. It can be seen from the metric that $\mathrm{d}l = x\,\mathrm{d}\phi$ is a *proper distance* perpendicular to the radial coordinate ϱ and that it is the correct expression for the length of a line segment which subtends the angle $\mathrm{d}\phi$ at geodesic distance ϱ from O. It is therefore what is known as an *angular diameter distance* since it is guaranteed to give the correct answer for the length of a line segment perpendicular to the line of sight. We can use either ϱ or x in our metric but notice that, if we use x, the increment of geodesic distance is $\mathrm{d}\varrho = \mathrm{d}x/(1 - \kappa x^2)^{1/2}$. We recall that the curvature $\kappa = 1/R_{\mathrm{c}}^2$ can be *positive* as in the spherical two-space discussed above, *zero* in which case we recover flat Euclidean space ($R_{\mathrm{c}} \to \infty$) and *negative* in which case the geometry becomes *hyperbolic* rather than spherical.

We can now write down the expression for the spatial increment in any isotropic, three-dimensional curved space. As mentioned above, the trick is that any two-dimensional section through an isotropic three-space must be an isotropic two-space for which the metric is (5.19) or (5.22). We note that, in spherical polar coordinates, the general angular displacement perpendicular to the radial direction is

$$\mathrm{d}\Phi^2 = \mathrm{d}\theta^2 + \sin^2\theta\,\mathrm{d}\phi^2, \tag{5.23}$$

and can be found by rotating the coordinate system about the radial direction. Thus, by a straightforward extension of the formalism we have derived already, we can write the spatial increment

$$\mathrm{d}l^2 = \mathrm{d}\varrho^2 + R_{\mathrm{c}}^2 \sin^2\left(\frac{\varrho}{R_{\mathrm{c}}}\right)[\mathrm{d}\theta^2 + \sin^2\theta\,\mathrm{d}\phi^2], \tag{5.24}$$

in terms of the three-dimenstional spherical polar coordinates (ϱ, θ, ϕ). An exactly equivalent form is obtained if we write the spatial increment in terms of x, θ, ϕ in which case we find

$$\mathrm{d}l^2 = \frac{\mathrm{d}x^2}{1 - \kappa x^2} + x^2[\mathrm{d}\theta^2 + \sin^2\theta\,\mathrm{d}\phi^2]. \tag{5.25}$$

We are now in a position to write down the *Minkowski metric* in any isotropic three-space. It is given by

$$ds^2 = dt^2 - \frac{1}{c^2}dl^2, \tag{5.26}$$

where dl is given by either of the above forms of the spatial increment, (5.24) or (5.25). Notice that we have to be careful about the meanings of the distance coordinates – x and ϱ are equivalent but physically quite distinct distance measures. We can now proceed to derive from this metric the *Robertson–Walker metric*.

5.4 The Robertson–Walker Metric

In order to apply this metric to isotropic, homogeneous world models, we need the *cosmological principle* and the concepts of *fundamental observers* and *cosmic time* which were introduced in Sect. 5.1. For uniform, isotropic world models, we define a set of *fundamental observers*, who move in such a way that the Universe always appears to be isotropic to them. Each of them has a clock and proper time measured by that clock is called *cosmic time*. There are no problems of synchronisation of the clocks carried by the fundamental observers because, according to Weyl's postulate, the geodesics of all observers meet at one point in the past and cosmic time can be measured from that reference epoch.

We can now write down the metric for such Universes from the considerations of Sect. 5.3. For reasons which will become apparent in a moment, I will use (5.24) to write the metric in the form

$$ds^2 = dt^2 - \frac{1}{c^2}[d\varrho^2 + R_c^2 \sin^2(\varrho/R_c)(d\theta^2 + \sin^2\theta\, d\phi^2)]. \tag{5.27}$$

Notice that, in this form, t is cosmic time and $d\varrho$ is an increment of proper distance in the radial direction.

There is a problem in using the metric in this form which is illustrated by the space-time diagram shown in Fig 5.5. Since light travels at a finite velocity, we observe all astronomical objects along our *past light cone* which is centred on the Earth at the present epoch t_0. Therefore, when we observe distant objects, we do not observe them at the present epoch but rather at an earlier epoch when the Universe was still homogeneous and isotropic but the distances between fundamental observers were smaller and the spatial curvature different. The problem is that we can only apply the metric (5.27) to an isotropic curved space defined *at a single epoch*.

To resolve this problem, we perform the following thought experiment. To measure a proper distance which can be included in the metric (5.27), we line up a set of fundamental observers between the Earth and the galaxy whose

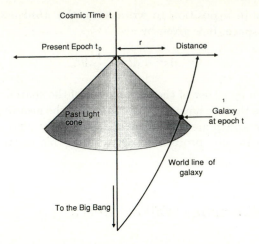

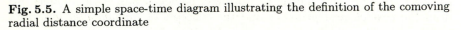

Fig. 5.5. A simple space-time diagram illustrating the definition of the comoving radial distance coordinate

distance we wish to measure. The observers are all instructed to measure the distance $d\varrho$ to the next fundamental observer at a particular cosmic time t which they read on their own clock. By adding together all the $d\varrho$s, we can find a proper distance ϱ which is measured at a single epoch and which can be used in the metric (5.27). Notice that ϱ is a *fictitious distance* in that we cannot actually measure distances in this way. We can only observe distant galaxies as they were at some epoch earlier than the present and we do not know how to project their positions relative to us forward to the present epoch without a knowledge of the kinematics of the expanding Universe. In principle, it is possible to determine the kinematics of the Universe observationally but, as we will discuss, this is not feasible at the moment. In practice, all we can do at the moment is to assume some suitable cosmological model for which the distance measure ϱ can be determined. In other words, *the distance measure ϱ depends upon the choice of cosmological model which we do not know*. We will show in Sect. 5.5 how to relate ϱ to measurable quantities.

Let us work out how the ϱ coordinates of galaxies change in a uniformly expanding Universe. The definition of a uniform expansion is that between two cosmic epochs, t_1 and t_2, the distances of any two fundamental observers, i and j, change such that

$$\frac{\varrho_i(t_1)}{\varrho_j(t_1)} = \frac{\varrho_i(t_2)}{\varrho_j(t_2)} = \text{constant}, \tag{5.28}$$

that is,

$$\frac{\varrho_i(t_1)}{\varrho_i(t_2)} = \frac{\varrho_j(t_1)}{\varrho_j(t_2)} = \dots = \text{constant} = \frac{R(t_1)}{R(t_2)}. \tag{5.29}$$

For isotropic world models, $R(t)$ is a universal function known as the *scale factor* which describes how the relative distances between *any* two fundamental observers change with cosmic time t. Let us therefore adopt the following definitions. We set $R(t)$ equal to 1 at the present epoch t_0 and let the value of ϱ at the present epoch be r, that is, we can rewrite (5.29) as

$$\varrho(t) = R(t)r. \tag{5.30}$$

r thus becomes a *distance label* which is attached to a galaxy or fundamental observer for all time and the variation in proper distance in the expanding Universe is taken care of by the scale factor $R(t)$; r is called the *comoving radial distance coordinate*.

Proper distances perpendicular to the line of sight must also change by a factor R between the epochs t and t_0 because of the isotropy of the world model.

$$\frac{\Delta l(t)}{\Delta l(t_0)} = R(t). \tag{5.31}$$

From the metric (5.27),

$$R(t) = \frac{R_c(t)\sin\left[\varrho/R_c(t)\right]\mathrm{d}\theta}{R_c(t_0)\sin[r/R_c(t_0)]\,\mathrm{d}\theta}. \tag{5.32}$$

Reorganising this equation and using (5.29), we see that

$$\frac{R_c(t)}{R(t)}\sin\left[\frac{R(t)r}{R_c(t)}\right] = R_c(t_0)\sin\left[\frac{r}{R_c(t_0)}\right]. \tag{5.33}$$

This is only true if

$$R_c(t) = R_c(t_0)R(t), \tag{5.34}$$

that is, the radius of curvature of the spatial sections is just proportional to $R(t)$. Thus, in order to preserve isotropy and homogeneity, *the curvature of space changes as the Universe expands as* $\kappa = R_c^{-2} \propto R^{-2}$. Notice that κ cannot change sign and so, if the geometry of the Universe was once, say, hyperbolic, it will always remain so.

Let us call the value of $R_c(t_0)$, that is, the radius of curvature at the present epoch, $\Re$. Then

$$R_c(t) = \Re R(t). \tag{5.35}$$

Substituting (5.30) and (5.35) into the metric (5.27), we obtain

$$\mathrm{d}s^2 = \mathrm{d}t^2 - \frac{R^2(t)}{c^2}[\mathrm{d}r^2 + \Re^2\sin^2(r/\Re)(\mathrm{d}\theta^2 + \sin^2\theta\,\mathrm{d}\phi^2)]. \tag{5.36}$$

This is the *Robertson–Walker metric* in the form we will use in much of our future analysis. Notice that it contains one unknown function $R(t)$, the scale factor, which describes the dynamics of the Universe and an unknown constant $\Re$ which describes the spatial curvature of the Universe at the present epoch.

It is possible to rewrite this metric in different ways. For example, if we use a *comoving angular diameter distance* $r_1 = \Re \sin(r/\Re)$, the metric becomes

$$ds^2 = dt^2 - \frac{R^2(t)}{c^2}\left[\frac{dr_1^2}{1 - \kappa r_1^2} + r_1^2(d\theta^2 + \sin^2\theta\, d\phi^2)\right], \qquad (5.37)$$

where $\kappa = 1/\Re^2$. By a suitable rescaling of the r_1 coordinate $\kappa r_1^2 = r_2^2$, the metric can equally well be written

$$ds^2 = dt^2 - \frac{R_1^2(t)}{c^2}\left[\frac{dr_2^2}{1 - k r_2^2} + r_2^2(d\theta^2 + \sin^2\theta\, d\phi^2)\right], \qquad (5.38)$$

with $k = +1, 0$ and -1 for universes with spherical, flat and hyperbolic geometries respectively. Notice that, in this rescaling, the value of $R_1(t_0)$ at the present epoch is $\Re$ and not unity. This is a rather popular form for the metric but I will normally use (5.36) because the r coordinate has a clear physical meaning.

The importance of the metrics (5.36), (5.37) and (5.38) is that they enable us to define the invariant interval ds^2 between events at any epoch or location in the expanding Universe. Let us recall the meanings of the various components and variables in the metric (5.36).

- t is cosmic time, that is, time as measured by a clock carried by a fundamental observer;
- r is the *comoving radial distance coordinate* which is fixed to a galaxy for all time and which is the proper distance the galaxy would have if its world line were projected forward to the present epoch t_0 and its distance measured at that time.
- $R(t)\, dr$ is the element of proper (or geodesic) distance in the radial direction at the epoch t;
- $R(t)[\Re \sin(r/\Re)]\, d\theta = R(t) r_1\, d\theta$ is the element of proper distance perpendicular to the radial direction subtended by the angle $d\theta$ at the origin;
- Similarly, $R(t)[\Re \sin(r/\Re)] \sin\theta\, d\phi = R(t) r_1 \sin\theta\, d\phi$ is the element of proper distance in the ϕ-direction.

Notice that so far we have specified nothing about the physics which determines the rate of expansion of the Universe – this has all been absorbed into the function $R(t)$. Note the key point that, whatever the physics which determines the function $R(t)$, only the three types of isotropic geometry described by the Robertson–Walker metric are allowed.

5.5 Observations in Cosmology

It is important to realise that many of the most important results which relate the intrinsic properties of distant objects to their observed properties are independent of the specific cosmological model. It is useful to produce a catalogue of results which describe how the observed properties of objects are related to their intrinsic properties and which are independent of the particular form of $R(t)$. First of all, let us elucidate the real meaning of redshift in cosmology.

5.5.1 Redshift

By redshift, we mean the shift of spectral lines to longer wavelength because of their recessional velocities from our Galaxy. If λ_e is the wavelength of the line as emitted and λ_0 the observed wavelength, the redshift z is defined to be

$$z = \frac{\lambda_0 - \lambda_e}{\lambda_e}. \tag{5.39}$$

According to special relativity, the radial velocity inferred from the redshift is given by the standard relation

$$1 + z = \left(\frac{1 + v/c}{1 - v/c}\right)^{1/2}. \tag{5.40}$$

In the limit of small redshifts, $v/c \ll 1$, (5.40) reduces to

$$v = cz. \tag{5.41}$$

This is the type of velocity which Hubble used in deriving the velocity–distance relation $v = H_0 r$.

Consider a wave packet of frequency ν_1 emitted between cosmic times t_1 and $t_1 + \Delta t_1$ from a distant galaxy. This wave packet is received by the observer at the present epoch in the cosmic time interval t_0 to $t_0 + \Delta t_0$. The signal propagates along null-cones, $ds^2 = 0$, and so, considering radial propagation from source to observer, $d\theta = 0$ and $d\phi = 0$, the metric (5.36) gives us the relation

$$dt = -\frac{R(t)}{c}\,dr \qquad \frac{c\,dt}{R(t)} = -dr. \tag{5.42}$$

The minus sign appears because the origin of the r coordinate is the observer at $t = t_0$. Considering first the leading edge of the wave packet, the integral of (5.42) is

$$\int_{t_1}^{t_0} \frac{c\,dt}{R(t)} = -\int_r^0 dr. \tag{5.43}$$

The end of the wave packet must travel the same distance in units of comoving distance coordinate since the r coordinate is fixed to the galaxy for all time. Therefore,

$$\int_{t_1+\Delta t_1}^{t_0+\Delta t_0} \frac{c\,dt}{R(t)} = -\int_r^0 dr,$$ (5.44)

that is,

$$\int_{t_1}^{t_0} \frac{c\,dt}{R(t)} + \frac{c\,\Delta t_0}{R(t_0)} - \frac{c\,\Delta t_1}{R(t_1)} = \int_{t_1}^{t_0} \frac{c\,dt}{R(t)}.$$ (5.45)

Since $R(t_0) = 1$, we find that

$$\Delta t_0 = \frac{\Delta t_1}{R(t_1)}.$$ (5.46)

This is the cosmological expression for the phenomenon of *time dilation*. Distant galaxies are observed at some earlier cosmic time t_1 when $R(t_1) < 1$ and so phenomena are observed to take longer in our frame of reference than they do in that of the source. The phenomenon is precisely the same as time dilation in special relativity, whereby, for example, relativistic muons, created at the top of the atmosphere, are observed to have longer lifetimes in the observer's frame as compared with their proper lifetimes. The expression (5.46) provides a direct test of the Robertson-Walker formalism and it has recently become possible to carry it out using the properties of supernovae of Type 1A. These objects, which are described in more detail in Sect. 8.4.2, have quite distinctive properties, specifically, there is a remarkable narrow dispersion in their absolute magnitudes and the nearby examples have exactly the same light curves, that is, the time-variation of their luminosities (Fig. 5.6a). These standard properties become even more precisely defined when account is taken of a weak correlation between luminosity and decline rate. These objects have such great luminosities at maximum light that they can be observed at large redshifts. Fig. 5.6b shows a plot of the duration of the initial outburst of the supernova as a function of redshift z, or, more precisely, $(1+z)$ (Goldhaber *et al.* 1997). It can be seen that the observations are in excellent agreement with the expectations of (5.46).

The result (5.46) provides us with an expression for *redshift*. If $\Delta t_1 = \nu_1^{-1}$ is the period of the emitted waves and $\Delta t_0 = \nu_0^{-1}$ the observed period, then

$$\nu_0 = \nu_1 R(t_1).$$ (5.47)

Rewriting this result in terms of redshift z,

$$z = \frac{\lambda_0 - \lambda_e}{\lambda_e} = \frac{\lambda_0}{\lambda_e} - 1 = \frac{\nu_1}{\nu_0} - 1,$$ (5.48)

that is,

$$1 + z = \frac{1}{R(t_1)}.$$ (5.49)

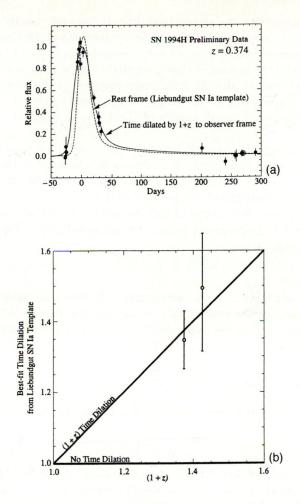

Fig. 5.6 (a) The time variation of the brightness of type 1A supernovae, showing the effect of time dilation for the supernova SN 1994H at redshift $z = 0.374$. (b) The correlation between the duration of the supernova outburst and redshift. The solid line shows the expectation of the cosmological time dilation formula (5.46) (Goldhaber *et al.* 1997).

This is one of the most important relations in cosmology and displays the real meaning of the redshifts of galaxies. *Redshift is a measure of the scale factor of the Universe when the radiation was emitted by the source.* When we observe a galaxy with redshift $z = 1$, the scale factor of the Universe when the light was emitted was $R(t) = 0.5$, that is, the distances between fundamental observers (or galaxies) were half their present values. Note, however, that we obtain no information about *when* the light was emitted. If we did, we could determine directly from observation the function $R(t)$. Unfortunately, we do not understand the astrophysics of galaxies and quasars well enough to be able to estimate times or ages from observation. Optimists, such as the present author, believe that we will eventually understand the physics of galaxies so well that certain of their properties can be used as cosmic clocks. In a few cases, we can already make some progress in this direction and we will discuss some examples in Chap. 20. At present, however, we have to adopt some theory of the dynamics of the Universe in order to determine $R(t)$.

Thus, although the redshift can be converted into a recession velocity using (5.40), its real meaning is much deeper as a measure of the scale factor $R(t) = (1 + z)^{-1}$. We will use (5.49) repeatedly throughout the rest of this text. It is a great pity that Hubble multiplied z by c.

One important consequence of this calculation is that we can now derive an expression for the comoving radial distance coordinate r. Equation (5.43) can be written

$$r = \int_{t_1}^{t_0} \frac{c\,dt}{R(t)}.$$ (5.50)

Thus, once we know $R(t)$ we can immediately find r by integration. This integral emphasises the point that r is an artificial distance which depends upon how the Universe has expanded between the emission and reception of the radiation.

5.5.2 Hubble's Law

In terms of proper distances, Hubble's Law can be written $v = H\varrho$ and so

$$\frac{d\varrho}{dt} = H\varrho.$$ (5.51)

We have written H rather than H_0 in Hubble's law since a 'Hubble's constant' H can be defined at any epoch as we show below. Substituting $\varrho = R(t)r$, we find that

$$r\frac{dR(t)}{dt} = H R(t)r,$$ (5.52)

that is,

$$H = \dot{R}/R.$$ (5.53)

Since we measure Hubble's constant H_0 at the present epoch, $t = t_0$, $R = 1$, we find

$$H_0 = (\dot{R})_{t_0}.$$ (5.54)

Thus, Hubble's constant H_0 defines the present expansion rate of the Universe. Notice that we can define a value of Hubble's constant at any epoch through the more general relation

$$H(t) = \dot{R}/R.$$ (5.55)

5.5.3 Angular Diameters

The great simplification which results from the use of the Robertson–Walker metric in the form (5.36) is apparent in working out the angular size of an object of proper length d perpendicular to the radial coordinate at redshift z. The relevant spatial component of the metric (5.36) is the term in $d\theta$. The proper length d of an object at redshift z, corresponding to scale factor $R(t)$, is

$$d = R(t)\,\Re\sin\left(\frac{r}{\Re}\right)\Delta\theta = R(t)D\,\Delta\theta = \frac{D\,\Delta\theta}{(1+z)}; \qquad (5.56)$$

$$\Delta\theta = \frac{d(1+z)}{D}, \qquad (5.57)$$

where we have introduced a *distance measure* $D = \Re\sin(r/\Re)$. For small redshifts, $z \ll 1$, $r \ll \Re$, (5.57) reduces to the Euclidean relation $d = r\Delta\theta$.

The expression (5.57) can also be written in the form

$$\Delta\theta = \frac{d}{D_A}, \qquad (5.58)$$

so that the relation between d and $\Delta\theta$ looks like the standard Euclidean relation. To achieve this, we have to introduce another distance measure $D_A = D/(1+z)$ which is known as the *angular diameter distance* and which is often used in the literature.

Another useful calculation is the angular diameter of an object which continues to partake in the expansion of the Universe. This is the case for infinitesimal perturbations in the expanding Universe. A good example is the angular diameter which large scale structures present in the Universe today would have subtended at an earlier epoch, say, the epoch of recombination, if they had simply expanded with the Universe. This calculation is used to work out physical sizes today corresponding to the angular scales of the fluctuations observed in the Cosmic Microwave Background Radiation. If the physical size of the object is $d(t_0)$ now and it expanded with the Universe, its physical size at redshift z was $d(t_0)/(1+z)$. Therefore, the object subtended an angle

$$\Delta\theta = \frac{d(t_0)}{D}. \qquad (5.59)$$

Notice that in this case the $(1+z)$ factor has disappeared from (5.56).

5.5.4 Apparent Intensities

Suppose a source at redshift z has luminosity $L(\nu_1)$ (measured in W Hz^{-1}), that is, the total energy emitted over 4π steradians per unit time per unit frequency interval. What is the flux density $S(\nu_0)$ of the source at the observing frequency ν_0, that is, the energy received per unit time, per unit area and per unit bandwidth (W m^{-2} Hz^{-1}) where $\nu_0 = R(t_1)\nu_1 = \nu_1/(1+z)$?

Suppose the source emits $N(\nu_1)$ photons of energy $h\nu_1$ in the bandwidth $\Delta\nu_1$ in the proper time interval Δt_1. Then the luminosity $L(\nu_1)$ of the source is

$$L(\nu_1) = \frac{N(\nu_1)\,h\nu_1}{\Delta\nu_1\Delta t_1}. \tag{5.60}$$

These photons are distributed over a 'sphere' centred on the source at epoch t_1 and, when the 'shell' of photons arrives at the observer at the epoch t_0, a certain fraction of them is intercepted by the telescope. The photons are observed at the present epoch t_0 with frequency $\nu_0 = R(t_1)\nu_1$, in a proper time interval $\Delta t_0 = \Delta t_1/R(t_1)$ and in the waveband $\Delta\nu_0 = R(t_1)\Delta\nu_1$.

We also need to know how the photons spread out over a sphere between the epochs t_1 and t_0, that is, we must relate the diameter of our telescope Δl to the angular diameter $\Delta\theta$ which it subtends at the source at epoch t_1. The metric (5.36) provides an elegant answer. The proper distance Δl refers to the present epoch at which $R(t) = 1$ and hence

$$\Delta l = D\Delta\theta, \tag{5.61}$$

where $\Delta\theta$ is the angle measured by a fundamental observer located at the source. We can understand this result by considering how the photons emitted by the source spread out over solid angle $d\Omega$, as observed from the source in the curved geometry. If the Universe were not expanding, the surface area over which the photons would be observed at a time t after their emission would be

$$dA = \Omega R_c^2 \sin^2 \frac{x}{R_c}, \tag{5.62}$$

where $x = ct$. In the expanding Universe, R_c changes as the Universe expands and so, in place of the expression x/R_c, we should write

$$\frac{1}{\Re}\int_{t_1}^{t_0} \frac{c\,dt}{R} = \frac{r}{\Re}, \tag{5.63}$$

where we recognise that r is the comoving radial distance coordinate. Thus,

$$dA = \Omega\Re^2 \sin^2 \frac{r}{\Re}. \tag{5.64}$$

Therefore, the diameter of the telescope as observed from the source is $\Delta l = D\Delta\theta$. Notice how the use of the comoving radial distance coordinate takes account of the changing geometry of the Universe in this calculation. Notice also the difference between (5.57) and (5.61). They correspond to angular diameters measured in opposite directions along the light cone. The factor of $(1 + z)$ difference between them is part of a more general relation concerning angular diameter measures along light cones which is known as the *reciprocity theorem*.

Therefore, the surface area of the telescope is $\pi\Delta l^2/4$ and the solid angle subtended by this area at the source is $\Delta\Omega = \pi\Delta\theta^2/4$. The number of photons incident upon the telescope in time Δt_0 is therefore

$$N(\nu_1)\Delta\Omega/4\pi, \tag{5.65}$$

but they are now observed with frequency ν_0. Therefore, the flux density of the source, that is, the energy received per unit time, per unit area and per unit bandwidth is

$$S(\nu_0) = \frac{N(\nu_1)\,h\nu_0\,\Delta\Omega}{4\pi\,\Delta t_0\,\Delta\nu_0\,(\pi/4)\Delta l^2}. \tag{5.66}$$

We can now relate the quantities in (5.66) to the properties of the source, using (5.46) and (5.47).

$$S(\nu_0) = \frac{L(\nu_1)R(t_1)}{4\pi D^2} = \frac{L(\nu_1)}{4\pi D^2(1+z)}. \tag{5.67}$$

If the spectra of sources are of power law form, $L(\nu) \propto \nu^{-\alpha}$, this relation becomes

$$S(\nu_0) = \frac{L(\nu_0)}{4\pi D^2(1+z)^{1+\alpha}}. \tag{5.68}$$

We can repeat the analysis for *bolometric* luminosities and flux densities. In this case, we consider the total energy emitted in a finite bandwidth $\Delta\nu_1$ which is received in the bandwidth $\Delta\nu_0$, that is

$$L_{\mathrm{bol}} = L(\nu_1)\Delta\nu_1 = 4\pi D^2 S(\nu_0)(1+z) \times \Delta\nu_0(1+z)$$
$$= 4\pi D^2(1+z)^2 S_{\mathrm{bol}}, \tag{5.69}$$

where the bolometric flux density is $S_{\mathrm{bol}} = S(\nu_0)\Delta\nu_0$. Therefore,

$$S_{\mathrm{bol}} = \frac{L_{\mathrm{bol}}}{4\pi D^2(1+z)^2} = \frac{L_{\mathrm{bol}}}{4\pi D_{\mathrm{L}}^2}. \tag{5.70}$$

The quantity $D_{\mathrm{L}} = D(1+z)$ is called the *luminosity distance* of the source since this definition makes the relation between S_{bol} and L_{bol} look like an inverse square law. The bolometric luminosity can be integrated over any suitable bandwidth so long as the corresponding redshifted bandwidth is used to measure the bolometric flux density at the present epoch.

$$\sum_{\nu_0} S(\nu_0)\Delta\nu_0 = \frac{\sum_{\nu_1} L(\nu_1)\Delta\nu_1}{4\pi D^2(1+z)^2} = \frac{\sum_{\nu_1} L(\nu_1)\Delta\nu_1}{4\pi D_{\mathrm{L}}^2}. \tag{5.71}$$

The formula (5.67) is the best expression for relating the observed intensity $S(\nu_0)$ to the intrinsic luminosity of the source $L(\nu_1)$. If the spectrum of the source is known over the appropriate range of frequencies (or wavelengths), it is straightforward to reconstruct $L(\nu_1)$ using (5.67). We can also write (5.67) in terms of the luminosity of the source at the observing frequency ν_0 as

$$S(\nu_0) = \frac{L(\nu_0)}{4\pi D_{\mathrm{L}}^2}\left[\frac{L(\nu_1)}{L(\nu_0)}(1+z)\right]. \tag{5.72}$$

The last term in square brackets is a form of what is known as the *K-correction*. The K-correction was introduced by the pioneer optical cosmologists in the 1930s in order to 'correct' the apparent magnitude of distant galaxies for the effects of redshifting their spectra when observations are made through standard filters with a fixed mean observing frequency ν_0. Taking logarithms and multiplying by -2.5, we can convert the terms in square brackets into a correction to the apparent magnitude of the galaxy and then we find

$$K(z) = -2.5 \log_{10} \left[\frac{L(\nu_1)}{L(\nu_0)} (1+z) \right]. \tag{5.73}$$

This form of K-correction is correct for *monochromatic* flux densities and luminosities. In the case of observations in the optical waveband, apparent magnitudes are measured through standard filters which usually have quite wide pass-bands. Therefore, to determine the appropriate K-corrections, the spectral energy distribution of the galaxy has to be convolved with the transmission function of the filter in the rest-frame and at the redshift of the galaxy. This is a straightforward calculation once the spectrum of the object is known.

Although I prefer to work directly with (5.67) and take appropriate averages, K-corrections are rather firmly established in the literature and it is often convenient to use the term to describe the effects of shifting the emitted spectrum into the observing wavelength window.

5.5.5 Number Densities

We often need to know the number of objects in a particular redshift interval, z to $z+dz$. Since there is a one-to-one relation between r and z, the problem is straightforward because, by definition, r is a radial proper distance coordinate defined *at the present epoch*. Therefore, the number of objects in the interval of radial comoving coordinate distance r to $r + dr$ is given by results already obtained in Sect. 5.3. The space-time diagram shown in Fig. 5.5 illustrates how we can evaluate the numbers of objects in the comoving distance interval r to $r + dr$ entirely by working in terms of *comoving volumes* at the present epoch. At the present epoch, the radius of curvature of the spatial geometry is $\Re$ and so the volume of a spherical shell of thickness dr at comoving distance coordinate r is

$$dV = 4\pi \Re^2 \sin^2(r/\Re) \, dr = 4\pi D^2 \, dr. \tag{5.74}$$

Therefore, if N_0 is the present space density of objects and their number is conserved as the Universe expands,

$$dN = N(z) \, dz = 4\pi N_0 D^2 \, dr. \tag{5.75}$$

The definition of comoving coordinates automatically takes care of the expansion of the Universe. Another way of expressing this result is to state that (5.75) gives the number density of objects in the redshift interval z to $z + \mathrm{d}z$, assuming the *comoving number density* of the objects is unchanged with cosmic epoch. If, for some reason, the comoving number density of objects changes with cosmic epoch as, say, $f(z)$ with $f(z = 0) = 1$, then the number of objects expected in the redshift interval $\mathrm{d}z$ is

$$\mathrm{d}N = N(z)\,\mathrm{d}z = 4\pi N_0\, f(z)\, D^2\, \mathrm{d}r. \tag{5.76}$$

5.5.6 The Age of the Universe

Finally, let us work out an expression for the age of the Universe, T_0 from a rearranged version of (5.42). The basic differential relation is

$$-\frac{c\,\mathrm{d}t}{R(t)} = \mathrm{d}r, \tag{5.42}$$

and hence

$$T_0 = \int_0^{t_0} \mathrm{d}t = \int_0^{r_{\max}} \frac{R(t)\,\mathrm{d}r}{c}, \tag{5.77}$$

where $r_{\max}$ is the comoving distance coordinate corresponding to $R = 0, z = \infty$.

5.6 Summary

The results we have derived can be used to work out the relations between intrinsic properties of objects and observables for any isotropic, homogeneous world model. Let us summarise the procedures described above:

1. First work out from theory, or otherwise, the function $R(t)$ and the curvature of space at the present epoch $\kappa = \Re^{-2}$. Once we know $R(t)$, we know the redshift–cosmic time relation.
2. Now work out the *radial comoving distance coordinate* r from the integral

$$r = \int_{t_1}^{t_0} \frac{c\,\mathrm{d}t}{R(t)}. \tag{5.51}$$

Recall what this expression means – the proper distance interval $c\,\mathrm{d}t$ at epoch t is projected forward to the present epoch t_0 by the scale factor $R(t)$. This integration yields an expression for r as a function of redshift z.
3. Next, work out the *distance measure D* from

$$D = \Re \sin \frac{r}{\Re}. \tag{5.78}$$

This relation determines D as a function of redshift z.

4. If so desired, the *angular diameter distance* $D_A = D/(1+z)$ and the *luminosity distance* $D_L = D(1+z)$ can be introduced to relate physical sizes and luminosities to angular diameters and flux densities respectively.

5. The number of objects dN in the redshift interval dz and solid angle Ω can be found from the expression

$$dN = \Omega N_0 D^2 \, dr, \tag{5.75}$$

where N_0 is the number density of objects at the present epoch which are assumed to be conserved as the Universe expands.

We will develop some explicit solutions for these functions in Chap. 7.

6 An Introduction to Relativistic Gravity

The standard world models which are used as the framework for astrophysical cosmology and for studying the problems of galaxy formation are based upon Einstein's General Theory of Relativity. General Relativity is a beautiful theory but it requires a thorough understanding of tensor calculus in four-dimensional non-Euclidean spaces to appreciate Einstein's epoch-making achievement. Since this is beyond the scope of the present text, Sects. 6.1 to 6.5 are intended to provide some flavour of the full theory and to introduce some key ideas which will be needed later. In Sect. 6.6, the current status of General Relativity is surveyed and it is shown that it is the best relativistic theory of gravity we possess. If you are happy to accept General Relativity at its face value, you may advance to Chap. 7.

6.1 The Principle of Equivalence

The key considerations which lead to Einstein's General Theory of Relativity are embodied in the *Principle of Equivalence* which follows from the null result of the Eötvös and Dicke experiments. These have shown that gravitational mass m_g is very precisely proportional to inertial mass m_i. In Braginsky's version of the Dicke experiment, the linearity of the relation between gravitational and inertial mass was established to better than one part in 10^{12} (for a review of this and many other aspects of experimental gravitation, see Will 1993). According to the Principle of Equivalence, the gravitational field **g** at any point in space can be precisely replaced by an accelerated frame of reference. In Einstein's own words,

- All local, freely falling, non-rotating laboratories are fully equivalent for the performance of all physical experiments.

An equivalent, more transparent statement of the principle is as follows:

- At any point in a gravitational field, in a frame of reference moving with the free-fall acceleration at that point, all the laws of physics have their usual special relativistic form, except for the force of gravity, which disappears locally.

By *free-fall*, we mean a frame of reference which is accelerated at the local gravitational acceleration at that point in space, $\mathbf{a} = \mathbf{g}$. These statements formally identifiy inertial and gravitational mass, since the force acting on a particle in a gravitational field depends upon the particle's *gravitational mass*, whereas the acceleration depends upon its *inertial mass*. The Principle of Equivalence has profound consequences for our understanding of the nature of space and time in a gravitational field. Let us illustrate these by two elementary examples.

6.2 The Gravitational Redshift

In the first example, we replace the stationary frame of reference located in a uniform gravitational field $\mathbf{g}$ by a frame of reference which is accelerated in the opposite direction. Let us consider a light wave of frequency ν propagating from the ceiling to the floor of a lift in a gravitational field $\mathbf{g} = -\mathbf{a}$ (Fig. 6.1). We assume that the acceleration is small. If the height of the lift is h, the light signal travels from the roof to the floor in a time $t = h/c$. In this time, the floor is accelerated to a velocity $u = at = |\mathbf{g}|t$. Hence,

$$u = \frac{|\mathbf{g}|h}{c}. \tag{6.1}$$

Therefore, the light wave is observed with a higher frequency when it arrives at the floor of the lift because of the Doppler effect. To first order in u/c, the observed frequency ν' is

$$\nu' = \nu \left(1 + \frac{u}{c}\right) = \nu \left(1 + \frac{|\mathbf{g}|h}{c^2}\right). \tag{6.2}$$

Let us now express this result in terms of the change in gravitational potential between the ceiling and floor of the lift. Since $\mathbf{g} = -\mathrm{grad}\,\phi$

$$|\mathbf{g}| = -\frac{\Delta\phi}{h}. \tag{6.3}$$

Notice that, because of the attractive nature of the gravitational force, ϕ is more negative at $h = 0$ than at the ceiling. Therefore,

$$\nu' = \nu \left(1 - \frac{\Delta\phi}{c^2}\right). \tag{6.4}$$

This is the formula for the *gravitational redshift* z_{g} in the 'Newtonian' limit. Recalling the definition of redshift,

$$z = \frac{\lambda_{\mathrm{obs}} - \lambda_{\mathrm{em}}}{\lambda_{\mathrm{em}}} = \frac{\nu - \nu'}{\nu}, \tag{6.5}$$

we find that

$$z_{\mathrm{g}} = \frac{\Delta\phi}{c^2}. \tag{6.6}$$

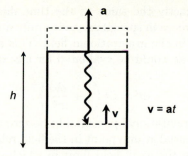

Fig. 6.1. Illustrating the gravitational redshift of an electromagnetic wave propagating from the ceiling to floor of a stationary lift in a gravitational field by application of the Principle of Equivalence.

Thus, the frequency of the waves depends upon to the *gravitational potential* in which the light waves are propagated.

The first observation which demonstrated the reality of the gravitational redshift was suggested by Eddington in 1924. He estimated that the gravitational redshift of the lines in the spectrum of the white dwarf star Sirius B should be $cz_g = 20$ km s^{-1}. The value measured by Adams in 1925 was 19 km s^{-1}. Eddington was jubilant:

> Prof. Adams has thus killed two birds with one stone. He has carried out a new test of Einstein's theory of general relativity, and has shown that matter at least 2000 times denser than platinum is not only possible, but actually exists in the stellar universe.

Laboratory experiments to measure the gravitational redshift were carried out by Pound and Rebka (1960) and by Pound and Snider (1965) who measured the difference in redshift of γ-ray photons moving up then down a tower 22.5 m high at Harvard University using the Mössbauer effect. In this effect, the recoil effects of the emission and absorption of the γ-ray photons are zero since the momentum is absorbed by the whole atomic lattice. The γ-ray resonance is very sharp indeed and only tiny Doppler shifts are needed to move off resonance absorption. In the Harvard experiment, the difference in redshifts for γ-ray photons moving up and down the tower were:

$$z_{\text{up}} - z_{\text{down}} = \frac{2gh}{c^2} = 4.905 \times 10^{-15}. \tag{6.7}$$

The measured value was $(4.900 \pm 0.037) \times 10^{-15}$, a precision of about 1%. Notice the key point that the gravitational redshift is incompatible with special relativity, according to which the observers at the top and bottom of the tower are at rest in the same inertial frame of reference.

Suppose we now write (6.4) in terms of the period of the waves T. Then,

$$T' = T\left(1 + \frac{\Delta\phi}{c^2}\right). \tag{6.8}$$

This expression is exactly the same as the time dilation formula between inertial frames of reference in special relativity, only now the expression refers to different locations in the gravitational field. This expression for the time dilation is exactly what would be evaluated for any time interval and so we can write in general

$$dt' = dt \left(1 + \frac{\Delta\phi}{c^2}\right). \tag{6.9}$$

Let us now take the gravitational potential to be zero at infinity and measure the gravitational potential at any point in the field relative to that value. We assume that we are in the weak field limit in which changes in the gravitational potential are small. Then, at any point in the gravitational field, we can write

$$dt'^2 = dt^2 \left[1 + \frac{\phi(r)}{c^2}\right]^2, \tag{6.10}$$

where dt is the time interval measured at $\phi = 0$, that is, at $r = \infty$. Since $\phi(r)/c^2$ is small, we can write this expression as

$$dt'^2 = dt^2 \left[1 + \frac{2\phi(r)}{c^2}\right]. \tag{6.11}$$

If we now adopt the Newtonian expression for the gravitational potential for a point mass M,

$$\phi(r) = -\frac{GM}{r}, \tag{6.12}$$

we find

$$dt'^2 = dt^2 \left(1 - \frac{2GM}{rc^2}\right). \tag{6.13}$$

Let us now introduce this expression for the time interval into the standard Minkowski metric of special relativity,

$$ds^2 = dt'^2 - \frac{1}{c^2}\, dl^2, \tag{6.14}$$

where dl is the differential element of proper distance. The metric of spacetime about the point mass can therefore be written as

$$ds^2 = dt^2 \left(1 - \frac{2GM}{rc^2}\right) - \frac{1}{c^2}\, dl^2. \tag{6.15}$$

This calculation shows how the metric coefficients become more complicated than those of Minkowski space when we attempt to derive a relativistic theory of gravity. Notice how careful we have to be about keeping track of time in general relativity. The time interval measured by an observer at a point in the gravitational field is dt'; the interval dt is a time interval at infinity. The gravitational redshift relates these differences in time keeping. Notice further that both of these are different from the time measured by an observer in free-fall in the gravitational field.

6.3 The Bending of Light Rays

Let us show how the expression for dl has to be changed as well. Consider the propagation of light rays in our lift but now travelling perpendicular to the gravitational acceleration. We again use the principle of equivalence to replace the stationary lift in a gravitational field by an accelerated lift in free space (Fig. 6.2).

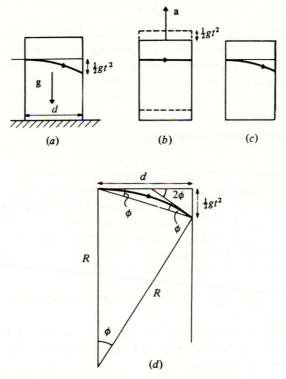

Fig. 6.2. Illustrating the application of the principle of equivalence to the propagation of a light ray in a gravitational field and in a uniformly accelerated lift. In the equivalent accelerated frame of reference, the light ray travels along a curved path.

In the time the light ray propagates across the lift, a distance l, the lift moves upwards a distance $\frac{1}{2}|\mathbf{g}|t^2$. Therefore, in the frame of reference of the accelerated lift, and also in the stationary frame in the gravitational field, the light ray follows a parabolic path as illustrated in the diagram. Let us approximate the light path by a circular arc of radius $\Re$. The length of the chord d across the circle is then

$$d^2 = \tfrac{1}{4}|\mathbf{g}|^2 t^4 + l^2. \tag{6.16}$$

Now, from the geometry of the diagram, it can be seen that $\theta = |\mathbf{g}|t^2/l$. Hence, since $\Re\theta = d$,

$$\Re^2 = \frac{d^2}{\theta^2} = \frac{l^2}{4} + \frac{l^4}{|\mathbf{g}|^2 t^4}.$$
(6.17)

Now, $\frac{1}{2}|\mathbf{g}|t^2 \ll l$, $l = ct$ and so

$$\Re = \frac{l^2}{|\mathbf{g}|t^2} = \frac{c^2}{|\mathbf{g}|}.$$
(6.18)

Thus, the radius of curvature of the path of the light rays depends only upon the local gravitational acceleration $|\mathbf{g}|$. Since $\mathbf{g}$ is determined by the gradient of the gravitational potential, it follows that the *curvature of the paths of light rays depends upon the mass distribution.*

6.4 Further Complications

The consequence of these two elementary calculations is that the rate at which clocks tick depends upon the gravitational potential in which they are located and the paths of light rays are bent by the gravitational influence of the mass-energy distribution in space. In other words, not only is space curved but, more generally, *space-time* is curved. Neither the space nor time coordinates take the simple 'Euclidean' values which appear in the Minkowski metric, which can be written in polar coordinates

$$\mathrm{d}s^2 = \mathrm{d}t^2 - \frac{1}{c^2}\left[\mathrm{d}r^2 + r^2(\mathrm{d}\theta^2 + \sin^2\theta\,\mathrm{d}\phi^2)\right].$$
(6.19)

It must be emphasised that the above arguments are *illustrative* of how Newtonian gravity has to be modified to incorporate gravity and there are many unsatisfactory steps in these.

To complicate matters further, any relativistic theory of gravity must be non-linear. This follows from Einstein's mass-energy relation $E = mc^2$ as applied to the gravitational field. The gravitational field of some mass distribution has a certain local energy density at each point in the field. Since $E = mc^2$, it follows that there is a certain inertial mass density in the gravitational field which is itself a source of gravitational field. This property contrasts with, say, the case of an electric field which possesses a certain amount of electromagnetic field energy (which has a certain inertial mass) but this does not generate additional electrostatic charge. Thus, relativistic gravity is intrinsically a non-linear theory and this accounts for a great deal of its complexity.

Finally, although we can eliminate the *acceleration* due to gravity at a particular point in space, we cannot eliminate completely the effects of gravity in the vicinity of that point. This is most easily seen by considering the

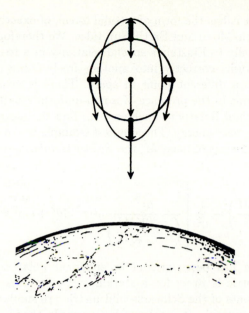

Fig. 6.3. Illustrating the 'tidal forces' which cannot be eliminated when the acceleration due to gravity **g** is replaced by an accelerated reference frame at a particular point in space. In this example, the observer is in centrifugal equilibrium in the field of a point mass. Initially, test masses are located on a sphere about the observer. At a later time, the sphere is distorted into an ellipsoid because of the tidal forces which cannot be eliminated by transforming away the gravitational field at the observer (Penrose 1997).

gravitational field at distance r in the vicinity of a point mass M (Fig. 6.3). It is apparent that we need different freely-falling lifts at different points in space in order to eliminate gravity everywhere. Even over very limited regions of space, if we make very precise measurements over a period of time, we will observe neighbouring particles beginning to move under the influence of the bits of gravity which were not exactly eliminated by transforming to a single accelerated reference frame. For example, consider a standard Euclidean (x, y, z) coordinate frame inside an orbiting Space Station, the z-coordinate being taken in the radial direction. It is a useful exercise to show that, if two test particles are released from rest, with an initial separation vector ξ, this separation vector varies with time as

$$\frac{\mathrm{d}^2}{\mathrm{d}t^2} \begin{bmatrix} \xi^x \\ \xi^y \\ \xi^z \end{bmatrix} = \begin{bmatrix} -\frac{GM}{r^3} & 0 & 0 \\ 0 & -\frac{GM}{r^3} & 0 \\ 0 & 0 & +\frac{2GM}{r^3} \end{bmatrix} \begin{bmatrix} \xi^x \\ \xi^y \\ \xi^z \end{bmatrix}. \tag{6.20}$$

The pleasant thing about this analysis is that it can be seen that the uncompensated forces depend upon r^{-3}. This is the part of the gravitational field which cannot be eliminated by transforming to a single freely-falling frame. Notice that it has the form of a 'tidal force', of exactly the same type which causes Earth–Moon and Earth–Sun tides. We therefore need a theory which reduces locally to Einstein's special relativity in a freely-falling frame and which transforms correctly into another freely-falling reference frame when we move to a different point in space. There is no such thing as a global Lorentz frame in the presence of a non-uniform gravitational field.

Einstein's General Relativity enables us to find the metric of space-time in the presence of mass-energy. The simplest example is the metric of space-time about a point mass of mass M, the *Schwarzschild metric*, which can be written

$$ds^2 = dt^2 \left(1 - \frac{2GM}{rc^2}\right) - \frac{1}{c^2}\left[\frac{dr^2}{\left(1 - \frac{2GM}{rc^2}\right)} + r^2(d\theta^2 + \sin^2\theta\, d\phi^2)\right], \quad (6.21)$$

where the metric has been written in spherical polar coordinates. The Schwarzschild metric is *exact* for a stationary point mass in general relativity. Some elements of the Schwarzschild metric are similar to those which we derived in our approximate analyses. For example, the increment of *proper time* is

$$dt' = dt\left(1 - \frac{2GM}{rc^2}\right)^{1/2}, \quad (6.22)$$

and has the same properties which we derived above, namely, the *coordinate time* t keeps track of how clocks measure time at infinity. Clocks closer to the origin run slower relative to clocks at infinity by the factor $(1 - 2GM/rc^2)^{1/2}$. This enables us to derive the general expression for *gravitational redshift*. The period of the light waves changes by precisely this factor as the light ray propagates from radius r from the point mass to infinity. Therefore, the change of frequency is

$$\nu' = \nu\left(1 - \frac{2GM}{rc^2}\right)^{-1/2}, \quad (6.23)$$

where ν is the frequency measured at infinity. Thus, the redshift z of the radiation is

$$z_{\rm g} = \frac{\lambda_{\rm obs} - \lambda_{\rm em}}{\lambda_{\rm em}} = \frac{\lambda_{\rm obs}}{\lambda_{\rm em}} - 1 = \left(1 - \frac{2GM}{rc^2}\right)^{-1/2} - 1, \quad (6.24)$$

or

$$1 + z_{\rm g} = \left(1 - \frac{2GM}{rc^2}\right)^{-1/2}. \quad (6.25)$$

Light rays emitted from the *Schwarzschild radius* $r_{\rm g} = 2GM/c^2$ are shifted to infinite wavelengths.

6.5 The Route to General Relativity

Einstein's great achievement was to understand how the features discussed in Sects. 6.1 to 6.4 could be incorporated into a self-consistent theory of relativistic gravity. The remarkable story of Einstein's struggles to find the theory is told by Abraham Pais in his splendid scientific biography of Einstein *'Subtle is the Lord ...'* (1982). This is not the place to go into the technical details of what Einstein did. My recommended approach would be to begin with Rindler's excellent introductory text *Essential Relativity* (1977) and then proceed to Weinberg's *Gravitation and Cosmology* (1972) which indicates clearly why General Relativity has to be as complex as it is – what I particularly like about his approach is that the physical content of the physics and the mathematics is elucidated at each stage. Another clear recommendation is d'Inverno's *Introducing Einstein's Relativity* (1995) which is particularly clear on the geometric aspects of the theory and goes much further than Weinberg's text in studying black hole solutions of Einstein's field equations. The understanding of the theory requires considerable effort. Let me outline some of the key steps in the formal development of the theory.

6.5.1 Four-tensors in Relativity

The first step is the realisation that all the laws of physics, with the exception of gravity, can be written in Lorentz invariant form, that is, they are *form-invariant* under Lorentz transformations. The simplest example of such a procedure is the introduction of *four-vectors* which are designed to be objects which are form-invariant under Lorentz transformations. Just for this section, we will use the notation used by professional relativists according to which the velocity v is measured in units of the speed of light, that is, it is taken to have the value $c = 1$. Then, the time coordinate $x^0 = t$ and the spatial components are $x = x^1$, $y = x^2$ and $z = x^3$. As a result, we can write the transformation of a four-vector V^α between two inertial frames of reference in standard configuration in the form

$$V^\alpha \to V'^\alpha = \Lambda^\alpha_\beta V^\beta, \tag{6.26}$$

where the matrix Λ^α_β is the standard Lorentz transformation

$$\Lambda^\alpha_\beta = \begin{bmatrix} \gamma & -\gamma v & 0 & 0 \\ -\gamma v & \gamma & 0 & 0 \\ 0 & 0 & 1 & 0 \\ 0 & 0 & 0 & 1 \end{bmatrix}, \tag{6.27}$$

where $\gamma = (1 - v^2)^{-1/2}$. The convention of summing over identical indices is adopted in (6.26).

Many physical quantities are naturally described in terms of tensors rather than vectors. Therefore, the natural extension of the idea of four-vectors in relativity is to *four-tensors* which are objects which transform according to the rule

$$T^{\alpha\gamma} \to T'^{\alpha\gamma} = \Lambda^{\alpha}_{\beta}\Lambda^{\gamma}_{\delta}T^{\beta\delta}. \tag{6.28}$$

For example, relativists call matter without any internal pressure 'dust' and the *energy–momentum tensor* for dust is $T^{\alpha\beta} = \varrho_0 u^{\alpha}u^{\beta}$, where ϱ_0 is the proper mass density of the dust, meaning the density measured by an observer moving with the flow, that is, a comoving observer; u^{α} is the velocity four-vector. $T^{\alpha\beta}$ can be written

$$T^{\alpha\beta} = \varrho_0 \begin{bmatrix} 1 & u_{x_1} & u_{x_2} & u_{x_3} \\ u_{x_1} & u_{x_1}^2 & u_{x_1}u_{x_2} & u_{x_1}u_{x_3} \\ u_{x_2} & u_{x_1}u_{x_2} & u_{x_2}^2 & u_{x_2}u_{x_3} \\ u_{x_3} & u_{x_1}u_{x_3} & u_{x_2}u_{x_3} & u_{x_3}^2 \end{bmatrix}. \tag{6.29}$$

The simplest case is to consider the transformation of the T^{00} component of this four-tensor, $T'^{00} = \Lambda^{0}_{\beta}\Lambda^{0}_{\delta}T^{\beta\delta}$. The result is $\varrho' = \gamma^2\varrho_0$, which has a natural interpretation in special relativity. The observed density of the dust ϱ' is increased by two powers of the Lorentz factor γ over the proper value ϱ_0. One of these is associated with the formula for the momentum of the dust $p, = \gamma m\mathbf{v}$ and the other is due to length contraction in the direction of motion of the dust, $l = l_0/\gamma$.

For a perfect fluid, the pressure cannot be neglected and then the energy–momentum tensor becomes $T^{\alpha\beta} = (\varrho_0 + p)u^{\alpha}u^{\beta} - pg^{\alpha\beta}$, where $g^{\alpha\beta}$ is the metric tensor. Then, it is a pleasant exercise to show that the equation

$$\partial_{\beta}T^{\alpha\beta} = 0, \tag{6.30}$$

expresses the laws of conservation of momentum and energy in relativity, where ∂_{β} means partial differentiation of the tensor components with respect to β, that is, the operator ∂_{β} has the form

$$[\partial/\partial x_0, \partial/\partial x_1, \partial/\partial x_2, \partial/\partial x_3]. \tag{6.31}$$

Maxwell's equations in a vacuum can be written in compact form in terms of the antisymmetric electromagnetic field tensor $F^{\alpha\beta}$

$$F^{\alpha\beta} = \begin{bmatrix} 0 & E_x & E_y & E_z \\ -E_x & 0 & B_z & -B_y \\ -E_y & -B_z & 0 & B_x \\ -E_z & B_y & -B_x & 0 \end{bmatrix}, \tag{6.32}$$

and the current density four-vector $j^{\alpha} = [\varrho_e, \mathbf{j}]$. I apologise for deviating from my normal practice of using strictly SI units. This elegant form of Maxwell's equations is written in Heaviside–Lorentz units with $c = 1$. The equation of continuity becomes

$$\partial_{\alpha}j^{\alpha} = 0. \tag{6.33}$$

Maxwell's equations for the relations between electric and magnetic fields and their sources become

$$\partial_\beta F^{\alpha\beta} = j^\alpha. \tag{6.34}$$

Thus, four-tensors provide the natural language for expressing the laws of physics in a form which guarantees that they transform correctly according to the Lorentz transformations.

6.5.2 What Einstein Did

The simple considerations of Sects. 6.1 to 6.4 indicate that the aim of General Relativity is to incorporate the influence of the mass-energy distribution upon space-time into the metric coefficients $g_{\mu\nu}$. The metric of space-time locally has to reduce to the standard Minkowski metric

$$ds^2 = dt^2 - \frac{1}{c^2}\,dl^2. \tag{6.35}$$

Therefore, the natural starting point for the development of general transformations between arbitrary four-dimensional spaces is the *Riemannian metric* of form

$$ds^2 = \sum_{\mu,\nu} g_{\mu\nu}\,dx^\mu dx^\nu = g_{\mu\nu}\,dx^\mu dx^\nu, \tag{6.36}$$

where the coordinates x^μ and x^ν define points in four-dimensional space and the interval ds^2 is given by a homogeneous quadratic differential form in these coordinates. The components of the *metric tensor* $g_{\mu\nu}$ vary from point-to-point in space-time and define its local curvature. Since the local curvature defines the properties of the gravitational field, the $g_{\mu\nu}$ can be thought of as being analogous to gravitational potentials.

We need to develop a way of relating the $g_{\mu\nu}$ to the mass-energy distribution, that is, the analogue of Poisson's equation in Newtonian gravity which involves second-order partial differential equations. For example, in the metric (6.11), we rationalised that the g_{00} component should have the form

$$g_{00} = \left(1 + \frac{2\phi}{c^2}\right), \tag{6.37}$$

(see also the Schwarzschild metric (6.21)). Poisson's equation for gravity is

$$\nabla^2\phi = 4\pi G\varrho, \tag{6.38}$$

and hence, using (6.29) and (6.37), we find that

$$\frac{c^2}{2}\nabla^2 g_{00} = 4\pi G T_{00}, \tag{6.39}$$

that is,

$$\nabla^2 g_{00} = \frac{8\pi G}{c^2} T_{00}. \tag{6.40}$$

This is a crude calculation but it shows why it is reasonable to expect a close relation between the derivatives of $g_{\mu\nu}$ and the energy–momentum tensor $T_{\mu\nu}$.

The tensor equivalent of this analysis involves the differentiation of tensors and this is where the complications begin – partial differentiation of tensors does not generally yield other tensors. Thus, the definitions of the equivalent vector operations of grad, div and curl are correspondingly more complicated for tensors as compared with vectors. How this problem was solved and how the components of the metric tensor $g_{\mu\nu}$ are related to the energy–momentum tensor took Einstein almost 10 years, although he was doing lots of other things at the same time.

What is needed is a tensor which involves the metric tensor $g_{\mu\nu}$ and its first and second derivatives and which is linear in its second derivatives. It turns out that there is a unique answer to this problem – it is the fourth-rank tensor $R^{\lambda}_{\mu\nu\kappa}$ which is known as the *Riemann–Christoffel tensor*. Other tensors can be formed from this tensor by contraction, the most important of these being the *Ricci tensor*

$$R_{\mu\kappa} = R^{\lambda}_{\mu\lambda\kappa}, \tag{6.41}$$

and the *curvature scalar*

$$R = g^{\mu\kappa} R_{\mu\kappa}. \tag{6.42}$$

This is the only time R will take this meaning. Throughout the rest of the book R is the scale factor.

Einstein's stroke of genius was to propose that these tensors are related to the energy–momentum tensor in the following way

$$R_{\mu\nu} - \tfrac{1}{2} g_{\mu\nu} R = -\frac{8\pi G}{c^2} T_{\mu\nu}. \tag{6.43}$$

This is the key relation which shows how the components of the metric tensor $g_{\mu\nu}$ are related to the mass-energy distribution in the Universe.

We will go no further along this route, except to note that Einstein realised that he could add an additional term to the left-hand side of (6.43). This is the origin of the famous cosmological constant Λ and was introduced specifically in order to construct a static closed model for the Universe. Equation (6.43) then becomes

$$R_{\mu\nu} - \tfrac{1}{2} g_{\mu\nu} R + \Lambda g_{\mu\nu} = -\frac{8\pi G}{c^2} T_{\mu\nu}. \tag{6.44}$$

In the discussion of Chap. 7, we will use the Newtonian equivalents of these equations but it must be appreciated that we can only do this with the reassurance that the complete Einstein equations give fully self-consistent world models, without any need to introduce *ad hoc* assumptions. Our Newtonian equivalents can however provide intuitive impressions of the physical content of the theory.

6.6 Experimental and Observational Tests of General Relativity

When Friedman first solved the field equations of General Relativity for isotropically expanding Universes, the evidence for the theory was good but not perhaps overwhelming. The history of the tests and their current status are comprehensively described by Will (1993) in his excellent book *Theory and Experiment in Gravitational Physics*. Traditionally, there are four tests of the theory. The *first* is the measurement of the *gravitational redshift* of electromagnetic waves in a gravitational field which we discussed in Sect. 6.2. There, we described the use of the Mössbauer effect to measure the redshift of γ-ray photons in terrestrial experiments and the observation of the gravitational redshift of the emission lines in white dwarfs. More recent versions of the test have involved placing hydrogen masers in rocket payloads and measuring very precisely the change in frequency with altitude. These experiments have demonstrated directly the gravitational redshift of light. In the rocket experiments, the gravitational redshift was measured with a precision of about 5 parts in 10^5.

The *second* and oldest test, and the first great triumph of General Relativity, was the explanation of the *perihelion shift* of the orbit of the planet Mercury. Mercury's orbit has ellipticity $e = 0.2$ and, in 1859, Le Verrier found that, once account is taken of the influence of all the other planets in the Solar System, there remained a small but significant advance of the perihelion of its orbit which amounted to about $\dot{\omega} \approx 43$ arcsec per century. The origin of this perihelion shift remained a mystery, possible explanations including the presence of a hitherto unknown planet close to the Sun, oblateness of the solar interior, deviations from the inverse square law of gravity near the Sun and so on. Continued observations of Mercury by radar ranging has established the advance of the perihelion of its orbit to about 0.1% precision with the result $\dot{\omega} = 42.98(1 \pm 0.001)$ arcsec per century (Shapiro 1990), once the perturbing effects of the other planets had been taken into account. Einstein's theory of General Relativity predicts a value of $\dot{\omega} = 42.98$ arcsec per century, in remarkable agreement with the observed value.

There has been some debate as to whether or not the agreement really is as good as this comparison suggests because there might be a contribution to the perihelion advance if the core of the Sun were rapidly rotating and so possessed a finite quadrupole moment. Observations of the vibrational modes of the Sun, or *helioseismology*, have shown that the core of the Sun is not rotating sufficiently rapidly to upset the excellent agreement between the predictions of general relativity and the observed perihelion advance. Specifically, the quadrupole moment of the Sun has now been measured and its contribution to the perihelion advance is less than 0.1% of the predicted advance.

The *third* test was the measurement of the deflection of light by the Sun. For light rays just grazing the limb of the Sun, the deflection amounts to $\Delta\theta_{\mathrm{GR}} = 4GM/R_\odot c^2 = 1.75$ arcsec, where $R_\odot$ is the radius of the Sun. Historically, this was a very important result. According to Newtonian theory, if we assume that the photon has a momentum $p = h\nu/c$ and then use the Rutherford scattering formula to work out the deviation of the light path, we find that the Newtonian deflection amounts to half the prediction of general relativity $\Delta\theta_{\mathrm{Newton}} = 2GM/R_\odot c^2$. This prediction led to the famous eclipse expeditions of 1919 led by Eddington and Crommelin to measure precisely the angular deflections of the positions of stars observed close to the limb of the Sun during a solar eclipse. One expedition went to Sobral in Northern Brazil and the other to the island of Principe, off the coast of West Africa. The Sobral result was 1.98 ± 0.012 arcsec and the Principe result 1.61 ± 0.3 arcsec. These were technically demanding observations and there was some controversy about the accuracy of the results.

The modern version of the test involves measuring very precisely the angular separations between compact radio sources as they are observed close to the Sun. By means of Very Long Baseline Interferometry (VLBI), angular deviations of less than a milliarcsecond are quite feasible. Fig. 6.4a shows how the precision of the test has improved over the years. In the most recent experiments, the VLBI technique is sensitive to deflections by the Sun over the whole sky. For example, at 90° to the direction of the Sun, the deflection still amounts to 4 milliarcsec, which is readily measurable by VLBI techniques. General Relativity agrees with the most recent observations within 0.1%.

The *fourth* of the traditional tests is closely related to the deflection of light by the Sun and concerns the time delay expected when an electromagnetic wave propagates through a varying gravitational potential. In 1964, Irwin Shapiro realised that the gravitational redshift of signals passing close to the Sun causes a small time delay which can be measured by very precise timing of signals which are reflected from planets or space vehicles as they are about to be occulted by the Sun. Originally, the radio signals were simply reflected from the surface of the planets, but later experiments used transponders on space vehicles which passed behind the Sun. The most accurate results were obtained using transponders on the *Viking* space vehicles which landed on Mars. These 'anchored' transponders have given the most accurate results displayed in Fig. 6.4b. General Relativity is in agreement with these experiments within 0.1%.

Some of the most remarkable results have come from radio observations of pulsars. These radio sources are rotating, magnetised neutron stars and they emit beams of radio emission from their magnetic poles, as shown schematically in Fig. 6.5. The typical parameters of a neutron star are given in the figure. Observations by Joseph Taylor and his colleagues using the Arecibo radio telescope have demonstrated that these are among the most stable clocks we know of in the Universe.

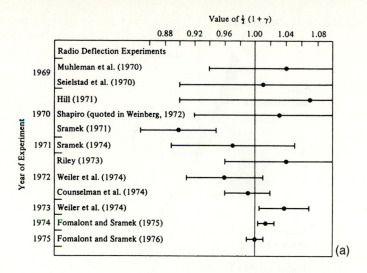

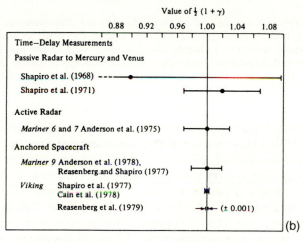

Fig. 6.4a-b. Measurements of the coefficient $(1+\gamma)/2$ from (**a**) the light deflection and (**b**) the time delay measurements. The coefficient γ measures how much space-curvature is produced by unit rest mass. In General Relativity, $\gamma = 1$. In the figures, the values predicted by General Relativity on the abscissa are unity. General Relativity is in agreement with the VLBI deflection experiments and the Shapiro time-delay experiments using the *Viking* spacecraft within 0.1% (Will 1993).

The pulsars have enabled a wide variety of very sensitive tests of General Relativity to be carried out and limits to be set to the possible existence of a background flux of gravitational radiation. The most intriguing systems are those pulsars which are members of binary systems. More than 20 of these are now known, the most important being those in which the other member of the binary system is also a neutron star and in which the two neutron

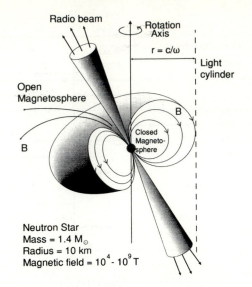

Fig. 6.5. A schematic diagram of a pulsar showing the displacement between the axis of the magnetic dipole and the rotation axis of the neutron star. The radio pulses are assumed to be due to beams of radio emission from the poles of the magnetic field distribution and are associated with the passage of the beam across the line of sight to the observer. Typical parameters of the neutron stars are indicated on the diagram.

stars form a close binary system. The first of these to be discovered was the binary pulsar PSR 1913+16 which is illustrated schematically in Fig. 6.6. The system has a binary period of only 7.75 hours and its orbital eccentricity is large, $e = 0.617$. This system is a pure gift for the relativist. To test General Relativity, we require a perfect clock in a rotating frame of reference and systems such as PSR 1913+16 are ideal for this purpose. The neutron stars are so inert and compact that the binary system is very 'clean' and so can be used for some of the most sensitive tests of General Relativity yet devised.

In Fig. 6.7, the determination of the masses of the neutron stars in the binary system PSR 1913+16 is shown assuming that General Relativity is the correct theory of gravity. Various parameters of the binary orbit can be measured very precisely and these provide different estimates of functions involving the masses of the two neutron stars. In Fig. 6.7, various parameters of the binary orbit are shown, those which have been measured with very good accuracy being indicated by an asterisk. It can be observed that the different loci intersect very precisely at a single point in the m_1/m_2 plane. Some measure of the precision with which the theory is known to be correct can be obtained from the accuracy with which the masses of the neutron stars are known as indicated in Fig. 6.7. These are the most accurately known masses for any extra solar-system object.

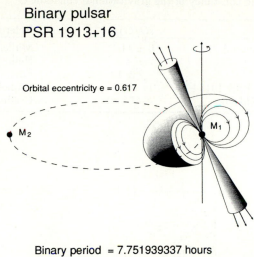

Binary pulsar
PSR 1913+16

Orbital eccentricity e = 0.617

M_2

M_1

Binary period = 7.751939337 hours
Pulsar period = 59 milliseconds
Neutron star mass M_1 = 1.4411(7) $M_\odot$
Neutron star mass M_2 = 1.3874(7) $M_\odot$

Fig. 6.6. A schematic diagram showing the binary pulsar PSR 1913+16. As a result of the ability to measure precisely many parameters of the binary orbit from ultra-precise pulsar timing, the masses of the two neutron stars have been measured with very high precision (Taylor 1992).

A second remarkable measurement has been the rate of loss of orbital rotational energy by the *emission of gravitational waves*. A binary star system loses energy by the emission of gravitational radiation and the rate at which energy is lost can be precisely predicted once the masses of the neutron stars and the parameters of the binary orbit are known. The rate of change of the angular frequency Ω of the orbit due to the emission of gravitational radiation is precisely known, $-\mathrm{d}\Omega/\mathrm{d}t \propto \Omega^5$. The change in orbital phase of the binary pulsar PSR 1913+16 has been observed over a period of 17 years and General Relativity is in precise agreement with the observed changes over that period (Fig. 6.8). Thus, although the gravitational waves themselves have not been detected, exactly the correct energy loss rate from the system has been measured – it is generally assumed that this is convincing evidence for the existence of gravitational waves and this observation acts as a spur to their direct detection by future generations of gravitational wave detectors. This is a very important result for the theory of gravitation since it alone enables a wide range of alternative theories of gravity to be eliminated. For example, since General Relativity predicts only quadrupolar emission of gravitational radiation, any theory which, say, involved the dipole emission of gravitational waves can be eliminated.

Table 6.1. The constancy of the gravitational constant G

Method	$(\dot{G}/G)/10^{-12}$ years	Reference
Lunar laser ranging	0 ± 11	Müller *et al.*(1991)
Viking radar	2 ± 4	Hellings *et al.* (1983)
Viking radar	-2 ± 10	Shapiro (1990)
Binary pulsar PSR 1913+16	11 ± 11	Damour and Taylor (1991)
Pulsar PSR 0655+64	< 55	Goldman (1990)

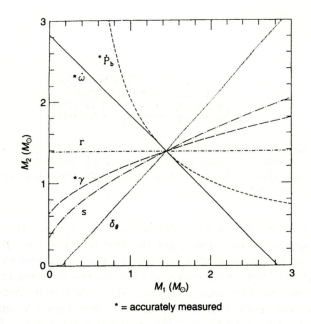

* = accurately measured

Fig. 6.7. The measurement of the masses of the neutron stars in the binary system PSR 1913+16 resulting from very precise timing of the arrival times of the pulses at the Earth. The different parameters of the neutron star's orbit depend upon different combinations of the masses m_1 and m_2 of the neutron stars. It can be seen that the lines intersect very precisely at a single point in the $m_1 - m_2$ plane (Taylor 1992).

An important question for cosmology is whether or not the gravitational constant G has varied with time. A summary of recent results by Will (1993) is shown in Table 6.1. The techniques of lunar laser ranging and the *Viking* radar ranging techniques have provided the best results to date. In these experiments, evidence is sought for steady changes in the lunar and planetary orbits which could be attributed to changes in the gravitational constant

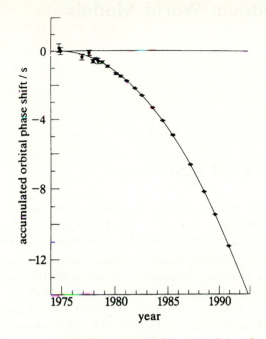

Fig. 6.8. The change of orbital phase as a function of time for the binary neutron star system PSR 1913+16 compared with the expected changes due to gravitational radiation energy loss by the binary system (Taylor 1992)

with time. The techniques of accurate pulsar timing have also been used to determine whether or not there is any evidence for the gravitational constant G changing with time. The latter tests are somewhat dependent upon the equation of state used to describe the interiors of the neutron stars. It can be seen that the best limits come from the *Viking* ranging experiments, largely because the transponders were anchored to Mars and the Lander 2 vehicle survived for 6 years. It can be seen that all the experiments are consistent with values of $\dot{G}/G$ are less than 10^{-11} year^{-1} and probably significantly less than this value. Thus, there can have been little change in the value of the gravitational constant over typical cosmological timescales which are about $(1-2) \times 10^{10}$ years.

In summary, General Relativity has passed every test which has been made of the theory and we can have confidence that it is the correct starting point for the development of theories of the large-scale dynamics of the Universe.

7 The Friedman World Models

7.1 Einstein's Field Equations

Einstein realised that, in General Relativity, he had discovered a theory which enabled fully self-consistent models for the Universe as a whole to be constructed. The standard models contain three essential ingredients:

- The *cosmological principle*, which, combined with the observations that the Universe is isotropic, homogeneous and uniformly expanding on a large scale, leads to the Robertson–Walker metric (5.36);
- *Weyl's postulate*, according to which the world lines of particles meet at a singular point in the finite or infinite past. This means that there is a unique world line passing through every point in space-time. The fluid moves along streamlines in the universal expansion and so behaves like a perfect fluid for which the energy–momentum tensor is given by the $T^{\alpha\beta}$ of (6.30);
- General Relativity, which enables us to relate the energy–momentum tensor to the geometrical properties of space-time through (6.43) or (6.44).

The assumptions of isotropy and homogeneity result in enormous simplifications of Einstein's field equations which reduce to the following pair of independent equations.

$$\ddot{R} = -\frac{4\pi G}{3} R \left(\varrho + \frac{3p}{c^2} \right) + \left[\tfrac{1}{3} \Lambda R^2 \right] ; \qquad (7.1)$$

$$\dot{R}^2 = \frac{8\pi G \varrho}{3} R^2 - \frac{c^2}{\Re^2} + \left[\tfrac{1}{3} \Lambda R^2 \right] . \qquad (7.2)$$

In these equations, R is the scale factor, ϱ is the total inertial mass density of the matter and radiation content of the Universe and p the associated total pressure. $\Re$ is the radius of curvature of the geometry of the world model at the present epoch and so the term $-c^2/\Re^2$ is simply a constant of integration. The *cosmological constant* Λ, which has been included in the terms in square brackets in (7.1) and (7.2), has had a chequered history since it was introduced by Einstein in 1917. We take up that story in Sect. 7.3.

Let us look more closely at the meanings of (7.1) and (7.2). Equation (7.2) is referred to as *Friedman's equation* and has the form of an energy equation, the term on the left-hand side corresponding to the kinetic energy of the expanding fluid and the first term on the right-hand side to its gravitational potential energy, as we will show in Sect. 7.2. An illuminating account of the Friedman's equation, its physical content and its solutions has been given by White (1990), whose presentation can be thoroughly recommended. The First Law of Thermodynamics can be built into Friedman's equation as follows. We write it in the usual form

$$dU = -p\,dV \tag{7.3}$$

We need to formulate the first law in such a way that it is applicable for relativistic and non-relativistic fluids and so we write the internal energy U as the sum of all the terms which can contribute to the total energy of the fluid in the relativistic sense. Thus, the total internal energy consists of the fluid's rest mass energy, its kinetic energy, its thermal energy and so on. If we write the sum of these energies as $\varepsilon_{tot} = \sum_i \varepsilon_i$, the internal energy is $\varepsilon_{tot}V$ and so, differentiating (7.3) with respect to R, it follows that

$$\frac{d}{dR}(\varepsilon_{tot}V) = -p\,\frac{dV}{dR}. \tag{7.4}$$

Now, $V \propto R^3$ and so, differentiating, we find

$$\frac{d\varepsilon_{tot}}{dR} + 3\frac{(\varepsilon_{tot} + p)}{R} = 0. \tag{7.5}$$

This result can be expressed in terms of the inertial mass density associated with the total energy $\varepsilon_{tot} = \varrho c^2$ and so (7.5) can also be written

$$\frac{d\varrho}{dR} + 3\frac{\left(\varrho + \dfrac{p}{c^2}\right)}{R} = 0. \tag{7.6}$$

This is the type of density ϱ which should be included in (7.1) and (7.2).

Let us show how (7.5) and (7.6) lead to a number of important results which we will use repeatedly in what follows. First of all, suppose the fluid is very 'cold' in the sense that $p \ll \varrho_0 c^2$, where ϱ_0 is its rest mass density. Then, setting $p = 0$ and $\varepsilon_0 = Nmc^2$, where N is the number density of particles of rest mass m, we find

$$\frac{dN}{dR} + \frac{3N}{R} = 0 \quad \text{and so} \quad N = N_0 R^{-3}, \tag{7.7}$$

that is, the continuity equation.

Next, the thermal pressure of non-relativistic matter can be included into (7.5). For essentially all our purposes, we will be dealing with monatomic gases or plasmas for which the thermal energy is $\varepsilon_{th} = \frac{3}{2}NkT$ and $p = NkT$. Then, substituting $\varepsilon_{tot} = \frac{3}{2}NkT + Nmc^2$ and $p = NkT$ into (7.5), we find

$$\frac{\mathrm{d}}{\mathrm{d}R}\left(\tfrac{3}{2}NkT + Nmc^2\right) + 3\left(\frac{\tfrac{5}{2}NkT + Nmc^2}{R}\right) = 0$$

$$\frac{\mathrm{d}(NkT)}{\mathrm{d}R} + \frac{5NkT}{R} = 0 \quad \text{and so} \quad NkT = N_0 kT_0 R^{-5}. \tag{7.8}$$

Since $N = N_0 R^{-3}$, we find the standard result for the adiabatic expansion of a monatomic gas with ratio of specific heats $\gamma = 5/3$, $T \propto R^{-2}$. More generally, if the ratio of specific heats of the gas is γ, the temperature changes as $T \propto R^{-3(\gamma-1)}$. We can deduce another important result from $T \propto R^{-2}$. If we write $\varepsilon_{\mathrm{th}} = \tfrac{1}{2}Nm\langle v^2\rangle$, we find that $\langle v^2\rangle \propto R^{-2}$. Thus, the random velocities of the particles of the gas decrease as $v \propto R^{-1}$. This result applies equally to the random motions of galaxies relative to the mean Hubble flow, what are known as the *peculiar velocities* of galaxies, v_{pec}. Therefore, as the Universe expands, we expect the peculiar velocities of galaxies to decrease as $v_{\mathrm{pec}} \propto R^{-1}$.

Finally, in the case of a gas of ultrarelativistic particles, or a gas of photons, we can write $p = \tfrac{1}{3}\varepsilon_{\mathrm{tot}}$. Therefore,

$$\frac{\mathrm{d}\varepsilon_{\mathrm{tot}}}{\mathrm{d}R} + \frac{4\varepsilon_{\mathrm{tot}}}{R} = 0 \quad \text{and so} \quad \varepsilon_{\mathrm{tot}} \propto R^{-4}. \tag{7.9}$$

In the case of a gas of photons, $\varepsilon_{\mathrm{rad}} = \sum Nh\nu$ and, since $N \propto R^{-3}$, we find $\nu \propto R^{-1}$. The purpose of these calculations is to show how (7.5) and (7.6) correctly describe the law of conservation of energy for both relativistic and non-relativistic gases.

Let us now return to the analysis of (7.2). Differentiating this equation with respect to time and dividing through by $\dot{R}$, we find

$$\ddot{R} = \frac{4\pi G R^2}{3}\frac{\mathrm{d}\varrho}{\mathrm{d}R} + \frac{8\pi G\varrho R^2}{3} + \left[\tfrac{1}{3}\Lambda R\right]. \tag{7.10}$$

Now, substituting the expression for $\mathrm{d}\varrho/\mathrm{d}R$ from (7.6), we find

$$\ddot{R} = -\frac{4\pi G}{3}R\left(\varrho + \frac{3p}{c^2}\right) + \left[\tfrac{1}{3}\Lambda R^2\right], \tag{7.1}$$

that is, we recover (7.1).

Equation (7.1) has the form of a force equation, but, as we have shown, it contains implicitly the First Law of Thermodynamics as well. An equation of this form can be derived from Newtonian considerations, but it does not contain the pressure term $3p/c^2$. This pressure term can be considered a 'relativistic correction' to the inertial mass density, but it is unlike normal pressure forces which depend upon the gradient of the pressure and, for example, hold up the stars. The term $\varrho + (3p/c^2)$ can be thought of as playing the role of an *active gravitational mass density*. What this analysis shows is that Einstein's equations of general relativity combined with the energy–momentum tensor of a perfect gas contain both the laws of energy conservation and of motion.

·The general solutions of (7.2) for expanding world models were discovered by Alexander Alexandrovich Friedman in two remarkable papers published in 1922 and 1924 (for translations, see *Cosmological Constants*, edited by Bernstein and Feinberg, 1986). In these papers, Friedman assumed that $\Lambda \neq 0$ and so it is appropriate to refer the complete set of models with and without the Λ-term as the *Friedman world models*.

7.2 The Standard Dust Models – the Friedman World Models with $\Lambda = 0$

By *dust*, cosmologists mean a pressureless fluid and so we set $p = 0$. In this section, the cosmological constant Λ will be taken to be zero. It is convenient to refer the density of fluid to its value at the present epoch ϱ_0. Because of conservation of mass, $\varrho = \varrho_0 R^{-3}$ and so (7.1) and (7.2) reduce to

$$\ddot{R} = -\frac{4\pi G \varrho_0}{3R^2} \qquad \dot{R}^2 = \frac{8\pi G \varrho_0}{3R} - \frac{c^2}{\Re^2}. \tag{7.11}$$

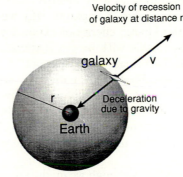

Velocity of recession
of galaxy at distance r

galaxy   v

r

Deceleration
due to gravity

Earth

Fig. 7.1. Illustrating the dynamics of Newtonian world models.

7.2.1 The Newtonian Analogue of the Friedman World Models

In 1934, Milne and McCrea showed that relations of this form can be derived using non-relativistic Newtonian dynamics. We will perform this calculation because the ideas implicit in the argument can be used to understand some of the problems which arise in the theory of galaxy formation. Consider a galaxy at distance x from the Earth and work out its deceleration due to the gravitational attraction of the matter inside the sphere of radius x centred on the Earth (Fig. 7.1). By Gauss's theorem, because of the spherical symmetry of the distribution of matter within x, we can replace that mass, $M = (4\pi/3)\varrho x^3$, by a point mass at the centre of the sphere and so the deceleration of the galaxy is

$$m\ddot{x} = -\frac{GMm}{x^2} = -\frac{4\pi x \varrho m}{3}. \tag{7.12}$$

The mass of the galaxy m cancels out on either side of the equation, showing that the deceleration refers to the sphere of matter as a whole rather than to any particular galaxy. We now make the same substitutions as before – replace x by the comoving value r using the scale factor R, $x = Rr$, and express the density in terms of its value at the present epoch, $\varrho = \varrho_0 R^{-3}$. Therefore,

$$\ddot{R} = -\frac{4\pi G \varrho_0}{3R^2}, \tag{7.13}$$

which is identical to (7.1) for dust models with $\Lambda = 0$. Multiplying (7.13) by $\dot{R}$ and integrating, we find

$$\dot{R}^2 = \frac{8\pi G \varrho_0}{3R} + \text{constant}. \tag{7.14}$$

This result is identical to (7.2) if we identify the constant with $-c^2/\Re^2$. This Newtonian calculation makes it apparent why we can identify the left-hand side of (7.2) with the kinetic energy of expansion of the fluid and the first term on the right-hand side with its gravitational potential energy.

The above analysis brings out a number of important features of the Friedman world models. First of all, note that, because of the assumption of isotropy, local physics is also global physics. This is why the Newtonian argument works. The same physics which defines the local behaviour of matter also defines its behaviour on the largest scales. For example, the curvature of space κ within one cubic metre is exactly the same as that on the scale of the Universe itself.

A second point is that, although we might appear to have placed the Earth in a rather special position in Fig. 7.1, an observer located on any galaxy would perform exactly the same calculation to work out the deceleration of any other galaxy relative to the observer's galaxy because of the cosmological principle which asserts that all fundamental observers should observe the same large scale features of the Universe at the same epoch. In other words, the Newtonian calculation applies for all observers who move in such a way that the Universe appears isotropic to them which is, by definition, for all fundamental observers.

Third, at no point in the argument did we ask over what physical scale the calculation was to be valid. It is a remarkable fact that this calculation describes correctly the dynamics of the Universe on scales which are greater than the *horizon scale* which, for the moment, we can take to be $r = ct$, that is, the maximum distance between points which can be causally connected at the epoch t. The reason for this is the same as for the first two points – local physics is also global physics and so, if the Universe were set up in such a way that it had uniform density on scales far exceeding the horizon scale, the dynamics on these very large scales would be exactly the same as the local dynamics. We will find this idea helpful in understanding the theory of the evolution of small perturbations in the expanding Universe.

Friedman died of typhoid during the civil war in Leningrad in 1925 and did not live to see what have become the standard models of the Universe bear his name (see Tropp, Frenkel and Chernin 1993). It is perhaps surprising that these papers did not attract more widespread interest at the time. This may have been partly due to a brief note published by Einstein in 1922 criticising some steps in Friedman's first paper. In the following year, Einstein graciously acknowledged that his criticism was based upon an error in his own calculations and that Friedman's solution was indeed correct (see the papers republished by Bernstein and Feinberg 1986). Georges Lemaître rediscovered Friedman's solutions in 1927 and brought Friedman's contributions to the wider notice of astronomers and cosmologists during the 1930s.

7.2.2 The Critical Density and the Density Parameter

It is convenient to express the density of the world models in terms of a *critical density* ϱ_c which is defined to be

$$\varrho_c = (3H_0^2/8\pi G) = 1.88 \times 10^{-26} \, h^2 \text{ kg m}^{-3}. \tag{7.15}$$

Then, the actual density of the model ϱ_0 at the present epoch can be referred to this value through a *density parameter* $\Omega_0 = \varrho_0/\varrho_c$. Thus, the density parameter is defined to be

$$\Omega_0 = \frac{8\pi G\varrho_0}{3H_0^2}. \tag{7.16}$$

The subscript 0 has been attached to Ω because the critical density ϱ_c changes with cosmic epoch, as does Ω. It is convenient to refer any cosmic density to ϱ_c. For example, we will often refer to the density parameter of baryons, Ω_B, or of visible matter, Ω_{vis}, or of dark matter, Ω_{dark}, and so on – these are convenient ways of describing the relative importance of different contributions to Ω_0.

The dynamical equations (7.1) and (7.2) therefore become

$$\ddot{R} = -\frac{\Omega_0 H_0^2}{2R^2}; \tag{7.17}$$

$$\dot{R}^2 = \frac{\Omega_0 H_0^2}{R} - \frac{c^2}{\Re^2}. \tag{7.18}$$

Several important results can be deduced from these equations. If we set the quantities in (7.18) equal to their values at the present epoch, $t = t_0$, $R = 1$ and $\dot{R} = H_0$, we find that

$$\mathfrak{R} = \frac{c/H_0}{(\Omega_0 - 1)^{1/2}} \quad \text{and} \quad \kappa = \frac{(\Omega_0 - 1)}{(c/H_0)^2}. \tag{7.19}$$

This last result shows that there is a one-to-one relation between the density of the Universe and its spatial curvature, one of the most beautiful results of the Friedman world models with $\Lambda = 0$.

7.2.3 The Dynamics of the Friedman Models with $\Lambda = 0$

To understand the solutions of (7.18), we substitute (7.19) into (7.18) to find the following expression for $\dot{R}$

$$\dot{R}^2 = H_0^2 \left[\Omega_0 \left(\frac{1}{R} - 1 \right) + 1 \right]. \tag{7.20}$$

In the limit of large values of R, $\dot{R}^2$ tends to

$$\dot{R}^2 = H_0^2 (1 - \Omega_0). \tag{7.21}$$

Thus,

- The models having $\Omega_0 < 1$ have open, hyperbolic geometries and expand to $R = \infty$. They continue to expand with a finite velocity at $R = \infty$ with $\dot{R} = H_0(1 - \Omega_0)^{1/2}$;
- The models with $\Omega_0 > 1$ have closed, spherical geometry and stop expanding at some finite value of $R = R_{\max}$ – they have 'imaginary expansion rates' at infinity. They reach the maximum value of the scale factor after a time

$$t_{\max} = \frac{\pi \Omega_0}{2 H_0 (\Omega_0 - 1)^{3/2}}. \tag{7.22}$$

These models collapse to an infinite density after a finite time $t = 2 t_{\max}$, an event sometimes referred to as the 'big crunch';

- The model with $\Omega_0 = 1$ separates the open from the closed models and the collapsing models from those which expand forever. This model is often referred to as the *Einstein–de Sitter* or the *critical model*. The velocity of expansion tends to zero as R tends to infinity. It has a particularly simple variation of $R(t)$ with cosmic epoch,

$$R = \left(\frac{t}{t_0} \right)^{2/3} \quad \kappa = 0, \tag{7.23}$$

where the present age of the world model is $t_0 = (2/3) H_0^{-1}$.

Some solutions of (7.20) are displayed in Fig. 7.2 which shows the well-known relation between the dynamics and geometry of the Friedman world models with $\Lambda = 0$. The abscissa in Fig. 7.2 is in units of $H_0 t$ and so the slope of the relations at the present epoch, $R = 1$, is always 1. The present age of the Universe is given by the intersection of each curve with the line $R = 1$.

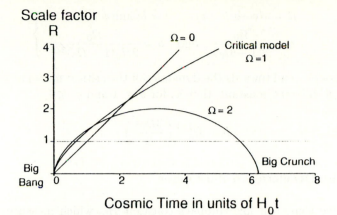

Fig. 7.2. The dynamics of the classical Friedman models parameterised by the density parameter $\Omega_0 = \varrho/\varrho_{\rm crit}$. If $\Omega_0 > 1$, the Universe collapses to $R = 0$ as shown; if $\Omega_0 < 1$, the Universe expands to infinity and has a finite velocity of expansion as R tends to infinity. In the case $\Omega_0 = 1$, $R = (t/t_0)^{2/3}$ where $t_0 = (2/3)H_0^{-1}$. The time axis is given in terms of the dimensionless time $H_0 t$. At the present epoch $R = 1$ and in this presentation, the three curves have the same slope of 1 at $R = 1$, corresponding to a fixed value of Hubble's constant. If t_0 is the present age of the Universe corresponding to $R = 1$, then for $\Omega_0 = 0$ $H_0 t_0 = 1$, for $\Omega_0 = 1$ $H_0 t_0 = 2/3$ and for $\Omega_0 = 2$ $H_0 t_0 = 0.57$.

Another useful result is the function $R(t)$ for the empty world model, $\Omega_0 = 0$, $R(t) = H_0 t$, $\kappa = -(H_0/c)^2$. This model is sometimes referred to as the *Milne model*. It is an interesting exercise to show why it is that, in the completely empty world model, the global geometry of the Universe is hyperbolic. The reason is that in the empty model, the galaxies partaking in the universal expansion are undecelerated and any particular galaxy always has the same velocity relative to the same fundamental observer. Therefore, the cosmic times measured in different frames of reference are related by the standard Lorentz transform $t' = \gamma(t - vr/c^2)$ where $\gamma = (1 - v^2/c^2)^{-1/2}$. The key point is that the conditions of isotropy and homogeneity apply at constant cosmic time t' in the frames of reference of all fundamental observers. The Lorentz transform shows that this cannot be achieved in flat space but it is uniquely satisfied in hyperbolic space with $\kappa = -(H_0/c)^2$. I have given a simple derivation of this result (Longair 1994).

The general solutions of (7.20) are most conveniently written in parametric form. For $\Omega_0 > 1$,

$$R = a(1 - \cos\theta) \qquad t = b(\theta - \sin\theta),$$
$$a = \frac{\Omega_0}{2(\Omega_0 - 1)} \quad \text{and} \quad b = \frac{\Omega_0}{2H_0(\Omega_0 - 1)^{3/2}}. \qquad (7.24a)$$

For $\Omega_0 < 1$,

$$R = a(\cosh\phi - 1) \qquad t = b(\sinh\phi - \phi),$$
$$\left. a = \frac{\Omega_0}{2(1 - \Omega_0)} \quad \text{and} \quad b = \frac{\Omega_0}{2H_0(1 - \Omega_0)^{3/2}}. \right\} \qquad (7.24b)$$

All the models tend towards the dynamics of the critical model at early times but with a different constant, that is, for $\theta \ll 1$ and $\phi \ll 1$,

$$R = \Omega_0^{1/3} \left(\frac{3H_0 t}{2} \right)^{2/3}. \qquad (7.25)$$

7.2.4 The Deceleration Parameter

We observe that, just like Hubble's constant H_0, which measures the local expansion rate of the distribution of galaxies, so we can define the local deceleration of the Universe at the present epoch, $\ddot{R}(t_0)$. It is conventional to define the *deceleration parameter* q_0 to be the dimensionless deceleration at the present epoch through the expression

$$q_0 = -\left(\frac{R\ddot{R}}{\dot{R}^2} \right)_{t_0}. \qquad (7.26)$$

Substituting into (7.17), we can show immediately that the deceleration parameter q_0 is directly proportional to the density parameter Ω_0,

$$q_0 = \Omega_0/2. \qquad (7.27)$$

Note that this result is only true if the cosmological constant Λ is zero. We consider the models with $\Lambda \neq 0$ in Sect. 7.3.

7.2.5 The Cosmic Time–Redshift Relation

An important result for many aspects of cosmology is the relation between cosmic time t and redshift z. Because $R = (1 + z)^{-1}$, it follows immediately from (7.20) that

$$\frac{dz}{dt} = -H_0(1 + z)^2(\Omega_0 z + 1)^{1/2}. \qquad (7.28)$$

Cosmic time t measured from the Big Bang follows by integration

$$t = \int_0^t dt = -\frac{1}{H_0} \int_\infty^z \frac{dz}{(1 + z)^2(\Omega_0 z + 1)^{1/2}}. \qquad (7.29)$$

For $\Omega_0 > 1$, we can write $x = (\Omega_0 - 1)R/\Omega_0 = (\Omega_0 - 1)/\Omega_0(1 + z)$, and then the cosmic time–redshift relation becomes

$$t(z) = \frac{\Omega_0}{H_0(\Omega_0 - 1)^{3/2}} \left[\sin^{-1} x^{1/2} - x^{1/2}(1 - x)^{1/2} \right]. \qquad (7.30a)$$

For $\Omega_0 < 1$, we write $y = (1 - \Omega_0)R/\Omega_0 = (1 - \Omega_0)/\Omega_0(1 + z)$, and then the cosmic time–redshift relation becomes

$$t(z) = \frac{\Omega_0}{H_0(1 - \Omega_0)^{3/2}} \left[y^{1/2}(1 + y)^{1/2} + \sinh^{-1} y^{1/2} \right]. \qquad (7.30b)$$

For large redshifts, $\Omega_0 z \gg 1$, (7.30a) and (7.30b) reduce to (7.25), which can be written

$$t(z) = \frac{2}{3H_0\Omega_0^{1/2}} z^{-3/2}. \qquad (7.30c)$$

We can then find the present age of the Universe for the different world models by integrating from $z = 0$ to $z = \infty$.

$$
\left.
\begin{aligned}
t_0 &= \frac{\Omega_0}{H_0(\Omega_0 - 1)^{3/2}} \left[\sin^{-1}\left(\frac{\Omega_0 - 1}{\Omega_0}\right)^{1/2} - \frac{(\Omega_0 - 1)^{1/2}}{\Omega_0} \right] \quad \text{if } \Omega_0 > 1; \\[2mm]
t_0 &= \frac{2}{3H_0} \quad \text{if } \Omega_0 = 1; \\[2mm]
t_0 &= \frac{\Omega_0}{H_0(1 - \Omega_0)^{3/2}} \left[\frac{(1 - \Omega_0)^{1/2}}{\Omega_0} - \sinh^{-1}\left(\frac{1 - \Omega_0}{\Omega_0}\right)^{1/2} \right] \quad \text{if } \Omega_0 < 1.
\end{aligned}
\right\}
\qquad (7.31)
$$

The age of the Universe is a monotonic function of Ω_0. The useful simple cases are those for the critical model $\Omega_0 = 1$ for which the present age of the Universe is $(2/3)H_0^{-1}$ and the empty model, $\Omega_0 = 0$, for which it is H_0^{-1}. For $\Omega_0 = 2$, the age of the Universe is $0.571\, H_0^{-1}$.

7.2.6 The Flatness Problem

We can find how Hubble's constant varies with cosmic epoch from (7.20) by setting $R = (1 + z)^{-1}$. Then,

$$H = \frac{\dot{R}}{R} = H_0(\Omega_0 z + 1)^{1/2}(1 + z). \qquad (7.32)$$

In the same way, we can define a density parameter Ω at any epoch through the definition $\Omega = 8\pi G\varrho/3H^2$. Since $\varrho = \varrho_0(1 + z)^3$, it follows that

$$\Omega = \frac{8\pi G}{3H^2} \varrho_0(1 + z)^3 = \Omega_0 \frac{1 + z}{\Omega_0 z + 1}. \qquad (7.33)$$

This relation can be rewritten

$$\left(1 - \frac{1}{\Omega}\right) = (1 + z)^{-1}\left(1 - \frac{1}{\Omega_0}\right). \qquad (7.34)$$

This is an important result because it shows that, whatever the value of Ω_0 now, because $(1 + z)^{-1}$ becomes very small at large redshifts, Ω tends to the value 1 in the distant past. There are two ways of looking at this result. On the one hand, it reaffirms our conclusion from (7.25) that the dynamics of all the world models with $\Lambda = 0$ tend to those of the Einstein–de Sitter model in their early stages. On the other hand, it is remarkable that the Universe is within roughly a factor of ten of the value $\Omega_0 = 1$ at the present day. If the value of Ω_0 were significantly different from 1 in the distant past, then it would be very different from 1 now as can be seen from (7.34). Notice that there is nothing in the standard models which requires Ω_0 to take any particular value. Ω_0 is simply a parameter which should be fixed as part of the initial conditions of our Universe.

The fact that the curvature of space κ must be close to zero now results in what is often referred to as the *flatness problem*, namely, that our Universe must have been very finely tuned indeed to the value $\Omega = 1$ in the distant past if it is to end up close to $\Omega_0 = 1$ now. Some argue that it is so remarkable that our Universe is within a factor of ten of $\Omega_0 = 1$ now the only reasonable value the Universe can have is Ω_0 precisely equal to 1. In Chap. 20, we will extend this argument back to the Planck era. Proponents of the inflationary picture of the early Universe have reasons why Ω_0 should be equal to one.

7.2.7 Distance Measures as a Function of Redshift

We can now complete our programme of finding expressions for the radial comoving distance coordinate r and the distance measure D. We recall that the increment of radial comoving coordinate distance is

$$\mathrm{d}r = \frac{c\,\mathrm{d}t}{R(t)} = -c\,\mathrm{d}t(1+z). \tag{7.35}$$

Therefore,

$$\mathrm{d}r = \frac{c\,\mathrm{d}z}{H_0(1+z)(\Omega_0 z + 1)^{1/2}}. \tag{7.36}$$

Integrating from redshifts 0 to z, we can find the expression for r.

$$r = \frac{2c}{H_0(\Omega_0 - 1)^{1/2}} \left[\tan^{-1}\left(\frac{\Omega_0 z + 1}{\Omega_0 - 1}\right)^{1/2} - \tan^{-1}(\Omega_0 - 1)^{-1/2} \right]. \tag{7.37}$$

Then, we can find D by evaluating $D = \Re \sin(r/\Re)$, where $\Re$ is given by (7.19). After some straightforward algebra, we find that

$$D = \frac{2c}{H_0 \Omega_0^2 (1+z)} \left\{ \Omega_0 z + (\Omega_0 - 2)[(\Omega_0 z + 1)^{1/2} - 1] \right\}. \tag{7.38}$$

This is the famous formula first derived by Mattig (1959). Although I have derived it using the formulae for spherical geometry, it has the great advantage of being correct for all values of Ω_0. In the limit of the empty, or Milne, world model, $\Omega_0 = 0$, (7.38) becomes

$$D = \frac{cz}{H_0} \frac{\left(1 + \frac{z}{2}\right)}{(1+z)}. \tag{7.39}$$

The variations of r and D with redshift for a range of standard world models are shown in Fig. 7.3. These diagrams are the basis for observational programmes to determine which of the standard Friedman models best describes the large scale dynamics of the Universe.

7.2.8 The Observed Properties of Standard Objects in the Friedman World Models with $\Lambda = 0$

It is now straightforward to work out how the observed properties of standard objects vary with redshift in the Friedman world models with $\Lambda = 0$. Three of the most useful examples are illustrated in Fig. 7.4a, b and c. In all three diagrams, $c/H_0 = 1$.

Angular Diameters. In Fig. 7.4a, the variation of the observed angular size of a rigid rod of unit proper length is shown based upon (5.57) and (7.38). Except for the empty world model with $\Omega_0 = 0$, there is a minimum in the angular diameter–redshift relation which occurs at $z = 1.25$ for the critical model, $\Omega_0 = 1$, and at $z = 1$ if $\Omega_0 = 2$. This type of angular diameter is known as a *metric angular diameter* and is distinct from the type of angular diameter which is often used to measure the sizes of galaxies. The latter are often defined to some limiting surface brightness and so, since bolometric surface brightnesses vary with redshift as $(1 + z)^{-4}$, angular sizes measured to a same limiting surface brightness at a wide range of redshifts are not rigid rods of fixed proper length. The angular diameter–redshift relation can be worked out for *isophotal angular diameters*, but this requires knowledge of the K-corrections to be applied as a function of radius within the galaxy.

Flux Densities. The observed flux density of a source of unit luminosity per unit frequency interval at frequency ν_0 with a power-law spectrum $L(\nu) \propto \nu^{-1}$ is given by (5.68) and (7.38) and is shown in Fig. 7.4b as a function of redshift. Inspection of (5.68) and (5.70) shows that this is the same as the variation of the bolometric flux density with redshift. For galaxies, the detailed form of the spectrum has to be taken into account and this is often done using the K-corrections described by (5.72) and (5.73).

The Comoving Volume Within Redshift z. This relation can be determined from (5.74) or (5.75) and (7.38). Alternatively, (5.74) can be integrated to give, for the case $\Omega_0 > 1$, $\Re = (c/H_0)(\Omega_0 - 1)^{-1/2}$,

$$V(r) = 2\pi\Re^3 \left(\frac{r}{\Re} - \frac{1}{2}\sin\frac{2r}{\Re}\right) = 2\pi\Re^3 \left(\sin^{-1}\frac{D}{\Re} - \frac{D}{\Re}\sqrt{1 - \frac{D^2}{\Re^2}}\right). \tag{7.40}$$

For the case $\Omega_0 < 1$, $\Re = (c/H_0)(1 - \Omega_0)^{-1/2}$,

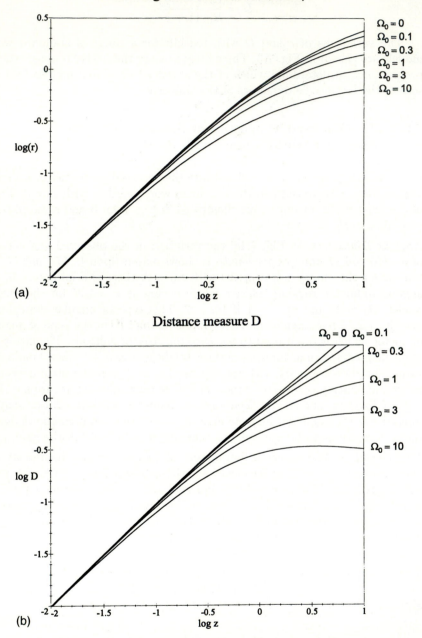

Fig. 7.3a-b. The variation with redshift of (a) the comoving radial distance coordinate r, and (b) the distance measure D for Friedman world models with $\Lambda = 0$ and $\Omega_0 = 0, 0.1, 0.3, 1, 3$ and 10. In this diagram, r and D are measured in units of c/H_0.

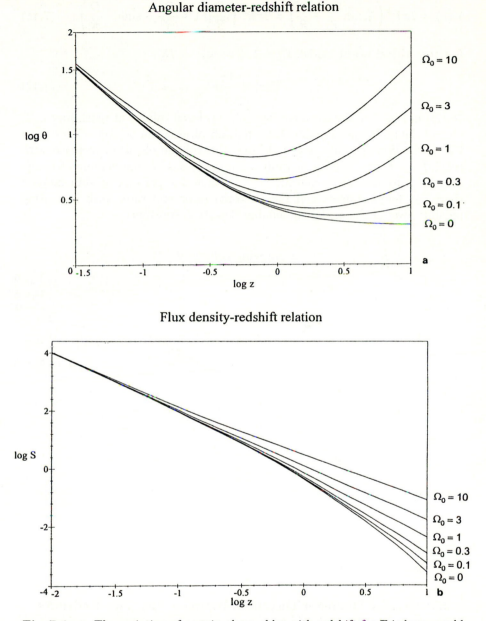

Angular diameter-redshift relation

Flux density-redshift relation

Fig. 7.4a-c. The variation of certain observables with redshift for Friedman world models with $\Lambda = 0$ and $\Omega_0 = 0$, 0.1, 0.3, 1, 3 and 10. In all three diagrams, c/H_0 has been set equal to unity. (a) The variation of the angular diameter of a rigid rod of unit proper length with redshift. (b) The variation of the flux density of a source of luminosity 1 W Hz^{-1} with a power-law spectrum $L(\nu) \propto \nu^{-1}$ with redshift. Inspection of (5.68) and (5.70) shows that this is the same as the variation of the bolometric flux densities with redshift. (c) The variation of the comoving volume within redshift z.

$$V(r) = 2\pi\Re^3 \left(\tfrac{1}{2}\sinh\frac{2r}{\Re} - \frac{r}{\Re} \right) = 2\pi\Re^3 \left(\frac{D}{\Re}\sqrt{1 + \frac{D^2}{\Re^2}} - \sinh^{-1}\frac{D}{\Re} \right). \quad (7.41)$$

For the critical world model, $\Omega_0 = 1, \Re = \infty, r = D$,

$$V(r) = \frac{4\pi}{3}r^3. \quad (7.42)$$

Notice in Fig. 7.4c the convergence of the enclosed volume at redshifts $z > 1$. One of the problems in finding large redshift objects can be appreciated from Fig. 7.4c. Whereas, at small redshifts, the volume elements increase with redshift as $z^2\,\mathrm{d}z$, at large redshifts, $\Omega_0 z > 1$, $z > 1$, the volume elements decrease with increasing redshift as $z^{-3/2}\,\mathrm{d}z$ and so per unit redshift interval there are fewer sources and they become rarer and rarer with increasing redshift, even if the comoving number density is constant.

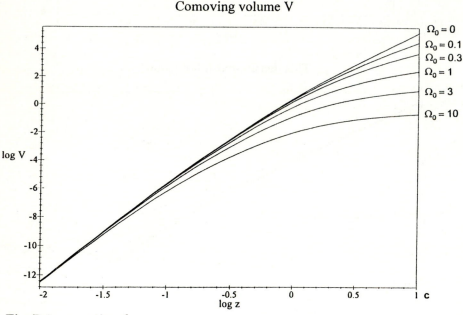

Comoving volume V

Fig. 7.4a-c. continued.

7.2.9 Angular Diameter Distances Between Any Two Redshifts

There are occasions when it is necessary to relate metric diameters as observed from locations other than the origin at $z = 0$. A good example is the geometry of gravitational lensing in which we need the angular diameter distance $D_A(z_i, z_j)$, meaning the angular diameter distance necessary to work out the physical size of an object which subtends an angle θ between redshifts z_i and z_j (see Sect. 4.3.4 and (4.31)). It is a useful exercise to derive the pleasant result presented by Blandford and Narajan (1992) for the appropriate angular diameter distance to be used in this case.

We begin with the expression for the radial comoving distance coordinate r_{ij} between redshifts z_i and z_j. By extension of (7.37), we find

$$
\begin{aligned}
r_{ij} &= \frac{2c}{H_0(\Omega_0 - 1)^{1/2}} \left[\tan^{-1} \left(\frac{\Omega_0 z_j + 1}{\Omega_0 - 1} \right)^{1/2} - \tan^{-1} \left(\frac{\Omega_0 z_i + 1}{\Omega_0 - 1} \right)^{1/2} \right] \\
&= \frac{2c}{H_0(\Omega_0 - 1)^{1/2}} \left[\tan^{-1} \frac{G_j}{(\Omega_0 - 1)^{1/2}} - \tan^{-1} \frac{G_i}{(\Omega_0 - 1)^{1/2}} \right]
\end{aligned}
\tag{7.43}
$$

where $G_j = (\Omega_0 z_j + 1)^{1/2}$ and $G_i = (\Omega_0 z_i + 1)^{1/2}$. Using the summation formulae for inverse tangents, this can be rewritten

$$
r_{ij} = \frac{2c}{H_0(\Omega_0 - 1)^{1/2}} \tan^{-1} \frac{N_1 (\Omega_0 - 1)^{1/2}}{N_2},
\tag{7.44}
$$

where $N_1 = (G_j - G_i)$ and $N_2 = (\Omega_0 - 1 + G_i G_j)$.

We now form the expression $\Re' \sin(r'_{ij}/\Re')$ in order to find the distance measure D_{ij} between the redshifts z_i and z_j. We need to use the radius of curvature of the spatial geometry $\Re'$ and the radial comoving distance coordinate r'_{ij} at the redshift z_1, but these both scale as $R(t)$ and so, since $\Re = (c/H_0)/(\Omega_0 - 1)^{1/2}$,

$$
\begin{aligned}
\sin \frac{r'_{ij}}{\Re'} = \sin \frac{r_{ij}}{\Re} &= \sin \left[2 \tan^{-1} \frac{N_1 (\Omega_0 - 1)^{1/2}}{N_2} \right] \\
&= \frac{2 N_1 N_2 (\Omega_0 - 1)^{1/2}}{N_1^2 (\Omega_0 - 1) + N_2^2}
\end{aligned}
\tag{7.45}
$$

Hence the distance measure $D(z_i, z_j)$ is

$$
D(z_i, z_j) = \Re' \sin \frac{r'_{ij}}{\Re'} = \frac{2c}{H_0(1 + z_i)} \frac{N_1 N_2}{N_1^2 (\Omega_0 - 1) + N_2^2}.
\tag{7.46}
$$

Expanding the denominator of the second term, we find

$$
N_1^2 (\Omega_0 - 1) + N_2^2 = \Omega_0^2 (1 + z_i)(1 + z_j),
\tag{7.47}
$$

and so the distance measure from z_i to z_j is

$$
D(z_i, z_j) = \frac{2c}{H_0 \Omega_0^2} \frac{(G_j - G_i)(\Omega_0 - 1 + G_i G_j)}{(1 + z_i)^2 (1 + z_j)}.
\tag{7.48}
$$

Therefore, following the argument which led to (5.57), but now between the scale factors $R(z_1)$ and $R(z_2)$, we find

$$
\theta = \frac{d(1 + z_j)}{D(z_i, z_j)(1 + z_i)} = \frac{d}{D_{\mathrm{A}}(z_i, z_j)},
\tag{7.49}
$$

where the angular diameter distance $D_A(z_i, z_j)$ is given by

$$
\begin{aligned}
D_A(z_i, z_j) &= \frac{D(z_i, z_j)(1 + z_i)}{(1 + z_j)} \\
&= \frac{2c}{H_0 \Omega_0^2} \frac{(G_j - G_i)(\Omega_0 - 1 + G_i G_j)}{(1 + z_i)(1 + z_j)^2}
\end{aligned}
\tag{7.50}
$$

This is the expression for the angular diameter distance between redshifts z_i and z_j quoted by Blandford and Narayan (1992) and can be used in the expression for the gravitational lens formula (4.31). Expression (7.50) has the attractive feature that it can be used in either direction along the light cone. Thus, the angular diameter distance from z_j to z_i is found by interchanging the indices i and j in (7.50),

$$
D_A(z_j, z_i) = \frac{2c}{H_0 \Omega_0^2} \frac{(G_i - G_j)(\Omega_0 - 1 + G_i G_j)}{(1 + z_j)(1 + z_i)^2},
\tag{7.51}
$$

and so we find

$$
\frac{D_A(z_i, z_j)}{D_A(z_j, z_i)} = \frac{(1 + z_i)}{(1 + z_j)}.
\tag{7.52}
$$

This is precisely the *reciprocity theorem* which we alluded to in Sect. 5.5.4. The angular diameter distances to be used in opposite directions along the light cone differ by the ratio of the scale factors corresponding to the redshifts z_1 and z_2. If we set $z_i = 0$ and $z_j = z$, we find

$$
\frac{D_A(0 \to z)}{D_A(z \to 0)} = \frac{1}{(1 + z)},
\tag{7.53}
$$

as we demonstrated in Sect. 5.5.4.

7.3 Models with Non-zero Cosmological Constant

Einstein's introduction of the cosmological constant predated Hubble's discovery of the expansion of the distribution of galaxies. In 1917, Einstein had no reason to seek non-stationary solutions of the field equations and the Λ-term was introduced to produce static solutions, which he believed, incorrectly, would enable Mach's principle to be incorporated into his model of the Universe. Once the expansion of the system of galaxies was discovered, Einstein is reported to have stated that the introduction of the cosmological constant was 'the greatest blunder of my life' (Gamow 1970). Since then it has been in and out of fashion in response to different cosmological problems. It was very popular during the 1930s when it seemed that the age of the Universe, as estimated by $T_0 \sim H_0^{-1}$, was significantly less than the age of the Earth. The Lemaître models were of special interest because the cosmological constant could be chosen so that the Universe had a long 'coasting phase' during which the attractive force of gravity was almost perfectly balanced against the repulsive effect of the Λ-term. The conflicting age estimates could be reconciled in this model. This particular problem evaporated when it was found that the value of Hubble's constant had been greatly overestimated. There would, however, be the same problem today if high values of Hubble's constant and large ages for the globular cluster stars were adopted. We review these data in Chap. 8.

To anticipate the discussion of that chapter, whilst there is no observational evidence that $\Lambda \neq 0$, we cannot exclude *a priori* that this is might not in fact be the case. The world models with $\Lambda \neq 0$ have some quite distinct properties which are not encountered in those with $\Lambda = 0$. For example, if 'ghost images' of the same objects in opposite directions on the sky were discovered, that would be evidence that the dynamics of the Universe cannot described by the Friedman models with $\Lambda = 0$.

7.3.1 The Cosmological Constant and the Vacuum Energy Density

Einstein's field equations with the inclusion of the cosmological constant can be written

$$\ddot{R} = -\frac{4\pi G R}{3}\left(\varrho + \frac{3p}{c^2}\right) + \tfrac{1}{3}\Lambda R; \tag{7.54}$$

$$\dot{R}^2 = \frac{8\pi G \varrho}{3}R^2 - \frac{c^2}{\Re^2} + \tfrac{1}{3}\Lambda R^2. \tag{7.55}$$

Considering dust-filled universes as in Sect. 7.2, we again set $3p/c^2 = 0$.

$$\ddot{R} = -\frac{4\pi G R \varrho}{3} + \tfrac{1}{3}\Lambda R = -\frac{4\pi G \varrho_0}{3R^2} + \tfrac{1}{3}\Lambda R. \tag{7.56}$$

Inspection of (7.56) gives some insight into the physical meaning of the cosmological constant. Even in an empty universe, with $\varrho = 0$, there is still a net force acting on a test particle. If Λ is positive, the term may be thought of as the 'repulsive force of a vacuum', in the words of Ya.B. Zeldovich (1968), the repulsion being relative to an absolute geometrical frame of reference. There is no obvious interpretation of this term in terms of classical physics. There is, however, a natural interpretation in the context of quantum field theory.

A key development has been the introduction of Higgs fields into the theory of weak interactions. These and other ideas of quantum field theory are clearly described by Zeldovich (1986) in an article aimed at astronomers and astrophysicists. The Higgs field was introduced into the electro-weak theory of elementary particles in order to eliminate singularities in the theory and to endow the $W^{\pm}$ and Z^0 bosons with masses. Precise measurement of the masses of these particles at CERN has confirmed the theory very precisely. The Higgs fields have the property of being *scalar* fields, unlike the vector fields of electromagnetism or the tensor fields of General Relativity. Zeldovich shows that scalar fields have negative energy equations of state

$p = -\varrho c^2$. Fields of this nature, associated with phase transitions when the strong force decoupled from the electro-weak force in the early Universe, are prime candidates for a cosmological negative-energy equation of state.

In the modern picture of the vacuum, there are zero-point fluctuations associated with the zero point energies of all quantum fields. The stress–energy tensor of a vacuum has a negative energy equation of state, $p = -\varrho c^2$. This pressure may be thought of as a 'tension' rather than a pressure. When such a vacuum expands, the work done $p\,dV$ in expanding from V to $V+dV$ is just $-\varrho c^2\,dV$ so that, during the expansion, the mass–density of the negative energy field remains constant. We can find the same result from (7.6). If the vacuum energy density is to remain constant, $\varrho_{\text{vac}} = $ constant, it follows from that equation that $p = -\varrho c^2$.

Carroll, Press and Turner (1992) describe how a theoretical value of Λ can be estimated using simple concepts from quantum field theory and they find the mass density of the repulsive field to be $\varrho_v = 10^{95}$ kg m^{-3}. This is quite a problem. This mass density is about 10^{120} times greater than allowable values at the present epoch which correspond to $\varrho_v \leq 10^{-27}$ kg m^{-3}. This represents a rather large discrepancy, but it is not one we should pass over lightly. If we take seriously the inflationary picture of the very early Universe, it is exactly this type of force which causes the inflationary expansion. We then have to explain why ϱ_v decreased by a factor of at least 10^{120} at the end of the inflationary era. Within this context, 10^{-120} looks remarkably close to zero, which would correspond to the standard Friedman picture with $\Lambda = 0$.

Thus, it is now quite natural to believe that there are indeed forces in nature which can provide Zeldovich's 'repulsion of the vacuum' and to associate a certain mass density ϱ_v with the energy density of the vacuum at the present epoch. It is convenient to rewrite the formalism in terms of a density parameter Ω_Λ associated with ϱ_v as follows. We begin with (7.1) in the form:

$$\ddot{R} = -\frac{4\pi GR}{3}\left(\varrho_m + \varrho_v + \frac{3p_v}{c^2}\right), \tag{7.57}$$

where, in place of the Λ-term, we have included the mass density ϱ_v and pressure p_v of the vacuum fields. Since $p_v = -\varrho_v c^2$, it follows that

$$\ddot{R} = -\frac{4\pi GR}{3}(\varrho_m - 2\varrho_v). \tag{7.58}$$

As the Universe expands, $\varrho_m = \varrho_0/R^3$ and $\varrho_v = $ constant. Therefore,

$$\ddot{R} = -\frac{4\pi G\varrho_0}{3R^2} + \frac{8\pi G\varrho_v R}{3}. \tag{7.59}$$

Equations (7.56) and (7.59) have precisely the same dependence of the 'cosmological term' upon the scale factor R and so we can formally identify the cosmological constant with the vacuum mass density.

$$\Lambda = 8\pi G \varrho_{\rm v}. \tag{7.60}$$

At the present epoch, $R = 1$ and so

$$\ddot{R}(t_0) = -\frac{4\pi G \varrho_0}{3} + \frac{8\pi G \varrho_{\rm v}}{3}. \tag{7.61}$$

A density parameter associated with $\varrho_{\rm v}$ can now be introduced, in exactly the same way as the density parameter Ω_0 was defined.

$$\Omega_\Lambda = \frac{8\pi G \varrho_{\rm v}}{3H_0^2} \quad \text{and so} \quad \Lambda = 3H_0^2 \Omega_\Lambda. \tag{7.62}$$

We can therefore find a new relation between the deceleration parameter q_0, Ω_0 and Ω_Λ from (7.60) and (7.62).

$$q_0 = \frac{\Omega_0}{2} - \Omega_\Lambda. \tag{7.63}$$

The dynamical equations (7.54) and (7.55) can now be written

$$\ddot{R} = -\frac{\Omega_0 H_0^2}{2R^2} + \Omega_\Lambda H_0^2 R; \tag{7.64}$$

$$\dot{R}^2 = \frac{\Omega_0 H_0^2}{R} - \frac{c^2}{\Re^2} + \Omega_\Lambda H_0^2 R^2. \tag{7.65}$$

We can now substitute the values of R and $\dot{R}$ at the present epoch, $R = 1$ and $\dot{R} = H_0$, into (7.65) to find the relation between the curvature of space, Ω_0 and Ω_Λ.

$$\frac{c^2}{\Re^2} = H_0^2[(\Omega_0 + \Omega_\Lambda) - 1], \tag{7.66}$$

or

$$\kappa = \frac{1}{\Re^2} = \frac{[(\Omega_0 + \Omega_\Lambda) - 1]}{(c^2/H_0^2)}. \tag{7.67}$$

Thus, the condition that the spatial sections are flat Euclidean space becomes

$$(\Omega_0 + \Omega_\Lambda) = 1. \tag{7.68}$$

We recall that the radius of curvature $R_{\rm c}$ of the spatial sections of these models change with scale factor as $R_{\rm c} = R\Re$ and so, if the space curvature is zero now, it must have been zero at all times in the past.

7.3.2 The Dynamics of World Models with $\Lambda \neq 0$

Let us now investigate the dynamics of world models with $\Lambda \neq 0$.

Models with $\Lambda < 0$; $\Omega_\Lambda < 0$ Models with negative cosmological constant are not of a great deal of interest because the net effect is to incorporate an attractive force in addition to gravity which slows down the expansion of the Universe. The one difference from the models with $\Lambda = 0$ is that, no matter how small the values of Ω_Λ and Ω_0 are, the universal expansion is eventually reversed as may be seen from inspection of (7.64).

Models with $\Lambda > 0$; $\Omega_\Lambda > 0$ These models are much more interesting because the positive cosmological constant leads to a repulsive force which opposes the attractive force of gravity. In each of these models, there is a minimum rate of expansion $\dot{R}_{\min}$ which occurs at a value of the scale factor

$$R_{\min} = (4\pi G \varrho_0 / \Lambda)^{1/3} = (\Omega_0 / 2\Omega_\Lambda)^{1/3}, \tag{7.69}$$

which is found immediately by setting $\ddot{R} = 0$ in (7.56). We then find the minimum rate of expansion

$$\dot{R}_{\min}^2 = \Lambda^{1/3} \left(\frac{3\Omega_0 H_0^2}{2} \right)^{2/3} - \frac{c^2}{\mathfrak{R}^2} = \frac{3H_0^2}{2} (2\Omega_\Lambda \Omega_0^2)^{1/3} - \frac{c^2}{\mathfrak{R}^2}. \tag{7.70}$$

If the right-hand side is greater than zero, the dynamical behaviour shown in Fig. 7.5a is found. For large values of R, the dynamics become those of the de Sitter universe

$$R(t) \propto \exp \left[\left(\frac{\Lambda}{3} \right)^{1/2} t \right] = \exp \left(\Omega_\Lambda^{1/2} H_0 t \right). \tag{7.71}$$

If the right-hand side of (7.70) is less than zero, there exists a range of scale factors for which no solution exists and it can be shown readily that the function $R(t)$ has two branches, as illustrated in Fig. 7.5b. For the branch B, the Universe never expanded to sufficiently large values of R that the repulsive effect of the Λ term can prevent the Universe collapsing. In the case of branch A, the dynamics are dominated by the Λ term – the repulsive force is so strong that the Universe never contracted to such a scale that the attractive force of gravity could overcome its effect. In this model, there was no initial singularity – the Universe 'bounced' under the influence of the Λ-term. In the limiting case in which the density of matter is zero, $\Omega_0 = 0$, the dynamics of the model are described by

$$\dot{R}^2 = H_0^2 [\Omega_\Lambda R^2 - (\Omega_\Lambda - 1)], \tag{7.72}$$

which has solution

$$R = \left(\frac{\Omega_\Lambda - 1}{\Omega_\Lambda} \right)^{1/2} \cosh \Omega_\Lambda^{1/2} H_0 \tau, \tag{7.73}$$

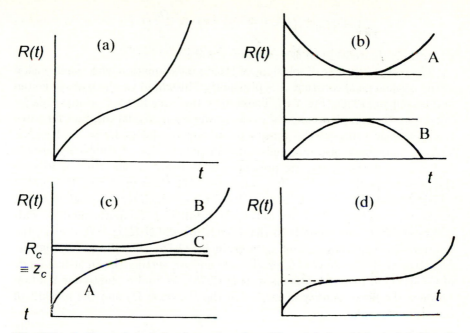

Fig. 7.5a-d. Examples of the dynamics of world models in which $\Lambda \neq 0$ (Bondi 1960). Models (**a**) and (**d**) are referred to as Lemaître models. (**c**) is the Eddington-Lemaître model. The Einstein static model is illustrated by the model for which $R(t)$ is constant for all time. The other models are described in the text.

where the time $\tau = t - t_{\min}$ is measured from the time at which the model 'bounced', that is, from the time at which $R = R_{\min}$. In all cases in which the models bounce, the variation of R with cosmic time is symmetrical about $R_{\min}$. Their asymptotic behaviour corresponds to exponentially collapsing and expanding de Sitter solutions

$$R = \left(\frac{\Omega_\Lambda - 1}{\Omega_\Lambda} \right)^{1/2} \exp \left(\pm \Omega_\Lambda^{1/2} H_0 \tau \right). \tag{7.74}$$

In these 'bouncing' Universes, the smallest value of R, $R_{\min}$, corresponds to the largest redshifts which objects could have.

The most interesting cases are those for which $\dot{R}_{\min} \approx 0$. The case $\dot{R}_{\min} = 0$ is known as the *Eddington–Lemaître model* and is illustrated in Fig. 7.5c. The literal interpretation of these models is either: A, the Universe expanded from an origin at some finite time in the past and will eventually attain a stationary state in the infinite future; B, the Universe is expanding away from a stationary solution in the infinite past. The stationary state C is unstable because, if it is perturbed, the Universe moves either onto branch B, or onto the collapsing variant of branch A. In Einstein's static Universe, the stationary phase occurs at the present day. In general, from (7.69), the value of Λ corresponding to $\dot{R}_{\min} = 0$ is

$$\Lambda = \tfrac{3}{2}\Omega_0 H_0^2 (1 + z_c)^3 \quad \text{or} \quad \Omega_\Lambda = \frac{\Omega_0}{2}(1 + z_c)^3 \qquad (7.75)$$

where z_c is the redshift of the stationary state.

The complete range of solutions of Friedman's equation with the inclusion of the cosmological constant Λ is pleasantly illustrated by the plots of Felten and Isaacman (1986) (Fig. 7.6). These show the scale factor – cosmic time relations for the complete range of values of the cosmological constant for three representative values of the density parameter of matter $\Omega_0 = 0.1, 1$ and 3. The relations have been normalised to the same value of Hubble's constant at the present epoch and are presented in terms of cosmic time relative to the present epoch $\tau = t - t_0$. It can be seen from these plots that the models which separate those which 'bounce' from those which have a singular origin in the past correspond to the stationary Eddington–Lemaître models. This stationary state corresponds to the models for which $\dot{R}_{\min} = 0$, that is, the relation between Ω_Λ, Ω_0 and z_c is given by (7.75). The static Eddington–Lemaître models have $\dot{R} = 0$ for all time and so, setting the right-hand side of (7.65) equal to zero and substituting (7.75), we find a one-to-one relation between the mean density of matter in the Universe Ω_0 and the redshift of the stationary phase z_c.

$$\Omega_0 = \frac{2}{(1 + z_c)^3 - 3(1 + z_c) + 2} = \frac{2}{z_c^2(z_c + 3)}. \qquad (7.76)$$

Fig. 7.6 repays some study. Table 7.1 shows the values of z_c, R_c, Ω_Λ and $\Lambda/H_0^2 = 3\Omega_\Lambda$ for the three illustrative models shown in Fig. 7.6.

Table 7.1. Parameters for stationary Eddington–Lemaître models with $\Omega_0 = 0.1, 1$ and 3.

Ω_0	z_c	R_c	Ω_Λ	$\Lambda/H_0^2 = 3\Omega_\Lambda$
0.1	2	1/3	1.35	4.05
1	$\sqrt{3} - 1 = 0.7321$	$1/\sqrt{3} = 0.5774$	$\tfrac{3}{2}\sqrt{3} = 2.5981$	$\tfrac{9}{2}\sqrt{3} = 7.7942$
3	0.4402	0.6943	4.4809	13.4428

The fact that quasars are already observed with redshifts as large as 4.89 (Schneider et al. 1991) indicates that large positive values of Ω_Λ can be excluded. For example, if we adopt a stationary redshift z_c of 5, the corresponding value of Ω_0 would be 0.01, which corresponds roughly to the mass present in the visible parts of galaxies, without taking account of the dark matter. The corresponding value of Ω_Λ would be 1.08.

The properties of the world models with non-zero cosmological constant are conveniently summarised in a plot of Ω_0 against $\Omega_0 + \Omega_\Lambda$ presented by Carroll, Press and Turner (1992) (Fig. 7.7). The world models with $\Lambda = 0$ lie along the 45° line passing through zero on both axes. As shown by (7.66), the

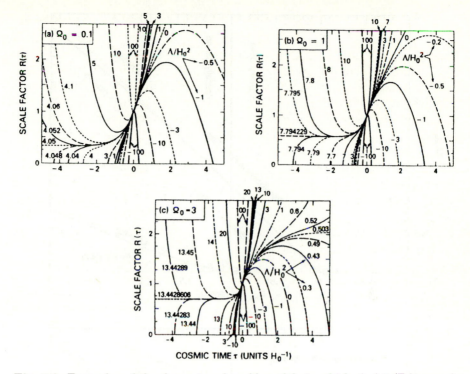

Fig. 7.6. Examples of the dynamics of world models in which $\Lambda \neq 0$ (Felten and Isaacman 1986). The scale factor – cosmic time relations are shown for the complete range of values of Λ and for three different representative values of the present matter density $\Omega_0 = 0.1, 1$ and 3. The time is measured in units of H_0^{-1} from the present epoch, $\tau = t - t_0$. Notice that both scales are linear.

spatial geometry of the world models depends upon the value of $\Omega_0 + \Omega_\Lambda$, the value unity separating the open from closed geometries. The models which were stationary in the past, corresponding to the dividing line between those models which had a singular past and those which 'bounced', are given by (7.75) and (7.76), the values of the stationary redshifts being indicated along the locus to the bottom right of the diagram. Finally, the diagram also shows the dividing line between those models which will eventually recollapse to a 'big crunch' in the future and those which will expand forever. This dividing line can also be found from (7.75) and (7.76) by requiring the models to tend to stationary phases in the future, for which the values of R are greater than one and the redshifts less than zero. For example, using (7.76), we find that the model which is stationary at a scale factor $R = 2$, corresponding to $(1 + z_c) = 0.5$, has $\Omega_0 = 3.2$. The corresponding value of Ω_Λ from (7.75) is 0.2, so that $\Omega_0 + \Omega_\Lambda = 3.4$, which lies on the solid line in Fig. 7.7.

The models with positive cosmological constant can have ages greater than H_0^{-1}. In the limiting cases of Eddington–Lemaître models with $\dot{R}_{\min} = 0$

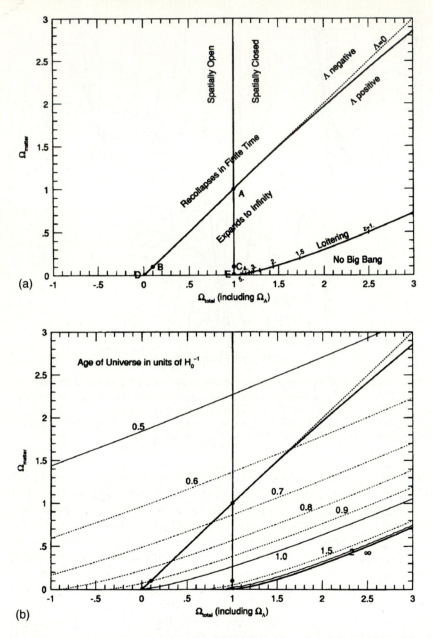

Fig. 7.7. (a) The classification of the Friedman world models with $\Omega_\Lambda \neq 0$ in a plot of Ω_0 against $\Omega_0 + \Omega_\Lambda$ (Carroll, Press and Turner 1992). The Eddington–Lemaître models lie along the line labelled 'loitering'. (b) The present age of the Universe for the range of world models displayed in (a) in units of H_0^{-1}.

in the infinite past, for example, the Universe is infinitely old. A closely related set of models, with ages which can be greater than H_0^{-1}, are the *Lemaître models* which have values of Ω_Λ such that the value of $\dot{R}_{\min}$ is just greater than zero. An example of this type of model is shown in Fig. 7.5d. There is a long 'coasting phase' when the velocity of expansion of the Universe was very small. To obtain cosmological time-scales significantly greater than H_0^{-1}, the models cannot be too far from the limiting stationary locus of Fig. 7.7, as can be seen by inspection of the models illustrated in Fig. 7.6.

To work out the age of the Universe for the Lemaître models, we follow the same procedure which led to (7.29) but now we need the expression for dz/dt which replaces (7.28). From (7.65), it is straightforward to show that the expression which replaces (7.28) is

$$\frac{dz}{dt} = -H_0(1+z)[(1+z)^2(\Omega_0 z + 1) - \Omega_\Lambda z(z+2)]^{1/2}. \qquad (7.77)$$

Cosmic time t measured from the Big Bang follows immediately by integration

$$t = \int_0^t dt = -\frac{1}{H_0}\int_\infty^z \frac{dz}{(1+z)[(1+z)^2(\Omega_0 z + 1) - \Omega_\Lambda z(z+2)]^{1/2}}. \qquad (7.78)$$

A convenient representation of the ages of the various world models with non-zero values of Ω_Λ is shown in Fig. 7.7b (Carroll, Press and Turner 1992).

The models with zero curvature are of particular interest for proponents of the inflationary picture of the early Universe and there is a simple analytic solution for the cosmic time-scale for these models. From (7.66), the condition that the curvature of space is zero, $\Re \rightarrow \infty$, is $\Omega_0 + \Omega_\Lambda = 1$. Then, either from (7.65) or (7.77), it is straightforward to show that

$$t = \int_0^t dt = -\frac{1}{H_0}\int_\infty^z \frac{dz}{(1+z)[\Omega_0(1+z)^3 + \Omega_\Lambda]^{1/2}}. \qquad (7.79)$$

The cosmic time–redshift relation becomes

$$t = \frac{2}{3H_0\Omega_\Lambda^{1/2}}\ln\left(\frac{1+\cos\theta}{\sin\theta}\right) \quad \text{where} \quad \tan\theta = \left(\frac{\Omega_0}{\Omega_\Lambda}\right)^{1/2}(1+z)^{3/2}. \qquad (7.80)$$

The present age of the Universe follows by setting $z = 0$

$$t_0 = \frac{2}{3H_0\Omega_\Lambda^{1/2}}\ln\left[\frac{1+\Omega_\Lambda^{1/2}}{(1-\Omega_\Lambda)^{1/2}}\right]. \qquad (7.81)$$

This relation illustrates how it is possible to find a Friedman model which has age greater than H_0^{-1} and yet has flat spatial sections. For example, if $\Omega_\Lambda = 0.9$ and $\Omega_0 = 0.1$, the age of the world model would be $1.28H_0^{-1}$.

It is also evident that, if our present Universe expanded from a stationary state at redshift z_c, it must now be accelerating. The deceleration parameter q_0 is negative and equal to

$$q_0 = -\frac{z_c^3 + 3z_c + 3}{z_c^2 + 3z_c}. \tag{7.82}$$

For example, if $z_c = 2$, $q_0 = -1.7$ and, if $z_c = 5$, $q_0 = -3.6$.

7.3.3 Observations in Lemaître World Models

To relate the physical properties of distant objects to observables such as angular sizes, flux densities, and so on, the apparatus developed in Sect. 5.5 can be used. The curvature of space at the present epoch is given by (7.66) and the comoving radial distance coordinate r is

$$r = \int_{t_1}^{t_0} \frac{c\,dt}{R} = \frac{c}{H_0} \int_0^z \frac{dz}{[(1+z)^2(\Omega_0 z + 1) - \Omega_\Lambda z(z+2)]^{1/2}}, \tag{7.83}$$

where dz/dt is given by (7.77) and $R = (1+z)^{-1}$. General solutions may be found in terms of elliptic functions but we do not wish to enter into that exercise here. It is generally easier to evaluate the distance measure $D = \Re \sin(r/\Re)$ numerically and then use the formulae derived in Sect. 5.5.

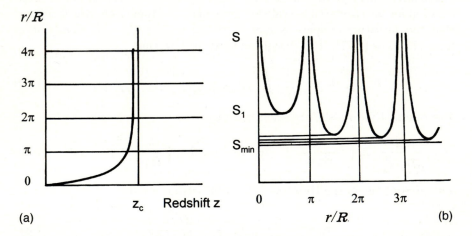

Fig. 7.8. (a) The function $r/\Re$ as a function of redshift z for the Eddington–Lemaître model which was stationary at redshift z_e. (b) The variation of the observed flux density S of sources of the same intrinsic luminosity as a function of distance coordinate $r/\Re$ for this Eddington–Lemaître model. The sources are assumed to have power-law spectra of the form $S \propto \nu^{-1}$.

One unique feature of the Lemaître world models is the possible appearance of *ghost images*. In the Eddington–Lemaître models, the curvature of space κ is positive, as may be seen from the 'loitering' line in Fig. 7.7a. Specifically, the spatial curvature of these models is

$$\kappa = \Re^{-2} = \frac{H_0^2}{c^2}\left[\frac{3\Omega_0}{2}(1+z_{\rm c})\right] = \frac{H_0^2}{c^2}\frac{3(1+z_{\rm c})}{(1+z_{\rm c})^3 - 3(1+z_{\rm c}) + 2}, \qquad (7.84)$$

where we have used (7.76) to relate Ω_0 and $z_{\rm c}$. Those models which have $\dot{R}_{\rm min}$ just greater than zero also have closed spherical spatial sections, as may be seen in Fig. 7.7a. Since the expansion almost stops at redshift $z_{\rm c}$, there is time for electromagnetic waves to propagate from the source to the observer a number of times around the closed geometry of the Universe. Therefore, the same object may be observed in diametrically opposite directions, or multiply in the same position on the sky, although at different redshifts and consequently at different times in its life-history. To illustrate this behaviour, we consider the case of the Eddington–Lemaître models for which analytic solutions for the comoving radial distance coordinate can be found from the integral 7.83. In these models, Ω_Λ is given by (7.75) and Ω_0 by (7.76). If we write $\eta = [2(1+z)/(1+z_{\rm c})] + 1$, we find that

$$r = \sqrt{3}\Re \int \frac{{\rm d}\eta}{\eta^{1/2}(3-\eta)}, \qquad (7.85)$$

and hence

$$r = \Re \ln\left(\frac{[\sqrt{3}+\eta^{1/2}][\sqrt{3}-\eta_0^{1/2}]}{[\sqrt{3}-\eta^{1/2}][\sqrt{3}+\eta_0^{1/2}]}\right), \qquad (7.86)$$

where $\eta_0 = [2/(1+z_{\rm c})] + 1$. This dependence of $r/\Re$ upon z is illustrated in Fig. 7.8a. The corresponding flux density–distance measure relation for a source with spectral index $\alpha = 1$, defined by $S \propto \nu^{-\alpha}$, is

$$S \propto \frac{1}{[\Re \sin(r/\Re)]^2(1+z)^2}, \qquad (7.87)$$

and is shown in Fig. 7.8b. The poles in the $S - (r/\Re)$ relation at $r/\Re = (2n+1)\pi$, where $n = 1, 2, 3, \ldots$, correspond to the antipodal points to the Earth in the closed geometry; the points $r/\Re = 2n\pi$ correspond to observing our own neighbourhood, but seen as it was at the epoch, or redshift, corresponding to $r/\Re = 2n\pi$. In this type of model, each fundamental observer acquires a number of comoving distance coordinates corresponding to the ranges of $r/\Re$, $0 - 2\pi$, $2\pi - 4\pi$, $\ldots$. In Fig. 7.8, we can read off the redshifts at which a galaxy or quasar would be observed in opposite directions on the sky, assuming that the object lived long enough to be observed at the various epochs corresponding to the different redshifts of observation.

Similar results are found in Lemaître models with $\dot{R}_{\rm min}$ just greater than zero (Fig. 7.9). The possibility of observing 'ghost' images does not occur

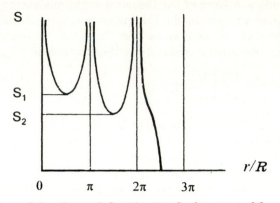

Fig. 7.9. The variation of the observed flux density S of sources of fixed intrinsic luminosity as a function of distance coordinate $r/\Re$ for a Lemaître model with cosmological constant $\Lambda = \Lambda_0(1 + \varepsilon)$ where Λ_0 is the value of the cosmological constant which would result in a static Universe at $z = 2$ and $\varepsilon = 0.02$ (Longair and Scheuer 1970). The sources are assumed to have power-law spectra of the form $S \propto \nu^{-1}$.

in the Friedman models with $\Omega_\Lambda = 0$. For all the closed Friedman models with $\Omega_\Lambda = 0$, $r/\Re$ never reaches the value π as z tends to infinity. Since $r/\Re = \pi$ corresponds to light rays propagating from the antipodal point of the closed geometry to the Earth, the possibility of observing the same source in opposite directions in the sky, or multiple images of the same object in the same direction, does not occur in these models. The reason for emphasising this aspect of the Lemaître models is that, if such repeated 'ghost' images were found, this would be positive evidence that the field equations of general relativity without the Λ-term were not an adequate description of satisfactory world models.

7.4 Inhomogeneous World Models

The results of Sects. 7.2 and 7.3 are exact for isotropic, homogeneous world models. The evidence which we assembled in Chap. 2 shows that the Universe is isotropic and homogeneous on the large scale but, on small scales, the Universe is very far from homogeneous. Matter is concentrated into stars and galaxies which are very large perturbations in the mean density. These perturbations cause deviations of the paths of light rays and it is important to evaluate their effect upon the results quoted above. We consider first the limiting case in which the matter distribution is so inhomogeneous that there is no matter within the light cone subtended by a distant object at the observer.

This problem has been treated elegantly by Zeldovich (1964) using simple physical arguments. Identical results are obtained more arduously from a general Riemannian approach to the propagation of light signals in inhomogeneous cosmological models. We consider the case of the Einstein–de Sitter world model, $\Omega_0 = 1$, for which the spatial geometry is flat, $\kappa = 0, \Re = \infty$.

If the Universe is so inhomogeneous that all the matter is condensed into point-like objects, there is only a small probability that there will be any matter within the light-cone subtended by a distant object of small angular size. Because of the long-range nature of gravitational forces, however, the background metric remains the standard flat Einstein–de Sitter metric and the overall dynamics of the Universe are unaltered. The Robertson–Walker metric can be written

$$ds^2 = dt^2 - \frac{R^2(t)}{c^2} \left[dr^2 + r^2(d\theta^2 + \sin^2\theta \, d\phi^2) \right]$$

$$= dt^2 - \frac{R^2(t)}{c^2} \left[dx^2 + dy^2 + dz^2 \right] \tag{7.88}$$

where $R(t) = (t/t_0)^{2/3}$ and $t_0 = \frac{2}{3}H_0$ is the present age of the Universe. r, x, y and z are co-moving coordinates referred to the present epoch t_0.

First, we consider the homogeneous case. Consider the events A and B which correspond to the emission of light signals at cosmic time t from either end of a standard rod of length L oriented at right angles to the line of sight at comoving radial distance coordinate r. Then, the interval between events is minus the square of a proper length

$$ds^2 = -R^2(t)r^2 \, d\theta^2 = -R^2(t) \, dy^2 = -L^2. \tag{7.89}$$

Since $R(t) = (1+z)^{-1}$, we recover the result of Sect. 7.2,

$$d\theta = \frac{L(1+z)}{r}, \tag{7.90}$$

where $r = D = (2c/H_0)\left[1 - (1+z)^{-1/2}\right]$ for the Einstein–de Sitter model. Notice that, in the homogeneous case, the angle between the light rays $d\theta$ remains a constant during propagation from the source to the observer. This fundamental result is true for all isotropic world models and is a consequence of the postulates of isotropy and homogeneity.

In the model of an inhomogeneous Universe, we consider the propagation of the light rays in this background metric, but include in addition the effect of the absence of matter within the light cone subtended by the source at the observer. As we discussed in Sect. 4.3.4, angular deflection of a light ray by a point mass, or by an axially symmetric distribution of mass at the same distance, is given by

$$\Delta\theta = \frac{4GM(<p)}{pc^2}, \tag{7.91}$$

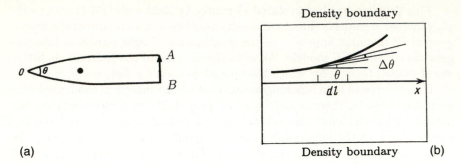

(b)

Fig. 7.10. (a) Illustrating the deflection of light beams by a point mass. (b) Illustrating the divergence of the light beam because of the absence of mass within the light cone (Zeldovich 1964).

where $M(< p)$ is the mass within 'collision parameter' p, that is, the distance of closest approach of the light ray to the point mass (Fig. 4.8a). Fig. 7.10 shows an idealised model for the propagation of the rays along the light cone, assuming the light paths are known. Because of the principle of superposition, the effect of the 'missing mass' within the light cone may be precisely found by supposing that the distribution of mass has negative density $-\varrho(t)$ within the light cone. The deviations of the light cones from the homogeneous result, $d\theta = dy/dx = $ constant, are due to the influence of the 'negative mass' within the light cone. As a result, the light rays bend *outwards* rather than inwards, as in the usual picture (Fig. 7.10).

Considering a small section of the light path of proper length $dl = R(t)\,dx$ in the radial direction, the change in θ due to the 'negative mass' inside the light cone is

$$\Delta\theta = \frac{4G\,dM(< p)}{pc^2}. \tag{7.92}$$

Now, $dM(< p) = \pi p^2 \varrho(t)\,dl$ and hence

$$\frac{d\theta}{dl} = \frac{4\pi G p \varrho(t)}{c^2}. \tag{7.93}$$

We now convert to comoving coordinates $\theta = dy/dx$, $l = R(t)x$, $p = R(t)y$. For the Einstein–de Sitter model, $\varrho(t) = \varrho_0(1+z)^3$ with $\varrho_0 = 3H_0^2/8\pi G$ and

$$x = \frac{2c}{H_0}\left[1 - (1+z)^{-1/2}\right]. \tag{7.94}$$

If we write $2c/H_0 = x_0$, then $(1+z) = x_0^2/(x_0 - x)^2$ and hence

$$\frac{d^2y}{dx^2} = \frac{6y}{(x_0 - x)^2}. \tag{7.95}$$

This equation can be solved using a series trial function $y = \sum_n a_n(x_0 - x)^n$, for which the solution can be written

$$y = a_3(x_0 - x)^3 + a_{-2}(x_0 - x)^{-2}. \tag{7.96}$$

Fitting the boundary conditions, namely that, at $x = y = 0$, the angle subtended by the source is $\Theta = dy/dx$, we find

$$y = \frac{2c\Theta}{5H_0}(1 + z)\left[1 - (1 + z)^{-5/2}\right]. \tag{7.97}$$

Therefore, since $L = R(t)y = y/(1 + z)$, the final result is

$$L = \frac{2c\Theta}{5H_0}\left[1 - (1 + z)^{-5/2}\right]. \tag{7.98}$$

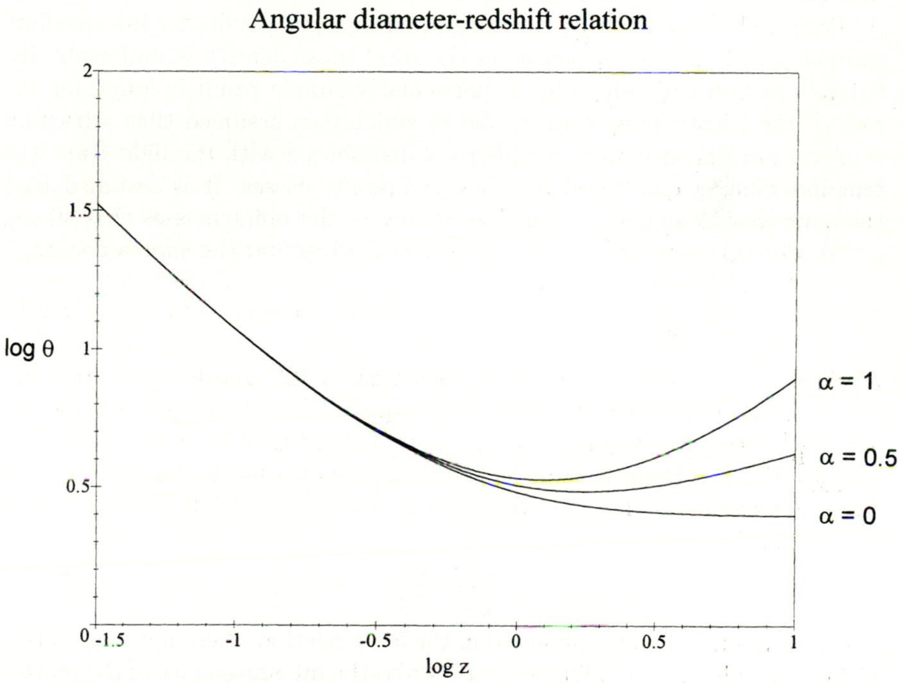

Angular diameter-redshift relation

Fig. 7.11. Comparison between the angular diameter–redshift relation in the homogeneous, uniform Einstein-de Sitter world model ($\alpha = 1$), the same background model in which there is no mass within the light cone subtended by the source ($\alpha = 0$) and the case in which half of the total mass is uniformly distributed and the rest is contained in point masses ($\alpha = 0.5$).

Corresponding results have been obtained for Friedman models with $\Omega_0 \neq 1$ by Dashevsky and Zeldovich (1964) and by Dyer and Roeder (1972). In these cases, if $\Omega_0 > 1$,

$$L = \frac{3c\Omega_0^2}{4H_0(\Omega_0 - 1)^{5/2}}\left[\sin^{-1}\left(\frac{\Omega_0 - 1}{\Omega_0}\right)^{1/2} - \sin^{-1}\left(\frac{\Omega_0 - 1}{\Omega_0(1 + z)}\right)^{1/2}\right]$$

$$- \frac{3c\Omega_0}{4H_0(\Omega_0 - 1)^2} \left[1 - \frac{(1 + \Omega_0 z)^{1/2}}{(1 + z)} \right] + \frac{1}{2(\Omega_0 - 1)} \left[1 - \frac{(1 + \Omega_0 z)^{1/2}}{(1 + z)^2} \right]$$

(7.99)

If $\Omega_0 < 1$, the inverse trigonometric functions are replaced by inverse hyperbolic functions according to the rule $\sin^{-1} ix = i \sinh^{-1} x$.

The $\theta - z$ relation (7.98) is compared with the standard result (7.90) in Fig. 7.11. It can be seen that the minimum in the standard $\theta - z$ relation disappears in the maximally inhomogeneous model. Thus, if no minimum is observed in the $\theta - z$ relation for a class of standard rods, it does not necessarily mean that the Universe must have $\Omega_0 \approx 0$. It might just mean that the Universe is of high density and is highly inhomogeneous.

Dyer and Roeder (1973) have presented analytic results for intermediate cases in which a certain fraction of the total mass density is uniformly distributed within the light cone. A particularly simple result is found for the case of the Einstein–de Sitter model in which it is assumed that a fraction α of the total mass density is uniformly distributed with the light cone, the remainder being condensed into discrete point masses. It is assumed that the light cone does not pass so close to any of the point masses that strong gravitational lensing distorts the light cones. They find the simple result:

$$L = \Theta D_{\mathrm{A}} = \Theta \frac{2}{\beta} (1 + z)^{(\beta - 5)/4} [1 - (1 + z)^{-\beta/2}],$$

(7.100)

where $\beta = (25 - 24\alpha)^{1/2}$. It can be seen that (7.100) reduces to (7.90) and (7.98) in the limits $\alpha = 1$ and $\alpha = 0$ respectively. The angular diameter–redshift relation for the case $\alpha = 0.5$ is included in Fig. 7.11. Finding the minimum of (7.100), Dyer and Roeder (1973) also show that, for the Einstein–de Sitter model, the minimum in the angular diameter–redshift relation occurs at a redshift

$$z_{\mathrm{min}} = \left(\frac{5 + \beta}{5 - \beta} \right)^{2/\beta} - 1.$$

(7.101)

Thus, if a minimum *is* observed in the $\theta - z$ relation, there must be matter within the light cone and limits can be set to the inhomogeneity of the matter distribution in the Universe. The effects upon the observed intensities of sources may be evaluated using the same approach as in Sect. 7.3. The $\theta - z$ relation may be used to work out the fraction of the total luminosity of the source incident upon the observer's telescope using the reciprocity theorem. These are not very different from the results of the standard models.

The case of strong gravitational lensing, in which the light cone subtended by the source at the observer passes close to a massive deflector, was discussed in Sect. 4.3.4. As shown in that section, strong gravitational lensing causes major distortions of the images of distant background sources, if they lie within roughly the Einstein angle θ_{E}, given by (4.31) and (4.35), of the deflector. The types of distortion, illustrated in Fig. 4.9, have been observed in a number of gravitationally lensed sources in the optical and radio wavebands. In addition, the flux densities of the background sources can be enhanced by factors of up to about 40 over their unlensed intensities. This type of flux density enhancement has been shown to account for the extraordinary luminosity of the galaxy IRAS F10214+4724. Assuming the galaxy were unlensed, its far infrared luminosity would be $\sim 3 \times 10^{14} \; L_\odot$. The image of the galaxy is, however, clearly distorted because of strong gravitational lensing and, once a best fitting mass model has been used to determine the flux density enhancement, the far-infrared luminosity of the galaxy is found to be $\sim 10^{13} \; L_\odot$, still a very large value, but not as extreme as once believed (Close *et al.* 1995).

As mentioned in Sect. 4.3.4, even in the weak lensing limit, the images of background objects are distorted by the presence of mass concentrations along the line of sight. Kaiser (1992) has shown how the statistics of the distortions of background objects by intervening mass concentrations can be used to determine the power spectrum of density fluctuations in the large-scale distribution of matter in the Universe.

8 The Determination of Cosmological Parameters

8.1 The Cosmological Parameters

We can summarise the results of the calculations of Chap. 7 for observational cosmology as follows. The standard uniform world models can be described by a small number of parameters:

- *Hubble's constant*, H_0, describes the present rate of expansion of the Universe

$$H_0 = \left(\frac{\dot{R}}{R} \right)_{t_0} = \dot{R}(t_0). \tag{8.1}$$

- The *deceleration parameter*, q_0, is defined to be the present dimensionless deceleration of the Universe.

$$q_0 = - \left(\frac{\ddot{R}}{\dot{R}^2} \right)_{t_0} = -\frac{\ddot{R}(t_0)}{H_0^2}. \tag{8.2}$$

- The *density parameter*, Ω_0, is defined to be the ratio of the present mass–energy density of the Universe ϱ_0 to the critical density $\varrho_c = 3H_0^2/8\pi G$

$$\Omega_0 = \frac{\varrho_0}{\varrho_c} = \frac{8\pi G \varrho_0}{3H_0^2}. \tag{8.3}$$

  For astrophysical cosmology, it is important to determine separately the density parameter in baryonic matter Ω_B and the overall density parameter Ω_0, which includes all forms of non-baryonic dark matter.
- The *curvature of space*, κ.
- The *cosmological constant* Λ which can be parameterised in terms of the density parameter of the vacuum fields $\Omega_\Lambda = 8\pi G \varrho_v / 3H_0^2 = \Lambda / 3H_0^2$.
- The *age of the Universe*, T_0,

$$T_0 = \int_0^{t_0} \frac{\mathrm{d}R}{\dot{R}}. \tag{8.4}$$

The observational determination of these parameters is a long and fascinating story, which has been delightfully told by Sandage (1995). The astronomical approach to the determination of these parameters can be appreciated from the graphs presented in Figs. 7.3 and 7.4 for the cosmological models with $\Lambda = 0$. The basic point is that the dependences of the comoving radial distance coordinate r and the distance measure D upon redshift are determined by the dynamics and geometry of the Universe. Thus, if the $D-z$ relation, or equivalently the $D_{\rm L}-z$ or $D_{\rm A}-z$ relations, can be defined precisely from observation, the parameters q_0 and Ω_0 can be estimated directly. The key problem is that, in order to estimate D, physical properties, such as dimensions or luminosities, should be measured *independent* of a knowledge of D. This procedure is possible in some cosmological tests, as we will discuss. The alternative is to select objects with the same intrinsic properties at different redshifts and then determine how the observed properties vary with redshift. This procedure is critically dependent upon our ability to identify reliably the same types of standard candle or rigid rod at different redshifts.

8.2 Testing the Friedman Models

Let us first study the relation between the cosmological parameters and how we might discriminate between them. The Einstein equations, which are the basis of the standard models, are:

$$\ddot{R} = -\frac{\Omega_0 H_0^2}{2R^2} + \Omega_\Lambda H_0^2 R; \tag{8.5}$$

$$\dot{R}^2 = \frac{\Omega_0 H_0^2}{R} + \Omega_\Lambda H_0^2 R^2 - \frac{c^2}{\Re^2}. \tag{8.6}$$

At the present epoch, $t = t_0$, $R = 1$, these reduce to

$$\ddot{R}(t_0) = -\frac{\Omega_0 H_0^2}{2} + \Omega_\Lambda H_0^2 \qquad q_0 = \frac{\Omega_0}{2} - \Omega_\Lambda; \tag{8.7}$$

$$\dot{R}^2(t_0) = \Omega_0 H_0^2 + \Omega_\Lambda H_0^2 - \frac{c^2}{\Re^2} \qquad \kappa \left(\frac{c}{H_0}\right)^2 = (\Omega_0 + \Omega_\Lambda) - 1. \tag{8.8}$$

Expressed in this way, the equations describe two different aspects of the cosmological models. Equations (8.5) and (8.7) are dynamical equations, describing the deceleration, or acceleration, of the Universe under the competing influences of gravity and the vacuum fields. Expression (8.7) shows that the deceleration parameter provides a measure of the difference between half the density parameter Ω_0 and the density parameter in the vacuum fields Ω_Λ. In contrast, (8.8) describes how the curvature of space, $\kappa = \Re^{-2}$, depends upon the total mass density in both the matter and the vacuum fields. Notice that (8.8) describes the basic feature of the isotropic, homogeneous world models of General Relativity that the curvature of space at any epoch is determined by the total mass–energy density at any epoch – as was shown in Sect. 5.4 (5.35), the curvature changes as $\kappa = R_{\rm c}^{-2}(t) = (\Re R)^{-2} \propto R^{-2}$, that is, its variation with cosmic epoch is independent of the details of the dynamics of the Universe. Let us look in a little more detail into how we can distinguish between these parameters observationally.

The deceleration of the Universe and its present mass density are separately measurable quantities, and so their determination provides a key test for the presence, or absence, of the cosmological constant in the field equations, as can be observed in (8.7), $q_0 = \Omega_0/2 - \Omega_\Lambda$. The density parameter Ω_0 can be found from the virial theorem in its various guises (Sect. 8.5), and so the question is how to measure the deceleration of the Universe from the classical cosmological tests, independent of the density parameter and the curvature of space. Inspection of the expressions for the comoving radial distance coordinate r and the distance measure D shows that the dynamical and geometrical properties of the models become entangled when we relate the intrinsic properties of objects at large redshifts to observables. It turns, however, out that at small redshifts the differences between the world models depend only upon the deceleration parameter and not upon the density parameter and the curvature of space. Let us demonstrate this first by a simple argument given by Gunn (1978).

It will be recalled that, in order to relate observables to intrinsic properties, we need to know how the distance measure D depends upon redshift and there are two steps involved in working out this dependence. First, we work out the dependence of the comoving radial distance coordinate r upon redshift z and then we form the distance measure $D = \Re \sin(r/\Re)$. Let us carry out this calculation first in terms of *kinematics* rather than through the dynamical equations proceeding from Einstein's field equations. We can write the variation of the scale factor R with cosmic epoch in terms of a Taylor series as follows

$$R = R(t_0) + \dot{R}(t_0)\,\mathrm{d}t + \tfrac{1}{2}\ddot{R}(t_0)(\mathrm{d}t)^2 + \dots$$
$$= 1 - H_0\tau - \tfrac{1}{2}q_0 H_0^2 \tau^2 + \dots. \qquad (8.9)$$

where we have introduced the look-back time $\tau = t_0 - t$ where t_0 is the present epoch and t is some earlier epoch. The above expansion can be written in terms of $x = H_0\tau$ and so, writing $R = (1+z)^{-1}$, we find

$$\frac{1}{1+z} = 1 - x - \frac{q_0}{2}x^2 + \dots. \qquad (8.10)$$

Now, we want to express the redshift z to second order in the look-back time τ. This is achieved by making a further Taylor expansion of $[1 - x - (1/2)q_0x^2]^{-1}$ to second order in x. Carrying out this expansion, we find

$$z = x + (1 + \frac{q_0}{2})x^2 + \dots. \qquad (8.11)$$

We can now find the expression for the comoving coordinate distance r by taking the integral

$$r = \int_0^\tau \frac{c\,d\tau}{R} = \int_0^\tau c(1+z)\,d\tau$$

$$= \frac{c}{H_0}\left[x + \frac{x^2}{2} + \left(1 + \frac{q_0}{2}\right)\frac{x^3}{3}\cdots\right]. \tag{8.12}$$

Finally, we can express r to second order in z by dividing (8.12) successively through by (8.11). We find

$$r = \left(\frac{c}{H_0}\right)\left[z - \frac{z^2}{2}(1+q_0) + \cdots\right]. \tag{8.13}$$

The last step is to evaluate $D = \Re\sin(r/\Re)$ but, since the expansion for small values of $r/\Re$ is

$$D = r\left(1 - \frac{1}{6}\frac{r^2}{\Re^2}\right), \tag{8.14}$$

the dependence upon the curvature only appears in third-order in z and so to second-order, we find

$$D = \left(\frac{c}{H_0}\right)\left[z - \frac{z^2}{2}(1+q_0)\right]. \tag{8.15}$$

Let us now evaluate the radial comoving distance coordinate r and the distance measure D to third order in redshift, using the general result (7.83) for r

$$r = \int_{t_1}^{t_0} \frac{c\,dt}{R} = \frac{c}{H_0}\int_0^z \frac{dz}{[(1+z)^2(\Omega_0 z + 1) - \Omega_\Lambda z(z+2)]^{1/2}}. \tag{7.83}$$

The denominator of the integral can be reorganised to show explicitly the dependence upon the deceleration parameter q_0 and the present matter density parameter Ω_0 using (8.7)

$$r = \frac{c}{H_0}\int_0^z \frac{dz}{[1 + (2 + \Omega_0 - 2\Omega_\Lambda)z + (1 + 2\Omega_0 - \Omega_\Lambda)z^2 + \Omega_0 z^3]^{1/2}}$$

$$= \frac{c}{H_0}\int_0^z \frac{dz}{[1 + 2(1+q_0)z + (1 + \frac{3}{2}\Omega_0 + q_0)z^2 + \Omega_0 z^3]^{1/2}} \tag{8.16}$$

Now expand the denominator to second order for small values of z

$$r = \frac{c}{H_0}\int_0^z\left[1 - z(1+q_0) + \frac{z^2}{2}\left(2 + 5q_0 + 3q_0^2 - \frac{3}{2}\Omega_0\right)\right]dz,$$

$$= \frac{c}{H_0}\left[z + \frac{z^2}{2}(1+q_0) + \frac{z^3}{6}\left(2 + 5q_0 + 3q_0^2 - \frac{3}{2}\Omega_0\right)\right]. \tag{8.17}$$

As before, we form $D = \Re\sin(r/\Re)$, where

$$\Re = \frac{c/H_0}{[(\Omega_0 + \Omega_\Lambda) - 1]^{1/2}} = \frac{c/H_0}{\left(\frac{3}{2}\Omega_0 - q_0 - 1\right)^{1/2}}. \tag{8.18}$$

Preserving quantities to third order in z, we find

$$D = \frac{c}{H_0}\left[z + \frac{z^2}{2}(1+q_0) + \frac{z^3}{6}\left(3 + 6q_0 + 3q_0^2 - 3\Omega_0\right)\right]. \qquad (8.19)$$

This is the result we have been seeking.

To second order in the redshift, we obtain exactly the same result as that obtained from the kinematic argument, (8.15). What this means is that, to second order in redshift, the distance measure D does not depend upon the density parameter Ω_0 at all – it only depends upon the deceleration. The physical meaning of this result is that, at the same redshift z or scale factor R, the look-back time is smaller in an $\Omega_0 = 1$ Universe than in an $\Omega_0 = 0$ model as can be observed by inspection of Fig. 7.2. Therefore, the same object is expected to be brighter in an $\Omega_0 = 1$ model as compared with an $\Omega_0 = 0$ model. Conversely, if the Universe accelerated between the redshift z and the present epoch, the look-back time would be greater and the galaxies fainter than in the $\Omega_0 = 0$ case. The significance of these remarks is that the prime reason for the difference between the world models at small redshifts is the deceleration, or acceleration, of the system of galaxies, whatever its cause, rather than the curvature of space. It can also be seen from (8.15) that, even to third order in the redshift, the dependence upon the density parameter is quite weak.

One approach is therefore straightforward, but observationally very demanding – we seek to determine the deceleration of the Universe by determining the distance measure D at small redshifts, say $z \leq 0.3$, at which there are small but appreciable differences between the world models. Alternatively, the $D - z$ relation can be determined very precisely to large redshifts and the two parameters Ω_0 and Ω_Λ found from the detailed shape of that relation. As a simple example, the variation of the comoving radial distance coordinate r as a function of redshift is shown in Fig. 8.1 for a range of flat world models, $\kappa = 0$, for which $\Omega_0 + \Omega_\Lambda = 1$. Since $\kappa = 0$, the distance measure $D = r$. It can be seen that the relations are of quite different shapes. This second procedure requires the use of high precision distance indicators.

The importance of these analyses is that they provide a test of General Relativity and the laws of physics on the largest scales accessible to us at the present epoch.

8.3 Hubble's Constant H_0

In order to measure Hubble's constant, the *recession velocities* v and *distances* r of galaxies which sample the mean Hubble flow have to be determined. This turns out to be a much more difficult task than might be imagined. Measuring the recession velocities of galaxies relative to the local standard of rest defined by the Cosmic Microwave Background Radiation is relatively straightforward,

Comoving radial distance coordinate, r

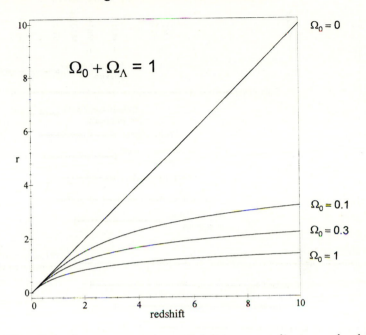

Fig. 8.1. The relation between comoving radial distance coordinate r and redshift z for the flat world models for which $\Omega_0 + \Omega_\Lambda = 1$. Since $\kappa = 0$, the distance measure $D = r$. The values of Ω_0 are 1, 0.3, 0.1 and 0. In the case $\Omega_0 = 0$, the relation is precisely $r = D \propto z$. In this diagram, distances are measured in units of c/H_0.

the radial velocities being obtained directly from their redshifts. It is now known, however, that galaxies possess significant *streaming motions* relative to the mean Hubble flow which can be as large as 500 km s^{-1} Mpc^{-1}. It is therefore necessary to extend the Hubble relation well beyond these local irregularities in the mean Hubble flow. In terms of our sponge picture, we need to get well beyond the local gravitational influence of the voids, walls and superclusters. The brightest galaxies in rich clusters of galaxies are good probes of the mean Hubble flow as illustrated by Fig. 2.9a.

The big problem is to measure accurate distances to galaxies which are independent of their redshifts. In the traditional approach, this is achieved by using the standard properties of various different types of *distance indicator* to take us from the nearby region of space within our Galaxy to the nearby galaxies. Within the local region of space in our own Galaxy, parallaxes provide direct geometric distances to nearby stars and star clusters. The local distance scale in our Galaxy is now known with much improved precision thanks to the magnificent observations made by the Hipparcos Astrometric Satellite of the European Space Agency. The measurement of the distances to nearby star clusters enables the properties of other secondary distance in-

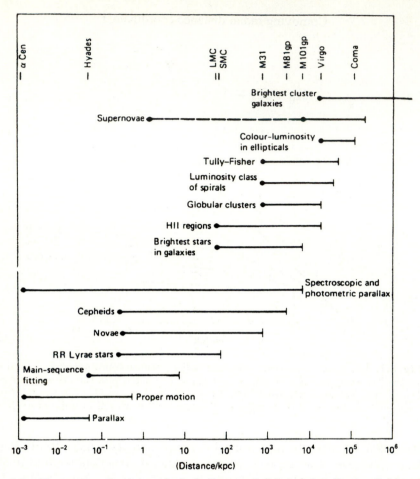

Fig. 8.2. Illustrating the 'cosmological distance ladder' (after Rowan-Robinson 1985). The diagram shows the range of distances over which different classes of object can be used to estimate the astronomical distances.

dicators to be calibrated and these can then be use to extend the distance scale to greater distances and to nearby galaxies. The use of a sequence of primary, secondary and tertiary distance indicators to extend the distance scale from within our Galaxy to extragalactic distances is indicated schematically in Fig. 8.2, which is discussed in detail by Rowan-Robinson (1985, 1988).

The great advance of the last 10 years has been that the distances to nearby galaxies are now known very much more precisely than they were in the past. Much of this improvement has been associated with observations made with the Hubble Space Telescope. There is now relatively little disagreement among the experts about the distances of those galaxies which have been studied in detail out to about the distance of the Virgo cluster.

This has involved an enormous amount of difficult and careful work. Where the differences come in are in how the data are analysed once these distances are known. What has been particularly difficult has been the elimination of systematic errors and biases in the observed samples of galaxies. One of the problems in interpreting much of the early work was that the estimates were quoted with small statistical errors, but with the caveat 'excluding systematic errors', which is where all the problems arise.

This is not the place to go into this enormous subject in any detail, but mention should be made of two of the most important distance indicators used in recent work. One class consists of the Cepheid variable stars which are luminous blue regular variable stars with characteristic light curves, that is, the variation of stellar luminosity with phase of the pulsation. It is found that there is a very tight relation between their luminosities and their periods of pulsation. With the Hubble Space Telescope, Cepheid variables can be recognised in galaxies as far away as the Virgo cluster of galaxies. By measuring their periods, their luminosities can be inferred from the period–luminosity relation and hence their distances derived from the inverse square law.

The second important class of standard candle consists of the supernovae of Type Ia, which all seem to have more or less the same luminosity at maximum light. The preferred interpretation of the origin of Type Ia supernovae is that they result from the explosion of accreting white dwarfs in binary systems. As matter is accreted onto their surfaces, the central temperature increases and, for carbon–oxygen white dwarfs, explosive nuclear burning in their centres is initiated as their masses approach the Chandrasekhar limit of 1.4 $M_\odot$. The attractive feature of the use of these objects as distance indicators is that this picture provides a natural physical explanation for the uniformity of this class of supernovae. Another great advantage of the supernovae of this type is that they can be observed in distant galaxies because they are so luminous (Branch and Tammann 1992).

The value of Hubble's constant has been a matter of controversy for many years, one school of thought, associated with the names of Sandage and Tammann, favouring values about 50 km s^{-1} Mpc^{-1}, while de Vaucouleurs, Aaronson and their colleagues have favoured values closer to 80 km s^{-1} Mpc^{-1}. Freedman and Tammann presented their determinations of Hubble's constant at the 1996 Princeton meeting on *Critical Dialogues in Cosmology* (1997) and quoted the values of $H_0 = 70 \pm 10$ km s^{-1} Mpc^{-1} and $H_0 = 55 \pm 10$ km s^{-1} Mpc^{-1} respectively. Both teams found much smaller statistical errors than the errors quoted above, which include an allowance for systematic errors as well. At the IAU Symposium No. 183 held in Kyoto in 1997, both teams presented their updated estimates: Freedman (1998) quoted a value $H_0 = 73 \pm 6(\text{stat}) \pm 8(\text{syst})$ km s^{-1} Mpc^{-1}, while Tammann (1998) quoted the value $H_0 = 58 \pm 3(\text{stat})$ km s^{-1} Mpc^{-1}. Thus, for the moment, the estimates by the traditional route seem to have stabilised. There is a problem in assigning formal errors to these estimates since the systematic errors are not symmetrical with respect to the best estimate and the systematic uncertainties should be combined with the random errors before adding them together statistically. Let me illustrate how I interpret these results. Suppose we adopt a compromise value of $H_0 = 62.5$ km s^{-1} Mpc^{-1} with a 1-σ random error of 15%. Then, there would be a 95% probability that the actual value of Hubble's constant lies between $H_0 = 44$ and 81 km s^{-1} Mpc^{-1}. The determination of Hubble's constant is one of the Key Projects of the Hubble Space Telescope and it is interesting that the goal is to measure its value to a 1-σ accuracy of 10%.

From these considerations, I draw a number of conclusions. First, I believe there is no longer any serious discrepancy between the estimates of Hubble's constant. Second, to do substantially better, say, to determine Hubble's constant to 5% will require an enormous effort. Third, although I personally dislike using it, it is only wise to continue to include the h^{-1} factor in all our estimates of distances and dimensions.

Fortunately, new *physical methods* of measuring H_0 are now becoming feasible and they have the advantage of eliminating the many of the steps in the cosmological distance ladder. These methods are based upon measuring a physical dimension l of a distant object and then measuring its angular size θ, so that an angular diameter distance D_A can be found from $D_A = l/\theta$ at a known redshift z. A beautiful example of the use of this technique was described by Panagia *et al.* (1991) who combined IUE observations of the time-variability of the emission lines from the supernova SN1987A in the Large Magellanic Cloud with Hubble Space Telescope observations of the emission-line ring observed about the site of the explosion to measure the physical size of the ring. The distance found for the Large Magellanic Cloud turned out to be as accurate as the value found by the traditional procedures. Another distance measure to the supernova SN1987A was provided by the *Baade–Wesselink method* applied to the expanding photosphere of the supernova. Schmidt *et al.* (1992) found a value in excellent agreement with the values found by the traditional approach and by Panagia and his colleagues. Extending the Baade–Wesselink technique to distant supernovae, they find values of H_0 of roughly 55 km s^{-1} Mpc^{-1}.

Another example of a physical method of measuring H_0 is to use gravitational lensing of distant objects by intervening galaxies or clusters (see, for example, Blandford and Narayan 1992). The key observations concern, not only the geometry of the lens but also the time variability of different images of the same background object. The time difference between the variability of the different images of the distant quasar arriving by the two different routes enables physical scales at the lensing galaxy to be determined – for example, in the double quasar 0957+561, a time-delay of 418 days has been measured. The main uncertainty in working out physical scales at the lensing galaxy results from uncertainties in modelling the mass distribution in the lens.

In the case of the double quasar 0957+561, Kundic *et al.* (1997) now claim that the mass distribution in the galaxy is sufficiently well determined that the model dependent errors are now small and they have derived a value of Hubble's constant of $H_0 = 64 \pm 13$ km s^{-1} Mpc^{-1} at the 95% confidence level from these observations.

A third approach involves the hot gas clouds observed in rich clusters of galaxies. As discussed in Sects. 4.3.2 and 4.3.3, clusters of galaxies contain vast quantities of hot gas which is detected by its X-ray bremsstrahlung. The X-ray surface brightness depends upon the electron density N_e and the electron temperature T_e through the relation $I_\nu \propto \int N_e^2 T_e^{-1/2} dl$. The electron temperature T_e can be found from the shape of the bremsstrahlung spectrum. Furthermore, a dip in the Microwave Background Radiation in the Rayleigh–Jeans region of the spectrum is expected due to Compton scattering of low energy photons as they transverse the cluster, a phenomenon known as the *Sunyaev–Zeldovich effect*. The decrement in the background is proportional to the Compton optical depth $y = \int (kT_e/m_e c^2)\sigma_T N_e dl \propto \int N_e T_e dl$. Thus, the physical properties of the hot gas are over-determined and the physical dimensions of the X-ray emitting gas can be found. Myers *et al.* (1997) have estimated a value of $H_0 = 54 \pm 14$ km s^{-1} Mpc^{-1} from detailed studies of the Abell clusters A478, 2142 and 2256. Similar values are found from studies with the Ryle Telescope at Cambridge, which has now measured Sunyaev-Zeldovich decrements in 12 rich clusters. According to Dr. Richard Saunders, the clusters used in these studies must be selected with care. There are complications in the interpretation if the clusters are irregular, possess cooling flows or contain diffuse, non-thermal radio emission. If the clusters pass the selection criteria, he estimates that typically Hubble's constant can be measured to about 30% accuracy for an individual cluster, consistent with the findings of the Myers *et al.* (1997).

The problem with the physical methods of measuring H_0 is that the astrophysics of the sources used must be well understood. Their advantage is that the methods are independent of the need to select identical types of object in different galaxies – the objects are used to measure directly an angular diameter distance at a known redshift.

8.4 The Deceleration Parameter q_0

We have already discussed in some detail in Sect. 8.2 how it is possible to determine separately the deceleration parameter q_0 and the density parameter Ω_0 by observations of distant objects. The hope of the pioneers of observational cosmology was that these could be found from the apparent magnitude–redshift relation or the angular diameter–redshift relation, examples of which were illustrated in Figs. 7.4a and b. The well-defined apparent magnitude–redshift relation for brightest cluster galaxies found by Sandage out to redshifts $z \sim 0.5$ (Fig. 2.9a) was an encouraging result, as was the infrared K magnitude–redshift relation for radio galaxies (Fig. 2.9b). The problem with these relations is that the impressive constancy of the intrinsic luminosities of these galaxies is an entirely empirical result without any strong astrophysical underpinning. Even worse was the fact that observational evidence was found for the evolution of the properties of these galaxies with cosmic epoch and so the astrophysics of these galaxies had to be understood as well if they were to be used in cosmological tests. There is now abundant evidence for the evolution of many classes of extragalactic object with cosmic epoch and we will survey the cosmological implications of these observations in Chap. 17. The net result is that we learn more about the evolution of the stellar populations of galaxies and their high energy astrophysical activity with cosmic epoch than about the value of q_0. The point is a very simple one, namely, that to measure the deceleration of the Universe, objects have to be observed at significant redshifts, but that automatically means that they are observed at epochs earlier than the present when they were at a younger stage in their evolution.

8.4.1 The Apparent Magnitude–Redshift Relation for Luminous Galaxies

To illustrate how tricky the traditional approach can be, let me discuss two examples of the use of the apparent magnitude–redshift relation for galaxies. The problem of using the brightest cluster galaxies is well illustrated by the discussion in Sandage's review of the problems of determining cosmological parameters (Sandage 1988). Although there is an excellent redshift–magnitude relation to redshifts $z \sim 0.5$, beyond that redshift, the colours of the galaxies diverge from the expectations of the standard world models. Part of the problem arises because Sandage used optical magnitudes, which are sensitive to star-formation activity in galaxies. According to Sandage (1995), all that can be stated from his data is that the value of q_0 is about 1 ± 1.

My own experience with this problem concerns the apparent magnitude-redshift relation at an infrared wavelength of 2.2 μm for the narrow-line 3CR radio galaxies, which are associated with the brightest extragalactic radio sources in the northern sky (Lilly and Longair 1984). It was known from studies in the 1960s and 1970s that the 3CR radio galaxies have more or less the same absolute optical magnitudes with a small dispersion about the mean value. In the late 1970s, it became feasible to make observations of faint galaxies in the 1 to 2.2 μm waveband using single element InSb detectors. We began a systematic programme of infrared photometry of a complete sample of the 3CR narrow-line radio galaxies in 1980 and by 1984 had derived their infrared $K-z$ relation. There are many advantages in using these observations for cosmological studies. First of all, the complete sample of galaxies extended with good statistics to redshifts $z = 1.8$, at which there

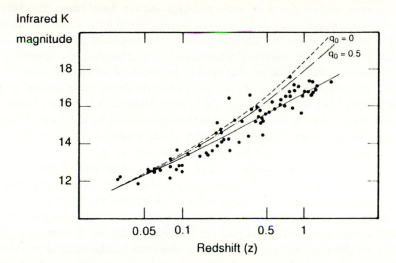

Fig. 8.3. The redshift – infrared magnitude relation for a complete sample of radio galaxies selected from radio sources in the 3CR catalogue (Lilly and Longair 1984). The observations were made in the infrared K waveband at 2.2 μm. The dashed lines show the expectations of world models having $q_0 = 0$ and 0.5. The solid line shows the expected relation when account is taken of the evolution of the stellar populations of the radio galaxies.

is a significant difference between the predicted K−z relations for different values of q_0. Second, the infrared magnitudes at 2.2 μm are dominated by the emission of the old stellar populations of galaxies, making them much less susceptible to contamination by star-bursts and ongoing star-formation than their optical magnitudes. We found a splendid correlation with a remarkably small dispersion about the mean relation, which is shown in Fig. 8.3 (Lilly and Longair 1984). If the evolution of the old stellar population was ignored, a best-fit value of q_0 of 3.5 was found, which was not a particularly popular conclusion. It turns out that the corrections for the stellar evolution of the old red giant populations of galaxies is remarkably model independent and, when we made the simplest correction for the evolution of their K luminosities, we found a best-fit value of $q_0 \approx 0.5 \pm 0.5$ (see Sect. 20.1.1).

This simple picture was thrown into some disarray in 1987 when Mc-Carthy, Chambers and their colleagues discovered that the optical images of the 3CR radio galaxies at redshifts $z \sim 1$ are aligned with the double radio structures, indicating that the optical luminosities of these galaxies are influenced by their radio properties. Optical images taken with the Hubble Space Telescope and ground based infrared images taken with the UK Infrared Telescope of five of the 3CR radio galaxies in the redshift interval $1 < z < 1.3$ are shown in Fig. 8.4 (Best, Longair and Röttgering 1996). The infrared images in the right-hand column look exactly like the classic picture of a double radio source associated with a giant elliptical galaxy. In the HST images in the left-hand column, however, a wide variety of high surface brightness structures is observed aligned with the radio structures and these bear little resemblance to giant elliptical galaxies. The spectrum of this optical radiation must be flat, but it is not clear precisely what emission mechanism is responsible for it. The linear polarisation of the optical emission observed in some of the sources suggests that scattering of the radiation from an obscured quasar nucleus must play some role, but the strong alignment with the radio jets indicates that the emission regions must be stimulated by the radio jet itself. It remains to be disentangled how much of the emission is associated with shocks, nebular emission and jet-induced star formation.

Analysis of the light distributions of the infrared images shows that they follow precisely that expected of giant elliptical galaxies. Thus, although the optical images are indeed strongly influenced by the presence of the radio source, the infrared images are dominated by the old stellar populations of the underlying giant elliptical galaxy. Therefore, the $K-z$ relation is little affected by the extraordinary structures observed in the optical waveband. We also investigated whether or not the infrared magnitudes are influenced by the presence of an active galactic nucleus and re-measured the K-magnitudes of most of the galaxies at $z \geq 0.6$ (Best *et al.* 1998).

The outcome of these analyses is a new determination of the $K-z$ relation for the 3CR radio galaxies which is shown in Fig. 8.5. The open triangles show the infrared magnitudes of the old stellar components of the 3CR radio galaxies. Also shown are various lines indicating the expected relations for different assumptions about the evolution of the stellar populations of the radio galaxies. The solid line shows the expected relation if the spectrum of a giant elliptical galaxy is redshifted in a standard $q_0 = 0.5$ world model, with no evolution of its stellar population. The dashed line shows the expected relation if the galaxies were formed at a redshift of 5 and the stellar populations allowed to evolve passively to the present epoch. It can be seen that the passive evolution model can account for the observed $K-z$ relation of the 3CR radio galaxies.

The story must, however, be more complicated that this. Also shown in Fig. 8.5 are the K-z relations for brightest cluster galaxies from the studies of Aragòn-Salamanca *et al.* (1993) (crossed circles) and a sample of 6C radio galaxies studied by Eales *et al.* (1997) (filled squared), both of which seem to follow the non-evolving $K-z$ relation. The $K-z$ relation for the brightest cluster galaxies is the infrared equivalent of Sandage's apparent magnitude–redshift relation, but now extended to redshifts $z \sim 1$. The sample of 6C radio galaxies is similar to the 3CR sample, but selected at about 5 times fainter flux densities. The fact that these $K-z$ relations follow the 'no evolution' locus is remarkable, since undoubtedly the stellar populations of both types of galaxy must have evolved significantly between a redshift $z \sim 1$ and the present epoch. Indeed, the $V-K$ colours of these galaxies are bluer at redshifts

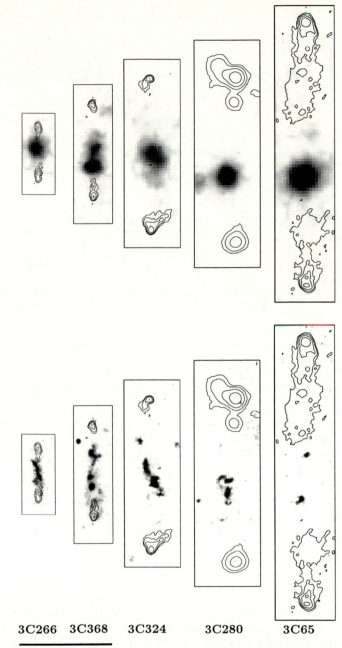

3C266   3C368   3C324   3C280   3C65

50 kpc

Fig. 8.4. HST and UKIRT images of the radio galaxies 3C 266, 368, 324, 280 and 65 with the VLA radio contours superimposed (Best, Longair and Röttgering 1996). The images are drawn on the same physical scale. The angular resolution of the HST images is 0.1 arcsec while that of the ground-based infrared images is about 1 arcsec.

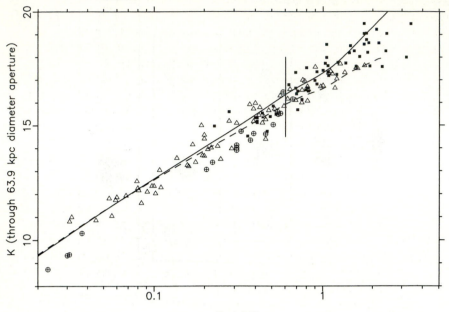

Fig. 8.5. The K−z relation for 3CR radio galaxies (open triangles), 6C radio galaxies (filled squares) and brightest cluster galaxies (crossed circles) (Best *et al.* 1998).
The data for the 6C galaxies are from Eales *et al.* (1997), the 3CR galaxies from
Lilly and Longair (1984) and Best *et al.* (1997), and the brightest cluster galaxies
from the sample of Aragòn-Salamanca *et al.* (1993), converted to a 63.9 kpc metric
aperture using a standard galaxy light curve. The solid line shows the relation expected for a non-evolving stellar population, whilst the dashed line shows that for a
passively evolving population which formed at redshift $z = 5$. These relations have
been normalised to match the magnitudes of the 3CR galaxies at low redshifts. The
vertical line at redshift $z \sim 0.6$ provides a useful division between the high and low
redshift populations.

$z \geq 0.5$ than they are nearby, consistent with passive evolution of their stellar
populations. As if this were not bad enough, the picture is further complicated
by the fact that a wide range of evidence now suggests that, whilst powerful
3CR radio galaxies at small redshifts tend to lie in isolated environments or in
small groups, the distant 3CR radio galaxies at redshifts $z \geq 1$ are associated
with the brightest galaxies in rich cluster, or proto-cluster, environments. This
dramatic change in the galactic environments of the 3CR radio galaxies casts
doubt on the 'uniform population' interpretation of the K−z relationship.

The greater K-band luminosities of the 3CR galaxies as compared with
the 6C galaxies at high redshifts must indicate that they contain a greater
mass of stars. It follows that, if there is a correlation between stellar mass
and central black hole mass, for which there is reasonably convincing evidence
(see, for example, Kormendy and Richstone 1995), the 6C galaxies possess

less powerful central engines than the 3CR galaxies at high redshifts, thus accounting for their lower radio luminosities. At redshifts $z \sim 1$, the 3CR galaxies and brightest cluster galaxies are equally luminous, consistent with the proposition that the distant 3CR galaxies are amongst the most massive galaxies in clusters. At low redshifts, however, brightest cluster galaxies are up to a magnitude brighter in absolute K-magnitude than the 3CR galaxies. Since the stellar populations of the brightest cluster galaxies must have evolved over this redshift interval in a similar way to those of the 3CR galaxies, their K$-z$ relation must reflect the fact that brightest cluster galaxies continue to accumulate matter through mergers with massive cluster galaxies, and through the infall of gas. Hierarchical clustering models for structure formation have suggested that the mass of the brightest cluster galaxies increases by as much as a factor of 5 between a redshift of one and the present epoch (Kaufmann 1995).

For the brightest cluster galaxies, and also for the 6C galaxies, this growth of the mass of the galaxies between $z \sim 1$ and $z = 0$ increases their absolute magnitudes, thereby compensating for the dimming of their stellar populations – the two effects conspire to give rise to apparently simple 'no evolution' tracks. The 3CR galaxies at high redshifts also lie at the centre of galaxy clusters and so their masses would be expected to increase with cosmic epoch just like the brightest cluster galaxies; the 'passively evolving' K$-z$ relation of these galaxies suggests, however, that this does not occur. We infer that the galaxies sampled at high and low redshifts do not form a uniform population, as is indicated by the dramatic change in their galactic environments with redshift. The apparent passive evolution disguises an important result: the most powerful 3CR radio sources at all redshifts $z \leq 1.5$ contain approximately the same mass of stars, a few times $10^{11} M_\odot$.

I have told this story in some detail, partly because it is a remarkable story in its own right, but also because it indicates clearly the problems of using galaxies as standard objects in cosmological tests. It is evident that the astrophysics of galaxies needs to be understood in considerable detail before real progress can be made.

8.4.2 Type 1A Supernovae

The discussion of Sect. 8.4.1 makes it clear that what is required is a set of standard candles which are not susceptible to the effects of stellar evolution. One of the most encouraging developments of the last few years has been the use of supernovae of Type 1a to extend their apparent magnitude-redshift relation to redshifts $z > 0.5$. This approach has a number of very attractive features. First of all, it has been shown empirically that these supernovae have a very small dispersion in absolute luminosity at maximum light (Branch and Tammann 1992). This dispersion can be further reduced if account is taken of the correlation found between the maximum luminosity of the Type 1a supernovae and their decay rates. This correlation, referred to as the *luminosity–width relation*, is in the sense that the supernovae with the slower decline rates from maximum light are more luminous than the faster supernovae. Second, there are good astrophysical reasons to suppose that these objects are likely to be good standard candles, despite the fact that they are observed at earlier cosmological epochs. The preferred picture is that these supernovae result from the explosion of white dwarfs which are driven over the limiting Chandrasekhar mass of 1.4 $M_\odot$ by accretion in a binary system. Nucleosynthesis in the degenerate carbon core results in a nuclear burning front which propagates out through the star, converting much of the inner material into iron group elements, including radioactive ^{56}Ni. The radioactive ejecta can account for the eventual exponential decline of the luminosity of the supernova.

Perlmutter *et al.* (1996) described the first results of a systematic search for Type 1a supernovae at redshifts $z \sim 0.5$ using an ingenious approach to detect them before they reach maximum light. Deep images of selected fields, including a number which contain distant clusters of galaxies, are taken during one period of new moon and the fields are imaged in precisely the same way during the next new moon. Using rapid image analysis techniques, any supernovae which appeared between the first and second epoch observations are quickly identified and are reobserved photometrically and spectroscopically over the succeeding weeks to determine their types and light curves. The pleasant aspect of this approach is that the effect of time dilation between a redshift of 0.5 and the present epoch means that a Type 1a supernova which exploded between the first and second epochs will not have reached maximum light by the time of the second epoch observation.

Using this search technique, Perlmutter *et al.* (1996) have discovered 27 supernovae of Type 1a between redshifts 0.4 and 0.6 in three campaigns in 1995 and 1996. These observations have a number of important consequences for cosmology. First of all, Goldhaber *et al.* (1997) have used these data to demonstrate directly the effects of cosmological time dilation for the light curves of Type 1a supernovae at redshifts $z \sim 0.4 - 0.6$ as compared with those of the same type at the present epoch, thus testing directly the basic cosmological relation (5.46) (see Fig. 5.6). Secondly, the same peak luminosity–width relation was found as that at small redshifts. When account is taken of this relation, the intrinsic spread in the luminosities of the Type 1a supernovae is only 0.21 magnitudes.

This same technique has been used to discover Type 1a supernovae at redshifts greater than $z = 0.8$ as a result of observations with the Hubble Space Telescope. In two independent programmes, Garnevich *et al.* (1998) and Perlmutter *et al.* (1998) have succeeded in discovering the Type 1a supernovae SN1997 ck at redshift $z = 0.97$ and SN1997ap at redshift $z = 0.83$ respectively. The great advantage of the HST observations is that their high angular resolution enables very accurate photometry to be carried out on

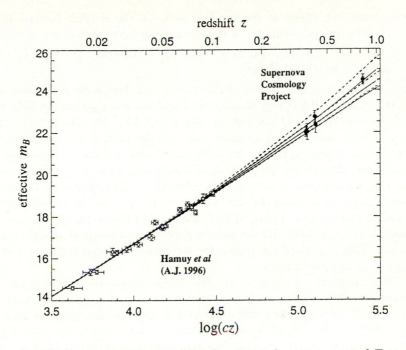

Fig. 8.6. The apparent magnitude–redshift relation for supernovae of Type 1a (Perlmutter *et al.* 1998). The SN1987ap at redshift $z = 0.83$ is plotted, as well as 5 of the first seven large redshift supernovae which can be corrected for the width-luminosity relation, and 18 lower redshift Type 1a supernovae from the Calan-Tololo survey. All the magnitudes plotted include the appropriate K-correction and have been corrected for the width-luminosity relation. The theoretical lines are for: $(\Omega_0, \Omega_\Lambda) = (0,0); (1,0); (2,0)$ from top to bottom – solid lines; $(\Omega_0, \Omega_\Lambda) = (0,1); (0.5, 0.5); (1,0); (1.5, -0.5)$ from top to bottom – dotted lines.

stellar objects in distant galaxies. These observations extend the redshift–apparent magnitude relation to such large redshifts that there are significant differences between the world models. The resulting redshift–apparent magnitude relation presented by Perlmutter *et al.* (1998) is shown in Fig. 8.6. The results of the two sets of observations are in excellent agreement.

- Garnavich *et al.* (1998) state that, if $\Omega_0 + \Omega_\Lambda = 1$, the value of Ω_0 is less than 1 at the 95% confidence level. If $\Omega_\Lambda = 0$, then $\Omega_0 = -0.1 \pm 0.5$.
- Perlmutter *et al.* (1998) state that, if $\Omega_0 + \Omega_\Lambda = 1$, then $\Omega_0 = 0.6 \pm 0.2$. If $\Omega_\Lambda = 0$, then $\Omega_0 = 0.2 \pm 0.4$.

These studies are advancing very rapidly at the time of writing. Since the above papers were published, Perlmutter and his colleagues have announced revised estimates of these cosmological parameters from a sample of 40 Type 1a supernovae discovered in the Supernova Cosmology Project, all of them with redshifts $z > 0.35$ and extending to redshifts of 0.9 (R.S. Ellis, personal communication). Many of the supernovae lie in the redshift interval $0.4 \leq z \leq 0.6$. The quoted estimates of the cosmological parameters are as follows.

- If $\Omega_0 + \Omega_\Lambda = 1$, $\Omega_0 = 0.25 \pm 0.06(\text{stat}) \pm 0.4(\text{syst})$. If $\Omega_\Lambda = 0$, $\Omega_0 = -0.4 \pm 0.1(\text{stat}) \pm 0.6(\text{syst})$.

There is one intriguing aspect of these analyses. The angle subtended by a supernova at $z \sim 0.5$ is very small indeed and the mass within the light-cone to the supernova would amount to only about $10^{-5} M_\odot$, if $\Omega_0 = 1$. Thus, the standard cosmological formulae can only be used if it is assumed that this amount of mass is indeed smoothly distributed within the light-cone. As pointed out by Lynden-Bell (personal communication), it is quite possible that this assumption is wrong, if the dominant dark matter is inhomogeneously distributed along the line of sight, as is suggested by the simulations of models involving cold dark matter discussed in Chap. 14. In this case, the formulae given in Sect. 7.4 for inhomogeneous cosmological models should be used. This is part of the reason for the relatively large systematic errors included in these estimates.

These are very encouraging results. They suggest that there is the real possibility of obtaining much improved estimates of q_0 over the next few years. We should also note that the physical methods described in Sect. 8.3 in connection with Hubble's constant can also be extended to cosmological distances and so enable cosmological parameters to be determined by quite independent means.

8.4.3 The Angular Diameter–Redshift Relation

It might seem that the angular diameter-redshift relation would provide a useful route to the determination of q_0 since it possesses the distinctive feature of a minimum angular size at redshift $z \sim 1$ if $q_0 \sim 0.5$ (Fig. 7.4a). The big problem has been to measure accurate proper distances at large redshifts. It is likely that the physical methods of measuring proper distances at large redshifts described in Sect. 8.3 will eventually provide the best means of using the angular diameter–redshift relations to estimate q_0. Thus, the Sunyaev–Zeldovich effect in conjunction with X-ray observations of the hot gas in clusters, gravitational lenses and the various versions of the Baade–Wesselink method all result in direct measurements of metric sizes at large redshifts and, if these could be determined for a range of redshifts, say from redshifts 0.3 to 1.5, the $D_A - z$ relation could be determined directly and the best fitting world model found from this relation. The only problem with this programme is the extent to which the predicted angular diameter–redshift relations are modified by inhomogeneities in the distribution of mass along the line of sight, as illustrated by relations shown in Fig. 7.11 for inhomogeneous world models.

The alternative approach is to use objects which may be considered to be 'rigid rods' and to use the shape of their angular diameter–redshift relation to find q_0. The big problem is to find suitable metric sizes which can be used in the test. A good example is the use of the separation of the radio components of double radio sources, such as those illustrated in Fig. 8.4. Large samples of these objects can be found spanning a wide range of redshifts, but, as illustrated in Fig. 8.7a, there appears to be no minimum in the median angular separation θ_m of the radio source components as a function of redshift (Kapahi 1987). The median angular separation of the source components is observed to be roughly inversely proportional to redshift and this has been interpreted as evidence that the median physical separation of the source components l_m was smaller in the past. Examples of fits to the observational data using evolution functions of the form $l_m \propto (1 + z)^{-n}$ are shown in Fig. 8.7a for world models with $q_0 = 0$ and 0.5. It can be seen that values of $n \approx 1.5 - 2.0$ can provide good fits to the data. There are, of course, many reasons why the separation of the radio source components might be smaller in the past – for example, the ambient interstellar and intergalactic gas was greater in the past and so the source components would not travel so far. Again, we learn more about astrophysical changes with cosmic epoch than about geometrical cosmology.

Another version of the same test was described by Kellermann (1993) and involved using only compact double radio structures studied by Very Long Baseline Interferometry. He argued that these are likely to be less influenced by changes in the properties of the intergalactic and interstellar gas, since the components are deeply embedded within the central regions of the host galaxy. His angular diameter–redshift relation is shown in Fig. 8.7b in which it can be seen that there is evidence for a minimum in the relation, which would be consistent with a value of $q_0 \sim 0.5$. The problem with this type of approach is that we cannot be certain that precisely the same types of double radio source are being selected at large and small redshifts. My own view is that this type of analysis is likely to tell us more about how the properties of these compact radio sources have changed with cosmic epoch than about the value of q_0.

8.5 The Density Parameter Ω_0

The big problem in estimating the average density of matter in the Universe ϱ_0, or equivalently the density parameter Ω_0, is that we know that there must be large amounts of *dark matter* present in the Universe. In Sect. 3.4, evidence for dark matter in spiral and elliptical galaxies was discussed and in Sect. 4.3 the corresponding results for clusters of galaxies. The issue of the total amount of dark matter present in the Universe will recur throughout much of the rest of this volume, since it strongly influences theories of the formation of galaxies, clusters and other large scale structures. This topic was the subject of a heated discussion at the 1996 Princeton meeting *Criti-*

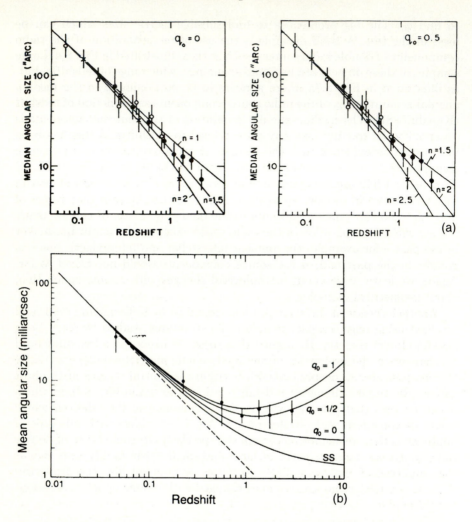

Fig. 8.7. (a) The angular diameter–redshift relation for double radio sources, in which the median angular separation of the double radio source components θ_m is plotted against redshift (Kapahi 1987). The observed relation follows closely the relation $\theta_m \propto z^{-1}$. The left-hand panel shows fits to the observations for a world model with $q_0 = 0$ and the right hand panel for a model with $q_0 = 0.5$, in both cases, the median separation of the components being assumed to change with redshift as $l_m \propto (1+z)^{-n}$. (b) The mean angular diameter–redshift relation for 82 compact radio sources observed by VLBI (Kellermann 1993). In addition to the standard Friedman models, the relation for steady state cosmology (SS) as well as the relation $\theta \propto z^{-1}$ (dashed line) are shown.

cal Dialogues in Cosmology between David Burstein and Avishai Dekel, the discussion being moderated by Simon White – this debate can be recommended as a good representation of the multiply-connected thorny issues to be resolved. The results can be summarised as follows:

- The matter contained in the visible parts of galaxies can be found by evaluating average values for the mass-to-light ratios of different types of galaxy and then, knowing the average luminosity per unit volume due to galaxies, the average density of visible matter in the Universe can be found. A number of independent estimates agree that this amounts to $\Omega_{\mathrm{gal}} \approx 0.01$.

- To the visible matter we have to add the *dark* or *hidden matter* found in galaxies and in clusters of galaxies. This amounts typically to a factor of about $10 - 30$ and so, assuming that this figure is representative of all the visible mass in the Universe, the value of the density parameter for bound systems would increase to $\Omega_{\mathrm{bound}} \sim 0.1 - 0.3$.

- On larger scales, estimates of the mass density in the general field can be found from what is known as the *cosmic virial theorem*. In this procedure, the random velocities of galaxies with respect to the mean Hubble flow are compared with the varying component of the gravitational acceleration due to large scale inhomogeneities in the distribution of galaxies. As in the other methods of mass determination, the mass density is found by comparing the random kinetic energy of galaxies with their gravitational potential energy, this comparison being carried out in terms of two-point correlation functions for both the velocities and positions of galaxies selected from the general field. The cosmic virial has suggested values of the density parameter of about 0.2 to 0.3.

- A similar argument involves studies of the infall of galaxies into the local supercluster of galaxies. As illustrated in Fig. 2.7, the Local Supercluster is an extensive region, roughly centred upon the Virgo cluster of galaxies, in which the galaxy density exceeds that in the general field by a factor of about 2. Therefore, galaxies in the vicinity of the supercluster should feel a gravitational acceleration towards it, thus providing a measure of the mean density of gravitating matter within the system. This method has also resulted in values of Ω_0 about 0.2 to 0.3.

In the last three arguments, it is assumed that the dark matter is distributed like the visible matter in galaxies. There is a problem in that the velocities induced by large-scale density perturbations depend upon the *density contrast* $\Delta\varrho/\varrho$ between the system studied and the mean background density and *not* upon the absolute value of the density ϱ itself. A typical formula for the infall velocity u of test particles into a density perturbation is

$$u \propto H_0 r \Omega_0^{0.6} \left(\frac{\Delta\varrho}{\varrho}\right)_0 , \qquad (8.20)$$

(see also Sect. 11.5 and Gunn (1978)). Thus, if, in addition to the observed distribution of galaxies, there existed a uniform background distribution of dark matter, this would have the effect of decreasing the value of $\Delta\varrho/\varrho$. The result would be that, for a given observed velocity of infall, u, a larger value of Ω_0 would be inferred. Thus, one can contrive models in which these dynamical estimates would be consistent with $\Omega_0 = 1$. The agreement is, however, obtained at the expense of assuming that there is a difference in the spatial distribution of the visible and the dark matter, that is, there is *biasing*, a topic to which we will return later.

- An analysis which has consistently suggested that the value of Ω_0 is close to 1 is the POTENT programme, in which the distribution of mass in the local Universe is determined entirely from the measured radial velocities and distances of complete samples of nearby galaxies. In this procedure, the objective is to determine a three-dimensional map of the velocity deviations from the mean Hubble flow. Then, from the application of Poisson's equation, the mass distribution can be reconstructed numerically with the requirement that the mean radial velocity deviations from the mean Hubble flow match the observed values throughout the volume surveyed. Fig. 8.8 shows an example of a reconstruction of the local density distribution using the POTENT procedure (Dekel 1995). It can be observed that, using only the velocities and distances, and *not* their number densities, many of the familiar features of our local Universe are recovered – the Virgo supercluster and the 'Great Attractor' can be seen as well as voids in the mean mass distribution. Dekel has consistently found best estimates of $\Omega_0 \sim 1$ by this procedure. He states that the density parameter is greater than 0.3 at the 95% confidence level.
- Finally, as we will show in Sect. 10.4, the mean density in baryons is constrained by the production of the light elements in the early stages of the Big Bang. To anticipate the discussion of that Section, a conservative upper limit to the density parameter in baryons is $\Omega_{\text{bar}} \leq 0.0375h^{-2}$ or else less than the observed abundance of deuterium is created primordially. Therefore, even if we adopt a value of Hubble's constant corresponding to $h = 0.5$, $\Omega_{\text{bar}} \leq 0.15$. Thus, even adopting this very conservative position, there cannot be sufficient baryons to close the Universe. As explained in Sect. 10.4, a good case can be made for adopting a best estimate of the baryon density parameter of $\Omega_{\text{bar}} = 6 \times 10^{-3}h^{-2}$, in which case, the total baryon density in the Universe cannot be much more than that of the luminous matter.

The upshot of these considerations is that most cosmologists believe that the value of Ω_0 is greater than 0.1 and a value of 0.2 to 0.3 would probably be consistent with all the data. It is somewhat worrying that the POTENT procedures reproduce the local large-scale distribution of visible matter, since we have argued that the mass-to-luminosity ratio for visible matter is less than the value needed to close the Universe, even when account is taken of the dark matter present in these systems. I believe there may be some

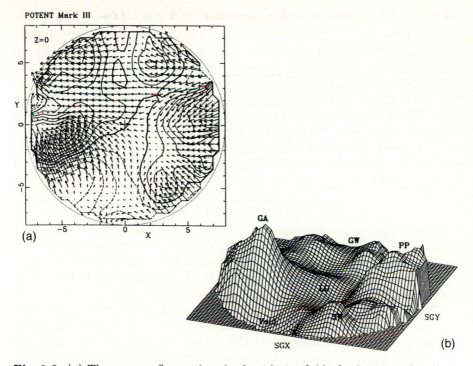

Fig. 8.8. (a) The average fluctuations in the velocity field of galaxies in the plane of the Local Supercluster with respect to the mean Hubble flow. Distances are measured in units of 1000 im s^{-1}. (b) The distribution of mass density in the plane of the Local Supercluster reconstructed by the POTENT procedure developed by Dekel, Bertschinger and their colleagues. Our Galaxy and the local group of galaxies lie at the centre of the diagram (LG). The large hump to the left of the diagram is the mass excess associated with the 'Great Attractor', which is associated with the Hydra–Centaurus supercluster. The fractional excess density (or density contrast) in the Great Attractor corresponds to $\delta\varrho/\varrho = 1.2 \pm 0.4$, if the density parameter $\Omega_0 = 1$. ϱ is the average density of matter throughout the whole region (Dekel 1995).

technical difference in the way in which the data are analysed which results in the discrepancies between the different methods of estimating Ω_0.

Values of $\Omega_0 \sim 0.1 - 0.3$ suggest that it is unlikely that most of the mass in the Universe is in baryonic form. On the other hand, there is also no evidence against the supposition that the density parameter Ω_0 is 1 and that most of the matter in the Universe is in some as yet undetermined non-baryonic form. In this case, the dark matter would not be distributed like the visible matter. In other words, if the Universe has $\Omega_0 = 1$, most of the dark matter has to be located where it can least readily be detected – in the space between clusters of galaxies.

8.6 The Cosmological Constant – Λ and Ω_Λ

It is perhaps surprising that, if the cosmological constant were to play an important rôle in cosmology, it has not made its presence felt already. For example, if it proved necessary to stretch out the cosmological time-scale by including the cosmological constant, the models should not be too far from the critical Eddington–Lemaître model which was stationary at some redshift z_c (see Sect. 7.3.2). If the Universe were stationary at redshift z_c, it is straightforward to show that there is a relation between the redshift of the stationary phase z_c and the density parameter Ω_0 of the Universe at the present epoch:

$$\Omega_0 = \frac{2}{(1+z_c)^3 - 3(1+z_c) + 2}. \tag{7.76}$$

Thus, if we adopt a reasonable lower limit of 0.1 for the value of Ω_0, $z_c \leq 2$. The important point is that the stationary phase could not have occurred at inaccessibly large redshifts, but rather at redshifts which are quite accessible by studies of luminous galaxies and quasars. It is remarkable that no trace of that quasi-stationary phase is present in the observations.

We described in Sect. 7.3.2 how limits to Ω_Λ can be found from the mere existence of galaxies and quasars at large redshifts and in Sect. 8.5.1 how important limits can already be found from the redshift–apparent magnitude relation for Type 1a supernovae. Another powerful method of setting limits to Ω_Λ comes from the numbers of gravitational lenses found in samples of distant quasars and radio sources. The power of this approach is illustrated by the following calculation (Fukugita *et al.* 1992). We assume, for simplicity, that we can represent the population of lensing galaxies by identical isothermal spheres which have constant comoving space density N_0. We can then use the calculation which resulted in (4.32) for the Einstein radius θ_E, within which strong distortions of the image of a background object are expected.

$$\theta_E = \frac{4\pi \langle v_\parallel^2 \rangle}{c^2} \frac{D_{ds}}{D_s}, \tag{4.44}$$

where $\langle v_\parallel^2 \rangle$ is the mean square velocity dispersion along the line of sight of the particles which make up each isothermal sphere, D_s is the angular diameter distance of the background quasar and D_{ds} the angular diameter distance from the deflector to the background source. We can therefore write the cross-section σ_E of the isothermal sphere for strong lensing as

$$\sigma_E = \pi D_d^2 \theta_E^2 = A \left(\frac{D_d D_{ds}}{D_s} \right)^2, \tag{8.21}$$

where D_d is the angular diameter distance of the isothermal sphere, or deflector. We now work out the probability that a background quasar at redshift z_s is observed to be strongly lensed. From (5.75), the number of isothermal spheres in the redshift interval z to $z + dz$ per steradian is

$$dN = N_0 D^2\, dr, \tag{8.22}$$

where D is the distance measure which is related to the angular diameter distance by $D_A = D/(1+z)$. Therefore, the probability of strong lensing in the redshift interval dz is given by the total solid angle subtended by all the isothermal spheres in the increment of comoving radial distance coordinate dr

$$p(z)\, dz = \frac{N_0 \sigma_E D^2}{D_A^2}\, dr = \frac{N_0 \sigma_E D^2}{[D/(1+z)]^2}\, dr = N_0 \sigma_E (1+z)^2\, dr. \tag{8.23}$$

We integrate this result from $z = 0$ to z_s to obtain the desired probability:

$$p(z_s) = AN_0 \int_0^{z_s} \left(\frac{D_d D_{ds}}{D_s}\right)^2 (1+z)^2\, dr. \tag{8.24}$$

In general, dr is given by the expression

$$dr = \frac{c\, dz}{H_0[(1+z)^2(\Omega_0 z + 1) - \Omega_\Lambda z(z+2)]^{1/2}}, \tag{8.25}$$

and so

$$p(z_s) = AN_0 \int_0^{z_s} \left(\frac{D_d D_{ds}}{D_s}\right)^2 \frac{c(1+z)^2\, dz}{H_0[(1+z)^2(\Omega_0 z + 1) - \Omega_\Lambda z(z+2)]^{1/2}}. \tag{8.26}$$

To integrate this expression we need to evaluate D_{ds}, the angular diameter distance from the deflector to the source (see also Sect. 7.2.9). The appropriate comoving distance coordinate between the epochs corresponding to z_d and z_s is

$$r(z_d, z_s) = -\int_{z_d}^{z_s} \frac{c\, dt}{R(t)}, \tag{8.27}$$

and then we form the distance measure $D(z_d, z_s)$

$$D(z_d, z_s) = \frac{\Re}{1+z_d} \sin \frac{r(z_d, z_s)}{\Re}, \tag{8.28}$$

recalling that the use of comoving coordinates takes account of the expansion of the system of coordinates and the radius of curvature of the geometry. The angular diameter distance is therefore

$$D_A(z_d, z_s) = D(z_d, z_s)\frac{1+z_d}{1+z_s} = \frac{\Re}{(1+z_s)} \sin \int_{z_d}^{z_s} \frac{dr}{\Re}. \tag{8.29}$$

Carroll *et al.* (1992) present the results of the integral (8.26) in a very pleasant format. They normalise the integral to the probability of lensing in the case of the Einstein–de Sitter model, $\Omega_0 = 1$, $\Omega_\Lambda = 0$ in which case, it is straightforward to show that the probability for any other model becomes

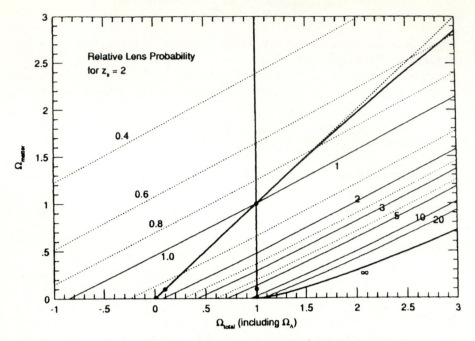

Fig. 8.9. The probability of observing strong gravitational lensing relative to that of the critical Einstein–de Sitter model, $\Omega_0 = 1, \Omega_\Lambda = 0$ for a quasar at redshift $z_s = 2$ (Carroll *et al* 1992). The contours show the relative probabilities derived from the integral (8.30) and are presented in the same format as in Fig. 7.7.

$$p(z_s) = \frac{15H_0^2}{4c^2} \left[1 - \frac{1}{(1 + z_s)} \right]^{-3} \int_0^{z_s} \left(\frac{D_d D_{ds}}{D_s} \right)^2$$

$$\times \frac{(1 + z)^2 \, dz}{[(1 + z)^2 (\Omega_0 z + 1) - \Omega_\Lambda z(z + 2)]^{1/2}}. \tag{8.30}$$

This function is plotted in Fig. 8.9 in the same format as Fig. 7.7, the contours showing the relative lensing probabilities for a quasar at a typical redshift $z_s = 2$. It can be seen that, if $\Omega_\Lambda = 0$, as represented by the solid diagonal line, there is little change in the probability of lensing. For low values of Ω_0, the probability increases by about a factor of 2. In contrast, the probability of lensing is very strongly dependent upon the value of Ω_Λ. For example, for the flat world models with $\Omega_0 + \Omega_\Lambda = 1$, the probability of lensing increases by almost a factor of ten as Ω_0 changes from 1 to 0.1. This dramatic increase occurs for a number of reasons. First of all, the distance measures D continue to increase with increasing redshift much more rapidly in models with low Ω_0 than in the model with $\Omega_\Lambda = 0$ (see Fig. 8.1) so that much greater volumes are encompassed at a given redshift. Second, the combination of parameters $D_d D_{ds}/D_s$ is very sensitive to the presence of the cosmological constant which stretches out the angular diameter distance for a given redshift.

To obtain limits to the value of Ω_Λ from the frequency and properties of gravitational lenses in complete samples of quasars and radio galaxies, modelling of the lens and background source populations needs to be carried out. The probabilities have to be averaged over the luminosity functions of lensing galaxies and the distribution of background sources. Models have to be adopted for the lenses which can account for the observed structures of the lensed images. The amplification of the brightness of the images as well as the detectability of the distorted structures need to be included in the computations. These complications are considered in some detail in the review by Carroll *et al.* (1992) and by Kochanek (1996).

A number of surveys have been carried out to search specifically for gravitationally lensed images among complete samples of quasars and distant radio sources. The quasar surveys include the HST Snapshot Survey (Moaz *et al.* 1993), the ESO/Liège survey (Surdej *et al.* 1993) and the Nordic Optical Telescope survey (Jaunsen *et al.* 1995). In Kochanek's analysis, he finds that combining these surveys and considering only quasars with redshifts $z > 1$, there are six gravitational lenses among 862 quasars. Similar analyses can be carried out for the large radio surveys carried out with the Very Large Array (see, for example, Burke *et al.* 1992), but the analysis of these data is more difficult because fewer redshifts are known. The data can be analysed in a number of ways, including estimating the absolute numbers of gravitational lenses, the statistics of the redshift distributions of the deflectors and sources and the separation of the gravitationally lensed images.

Although there are many complications in making reliable estimates, the best-fit cosmological models generally turn out to be those with $\Omega_\Lambda = 0$ and $\Omega_0 = 1$. More precisely, following an exhaustive analysis of the various sources of error and bias, Kochanek (1996) finds an upper limit of $\Omega_\Lambda = 0.65$ at the 95% confidence limit for flat cosmological models, that is, those in which $\Omega_0 + \Omega_\Lambda = 1$. In case of models with $\Omega_\Lambda = 0$, he finds $\Omega_0 \geq 0.15$ at the 90% confidence level.

Taken together with the preliminary results of the redshift–apparent magnitude relation for Type 1a supernovae, these results suggest that we cannot be living in a flat Universe which is currently dominated dynamically by the energy density of the vacuum fields. There is also a fine-tuning problem – if the cosmological constant were to play an important role, why did Ω_Λ turn out to have a value which is of the same order of magnitude as the density parameter? Perhaps, as has been suggested by Weinberg (1997), there is only an anthropic answer to this problem.

8.7 The Cosmic Time–scale T_0

Lower limits to the age of the Universe can be derived from nucleocosmochronology and from studies of the ages of the oldest globular clusters. A secure lower limit to the age of the Universe can be derived from the abundances of long-lived radioactive species. Anders (1963) used these to determine an accurate age for the Earth of 4.6×10^9 years. Some pairs of long-lived radioactive species, such as $^{232}\mathrm{Th}$–$^{238}\mathrm{U}$, $^{235}\mathrm{U}$–$^{238}\mathrm{U}$ and $^{187}\mathrm{Re}$–$^{187}\mathrm{Os}$ can provide information about nucleosynthetic timescales before the formation of the Solar System (Schramm and Wasserburg 1970). These pairs of elements are all produced by the r-process in which the time scale for neutron capture is less than the β-decay lifetime. The production abundances of these elements can be predicted and compared with their present observed ratios (see Cowan *et al.* 1991). A conservative lower bound to the cosmological timescale can be found by assuming that all the elements were formed promptly at the beginning of the Universe. From this line of reasoning, Schramm (1990) found the lower limit to the age of the Galaxy to be 9.6×10^9 years. The best estimates of the age of the Galaxy are somewhat model-dependent, but typically ages of about $(12 - 14) \times 10^9$ years are found (Cowan *et al.* 1991).

The oldest stars in our Galaxy are found in the globular clusters and ages can be found from the forms of their Hertzsprung–Russell diagrams. The feature of these diagrams which is particularly sensitive to the age of the cluster is the *main sequence termination point* which can be measured with some accuracy (Fig. 8.10). In the oldest globular clusters, the main sequence termination point has reached a mass of about $0.9\ M_\odot$ and in the most metal-poor, and presumably oldest, clusters the abundances of the elements with $Z \geq 3$ are about 150 times lower than their Solar System values. These facts make the determination of stellar ages much simpler than might be imagined. As Bolte (1997) points out, these stars have radiative cores and so are unaffected by the convective mixing of unprocessed material into the core. Secondly, the corrections to the perfect gas law equation of state are relatively small throughout most $1\ M_\odot$ stars. Thirdly, the surface temperatures of these stars are high enough that molecules are rare, simplifying the conversion of the effective temperature to a predicted colour. Examples of the fitting of isochrones to the Hertzsprung–Russell diagram of the globular cluster 47 Tucanae are shown in Fig. 8.10.

Maeder (1994) reported evidence that the ages of the oldest globular clusters are $\approx 16 \times 10^9$ years. Similar results were reported by Sandage at the 1993 Saas-Fee meeting. Bolte (1997) reviewed these data at the Princeton meeting on *Critical Dialogues in Cosmology* and concluded that the ages of the oldest globular clusters were

$$T_0 = 15 \pm 2.4 (\text{stat}) \, {}^{+4}_{-1} \, (\text{syst}) \ \text{Gy}. \tag{8.31}$$

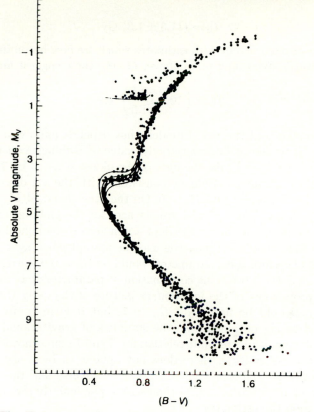

Fig. 8.10. The Hertzspring–Russell diagram for the globular cluster 47 Tucanae (Hesser *et al.* 1989). The scatter in the points increases at faint magnitudes because of the increase in the observational error associated with the photometry of faint stars. The solid lines show fits to the data using theoretical models of the evolution of stars of different masses from the main sequence to the giant branch due to VandenBerg. The isochrones shown have ages of 10, 12, 14 and 16 $\times 10^9$ years, the best fitting values lying in the range $(1.2 - 1.4) \times 10^{10}$ years. The cluster is metal-rich relative to other globular clusters, the metal abundance corresponding to about 20% of the solar value.

Since that meeting, the first results of the determination of the local distance scale from the Hipparcos survey have been announced (Feast and Catchpole 1997). The local distance scale has been increased by about 10%, which has two important effects. First of all, all extragalactic distances should become greater by this factor. Second, the stars in globular clusters become more luminous which has the effect of reducing their ages. Thus, this recalibration of the distance scale has two effects, both of which tend to improve the consistency of the ages of the oldest stars with the expansion age of the Universe. Chaboyer (1998) reviewed the recent age determinations of globular clusters in the light of the revision of the distance scale and derived a best-estimate of

$$T_0 = (11.5 \pm 1.3) \text{ Gy.} \qquad (8.32)$$

This value is also consistent with estimates which are now becoming available for white dwarfs. According to Chaboyer (1998), these suggest an age of

$$T_0 = \left(10.5 {+2.5 \atop -1.5}\right) \text{ Gy.} \qquad (8.33)$$

The consistency of the world models thus depends upon one's preferred value of H_0. If we adopt our compromise value of Hubble's constant, $H_0 = 62.5$ km s^{-1} Mpc^{-1} and take the ages of the oldest stars to be $T_0 = 11.5 \times 10^9$ years, these numbers would be consistent with the set of cosmological parameters $\Omega_0 = 2q_0 = 1$ and $\Omega_\Lambda = 0$. On the other hand, if H_0 were taken to be, say, 80 km s^{-1} Mpc^{-1} and greater ages for the globular cluster stars adopted, it would be necessary to adopt a non-zero cosmological constant Λ to reconcile the expansion time-scale and the astrophysical ages. A possible combination of cosmological parameters would be $\Omega_\Lambda = 0.9$, corresponding to $\Lambda = 2.7H_0^2$, and $\Omega_0 = 0.1$. This combination of parameters has the desirable feature for proponents of the inflationary picture of the early Universe that the geometry would be flat, $\Omega_0 + \Omega_\Lambda = 1$, but it is in conflict with the upper bound to Ω_Λ obtained from the statistics of gravitational lenses and the apparent magnitude–redshift relation for Type 1a supernovae.

At the time of writing, there does not appear to be a significant discrepancy between expansion time-scale of the Universe and the ages of the oldest stars, or nucleocosmological time-scales, particularly if realistic errors are placed upon the estimates.

9 The Thermal History of the Universe

We now have to set the scene for our attack on the major problems of astrophysical cosmology – how do we account for the origin and evolution of the galaxies and the large-scale structure of our Universe? This is one of the most exciting areas of modern cosmology. The first thing we have to do is to work out the thermal history of the matter and radiation content of the standard Big Bang picture. In this and the succeeding Chap. 10, we will develop many concepts which are essential for understanding the problems of galaxy formation. Let us first consider the dynamics of radiation-dominated universes.

9.1 Radiation-Dominated Universes

For a gas of photons, massless particles or a relativistic gas in the ultrarelativistic limit $E \gg mc^2$, pressure p is related to energy density ε by $p = \frac{1}{3}\varepsilon$ and the inertial mass density of the radiation ϱ_r is related to its energy density ε by $\varepsilon = \varrho_r c^2$. If $N(\nu)$ is the number density of photons of energy $h\nu$, then the energy density of radiation is found by summing over all frequencies

$$\varepsilon = \sum_\nu h\nu N(\nu). \tag{9.1}$$

If the number of photons is conserved, their number density varies as $N = N_0 R^{-3} = N_0 (1+z)^3$ and the energy of each photon changes with redshift by the usual redshift factor $\nu = \nu_0(1+z)$. Therefore, the variation of the energy density of radiation with cosmic epoch is

$$\varepsilon = \sum_{\nu_0} h\nu_0 N_0(\nu_0)(1+z)^4; \tag{9.2}$$

$$\varepsilon = \varepsilon_0(1+z)^4 = \varepsilon_0 R^{-4}. \tag{9.3}$$

In the case of black-body radiation, the energy density of the radiation is given by the Stefan–Boltzmann law $\varepsilon = aT^4$ and its spectral energy density, that is, its energy density per unit frequency range, by the Planck distribution

$$\varepsilon(\nu)\,d\nu = \frac{8\pi h\nu^3}{c^3}\frac{1}{e^{h\nu/k_B T}-1}\,d\nu. \tag{9.4}$$

It immediately follows that, for black-body radiation, the radiation temperature T_r varies with redshift as $T_r = T_0(1+z)$ and the spectrum of the radiation changes as

$$\varepsilon(\nu_1)\,d\nu_1 = \frac{8\pi h\nu_1^3}{c^3}[(e^{h\nu_1/k_B T_1}-1)]^{-1}\,d\nu_1$$

$$= \frac{8\pi h\nu_0^3}{c^3}[e^{h\nu_0/k_B T_0}-1)^{-1}](1+z)^4\,d\nu_0$$

$$= (1+z)^4\,\varepsilon(\nu_0)\,d\nu_0. \tag{9.5}$$

Thus, a black-body spectrum preserves its form as the Universe expands but the radiation temperature changes as $T_r = T_0(1+z)$ and the frequency of each photon as $\nu = \nu_0(1+z)$. Another way of looking at these results is in terms of the adiabatic expansion of a gas of photons. The ratio of specific heats γ for radiation and a relativistic gas in the ultrarelativistic limit is $\gamma = 4/3$. It is a simple exercise to show that, in an adiabatic expansion, $T \propto V^{-(\gamma-1)} = V^{-1/3} \propto R^{-1}$, which is exactly the same as the above result.

A key test of the standard Big Bang picture is whether or not the temperature of the Cosmic Microwave Background Radiation has followed this predicted relation $T_r = T_0(1+z)$. The fine-structure splitting of the ground state of neutral carbon atoms CI enables this test to be carried out. The photons of the background radiation excite the fine-structure levels of the ground state of the neutral carbon atoms and the relative strengths of the absorption lines originating from the ground and first excited states are determined by the energy density and temperature of the background radiation. The test has been carried out using the absorption lines observed in damped Lyman-α clouds observed in the spectra of quasars (see Sect. 18.3). This is a difficult experiment since the lines are weak, but there have now been two successful observations made with the Keck 10-metre Telescope. Cowie *et al.* (1994) observed the CI absorption lines in a damped Lyman-α cloud at redshift $z = 1.776$ in the spectrum of the quasar Q1331+170 and derived a background temperature $T_{rad} = 7.4 \pm 0.8\,\mathrm{K}$, consistent with expected temperature, $T(z) = T_0(1+z) = 7.58$ K. More recently, the experiment has been repeated by Ge *et al.* (1997) who studied the CI absorption lines in a damped Lyman-α cloud at redshift $z = 1.9731$ in the spectrum of the quasar QSO 0013-004 and found $T_{rad} = 7.9 \pm 1.0\,\mathrm{K}$, consistent with the predicted temperature of $T(z) = T_0(1+z) = 8.105$ K. Strictly speaking, these are only upper limits to the temperature of the background radiation, since there might be other local sources of excitation of the fine-structure lines. This is regarded as unlikely and the satisfactory agreement of the two independent experiments suggests that a genuinely background radiation field has been measured.

The variations of p and ϱ with R can now be substituted into Einstein's field equations:

$$\ddot{R} = -\frac{4\pi G R}{3}\left(\varrho + \frac{3p}{c^2}\right) + \left[\tfrac{1}{3}\Lambda R\right]; \tag{7.1}$$

$$\dot{R}^2 = \frac{8\pi G \varrho}{3}R^2 - \frac{c^2}{\Re^2} + \left[\tfrac{1}{3}\Lambda R^2\right]. \tag{7.2}$$

Therefore, setting the cosmological constant $\Lambda = 0$, we find

$$\ddot{R} = \frac{8\pi G \varepsilon_0}{3c^2}\frac{1}{R^3} \qquad \dot{R}^2 = \frac{8\pi G \varepsilon_0}{3c^2}\frac{1}{R^2} - \frac{c^2}{\Re^2}. \tag{9.6}$$

At early epochs we can neglect the constant term $c^2/\Re^2$ and integrating

$$R = \left(\frac{32\pi G \varepsilon_0}{3c^2}\right)^{1/4} t^{1/2} \quad \text{or} \quad \varepsilon = \varepsilon_0 R^{-4} = \left(\frac{3c^2}{32\pi G}\right) t^{-2}. \tag{9.7}$$

The dynamics of the radiation-dominated models, $R \propto t^{1/2}$, depend only upon the *total inertial mass density in relativistic and massless forms*. Thus, to determine the dynamics of the early Universe, we have to include all the massless and relativistic components in the total energy density. The force of gravity acting upon the sum of these determines the rate of deceleration of the early Universe.

9.2 The Matter and Radiation Content of the Universe

A schematic representation of the intensity I_ν of the extragalactic background radiation from radio to γ-ray wavelengths is shown in Fig. 9.1a (Longair and Sunyaev 1971). Although this is rather old picture, to which I am rather attached, it still gives a good representation of the intensity of the background radiation in those regions in which it has been measured (solid lines) and those regions in which only estimates of the extragalactic background have been made (dashed lines). A more recent representation plotted in terms of νI_ν is shown in Fig. 9.1b. Since νI_ν provides an estimate of the energy in the background radiation at each frequency, it is more useful for our present purpose, which is to find out the main contributors to the extragalactic background radiation.

I have discussed measurements of the background radiation in all wavebands recently (Longair 1995) and I refer the interested reader to that volume. Suffice to say that, from the far-infrared to the soft X-ray background, only upper limits to the background intensity are available. In the radio, centimetre and millimetre wavebands and in the X- and γ-ray wavebands, diffuse background radiation of cosmological origin has been detected. The upper limits to the background radiation shown in Fig. 9.1b are somewhat conservative, generally corresponding to the minimum sky brightnesses observed in

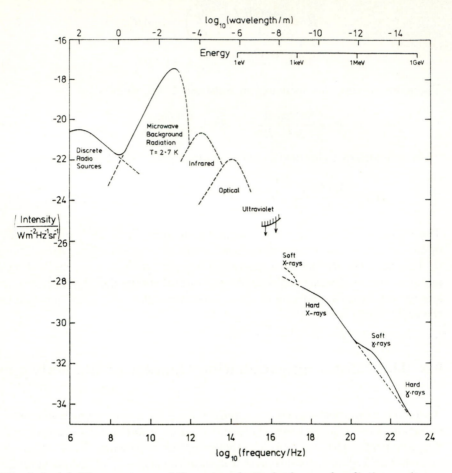

Fig. 9.1. (a) The spectrum of the extragalactic background radiation as it was known in 1969 (Longair and Sunyaev 1971). This figure still provides a good indication of the overall spectral energy distribution of the background radiation. The solid lines indicate regions of the spectrum in which extragalactic background radiation has been detected. The dashed lines were theoretical estimates of the background intensity due to discrete sources and should not be taken too seriously. **(b)** The spectrum of the extragalactic background radiation plotted as $I = \nu I_\nu = \lambda I_\lambda$ (Longair 1995, courtesy of Dr. Andrew Blain). This presentation shows the amount of energy $\varepsilon = 4\pi I/c$ present per unit volume throughout the Universe at the present epoch. The solid lines in the radio, millimetre, X- and γ-ray wavebands show the observed background intensities. The circles, crosses and squares correspond to upper limits to the background intensity in the far infrared to ultraviolet wavebands and are usually conservative upper limits. We will return to discuss the significance of these limits in Sect. 18.2

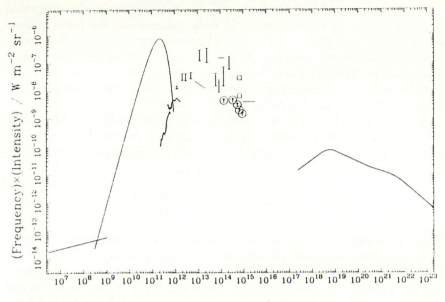

Frequency / Hz

Fig. 9.1. continued

Table 9.1. The energy densities and photon number densities of the extragalactic background radiation in different regions of the electromagnetic spectrum. Note that these are usually rough estimates which are only useful for making order of magnitude calculations.

Waveband	Energy density of radiation eV m^{-3}	Number density of photons m^{-3}
Radio	$\sim 5 \times 10^{-2}$	$\sim 10^6$
Microwave	3×10^5	10^8
Infrared	?	?
Optical	$\sim 2 \times 10^3$	$\sim 10^3$
Ultraviolet	?	?
X-ray	75	3×10^{-3}
γ-ray	25	3×10^{-6}

various sky surveys. In Table 9.1, typical energy densities and number densities of the photons in each of the wavebands in which a positive detection of the extragalactic background radiation has been made are listed. It must be emphasised that these are rough estimates and, for precise work, integrations should be taken over the appropriate regions of the spectrum. These figures are, however, often useful for making order of magnitude estimates.

In the radio waveband, the background radiation is due to the integrated emission of discrete radio sources, the radio galaxies and the radio quasars and it is likely that the X- and γ-ray backgrounds are also due to discrete X- and γ-ray sources (see Longair (1995) for a detailed discussion). The Cosmic Microwave Background Radiation, however, cannot be associated with discrete sources and is convincingly identified with the cooled remnant of the hot early phases of the Big Bang, as we will show.

It can be seen from Fig. 9.1b and Table 9.1 that the Cosmic Microwave Background Radiation provides by far the greatest contribution to the energy density of radiation in intergalactic space. Comparing the inertial mass density in the radiation and the matter, we find

$$\frac{\varrho_r}{\varrho_m} = \frac{aT^4(z)}{\Omega_0\varrho_c(1+z)^3c^2} = \frac{2.48 \times 10^{-5}(1+z)}{\Omega_0 h^2}. \tag{9.8}$$

Thus, at redshifts $z \geq 4 \times 10^4 \Omega_0 h^2$, the Universe was certainly *radiation-dominated*, even before we take account of the contribution of the three types of neutrino to the inertial mass density during the radiation-dominated phase, and the dynamics are described by the relation, $R \propto t^{1/2}$. According to this analysis, the Universe is *matter-dominated* at redshifts $z \leq 4 \times 10^4 \Omega_0 h^2$ and the dynamics are described by the standard Friedman models, $R \propto t^{2/3}$ provided $\Omega_0 z \gg 1$. We will provide a more precise estimate of the epoch at which the inertial mass densities in the massless particles and non-relativistic matter were equal in Sect. 10.5.

The present *photon-to-baryon ratio* is another key cosmological parameter. Assuming $T = 2.728$ K,

$$\frac{N_\gamma}{N_B} = \frac{3.6 \times 10^7}{\Omega_B h^2}. \tag{9.9}$$

If photons are neither created nor destroyed during the expansion of the Universe, this number is an invariant. This ratio is a measure of the factor by which the photons outnumber the baryons in the Universe at the present epoch and is also proportional to the specific entropy per baryon during the radiation-dominated phases of the expansion (see Sect. 10.5). The resulting thermal history of the Universe is shown in Fig. 9.2. Certain epochs are of special significance in the temperature history of the Universe and we now deal with some of these in more detail.

9.3 The Epoch of Recombination

At a redshift $z \approx 1500$, the radiation temperature of the Cosmic Microwave Background Radiation was $T \approx 4,000$ K and then there were sufficient photons with energies $h\nu \geq 13.6$ eV in the tail of the Planck distribution to ionise most of the neutral hydrogen present in the intergalactic medium. It

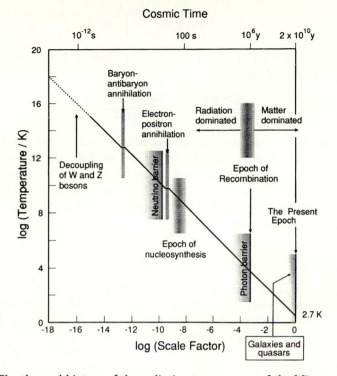

Fig. 9.2. The thermal history of the radiation temperature of the Microwave Background Radiation according to the standard Big Bang picture. The radiation temperature decreases as $T_r \propto R^{-1}$ except for abrupt jumps as different particle–antiparticle pairs annihilate at $k_B T \approx mc^2$. Various important epochs in the standard model are indicated. An approximate time scale is indicated along the top of the diagram. The neutrino and photon barriers are indicated. In the standard model, the Universe is optically thick to neutrinos and photons prior to these epochs.

may at first appear strange that the temperature is not closer to 150,000 K, the temperature at which $\langle h\nu \rangle = k_B T = 13.6$ eV, the ionisation potential of neutral hydrogen. The important points to note are that the photons far outnumber the baryons in the intergalactic medium and there is a broad range of photon energies present in the Planck distribution.

It is a useful calculation to work out the fraction of photons in the high frequency tail of the Planck distribution, that is, in the *Wien region* of the spectrum, with energies $h\nu \geq E$ in the limit $h\nu \gg k_B T$.

$$n(\geq E) = \int_{E/h}^{\infty} \frac{8\pi\nu^2}{c^3} \frac{\mathrm{d}\nu}{\mathrm{e}^{h\nu/k_B T}} = \frac{1}{\pi^2} \left(\frac{2\pi k_B T}{hc} \right)^3 \mathrm{e}^{-x}(x^2 + 2x + 2), \quad (9.10)$$

where $x = h\nu/k_B T$. Now, the total number density of photons in a black-body spectrum at temperature T is

$$N = 0.244 \left(\frac{2\pi k_B T}{hc} \right)^3 \ \mathrm{m}^{-3}. \qquad (9.11)$$

Therefore, the fraction of the photons of the black-body spectrum with energies greater than E is

$$\frac{n(\geq E)}{n_{\mathrm{ph}}} = \frac{e^{-x}(x^2 + 2x + 2)}{0.244\pi^2}. \qquad (9.12)$$

Roughly speaking, the intergalactic gas will be ionised, provided there are as many ionising photons with $h\nu \geq 13.6$ eV as there are hydrogen atoms, that is, we need only one photon in $3.6 \times 10^7 / \Omega_0 h^2$ of the photons of the Cosmic Microwave Background Radiation to have energy greater than 13.6 eV to ionise the gas. For illustrative purposes, let us take the ratio to be one part in 10^9. Then, we need to solve

$$\frac{1}{10^9} = \frac{e^{-x}(x^2 + 2x + 2)}{0.244\pi^2}. \qquad (9.13)$$

We find $x = E/k_B T \approx 26.5$. This is a very important result. There are so many photons relative to hydrogen atoms that the temperature of the radiation can be 26.5 times less than that found from setting $E = k_B T$ and there are still sufficient photons with energy $E \geq 13.6$ eV to ionise the gas. Therefore, the intergalactic gas was largely ionised at a temperature of $150,000/26.5$ K $\approx 5,600$ K. Since the present temperature of the Cosmic Microwave Background Radiation is 2.728 K, this means that the Universe became ionised at a scale factor $R = 2.728/5,600 = 5 \times 10^{-4}$ or a redshift $z \approx 2,000$. Note that this type of calculation appears in a number of different guises in astrophysics – the nuclear reactions which power the Sun take place at a much lower temperature than expected, the temperature at which regions of ionised hydrogen become fully ionised is only about 10,000 K, light nuclei are destroyed in the early Universe at much lower temperatures than would be expected. In all these cases, the tails of the Planck and Maxwell distributions contain large numbers of photons and particles respectively with energies very much greater than the mean.

Detailed calculations show that the pregalactic gas was 50% ionised at a redshift $z_r \approx 1,500$ and this epoch is referred to as the *epoch of recombination*, since the pregalactic gas was ionised prior to this epoch and, when we run the clocks forward, the universal plasma recombined at this time. At earlier epochs, $z \approx 6000$, helium was 50% ionised and rapidly became fully ionised before this time. The most important consequence is that, at redshifts greater than about 1000, the Universe became opaque to *Thomson scattering*. This is the simplest of the scattering processes which impede the propagation of photons from their sources to the Earth through an ionised plasma (see for example, Longair 1992). The photons are scattered without any loss of energy by free electrons. The optical depth of the intergalactic gas to Thomson scattering can be written

$$\mathrm{d}\tau_T = \sigma_T N_e(z)\, \mathrm{d}x = \sigma_T N_e(z)\frac{\mathrm{d}r}{(1+z)} = \sigma_T N_e(z)c\,\frac{\mathrm{d}t}{\mathrm{d}z}\, \mathrm{d}z, \qquad (9.14)$$

where σ_T is the Thomson scattering cross-section $\sigma_T = 6.653 \times 10^{-29}$ m^2. Notice that $\mathrm{d}x = \mathrm{d}r/(1+z)$ is an increment of proper distance at redshift z. Let us evaluate this integral in the limit of large redshifts, assuming that the Universe was matter-dominated at the epoch of recombination. Then, from (7.30a), the cosmic time–redshift relation can be written $\mathrm{d}z/\mathrm{d}t = -H_0\Omega_0^{1/2}z^{5/2}$. It is important to distinguish between the total mass density ϱ_0 and the mass density in baryons ϱ_B. Assuming that 25% of the primordial material is helium (see Sect. 10.4), we find that $N_H = (3/4)\varrho_B/m_p$ and so we can write the density parameter in baryons $\Omega_B = 8\pi G\varrho_B/3H_0^2 = 32\pi G m_p N_H/9H_0^2$. Then, if $x(z)$ is the fractional ionisation of hydrogen, the number density of electrons is $N_H x(z)(1+z)^3$ and so the optical depth for Thomson scattering in the limit $z \gg 1$ is

$$\tau_T = \frac{9\sigma_T H_0 c}{32\pi G m_p}\frac{\Omega_B}{\Omega_0^{1/2}}\int \frac{z^3 x(z)}{z^{5/2}}\, \mathrm{d}z = 0.052\frac{\Omega_B}{\Omega_0^{1/2}}h\int x(z)\, z^{1/2}\, \mathrm{d}z. \qquad (9.15)$$

It can be seen that, as soon as the pregalactic hydrogen was fully ionised at $z \approx 1500$, the optical depth to Thomson scattering became very large. For example, if we assume the intergalactic gas was more or less fully ionised at $z > 1000$, the optical depth at larger redshifts was

$$\tau = 0.035\frac{\Omega_B}{\Omega_0^{1/2}}hz^{3/2}. \qquad (9.16)$$

For reasonable values of Ω_B, Ω_0 and h, $\tau_T \gg 1$. Detailed calculations of the ionisation state of the intergalactic gas with redshift are discussed in Sect. 15.1 where it is shown that the optical depth of the intergalactic gas became unity at a redshift very close to 1000. Therefore, the Universe beyond a redshift of 1000 is unobservable. Any photons originating from larger redshifts were scattered many times before they propagated to the Earth and consequently all the information they carry about their origin is rapidly lost. Exactly the same process prevents us observing inside the Sun. There is therefore a *photon barrier* at a redshift of 1000 beyond which we cannot obtain information directly using photons.

We will return to the process of recombination and the variation of the optical depth to Thomson scattering in Sect. 15.1 because this is a crucial topic in evaluating the observability of fluctuations in the Cosmic Microwave Background Radiation. If there is no further scattering of the photons of the background radiation, the redshift of 1000 becomes the *last scattering surface* and therefore the fluctuations imprinted on the radiation at this epoch determine the spatial fluctuations in the intensity of the background radiation observed today.

9.4 The Radiation-Dominated Era

At redshifts $z \gg 4 \times 10^4 \Omega_0 h^2$, the Universe was radiation-dominated. If the matter and radiation were not thermally coupled, they would cool independently, the hot gas having ratio of specific heats $\gamma = 5/3$ and the radiation $\gamma = 4/3$. These result in adiabatic cooling which depends upon the scale factor R as $T_B \propto R^{-2}$ and $T_r \propto R^{-1}$ for the diffuse baryonic matter and radiation respectively. We therefore expect the matter to cool more rapidly than the radiation and this is indeed what is expected to take place during the post-recombination era. This is not the case, however, during the pre-recombination and immediate post-recombination eras because the matter and radiation are strongly coupled by *Compton scattering*. As shown in the last section, the optical depth of the pre-recombination plasma for Thomson scattering is very large, so large that we can no longer ignore the small energy transfers which take place between the photons and the electrons in Compton collisions. It turns out that these Compton scatterings are sufficient to maintain the matter at the same temperature as the radiation.

The exchange of energy between photons and electrons is an enormous subject and has been treated by Weymann (1965), Sunyaev and Zeldovich (1980) and Pozdnyakov *et al.* (1983). The equation for the rate of exchange of energy between a thermal radiation field at radiation temperature T_r and a plasma with electron temperature T_e interacting solely by Compton scattering was derived by Weymann in 1965.

$$\frac{d\varepsilon_r}{dt} = -\frac{d\varepsilon_m}{dt} = 4N_e \sigma_T c \varepsilon_r \left(\frac{k_B T_e - k_B T_r}{m_e c^2} \right), \qquad (9.17)$$

where ε_r and ε_m are the energy densities of radiation and matter. We can understand the form of this equation by considering the case in which the temperature of the electrons is greater than that of the radiation. The number of collisions per electron per second with the photon field is $N_\gamma \sigma_T c$, where N_γ is the number density of photons. In each collision, the average energy transfer to the photon field is $(4/3)(v^2/c^2)h\bar{\nu}$, where $h\bar{\nu}$ is the mean energy of the photons (see, for example, Longair 1992). Since the average energy of the electrons is $\frac{1}{2}m_e v^2 = \frac{3}{2}k_B T_e$, the rate of loss of energy per electron is

$$-\frac{d\varepsilon_m}{dt} = 4\sigma_T c N_\gamma h\bar{\nu} \left(\frac{k_B T_e}{m_e c^2} \right) = 4\sigma_T c \varepsilon_r \left(\frac{k_B T_e}{m_e c^2} \right). \qquad (9.18)$$

Equation (9.17) expresses the fact that, if the electrons are hotter than the radiation, the radiation is heated up by the matter and, contrariwise, if the radiation is hotter than the matter, the matter is heated by the radiation. The astrophysical difference between the two cases arises from the enormous difference in the number densities of the photons and electrons $N_\gamma/N_e = 3.6 \times 10^7 (\Omega_B h^2)^{-1}$. Let us look at this difference from the point of view of the optical depths for the interaction of an electron with the radiation field and of a photon with the electrons of the intergalactic gas. In the first case, the optical depth for interaction of an electron with the radiation field is $\tau_e = \sigma_T c N_\gamma t$, whereas that of the photon with the electrons is $\tau_\gamma = \sigma_T c N_e t$ where σ_T is the Thomson cross-section and t is the age of the Universe. Thus, because $N_\gamma \gg N_e$, it is much more difficult to modify the spectrum of the photons as compared with the energy distribution of the electrons because, in the time any one photon is scattered by an electron, the electron has been scattered $3.6 \times 10^7 (\Omega_B h^2)^{-1}$ times by the photons. Another way of expressing this is to say that the heat capacity of the radiation is very much greater than that of the matter.

We consider two important applications of these results. In the first, we consider the heating of the electrons by Compton scattering of the photons of the Microwave Background Radiation. The collision times between electrons, protons and atoms are always much shorter than the age of the Universe and hence, when energy is transferred from the radiation field to the electrons, it is rapidly communicated to the matter as a whole. This is the process by which the matter and radiation are maintained at the same temperature in the early Universe. Let us work out the redshift to which Compton scattering can maintain the matter and radiation at the same temperature.

We rewrite (9.17) for the case in which the plasma is heated by the radiation field. The thermal energy density of the plasma is $\varepsilon_m = 3 N_e k_B T_e$, since both the electrons and protons are maintained at the same temperature, and then

$$\frac{dT_e}{dt} = \tfrac{4}{3}\sigma_T \varepsilon_r \left(\frac{T_r - T_e}{m_e c} \right). \tag{9.19}$$

As pointed out above, because of the enormous heat capacity of the radiation, T_r scarcely changes at all and so (9.19) defines the characteristic exponential time-scale τ for the exchange of energy between the radiation and the plasma. Assuming $z \gg 1$,

$$\tau = \frac{3 m_e c}{4 \sigma_T a T_r^4} = \frac{3 m_e c}{4 \sigma_T a T_0^4} (1+z)^{-4} = 7.4 \times 10^{19} z^{-4} \text{ s}. \tag{9.20}$$

Thus, when the plasma was fully ionised at $z \gg 1000$, the time scale τ was 7.4×10^7 s $= 2.7$ years, that is, very much shorter than the age of the Universe and so the matter and radiation were maintained at the same temperature throughout the radiation-dominated era.

As the temperature fell below 4,000 K, most of the protons recombined with electrons to form neutral hydrogen, but there remained a small, but finite, fraction of free electrons which did not recombine, $x \approx 2.5 \times 10^{-5}$, at redshifts $z < 700$. These enabled energy to be transferred from the photons to the gas even in the post-recombination era. The energy density of the gas was predominantly associated with the kinetic energy of hydrogen atoms, $\varepsilon_H = \tfrac{3}{2} N_H k_B T$, and the number density of free electrons was $x(z) N_H$. Repeating the above calculation for the post-recombination era, the rate of change of the temperature of the neutral hydrogen is given by

$$\frac{dT_H}{dt} = \frac{8}{3}\sigma_T \varepsilon_r x(z) \left(\frac{T_r - T_H}{m_e c}\right), \tag{9.21}$$

and the characteristic time-scale τ for the exchange of energy was

$$\tau = \frac{3m_e c}{8\sigma_T x(z) a T_r^4} = \frac{3m_e c}{8\sigma_T x(z) a T_0^4}(1+z)^{-4} = 1.47 \times 10^{24} z^{-4} \text{ s}. \tag{9.22}$$

During the matter-dominated epochs, when $\Omega_0 z \gg 1$, the age of the Universe was

$$t = \frac{2.06 \times 10^{17}}{h\Omega_0^{1/2}} z^{-3/2} \text{ s}. \tag{9.23}$$

Therefore, equating (9.22) and (9.23), the time-scale τ for the exchange of energy from the background radiation to the neutral hydrogen was equal to the age of the Universe at a redshift

$$z = 550 \, h^{2/5} \Omega_0^{1/5}. \tag{9.24}$$

Thus, there was a period after the epoch of recombination when the matter and radiation cooled at the same rate but, at redshifts less than $550 \, h^{2/5}\Omega_0^{1/5}$, the matter and the radiation cooled independently, the matter temperature as $T_H \propto R^{-2}$ and the radiation temperature as $T_r \propto R^{-1}$.

In the second application, we derive the necessary condition for significant distortions of the spectrum of the Microwave Background Radiation to take place. Suppose the electrons are heated to a temperature greater than the radiation temperature by some process. This might be, for example, the dissipation of primordial sound waves or turbulence, matter–antimatter annihilation, the evaporation of primordial black holes by the Hawking mechanism or the decay of heavy unstable leptons. If no photons were created, the spectrum of the radiation would be distorted from its black-body form by Compton scattering. The interaction of the hot electrons with the photons results in an average frequency change of $\Delta\nu/\nu = 4k_B T_e/m_e c^2$. Thus, to obtain a significant change in the energy of the photon, $\Delta\nu/\nu \approx 1$, not only must the optical depth for Thomson scattering be very much greater than one, there must also be sufficient Compton scatterings to change the energies of the photons significantly so that $\Delta\nu/\nu \approx 1$. Evidently, the *Compton optical depth*

$$\tau_C = \int \left(\frac{k_B T_e}{m_e c^2}\right) \sigma_T c N_e dt, \tag{9.25}$$

should be greater or equal to one. We know that the Thomson scattering condition was satisfied during the pre-recombination era and also that the temperature of the electrons was maintained at that of the thermal background radiation. Therefore, assuming $T_e = T_0(1 + z)$ and $N_e = N_0(1 + z)^3$, the Compton optical depth is

$$\tau_C = \left(\frac{k_B T_0}{m_e c^2}\right) \frac{\sigma_T c N_0}{H_0} \int \frac{(1+z)^2}{(\Omega_0 z + 1)^{1/2}} \, dz. \tag{9.26}$$

For $z \gg 1$,

$$\tau_C = \left(\frac{k_B T_0}{m_e c^2}\right) \frac{2\sigma_T c N_0}{5\Omega_0^{1/2} H_0} z^{5/2}. \tag{9.27}$$

It is convenient to express this result in terms of the density parameter in baryons at the present epoch Ω_B. Assuming that 25% of the baryonic mass is in the form of helium and that the plasma is fully ionised in the relevant range of redshifts, we find that $\Omega_B = (64\pi G m_p / 21 H_0^2) N_e$ and so

$$\tau_C = \left(\frac{k_B T_0}{m_e c}\right) \frac{21\sigma_T}{160\pi G m_p} \frac{H_0 \Omega_B}{\Omega_0^{1/2}} z^{5/2} = 1.1 \times 10^{-11} \frac{h\Omega_B}{\Omega_0^{1/2}} z^{5/2}. \tag{9.28}$$

Thus, the optical depth was unity at redshift

$$z = 2.4 \times 10^4 \left(\frac{\Omega_0}{h^2 \Omega_B^2}\right)^{1/5}. \tag{9.29}$$

To observe a distortion of the black-body spectrum, the temperature of the electrons must be raised significantly above the temperature $T_e = T_0(1+z)$. If this were to occur, the spectrum would relax from a Planck to a Bose–Einstein spectrum with a finite dimensionless chemical potential μ.

$$I_\nu = \frac{2h\nu^3}{c^2} \left[\exp\left(\frac{h\nu}{k_B T_r} + \mu\right) - 1\right]^{-1}. \tag{2.1}$$

This is the form of equilibrium spectrum expected when there is a mismatch between the number of photons and the energy to be distributed among them to create a Planck spectrum. In the present instance, the photon energies are redistributed by Compton scattering. At early enough epochs, $z \sim 10^7$, Compton double scattering can create additional low energy photons and so, at these very large redshifts, the black-body spectrum is re-established. As discussed in Sect. 2.1.1, there are now very good upper limits to the value of μ from the COBE spectral observations of the Cosmic Microwave Background Radiation, $|\mu| \leq 10^{-4}$. In general terms, this means that there cannot have been major injections of energy into the intergalactic gas in the redshift interval $10^7 \geq z \geq 2 \times 10^4$. Useful limits to the significance of the processes listed above can be found from these observations (see Sunyaev and Zeldovich 1980). These authors have also surveyed the types of distortion which would result from large injections of thermal energy into the intergalactic gas during the post recombination era.

9.5 The Speed of Sound as a Function of Cosmic Epoch

One crucial result for understanding the physics of the formation of structure in the Universe is the variation of the speed of sound with cosmic epoch, particularly through the epochs when Universe changed from being radiation- to matter-dominated. All sound speeds are proportional to the square root of the ratio of the pressure which provides the restoring force to the inertial mass density of the medium. More precisely, the speed of sound c_s is given by

$$c_s^2 = \left(\frac{\partial p}{\partial \varrho}\right)_S, \tag{9.30}$$

where the subscript S means 'at constant entropy', that is, we consider adiabatic sound waves. The complication is that, from the epoch when the energy densities of matter and radiation were equal to beyond the epoch of recombination the dominant contributors to p and ϱ change dramatically as the Universe changes from being radiation- to matter-dominated. The key point is that the matter and radiation are closely coupled throughout the pre-recombination era and the square of the sound speed can then be written

$$c_s^2 = \frac{(\partial p/\partial T)_r}{(\partial \varrho/\partial T)_r + (\partial \varrho/\partial T)_m}, \tag{9.31}$$

where the partial derivatives are taken at constant entropy. It is straightforward to show that this reduces to the following expression:

$$c_s^2 = \frac{c^2}{3} \frac{4\varrho_r}{4\varrho_r + 3\varrho_m}. \tag{9.32}$$

Thus, in the radiation-dominated era, $z \gg 4 \times 10^4 \Omega_0 h^2$, $\varrho_r \gg \varrho_m$ and the speed of sound tends to the relativistic sound speed, $c_s = c/\sqrt{3}$. At smaller redshifts, the sound speed decreases as the contribution of the inertial mass density of the matter becomes more important. Specifically, between the epoch of equality of the matter and radiation energy densities and the epoch of the recombination, the pressure of sound waves is provided by the radiation, but the inertia is provided by the matter. Thus, the speed of sound decreases from the relativistic value of $c_s = c/\sqrt{3}$ to

$$c_s = \left(\frac{4c^2}{9}\frac{\varrho_r}{\varrho_m}\right)^{1/2} = \left[\frac{4aT_0^4(1+z)}{9\Omega_m\varrho_c}\right]^{1/2} = \frac{10^6 z^{1/2}}{(\Omega_m h^2)^{1/2}} \text{ m s}^{-1}. \tag{9.33}$$

After recombination, the sound speed is just the thermal sound speed of the matter which, because of the close coupling between the matter and the radiation, has temperature $T_r = T_m$ at redshifts $z \geq 550\, h^{2/5}\Omega_0^{1/5}$, as explained above. Thus, at a redshift of 500, the temperature of the gas was 1300 K. If nothing else happened to the matter, we would expect it to be very cold at the present epoch, about 500 times colder than 2.728 K. In fact, whatever intergalactic gas there is at the present day must be very much hotter than this – it must have been heated up once the process of galaxy formation got underway. We will take up this intriguing story in Chap. 19.

9.6 Early Epochs

Let us complete this brief thermal history of the Universe by outlining what happens at earlier times.

- We can extrapolate back to redshifts $z \approx 3 \times 10^8$ when the radiation temperature was about $T = 10^9$ K. This temperature is sufficiently high for the background photons to have γ-ray energies, $\varepsilon = k_B T = 100$ keV. At this high temperature, the high energy photons in the tail of the Planck distribution are energetic enough to dissociate light nuclei such as helium and deuterium. At earlier epochs, all nuclei are dissociated in protons and neutrons. When we run the clocks forward, it is at about this epoch that the process of primordial nucleosynthesis of the light elements takes place. This is a key topic for the whole of cosmology and we will discuss it in some detail in Chap. 10.
- At redshift, $z \approx 10^9$, electron–positron pair production from the thermal background radiation can take place and the Universe was then flooded with electron–positron pairs. When we run the clocks forward from an early epoch, the electrons and positrons annihilate at about this epoch and their energy is transferred to the photon field – this accounts for the little discontinuity in the temperature history at the epoch when the electrons and positrons were annihilated. The change in the energy density of the background radiation associated with this process is evaluated in Sect. 10.5.
- At a slightly earlier epoch the opacity of the Universe for weak interactions became unity (see Sect. 10.2). This results is a *neutrino barrier*, similar to the photon barrier at $z \sim 1,000$.
- We can extrapolate even further back in time to $z \approx 10^{12}$ when the temperature of the background radiation was sufficiently high for baryon–antibaryon pair production to take place from the thermal background. Just as in the case of the epoch of electron–positron pair-production, the Universe was flooded with baryons and antibaryons. Again, there is a little discontinuity in the temperature history at this epoch. These considerations lead to one of the great cosmological problems, the *baryon asymmetry problem*. In order to produce the matter-dominated Universe we live in today, there must have been a tiny asymmetry between matter and antimatter in the very early Universe. For every 10^9 antibaryons, there must have been $10^9 + 1$ baryons. When we run the clocks forward, 10^9 baryons annihilate with the 10^9 antibaryons, leaving one baryon which

becomes the Universe as we know it with the correct photon-to-baryon ratio. If the early Universe were completely symmetric with respect to matter and antimatter, the photon-to-baryon ratio would be about 10^9 times greater than it is today and there would be equal amounts of matter and antimatter in the Universe; we return to this topic in Sect. 10.6. The baryon asymmetry must originate in the very early Universe. Fortunately, we know that there is a slight asymmetry between matter and antimatter because of CP violation observed in the decays of K^0 mesons – this is a major industry for the early Universe theorists.

We can carry this process of extrapolation further and further back into the mists of the early Universe, as far as we believe we understand high energy particle physics. Probably most particle physicists would agree that the standard model of elementary particles has been tried and tested to energies of at least 100 GeV and so we can probably trust laboratory physics back to epochs as early as 10^{-6} s, although the more conservative among us would probably be happier to accept 10^{-3} s. How far back one is prepared to extrapolate is largely a matter of taste. The most ambitious theorists have no hesitation in extrapolating back to the very earliest Planck eras, $t_P \sim (Gh/c^5)^{1/2} = 10^{-43}$ s, when the relevant physics was certainly very different from the physics of the Universe from redshifts of about 10^{12} to the present day. Kolb and Turner's splendid text *The Early Universe* (1990) can be thoroughly recommended for those who wish to take the study of these very early epochs seriously.

10 Nucleosynthesis in the Early Universe

One of the reasons why the standard Big Bang picture is taken so seriously is its remarkable success in accounting for the observed abundances of the light elements by *primordial nucleosynthesis* which took place during the first 10 minutes of our Universe. The results of this analysis are crucial for a number of aspects of galaxy formation. In particular, primordial nucleosynthesis provides one of the most important constraints upon the density parameter in the form of baryons Ω_B and this is a key part of our story. In the process of developing these results, we need to study in somewhat more detail the rôle of neutrinos in the early Universe, how they change the dynamics of the expanding Universe and how they decouple from the electrons and positrons. The neutrinos provide an example of the decoupling processes which may be important for other unknown types of weakly interacting particle. We will find a qualitatively similar example when we study possible forms of the dark matter.

10.1 Equilibrium Abundances in the Early Universe

Consider a particle of mass m at very high temperatures such that its total energy is much greater than its rest mass energy, $k_B T \gg mc^2$. If the timescales of the interactions which maintain this species in thermal equilibrium with all the other species present at temperature T are shorter than the age of the Universe at that epoch, statistical mechanics tells us that the equilibrium number densities of the particle and its antiparticle are

$$N = \bar{N} = \frac{4\pi g}{h^3} \int_0^\infty \frac{p^2 dp}{e^{E/k_B T} \pm 1}, \tag{10.1}$$

where g is the statistical weight of the particle, p its momentum and the $\pm$ sign depends upon whether the particles are fermions ($+$) or bosons ($-$). The photons are massless bosons for which $g = 2$, nucleons, antinucleons, electrons and positrons are fermions with $g = 2$ and the electron, muon and tau neutrinos are fermions with helicity for which $g = 1$. The equilibrium number densities for these particles and their antiparticles $N = \bar{N}$ and their energy densities ε can be found from this expression. For (a) photons, (b) nucleons, electrons and their antiparticles and (c) neutrinos and antineutrinos respectively, these are:

$$g = 2 \quad N = 0.244 \left(\frac{2\pi k_B T}{hc} \right)^3 \mathrm{m}^{-3} \quad \varepsilon = aT^4; \tag{10.2a}$$

$$g = 2 \quad N = \bar{N} = 0.183 \left(\frac{2\pi k_B T}{hc} \right)^3 \mathrm{m}^{-3} \quad \varepsilon = \tfrac{7}{8} aT^4; \tag{10.2b}$$

$$g = 1 \quad N = \bar{N} = 0.091 \left(\frac{2\pi k_B T}{hc} \right)^3 \mathrm{m}^{-3} \quad \varepsilon = \tfrac{7}{16} aT^4. \tag{10.2c}$$

To find the total energy density, we have to add all the equilibrium energy densities, that is

$$\text{Total energy density} = \varepsilon = \chi(T)\, aT^4. \tag{10.3}$$

When the particles become non-relativistic, $k_B T \ll mc^2$, and the abundances of the different species are still maintained by interactions between the particles, the non-relativistic limit of the integral (10.1) gives an equilibrium number density

$$N = g \left(\frac{m k_B T}{2\pi \hbar^2} \right)^{3/2} \exp\left(-\frac{mc^2}{k_B T} \right). \tag{10.4}$$

Thus, once they become non-relativistic, the species no longer contribute to the inertial mass density which determines the rate of deceleration of the Universe.

Let us consider the abundances of protons and neutrons in the early Universe. At redshifts less than 10^{12}, the neutrons and protons are non-relativistic, $k_B T \ll mc^2$, and their equilibrium abundances are maintained by the electron–neutrino weak interactions

$$\mathrm{e}^+ + \mathrm{n} \leftrightarrow \mathrm{p} + \bar{\nu}_\mathrm{e} \qquad \nu_\mathrm{e} + \mathrm{n} \leftrightarrow \mathrm{p} + \mathrm{e}^-. \tag{10.5}$$

The values of g for neutrons and protons are the same and so the relative abundances of neutrons to protons is

$$\left[\frac{\mathrm{n}}{\mathrm{p}} \right] = \exp\left(-\frac{\Delta mc^2}{k_B T} \right), \tag{10.6}$$

where Δmc^2 is the mass difference between the neutron and the proton.

10.2 The Decoupling of Neutrinos and the Neutrino Barrier

The abundance ratio of neutrons to protons freezes out when the neutrino interactions can no longer maintain the equilibrium abundances of neutrons and protons. The condition for 'freezing out' is that the timescale of the weak interactions becomes greater than the age of the Universe. The relevant processes which prevent the neutrinos escaping freely are interactions with the electrons and positrons which are present in the abundances given by (10.2b). These interactions are:

$$e^- + e^+ \leftrightarrow \nu_e + \bar{\nu}_e \quad e^\pm + \nu_e \leftrightarrow e^\pm + \nu_e \quad e^\pm + \bar{\nu}_e \leftrightarrow e^\pm + \bar{\nu}_e. \tag{10.7}$$

The timescale for these interactions is $t_{\mathrm{weak}} = (\sigma N c)^{-1}$, where the cross-section for the weak neutrino interactions, $\sigma = 3 \times 10^{-49}(E/m_e c^2)^2$ m^2, is proportional to the square of the neutrino's energy. N is the total number density of electrons and positrons which decreases as R^{-3} and so, since $\bar{E} = 3k_B T$ for relativistic particles, it follows that the timescale for the weak interactions changes with temperature as T^{-5}. Specifically, inserting (10.2b) for N and the above estimate for σ, we find

$$t_{\mathrm{weak}} \approx \frac{3 \times 10^{40}}{\left(\frac{3k_B}{m_e c^2}\right)^2 \left(\frac{2\pi k_B}{hc}\right)^3} \frac{1}{T^5} = \left(\frac{1.7 \times 10^{10}}{T}\right)^5 \text{ s.} \tag{10.8}$$

This timescale increases with decreasing temperature much more rapidly than the expansion time-scale of the Universe, which is given by (9.7). We have to modify (9.7) to take account of all the types of elementary particles which can contribute to the energy density ε during these early epochs, that is, we have to use the (10.3). During these epochs, the particles which contribute to the total energy density are the photons, the electrons, the electron-, muon- and tau- neutrinos and their antiparticles. Adding together these contributions, we find from (10.2) that

$$\chi = 1 + 2 \times \tfrac{7}{8} + 2N_\nu \times \tfrac{7}{16}, \tag{10.9}$$

if there are N_ν neutrino species. If we adopt the known numbers of neutrino species, $N_\nu = 3$, we find that $\chi = 43/8$ and so

$$\varepsilon = \chi a T^4 = \frac{3c^2}{32\pi G} t^{-2} \quad T = \left(\frac{3c^2}{32\pi G \chi a}\right)^{1/4} t^{-1/2} = 10^{10} t^{-1/2} \text{ K,} \tag{10.10}$$

where the time t is measured in seconds. It can be seen that the timescales for the expansion of the Universe and the decoupling of the neutrinos are the same when the Universe had a temperature of 1.2×10^{10} K at $t = 0.7$ s; at this epoch, $k_B T$ was almost precisely 1 MeV. Notice the important point that this time-scale and temperature are determined by fundamental constants of physics.

As a by-product of the above analysis, we have derived the epoch at which the Universe became opaque to neutrinos, namely, the epoch when the neutrinos could no longer maintain the neutrons and protons in thermodynamic equilibrium. Just as there is a barrier for photons at a redshift of about 1,500, so there is a *neutrino barrier* at an energy $k_B T = 1$ MeV. This means that, if it were possible to undertake neutrino astronomy, we would expect the background neutrinos to be last scattered at the epoch corresponding to $k_B T = 1$ MeV, about 1 second from the origin of the Big Bang.

10.3 The Synthesis of the Light Elements

At that time $t = 1$ s, the neutron fraction, as determined by (10.6), 'freezes out' at a ratio

$$\left[\frac{n}{n+p} \right] = 0.21. \tag{10.11}$$

After this time, the neutron fraction decreases slowly. The protons and neutrons can now begin to interact to form the light elements through the nuclear reactions:

$$p + n \rightarrow {}^3He + \gamma \quad n + D \rightarrow {}^3H + \gamma \quad p + {}^3H \rightarrow {}^4He + \gamma$$
$$n + {}^3He \rightarrow {}^4He + \gamma \quad d + d \rightarrow {}^4He + \gamma \quad {}^3He + {}^3He \rightarrow {}^4He + 2p. \tag{10.12}$$

The net result is that almost all the neutrons are combined with protons to form ${}^4He$ nuclei so that, for every pair of neutrons which survives, a helium nucleus is formed. Most of the nucleosynthesis takes place at a temperature less than about 10^9 K since, at higher temperatures, the deuterons are destroyed by the γ-rays of the background radiation. The binding energy of deuterium is $E_B = 2.23$ MeV and so this energy is equal to $k_B T$ at $T = 2.6 \times 10^{10}$ K. Just as in the case of the recombination of the intergalactic gas (Sect. 9.3), the photons far outnumber the nucleons and it is only when the temperature of the expanding gas has decreased to about 26 times less than this temperature that the number of dissociating photons is less than the number of nucleons. Thus, the bulk of the nucleosynthesis only takes place after about 100 s when the temperature of the radiation has fallen to about 10^9 K.

The detailed temperature history and evolution of the light elements during the epoch of nucleosynthesis were worked out in a classic paper by Wagoner (1973), following earlier pioneering computations by Wagoner, Fowler and Hoyle (1967) (Fig. 10.1). It can be seen that the bulk of the synthesis of the light elements occurs when the Universe was about 300 seconds old. Although the neutrons begin to decay spontaneously at this time, the bulk of them survive. Detailed calculations show that after 300 s the neutron mass fraction had fallen to 0.123 (see, for example, Weinberg 1972) and so the predicted mass fraction of helium Y_p is expected to be twice the neutron fraction

$$Y_p = \left[\frac{{}^4\text{He}}{\text{H} + {}^4\text{He}} \right] \approx 0.25. \qquad (10.13)$$

In addition to ^{4}He, which is always produced with an abundance of about 23 to 25%, traces of the light elements deuterium (D), helium-3 (^{3}He) and lithium-7 (^{7}Li) are created. Heavier elements are not synthesised because of the absence of stable isotopes with mass numbers 5 and 8. The heavier elements are synthesised during the course of stellar evolution, the key step being the rare triple-α process, which enables carbon nuclei to be formed when three helium nuclei come together.

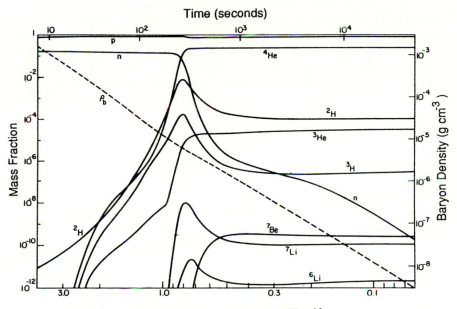

Fig. 10.1. An example of the time and temperature evolution of the abundances of different light elements in the standard Big Bang Model of the Universe from detailed computer calculations by Wagoner (1973). Before about 100 s from the origin of the model, no significant synthesis of the light elements takes place because deuterium ^{2}H is destroyed by hard γ–rays in the high energy tail of the black-body spectrum. As the temperature decreases, more and more of the deuterium survives and the synthesis of heavier light elements becomes possible through the reactions (10.12). Notice that the synthesis of elements such as D, ^{3}He, ^{4}He, ^{7}Li and ^{7}Be is completed after about 15 minutes. The neutrons decay with a half-life of about 10.25 minutes and the ^{3}H with a half-life of 12.46 years

These are remarkable results. It has always been a great problem to understand why the abundance of helium is so high wherever it can be observed in the Universe. Its chemical abundance always appears to be greater than about 23%. In addition, it has always been a mystery where the deuterium in the Universe could have been synthesised. It is a very fragile nucleus and is destroyed rather than created in stellar interiors. The same argument applies to a lesser extent to the lighter isotope of helium, ^{3}He, and to ^{7}Li. Precisely these elements are synthesised in the early stages of the Hot Big Bang. The difference between these modes of element formation is that, in stellar interiors, nucleosynthesis takes place in roughly thermodynamic equilibrium over very long timescales, whereas in the early stages of the Big Bang 'explosive' nucleosynthesis is all over in a few minutes.

The deuterium and ^{3}He abundances provide strong constraints upon the present baryon density of the Universe. Note that the physics which determines the abundance of ^{4}He is different from that which determines the abundance of the other light elements. The above analysis shows that the synthesis of ^{4}He is essentially thermodynamic, in that it is fixed by the ratio of neutrons to protons when the neutrinos decouple from the nuclear reactions which maintain them in their equilibrium abundances. In other words, the ^{4}He abundance is a measure of the *temperature* at which the decoupling of the neutrinos took place. On the other hand, the abundances of D, ^{3}He and ^{7}Li are determined by the extent to which the sequence of reactions (10.12) can convert all the neutrons into ^{4}He before the temperature falls below that at which nucleosynthesis can continue to take place. Thus, in high density Universes, there is time for essentially all the neutrons to be combined into deuterium nuclei which then combine to form ^{4}He nuclei and the predicted deuterium abundance is low. On the other hand, if the baryon density is low, there is not time for all the intermediate stages in the synthesis of helium to be completed and the result is a much higher abundance of deuterium and ^{3}He. Thus, the abundances of the deuterium and ^{3}He are measures of the *present baryon density* of the Universe.

Improved versions of his computer codes used by Wagoner (1973) using updated data on the nuclear cross-sections are described by Kolb and Turner (1990) (Fig. 10.2). The predicted abundances are displayed as a function of the baryon-to-photon ratio $\eta = n_B/n_\gamma = 2.74 \times 10^{-8}\, \Omega_B h^2$, which is conserved as the Universe expands from the epoch of nucleosynthesis to the present day. Y_p is the abundance of helium by mass and is plotted on a linear scale, whereas the abundances of D, ^{3}He and ^{7}Li are plotted as the ratios by number relative to hydrogen and are plotted logarithmically. The predictions are plotted for different assumed numbers of neutrino species, the standard Big Bang corresponding to $N_\nu = 3$. Also shown are the uncertainties in the predictions because of uncertainties in the half-life of the neutron. The error bar on the $N_\nu = 3$ line corresponds to a ± 0.2 minute uncertainty relative to the adopted value of 10.6 minutes. Note that more recent determinations

of the half-life of the neutron have found a value of 10.25 minutes. Kolb
and Turner (1990) give a useful analytic relation for the ^{4}He abundance as
functions of the baryon-to-photon ratio η, the value of χ and the half-life of
the neutron $\tau_{1/2}$,

$$Y_p = 0.230 + 0.25 \log \left(\frac{\eta}{10^{-10}} \right) + 0.015(\chi - 5.375) + 0.014(\tau_{1/2} - 10.6).$$

Thus, with the revised neutron half-life, the predicted ^{4}He abundance would
be slightly less than the values shown in Fig. 10.2. It can be seen that for
the standard Hot Big Bang, the ^{4}He abundance is relatively insensitive to
the present baryon density in the Universe, in contrast to those of the other
light elements.

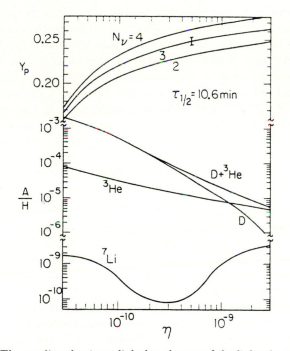

Fig. 10.2. The predicted primordial abundances of the light elements as a function
of the present baryon-to-photon ratio $\eta = n_B/n_\gamma = 2.74 \times 10^{-8} \, \Omega_B h^2$. Y_p is the
abundance of helium by mass and is plotted on a linear scale, whereas the abun-
dances for D, ^{3}He and ^{7}Li are plotted as the ratios by number relative to hydrogen
and are plotted logarithmically. The predictions are plotted for different assumed
numbers of neutrino species, $N_\nu = 3$ corresponding to the standard Big Bang in
which it is assumed that the three known neutrino species are massless. The error
bar on the to $N_\nu = 3$ line corresponds to a ± 0.2 minute uncertainty in the adopted
value of the half-life of the neutron of 10.6 minutes. Note that more recent estimates
of the half-life of the neutron give a value of 10.25 minutes. (Kolb and Turner 1990)

10.4 The Abundances of the Light Elements

These predictions of primordial nucleosynthesis are of the greatest importance for astrophysical cosmology and a great deal of effort has been devoted to the observational determination of the primordial abundances of the light elements. This is a far from trivial exercise and, whilst there is good general agreement between the observations and the predictions of the standard model, there are discrepancies which should not be ignored (see the exchange of views between Hogan and Steigman 1997). The present situation can be broadly summarised as follows:

Helium-4 ^{4}He. Helium can only be observed in hot stars and in regions of ionised hydrogen. Helium is synthesised in the course of stellar evolution and so it is important to determine the helium abundance in systems which are relatively uncontaminated by the effects of stellar nucleosynthesis. This can be achieved by determining the helium abundance as a function of the metallicity of the stars studied and extrapolating to zero metallicity. The results of independent analyses by a number of workers are summarised by Hogan (1997) and are in good agreement. For example, Olive and Steigman (1995) find a value $Y_p = 0.232 \pm 0.003(\text{stat}) \pm 0.005(\text{syst})$. Hogan remarks that the systematic error of 0.005 is due to uncertainties in the modelling of the HII regions, for example, in estimating the amount of neutral helium and in modelling the collisional excitation of HI, as well as the effect of temperature inhomogeneities. Note that this estimate results in a lower as well as an upper limit to the value of η.

Deuterium D. The abundance of deuterium is crucial cosmologically because of its strong dependence upon the present baryon density. The local interstellar abundance of deuterium in our Galaxy has been well determined by observations of the resonance absorption lines of D in interstellar clouds by the Copernicus satellite and by the Hubble Space Telescope (see, for example, Linsky *et al.* 1994). The values found amount to $[\text{D/H}] = (1.5 \pm 0.2) \times 10^{-5}$. This is very much a lower limit to the primordial deuterium abundance since much of the primordial deuterium has been destroyed as it is circulated through the hot interiors of stars. The key issue is the extent to which this process of deuterium destruction has taken place during the course of the chemical evolution of interstellar material.

The most important recent results have come from estimates of the deuterium abundance in the Lyman-α absorbers observed in the spectra of high redshift quasars. Absorption-line systems in the spectra of quasars with redshifts $z \geq 2.5$ have sufficiently large redshifts that the deuterium Lyman resonance lines are redshifted into the optical region of the spectrum and so are accessible to ground-based high-resolution spectroscopy with large telescopes (see also Chap. 18). These are highly non-trivial observations and the interpretation depends upon understanding the processes responsible for the line broadening. Hogan (1997) has reported that there seems to be a ceiling to the deuterium abundance of about $[D/H] = 2 \times 10^{-4}$ in the absorption line systems at redshifts about 3. On the other hand, Tytler and his colleagues (1996) reported observations of two systems at similar redshifts in which the deuterium abundance is almost an order of magnitude smaller than this value, $[D/H] = [2.4 \pm 0.3(\text{stat}) \pm 0.3(\text{syst})] \times 10^{-5}$, although this inference has been questioned by Songaila *et al.* (1997), who find rather larger values for the same objects. In Hogan's interpretation, the primordial $[D/H]$ ratio is high $[D/H] = 2 \times 10^{-4}$, and the low values found in our Galaxy today and in some of the other large redshift absorption systems are the result of stellar processing of primordial deuterium. This does, however, require that the primordial deuterium abundance has been depleted by an order of magnitude in the course of the chemical evolution of the Galaxy. It is a challenge to construct such models which are consistent with all the other constraints on the chemical evolution of the interstellar medium. A further complication is provided by the observations by Webb *et al.* (1997) who discovered a Lyman-α absorption system at redshift $z = 0.7$ with the high deuterium abundance $[D/H] = 2.0 \times 10^{-4}$. They suggested that it may be necessary to invoke spatially inhomogeneous models for primordial nucleosynthesis in order to account for the spread in the observed values of $[D/H]$.

Helium-3 ^{3}He. Helium-3 is observed in the oldest meteorites, the carbonaceous chondrites, with an abundance $[^3\text{He}/H] = 1.4 \pm 0.4 \times 10^{-5}$. This value is taken to be representative of the ^{3}He abundance about 4.6×10^9 years ago when the solar system formed. It has also been observed at radio wavelengths through the equivalent of the 21-cm hyperfine line of neutral hydrogen and the abundances in interstellar clouds lie in the range $[^3\text{He}/H] = 1.2$ to 15 $\times 10^{-5}$. ^{3}He is destroyed inside stars but it is a more robust isotope than deuterium. There are two important aspects to its cosmic abundance. First of all, when deuterium is burned, ^{3}He is one of the products. Second, when ^{3}He is destroyed, it creates ^{4}He which is then burned to form heavier elements. Thus, the interpretation of the ^{3}He abundances is more complicated than the deuterium abundance.

Lithium-7 ^{7}Li. In 1982, Spite and Spite made the first estimates of the ^{7}Li abundance for metal-poor halo stars. ^{7}Li is also a fragile element and it can be depleted by circulation through the centres of these stars. It is therefore important to use metal poor stars in which the convective zones are not too deep, which would result in depletion of the ^{7}Li from the atmosphere. According to Hogan (1997), the primordial abundance of ^{7}Li is estimated to be $\log[Li/H] = -9.80 \pm 0.16$. As he emphasises, the interpretation of these data is more uncertain than that of deuterium and there is also about a factor of two uncertainty in the predictions of the standard Big Bang nucleosynthesis predictions because of uncertainties in the cross-sections.

How is one to interpret these data? Hogan (1997) has suggested two possible positions. On the one hand, if we are very conservative, we can suggest that the data would be broadly consistent with values of η which lie in the range 10^{-9} to 10^{-10}. The value $\eta = 10^{-9}$ is a very firm upper limit since this results in a low value of the deuterium abundance of $[D/H] = 10^{-5}$. Thus, since $\eta = n_B/n_\gamma = 2.74 \times 10^{-8}\,\Omega_B h^2 \le 10^{-9}$ we find that $\Omega_B h^2 \le 3.6 \times 10^{-2}$ and so, even adopting a low value of $h = 0.5$, $\Omega_B \le 0.15$. Thus, even adopting this very conservative position, *there cannot be sufficient baryonic matter to close the Universe*. This will prove to be a key part of the story of the problems of galaxy formation.

Hogan also suggests a more somewhat bolder approach. If we adopt the upper limit he finds to the abundance of deuterium in the large redshift absorption line systems as the primordial deuterium abundance, $[D/H] = (1.9\pm0.5) \times 10^{-4}$, then we find that $\eta = (1.7\pm0.3) \times 10^{-10}$, $Y_p = 0.233\pm0.003$ and $\log[Li/H] = -9.75 \pm 0.2$. On other words, we obtain excellent agreement with the primordial abundances inferred from observation. If we adopt this more ambitious position, then we have already determined rather precisely the present baryon density in the Universe

$$\Omega_B h^2 = (6.2 \pm 1.1) \times 10^{-3}. \tag{10.14}$$

If we adopt a compromise value of Hubble's constant of 62.5 km s^{-1} Mpc^{-1}, $h = 0.625$ and $\Omega_B = 1.6 \times 10^{-2}$. In this case, the baryon density of the Universe would not be much greater than that attributable to the visible mass in galaxies. Since we have argued that the overall density parameter $\Omega_0 \ge 0.1$, it would follow that, even if $\Omega_0 \ne 1$, the Universe must be dominated dynamically by non-baryonic dark matter.

It is wise to be cautious about this line of reasoning, but it does indicate clearly the major issues which are at stake and which can be resolved by further observational studies, particularly with the new generation of 8 – 10 m optical-infrared telescopes. For example, the bolder line of reasoning would have major implications for the recycling of the light elements through stars. Many of these potential 'crises' for the standard Big Bang picture are described by Steigman (1997).

10.5 Electron–Positron Annihilation, the Value of χ and Other Considerations

There are several important aspects the above analysis which are worth noting since we will need these in the subsequent chapters. First, the type of decoupling process described in Sect. 10.2 is also used in calculating the abundances of massive neutrino-like particles which might have been present in equilibrium in the early Universe. For example, it is conceivable that particles with masses $m \gg 1 - 10$ GeV could have been present in the early

Universe and that, making reasonable assumptions about the cross-sections for the interaction of these particles and their antiparticles, sufficient of them could have survived to close the Universe now. We will return to this topic later.

One of the more remarkable results of these studies is that they enable limits to be set to the number of neutrino species which could have been present during the epochs when the light elements were being synthesised. If there had been more than three species of neutrino present, they would have contributed to the inertial mass density of massless particles and so would have speeded up the early expansion of the Universe (see 9.7 and 10.10). The decoupling of the neutrinos would have taken place at a higher temperature resulting in the overproduction of helium (see Fig. 10.2). From this type of cosmological argument, it was shown that there could not be more than three families of neutrinos, a result subsequently confirmed by the width of the decay spectrum of the $W^\pm$ and Z^0 bosons measured by LEP at CERN.

There is another important piece of physics associated with the determination of the temperature of the neutrino background radiation. We showed in Sect. 10.2 that the weak interactions maintain the equilibrium abundances of the neutrons and protons until $k_\mathrm{B}T \approx 1$ MeV. Prior to this epoch, the photons, neutrinos, electrons and their antiparticles were the only species which are relativistic and they all have the same thermal temperature. Just after neutrino decoupling, at an energy $k_\mathrm{B}T \approx 0.5$ MeV, the electrons and positrons annihilate creating γ-ray photons. These high energy photons are rapidly thermalised by Compton scattering and so the thermal temperature of the radiation becomes greater than that of the neutrinos. The expansion is adiabatic and the entropy per baryon is conserved during the expansion. Since the electrons and neutrinos are no longer coupled at $k_\mathrm{B}T \approx 0.5$ MeV, the temperatures of the neutrinos and photons can be worked out assuming the entropy of the electrons and positrons is transferred to the radiation.

As shown by Kolb and Turner (1990), the entropy per unit comoving volume s is conserved as the Universe expands

$$\mathrm{d}s = \mathrm{d}\left[\frac{(\varepsilon + p)R^3}{T}\right] = \mathrm{d}\left[\sum_i g_i a T^3 R^3\right] = 0, \tag{10.15}$$

where ε and p are the equilibrium energy density and pressure respectively. Before the epoch at which the electrons and positrons annihilated, the ratios of the energy densities ε and entropies s of the various equilibrium relativistic components were as given in Table 10.1.

After annihilation, the energy of the electron–positron pairs is transferred to the radiation field and so its entropy increases by a factor

$$\frac{s_\gamma + s_{e^+} + s_{e^-}}{s_\gamma} = \frac{11}{4}. \tag{10.16}$$

Therefore, since $\sum_i g_i T^{-3}$ is conserved and the neutrinos are decoupled from equilibrium, the temperature of the radiation is increased relative to that

Table 10.1. The ratios of the energy densities and entropies of photons, electrons, positrons and neutrinos prior to the epoch of electron–positron annihilation.

Energy density	ε_γ	$(\varepsilon_{e^-} + \varepsilon_{e^+})$	$(\varepsilon_\nu + \varepsilon_{\bar\nu})$	$(\varepsilon_{\nu_\mu} + \varepsilon_{\bar\nu_\mu})$	$(\varepsilon_{\nu_\tau} + \varepsilon_{\bar\nu_\tau})$
Entropy per baryon	s_γ	$(s_{e^-} + s_{e^+})$	$(\varepsilon_\nu + \varepsilon_{\bar\nu})$	$(s_{\nu_\mu} + s_{\bar\nu_\mu})$	$(s_{\nu_\tau} + s_{\bar\nu_\tau})$
$\sum_i g_i$	1	7/4	7/8	7/8	7/8

of the neutrinos by a factor $(11/4)^{1/3}$. Since the entropies per baryon are conserved throughout the subsequent evolution of the standard Big Bang, this is also the ratio of the temperatures of the photons and neutrinos at the present day. Adopting $T_{\rm rad} = 2.728$ K, the temperature of the neutrinos is expected to be $T_\nu = (4/11)^{1/3} T_{\rm rad} = 1.95$ K. This is the predicted temperature of the neutrino background which should accompany the Cosmic Microwave Background Radiation and which was last scattered at the epoch when $k_{\rm B} T \approx 1$ MeV, that is $t \approx 1$ s. Unfortunately, this background is far beyond the capabilities of the present generation of neutrino detectors. This is the process which accounts for the little 'kink' in the temperature history of the Big Bang illustrated in Fig. 9.1 at the epoch when the electrons and positrons annihilate.

There is one other important point about the influence of the neutrinos upon the dynamics of the Universe. The dynamics during the radiation-dominated phase is determined by the total inertial mass density of massless particles. For massless particles, even when they are decoupled from the electrons, their energy density decreases as $\varepsilon_\nu \propto R^{-4}$, just like the radiation field, and so they still contribute to the total inertial mass density which retards the expansion of the Universe. The energy densities in photons and neutrinos after the annihilation of the electron–positron pairs is shown in Table 10.2.

Table 10.2. The energy densities of photons and neutrinos in units of $a T_\nu^4$ after the epoch of electron–positron annihilation.

Energy density	ε_γ	$(\varepsilon_\nu + \varepsilon_{\bar\nu})$	$(\varepsilon_{\nu_\mu} + \varepsilon_{\bar\nu_\mu})$	$(\varepsilon_{\nu_\tau} + \varepsilon_{\bar\nu_\tau})$
Contribution to energy density ε	$(11/4)^{4/3}$	7/8	7/8	7/8

Thus, the total energy density ε is

$$\varepsilon = \left[\left(\tfrac{11}{4}\right)^{4/3} + \tfrac{21}{8} \right] a T_\nu^4 = 6.48 \, a T_\nu^4$$

$$= 6.48 \left(\tfrac{4}{11}\right)^{4/3} a T_{\rm rad}^4 = 1.68 \, a T_{\rm rad}^4. \tag{10.17}$$

Thus, during the radiation-dominated era after electron–positron annihilation, we should use a value $\chi = 1.68$ in the expression for the dynamics of the Universe,

$$\varepsilon = \chi a T^4 = \frac{3c^2}{32\pi G} t^{-2};$$ (10.18a)

$$T = \left(\frac{3c^2}{32\pi G \chi a}\right)^{1/4} t^{-1/2} = 1.33 \times 10^{10} t^{-1/2} \text{ K}.$$ (10.18b)

We can therefore reformulate the expression for the epoch at which the Universe changed from being 'radiation-dominated' to 'matter-dominated', in the sense that the dynamics changed from $R \propto t^{1/2}$ to $R \propto \Omega_0^{1/3} t^{2/3}$. Using (9.8) and (10.18), the ratio of energy densities of massless to non-relativistic matter becomes

$$\frac{\varrho_{\text{massless}}}{\varrho_{\text{matter}}} = \frac{1.68 \, a T_{\text{rad}}^4(z)}{\Omega_0 \varrho_c (1+z)^3 c^2} = \frac{4.2 \times 10^{-5}(1+z)}{\Omega_0 h^2}.$$ (10.19)

Therefore, the change-over in the dynamics takes place at a redshift $z = 2.4 \times 10^4 \Omega_0 h^2$.

10.6 Baryon-symmetric Universes

The physics of the formation of the light elements is similar in some ways to the physical processes involved in particle-antiparticle symmetric Universes and this has general significance for the physics of relic particles. Let us consider the processes of particle-antiparticle annihilation in a little more detail. The analyses of Zeldovich and Novikov (1983) and, in more detail, Kolb and Turner (1990) can be recommended. It might be imagined that there would be negligible abundances of baryons and antibaryons in the Universe now if there were equal numbers of baryons and antibaryons to begin with, but this is not correct. As discussed in Section 10.1, the equilibrium abundance of any species of mass m in thermal equilibrium at temperature T in the non-relativistic limit, $k_B T \ll mc^2$, is

$$N = \bar{N} = g \left(\frac{m k_B T}{2\pi \hbar^2}\right)^{3/2} \exp\left(-\frac{mc^2}{k_B T}\right),$$ (10.4)

where, for protons, antiprotons, neutrons and antineutrons $g = 2$. In the present case, the mass m is the mass of the nucleon and so it might appear that, since $T \sim 3$ K at the present day, the abundances of nucleons and antinucleons would be of order $10^{-10^{12}}$. This is not correct, however, since the abundances 'freeze out' when the time-scale of the processes which maintain the nucleons and antinucleons in their equilibrium abundances becomes greater than the age of the Universe. Zeldovich and Novikov present a simple solution for relic abundances of nucleons and antinucleons in baryon-symmetric Universes. The problem is to solve for the nucleon-to-photon ratio

η in the non-relativistic regime. The annihilation cross-section for nucleons and antinucleons in the non-relativistic regime varies as $\sigma_{\mathrm{p\bar{p}}} = \sigma_0 c/v$ where $\sigma_0 \approx 2 \times 10^{-30}$ m^2 and so the probability of annihilation of one nucleon with an antinucleon per unit time is $\sigma_{\mathrm{p\bar{p}}} v N = \sigma_0 c N$. Therefore, the rate of change of the number density of nucleons is

$$\frac{\mathrm{d}N}{\mathrm{d}t} = -\sigma_0 c N^2 + \psi(t) - \frac{3N}{2t} \qquad (10.20)$$

where $\psi(t)$ is the pair-creation rate and the term $-3N/2t$ describes the decrease in number density due to the expansion of the Universe in the radiation-dominated phase. This rate of change of nucleons can be compared with the rate at which the number density of photons changes with time. The number of photons is conserved as the Universe expands during the relevant epochs and so we write

$$\frac{\mathrm{d}N_\gamma}{\mathrm{d}t} = -\frac{3N_\gamma}{2t} \qquad (10.21)$$

Therefore, since $\eta = N/N_\gamma$, differentiating N/N_γ, we find that

$$\frac{\mathrm{d}\eta}{\mathrm{d}t} = -\sigma_0 c \eta^2 + \frac{\psi(t)}{N_\gamma} \qquad (10.22)$$

As expected, if there were no annihilation or creation of proton-antiproton pairs, the right-hand side would be zero and this ratio would remain unchanged. Following Zeldovich and Novikov, it is convenient to introduce the equilibrium value of $\eta = \eta_{\mathrm{eq}}$ which would be found if the nucleons and antinucleons remained in equilibrium at all times. In their calculations, they consider the total numbers of baryons, antibaryons, neutrons and antineutrons. Then, η_{eq} would be given by the ratio of (10.4) to (10.2a),

$$\eta_{\mathrm{eq}} = A\theta^{-3/2} \exp(-1/\theta) \qquad (10.23)$$

where $\theta = k_{\mathrm{B}}T/mc^2$ and A is a constant of order unity.

Now, in equilibrium, the rate of annihilations must equal the rate of pair production and so the right-hand side of (19.22) would be zero, that is,

$$-\sigma_0 c N_\gamma \eta_{\mathrm{eq}}^2 + \frac{\psi}{N_\gamma} = 0. \qquad (10.24)$$

Therefore, the equation for η can be written

$$\frac{\mathrm{d}\eta}{\mathrm{d}t} = -\sigma_0 c N_\gamma (\eta^2 - \eta_{\mathrm{eq}}^2) \qquad (10.25)$$

We are interested in finding the time at which the value of η begins to depart from its equilibrium value and this occurred when $\eta - \eta_{\mathrm{eq}} \approx \eta_{\mathrm{eq}}$, that is, when

$$\frac{\mathrm{d}\eta_{\mathrm{eq}}}{\mathrm{d}t} \approx -\sigma_0 c N_\gamma (\eta - \eta_{\mathrm{eq}})(\eta + \eta_{\mathrm{eq}}) \approx -2\sigma_0 c N_\gamma \eta_{\mathrm{eq}}^2 \qquad (10.26)$$

Therefore,

$$\frac{1}{\eta_{eq}}\frac{d\eta_{eq}}{dt} = \frac{d(\ln \eta_{eq})}{dt} \approx -2\sigma_0 c N_\gamma \eta_{eq} \qquad (10.27)$$

Now,

$$\ln \eta_{eq} = -\frac{3}{2}\ln \theta - 1/\theta \qquad (10.28)$$

and so, since we are well into the regime in which $\theta \ll 1$, we can approximate $\ln \eta_{eq} = -1/\theta$. Furthermore, during these radiation-dominated phases, $\theta \propto R^{-1} \propto t^{-1/2}$ and so

$$\frac{d(\ln \eta_{eq})}{dt} = \frac{d(\ln \eta_{eq})}{d\theta}\frac{d\theta}{dt} = \frac{1}{2\theta t} \approx 2\sigma_0 c N_\gamma \eta_{eq} \qquad (10.29)$$

Thus, the epoch at which the decoupling took place is given by the solution of

$$4\sigma_0 c \eta_{eq} N_\gamma \theta t \approx 1 \quad \text{or} \quad 4\sigma_0 c t\,\theta^{-1/2}\,e^{-1/\theta} N_\gamma \approx 1 \qquad (10.30)$$

Now, from (10.2a) and (10.10), the dependences of N_γ and θ as a function of cosmic time t are known,

$$N_\gamma = 0.244 \left(\frac{2\pi k_B T}{hc}\right)^3 = 2.6 \times 10^{46}\,\theta^3\,\mathrm{m}^{-3} \qquad (10.30a)$$

$$T = 10^{10}\,t^{-1/2}\,K \qquad \theta = \frac{k_B T}{mc^2} = 10^{-3}t^{-1/2} \qquad (10.30b)$$

Substituting these values into (10.30), we find the critical value of $\theta = \theta_d$ is given by the solution of

$$e^{1/\theta_d} = 6.4 \times 10^{19}\,\theta_d^{1/2} \qquad (10.31)$$

The solution is found by repeated approximation, with the result

$$\theta_d = \frac{k_B T_d}{mc^2} \approx \tfrac{1}{44}; \quad t_d = 2.5 \times 10^{-3}\,\mathrm{s}; \quad \eta_d = 2 \times 10^{-18} \qquad (10.32)$$

Before this time, the particles and antiparticles were maintained in thermal equilibrium by the annihilation and pair production mechanisms. At later times, the mean free path for $p\bar{p}$ collisions exceeded the horizon scale at that time as may be appreciated by evaluating the optical depth for these collision $\tau_{p\bar{p}} = \sigma_0 N c t$. At that time, the nucleon number density was $N \approx 6 \times 10^{23}\,\mathrm{m}^3$ and so $\tau_{p\bar{p}} \approx 1$. Thus, at later times, there were insufficient interactions to maintain the equilibrium abundances.

We can now work out how the value of η changes after this epoch. Inspection of (10.23) shows that at later times η_{eq} decreased exponentially and so (10.25) for η reduces to

$$\frac{d\eta}{dt} = -\sigma_0 c N_\gamma \eta^2 \qquad (10.33)$$

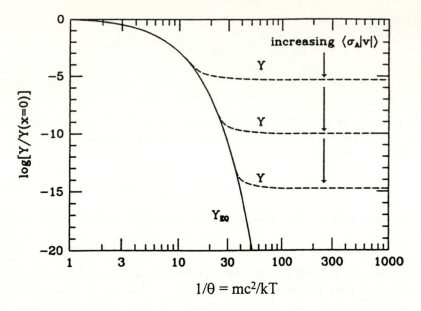

$$1/\theta = mc^2/kT$$

Fig. 10.3 Illustrating the 'freeze-out' of massive species due to particle-antiparticle annihilation in the early Universe (Kolb and Turner 1990). The parameter Y is the specific entropy of the particle, which is proportional to the number density of particles per unit comoving volume and so equivalent to the quantity η. The quantity plotted on the abscissa is $mc^2/k_BT = \theta^{-1}$. It can be seen that the freeze-out abundance corresponding to $\theta^{-1} = 44$ agrees with the calculations presented in the text.

Taking the initial conditions to be given by (10.32), the solution is

$$\frac{\eta}{\eta_d} \approx \frac{1}{[1 + 10^{19}\eta_d(1 - \tau^{-1/2})]} \tag{10.33}$$

where $\tau = t/t_d$. Thus, when $\tau \to \infty$,

$$\frac{\eta_\infty}{\eta_d} \approx \frac{1}{[1 + 10^{19}\eta_d]} \approx 0.05 \tag{10.34}$$

Thus, from the time the nucleons decoupled, the value of η only decreased by a factor of about 20 and so the value of η at the present day should be about 10^{-19}, that is, about 10^9 times smaller than the observed value. This is the origin of the baryon-asymmetry problem discussed in Sect. 9.6.

The reason for carrying out this calculation in some detail is that it gives physical insight into the decoupling of massive particles in the early Universe, a subject dealt with in some detail by Kolb and Turner (1990) and to which we return in Sect. 13.2. Fig. 10.3 shows results of their calculations for the 'freeze-out' of massive particle species in the early Universe for different values of the cross-section $\langle\sigma v\rangle$. The solid line shows the equilibrium abundances

and the dashed lines the actual abundances when account is taken of the decoupling of the particles from equilibrium. These results can be understood in
terms of the above calculations. After decoupling takes place, the abundances
freeze out with more or less the value at decoupling. As the cross-section decreases, the freeze-out occurs at higher abundances of the massive species, as
illustrated.

Part III

The Development of Primordial Fluctuations under Gravity

11 The Evolution of Fluctuations in the Standard Big Bang

At last, we can begin the study of the origin and formation of galaxies and the large-scale structure of the Universe. The first two parts of this text have presented the essential framework within which these problems have to be tackled. It will turn out that much of this chapter, and indeed most of Part 3, is concerned with understanding how the process of galaxy and structure formation gets underway. The details of what happened once these structures began to form are the subject of Part 4.

11.1 What the Theorists are Trying to Do

Galaxies and clusters of galaxies are complex systems, as the analyses of Chaps. 3 and 4 show, but the aim of the cosmologist is not to explain all their detailed features – that is the job of the astrophysicist. The aim of the cosmologist is to explain how large-scale structures formed in the expanding Universe in the sense that, if $\delta\varrho$ is the enhancement in density of some region over the average background density ϱ, the *density contrast* $\Delta = \delta\varrho/\varrho$ reached amplitude 1 from initial conditions which must have been remarkably isotropic and homogeneous. Once the initial perturbations have grown in amplitude to $\Delta = \delta\varrho/\varrho \approx 1$, their growth becomes non-linear and they rapidly evolve towards bound structures in which star formation and other astrophysical processes lead to the formation of galaxies and clusters of galaxies as we know them. The cosmologist's objectives are therefore twofold – first, to understand how density perturbations evolve in the expanding Universe and, second, to derive the initial conditions necessary for the formation of structure in the Universe. These may appear to be rather modest goals but they turn out to result in some of the most profound problems of modern cosmology. As we will show, the origin of the fluctuations must lie in the very early Universe.

Galaxies, clusters of galaxies and other large-scale features of our local Universe must have formed relatively late in the history of the Universe, as can be deduced from the following argument. As discussed in Sect. 8.5, the average density of matter in the Universe today corresponds to a density parameter $\Omega_0 \sim 0.1 - 1$. The analyses of Chaps. 3 and 4 show that the average densities of gravitationally bound systems such as galaxies and clusters of galaxies are much greater than this value, typically their densities being about 10^6 and 1000 times greater than the mean background density respectively. Superclusters have mean densities a few times the background density. Therefore, the density contrasts $\Delta = \delta\varrho/\varrho$ for galaxies, clusters of galaxies and superclusters at the present day are about $\sim 10^6$, 1000 and a few respectively. Since the average density of matter in the Universe ϱ changes as $(1 + z)^3$, it follows that typical galaxies must have had $\Delta = \delta\varrho/\varrho \approx 1$ at a redshift $z \approx 100$. They could not have separated out as discrete objects at greater than redshifts, or else their mean densities would be much greater than those observed at the present epoch. The same argument applied to clusters and superclusters suggests that they could not have separated out from the expanding background at redshifts greater than $z \sim 10$ and 1 respectively. As we will discuss in more detail in Sect. 16.1, these are upper limits to the redshifts at which these objects could have formed.

We conclude that galaxies and larger scale structures must have separated out from the expanding gas at redshifts significantly less than 100. This epoch occurred long after the epoch of recombination at $z \approx 1000$ and well into the matter-dominated phase of the standard Big Bang. These are important conclusions since it means that these structures were not formed in the inaccessibly remote past, but at redshifts which are in principle accessible by observation. These considerations indicate why it is natural to begin this study with an analysis of how small density perturbations develop in the expanding Universe.

11.2 The Non-relativistic Wave Equation for the Growth of Small Perturbations in the Expanding Universe

The analysis which follows is one of the classics of theoretical astrophysics. The problem of the growth of small perturbations under gravity dates back to the work of Jeans in the first decade of this century and then to a classic paper by Lifshitz (1946). The problem gets off to a very bad start. Let us first write down the standard equations of gas dynamics for a fluid in a gravitational field. These consist of three partial differential equations which describe (i) the conservation of mass, or the equation of continuity, (ii) the equation of motion for an element of the fluid, Euler's equation, and (iii) the equation for the gravitational potential in the presence of a density distribution ϱ, Poisson's equation.

$$\textit{Equation of Continuity}: \frac{\partial\varrho}{\partial t} + \nabla \cdot (\varrho\mathbf{v}) = 0; \tag{11.1}$$

$$\text{Equation of Motion} : \frac{\partial \mathbf{v}}{\partial t} + (\mathbf{v} \cdot \nabla)\mathbf{v} = -\frac{1}{\varrho}\nabla p - \nabla \phi; \qquad (11.2)$$

$$\text{Gravitational Potential} : \nabla^2 \phi = 4\pi G \varrho. \qquad (11.3)$$

Let us recall the meaning of these equations. They describe the dynamics of a fluid of density ϱ and pressure p in which the velocity distribution is $\mathbf{v}$. The gravitational potential ϕ at any point is given by Poisson's equation (11.3) in terms of the density distribution ϱ. It is important to remember exactly what the partial derivatives mean. In (11.1), (11.2) and (11.3), the partial derivatives describe the variations of these quantities *at a fixed point in space*. This coordinate system is often referred to as *Eulerian coordinates*. There is another way of writing the equations of fluid dynamics in which the motion of a particular fluid element is followed. These are known as *Lagrangian coordinates*. Derivatives which follow a particular fluid element are written as total derivatives d/dt and it is straightforward to show that

$$\frac{\mathrm{d}}{\mathrm{d}t} = \frac{\partial}{\partial t} + (\mathbf{v} \cdot \nabla), \qquad (11.4)$$

(see, for example, Longair 1991, Appendix to Chap. 5). Notice also the significance of the operator $(\mathbf{v} \cdot \nabla)$. There is no ambiguity when this operator is used with a scalar quantity. When it operates upon a vector quantity, it means that the derivative $v_x \partial/\partial x + v_y \partial/\partial y + v_z \partial/\partial z$ should be taken for each component of the vector. The equations of motion can therefore be written in Lagrangian form in which we follow the behaviour of an element of the fluid.

$$\frac{\mathrm{d}\varrho}{\mathrm{d}t} = -\varrho \nabla \cdot \mathbf{v}; \qquad (11.5)$$

$$\frac{\mathrm{d}\mathbf{v}}{\mathrm{d}t} = -\frac{1}{\varrho}\nabla p - \nabla \phi; \qquad (11.6)$$

$$\nabla^2 \phi = 4\pi G \varrho. \qquad (11.7)$$

In deriving (11.5), we have used the identity $\nabla \cdot (\varrho \mathbf{v}) = \varrho \nabla \cdot \mathbf{v} + \mathbf{v} \cdot \nabla \varrho$. For the cosmological problem we are analysing, (11.5), (11.6) and (11.7) can be thought of as being written in comoving form, that is, the properties of a fluid element expanding with the Universe is followed, rather than what would be observed if one were located at a fixed point in space and watched the Universe expand past it. Thus, writing Hubble's law in the form $\mathbf{v} = H_0 \mathbf{r}$, (11.5) becomes

$$\frac{\mathrm{d}\varrho}{\mathrm{d}t} = -3H_0 \varrho. \qquad (11.8)$$

Then, since $H_0 \, \mathrm{d}t = (\dot{R}/R) \, \mathrm{d}t$, $\varrho = \varrho_0 R^3$ as expected.

It is standard practice to establish first of all the solution for the unperturbed medium, that is, a uniform state in which ϱ and p are the same everywhere and $\mathbf{v} = 0$. Unfortunately this solution does not exist. Equations

(11.5) to (11.7) show that, if everything is uniform and the velocity is zero, we only obtain solutions if $\varrho = 0$. This is a problem since it means that there is no static solution with finite density and pressure about which to perturb the medium. Fortunately, we need to treat the growth of fluctuations in a uniformly expanding medium and this eliminates the problem. We first derive the unperturbed solutions for the velocity $\mathbf{v}_0$, density ϱ_0, pressure p_0 and gravitational potential ϕ_0 which satisfy (11.5), (11.6) and (11.7).

$$\frac{d\varrho_0}{dt} = -\varrho_0 \nabla \cdot \mathbf{v}_0; \tag{11.9}$$

$$\frac{d\mathbf{v}_0}{dt} = -\frac{1}{\varrho_0} \nabla p_0 - \nabla \phi_0; \tag{11.10}$$

$$\nabla^2 \phi_0 = 4\pi G \varrho_0. \tag{11.11}$$

The next step is to write down the equations including first order perturbations and so we write

$$\mathbf{v} = \mathbf{v}_0 + \delta\mathbf{v}, \quad \varrho = \varrho_0 + \delta\varrho, \quad p = p_0 + \delta p, \quad \phi = \phi_0 + \delta\phi. \tag{11.12}$$

These are substituted into (11.5), (11.6) and (11.7), which are expanded to first order in small quantities and then (11.9), (11.10) and (11.11) are subtracted from each of them in turn. From the subtraction of (11.9) from (11.5), we find

$$\frac{d}{dt}\left(\frac{\delta\varrho}{\varrho_0}\right) = \frac{d\Delta}{dt} = -\nabla \cdot \delta\mathbf{v}, \tag{11.13}$$

where $\Delta = \delta\varrho/\varrho_0$ is the density contrast. This is an important equation since it relates the rate at which the density contrast develops to the peculiar velocity $\delta\mathbf{v}$ associated with the collapse of the perturbation.

To make progress with (11.6), we expand $d\mathbf{v}/dt$ to first order in small quantities using (11.4).

$$\frac{d(\mathbf{v}_0 + \delta\mathbf{v})}{dt} = \frac{\partial\mathbf{v}_0}{\partial t} + (\mathbf{v}_0 \cdot \nabla)\mathbf{v}_0 + \frac{d(\delta\mathbf{v})}{dt} + (\delta\mathbf{v} \cdot \nabla)\mathbf{v}_0. \tag{11.14}$$

In expanding the right hand side of (11.10), we assume that the initial state is *homogeneous* and *isotropic* so that $\nabla p_0 = 0$ and $\nabla \varrho_0 = 0$. We then find, when we subtract (11.10) from (11.14), that

$$\frac{d(\delta\mathbf{v})}{dt} + (\delta\mathbf{v} \cdot \nabla)\mathbf{v}_0 = -\frac{1}{\varrho_0} \nabla \delta p - \nabla \delta\phi. \tag{11.15}$$

The third equation results from the subtraction of (11.11) from (11.7). Because of the linearity of Poisson's equation (11.7), we find

$$\nabla^2 \delta\phi = 4\pi G \, \delta\varrho. \tag{11.16}$$

Equations (11.13), (11.15) and (11.16) are the key differential equations in the present analysis.

In the cosmological case, the background is expanding uniformly and so it is convenient to use comoving coordinates by writing in the usual way $\mathbf{x} = R(t)\mathbf{r}$, where $\mathbf{r}$ is comoving coordinate distance and $R(t)$ is the scale factor. Therefore,

$$\delta\mathbf{x} = \delta[R(t)\mathbf{r}] = \mathbf{r}\,\delta R(t) + R(t)\,\delta\mathbf{r}. \tag{11.17}$$

The velocity can therefore be written

$$\mathbf{v} = \frac{\delta\mathbf{x}}{\delta t} = \frac{\mathrm{d}R}{\mathrm{d}t}\mathbf{r} + R(t)\frac{\mathrm{d}\mathbf{r}}{\mathrm{d}t}. \tag{11.18}$$

Thus, we can identify $\mathbf{v}_0$ with the Hubble expansion term $(\mathrm{d}R/\mathrm{d}t)\mathbf{r}$ and the perturbation to the Hubble flow $\delta\mathbf{v}$ with the term $R(t)(\mathrm{d}\mathbf{r}/\mathrm{d}t)$. It is convenient to write the perturbed velocity as $R(t)\mathbf{u}$ so that $\mathbf{u}$ is the perturbed comoving velocity. Equation (11.15) therefore becomes

$$\frac{\mathrm{d}}{\mathrm{d}t}(R\mathbf{u}) + (R\mathbf{u}\cdot\nabla)\dot{R}\mathbf{r}_0 = -\frac{1}{\varrho_0}\nabla\delta p - \nabla\delta\phi. \tag{11.19}$$

It will prove convenient to write the derivatives with respect to the comoving coordinate $\mathbf{r}$ rather than $\mathbf{x}$ so that $\mathrm{d}/\mathrm{d}x = (1/R)\mathrm{d}/\mathrm{d}r$. Differentials with respect to comoving coordinates will be written ∇_{c}. Therefore, since $(R\mathbf{u}\cdot\nabla)\dot{R}\mathbf{r} = \mathbf{u}\dot{R}$, (11.19) becomes

$$\frac{\mathrm{d}\mathbf{u}}{\mathrm{d}t} + 2\left(\frac{\dot{R}}{R}\right)\mathbf{u} = -\frac{1}{\varrho_0 R^2}\nabla_{\mathrm{c}}\delta p - \frac{1}{R^2}\nabla_{\mathrm{c}}\delta\phi. \tag{11.20}$$

Now, let us consider *adiabatic perturbations* in which the perturbations in pressure and density are related to the adiabatic sound speed c_{s}^2 by $\partial p/\partial\varrho = c_{\mathrm{s}}^2$. Thus, δp can be replaced by $c_{\mathrm{s}}^2\,\delta\varrho$ in (11.20). We now combine (11.13) and (11.19) by taking the divergence in comoving coordinates of (11.17) and the time derivative of (11.13).

$$\nabla_{\mathrm{c}}\cdot\dot{\mathbf{u}} + 2\left(\frac{\dot{R}}{R}\right)\nabla_{\mathrm{c}}\cdot\mathbf{u} = -\frac{c_{\mathrm{s}}^2}{\varrho_0 R^2}\nabla_{\mathrm{c}}^2(\delta\varrho) - \frac{1}{R^2}\nabla_{\mathrm{c}}^2(\delta\phi). \tag{11.21}$$

$$\frac{\mathrm{d}^2}{\mathrm{d}t^2}\left(\frac{\delta\varrho}{\varrho}\right) = -\nabla_{\mathrm{c}}\cdot\dot{\mathbf{u}}. \tag{11.22}$$

Therefore,

$$\frac{\mathrm{d}^2\Delta}{\mathrm{d}t^2} + 2\left(\frac{\dot{R}}{R}\right)\frac{\mathrm{d}\Delta}{\mathrm{d}t} = \frac{c_{\mathrm{s}}^2}{\varrho_0 R^2}\nabla_{\mathrm{c}}^2\delta\varrho + 4\pi G\delta\varrho. \tag{11.23}$$

We now seek wave solutions for Δ of the form $\Delta \propto \exp\mathrm{i}(\mathbf{k}_{\mathrm{c}}\cdot\mathbf{r} - \omega t)$ and hence derive a wave equation for Δ.

$$\frac{\mathrm{d}^2\Delta}{\mathrm{d}t^2} + 2\left(\frac{\dot{R}}{R}\right)\frac{\mathrm{d}\Delta}{\mathrm{d}t} = \Delta(4\pi G\varrho_0 - k^2 c_{\mathrm{s}}^2), \tag{11.24}$$

where $\mathbf{k}_c$ is the wavevector in comoving coordinates. The proper wavevector $\mathbf{k}$ is related to $\mathbf{k}_c$ by $\mathbf{k}_c = R\mathbf{k}$. Equation (11.24) is the result we have been seeking and a number of important conclusions follow from it. I make no apology for deriving (11.24) in somewhat gruesome detail because it is as important as any equation in astrophysical cosmology.

11.3 The Jeans' Instability

Let us return first of all to the problem originally studied by Jeans. The differential equation for gravitational instability in a static medium is obtained by setting $\dot{R} = 0$ in (11.24). Then, for waves of the form $\Delta = \Delta_0 \exp i(\mathbf{k} \cdot \mathbf{r} - \omega t)$, the dispersion relation,

$$\omega^2 = c_s^2 k^2 - 4\pi G\varrho_0, \tag{11.25}$$

is found. This relation was first derived by Jeans in 1902. The corresponding equation for the electrostatic case was only derived after the discovery of plasma oscillations by Langmuir and Tonks in the 1920s, and describes the dispersion relation for longitudinal plasma oscillations, or Langmuir waves:

$$\omega^2 = c_s^2 k^2 + \frac{N_e e^2}{m_e \varepsilon_0}, \tag{11.26}$$

where N_e is the electron density and m_e is the mass of the electron. The formal similarity of the physics may be appreciated by comparing the attractive gravitational acceleration of a region of mass density ϱ_0 and the repulsive electrostatic acceleration of a region of electron charge density $N_e e$. The equivalence of $-G\varrho_0$ and $N_e e^2 / 4\pi\varepsilon_0 m_e$ is apparent.

The dispersion relation (11.25) describes oscillations or instability depending upon the sign of its right-hand side.

- If $c_s^2 k^2 > 4\pi G\varrho_0$, the right-hand side is positive and the perturbations are oscillatory, that is, they are sound waves in which the pressure gradient is sufficient to provide support for the region. Writing the inequality in terms of wavelength, stable oscillations are found for wavelengths less than the critical *Jeans' wavelength* λ_J

$$\lambda_J = \frac{2\pi}{k_J} = c_s \left(\frac{\pi}{G\varrho} \right)^{1/2}. \tag{11.27}$$

- If $c_s^2 k^2 < 4\pi G\varrho_0$, the right-hand side of the dispersion relation (11.25) is negative, corresponding to unstable modes. The solutions can be written

$$\Delta = \Delta_0 \exp(\Gamma t + i\mathbf{k} \cdot \mathbf{r}), \tag{11.28}$$

where

$$\Gamma = \pm \left[4\pi G \varrho_0 \left(1 - \frac{\lambda_J^2}{\lambda^2} \right) \right]^{1/2}. \tag{11.29}$$

The positive solution corresponds to exponentially growing modes. For wavelengths much greater than the Jeans' wavelength, $\lambda \gg \lambda_J$, the growth rate Γ becomes $(4\pi G \varrho_0)^{1/2}$. In this case, the characteristic growth time for the instability is

$$\tau = \Gamma^{-1} = (4\pi G \varrho_0)^{-1/2} \sim (G \varrho_0)^{-1/2}. \tag{11.30}$$

This is the famous *Jeans' Instability* and the time scale τ is the typical collapse time for a region of density ϱ_0.

The physics of this result is very simple. The instability is driven by the self-gravity of the region and the tendency to collapse is resisted by the internal pressure gradient. We can derive the Jeans' instability criterion by considering the pressure support of a region with pressure p, density ϱ and radius r. The equation of hydrostatic support for the region is

$$\frac{dp}{dr} = -\frac{G \varrho M(< r)}{r^2}. \tag{11.31}$$

The region becomes unstable when the self-gravity of the region on the right-hand side of (11.31) overwhelms the pressure forces on the left-hand side. To order of magnitude, we can write $dp/dr \sim -p/r$ and $M \sim \varrho r^3$. Therefore, since $c_s^2 \sim p/\varrho$, the region becomes unstable if $r > r_J \sim c_s/(G\varrho)^{1/2}$. Thus, the Jeans' length is the scale which is just stable against gravitational collapse. Notice that the expression for the Jeans' length is just the distance a sound wave travels in a collapse time. The Jeans' instability is of central importance for the processes of star formation in giant molecular clouds.

11.4 The Jeans' Instability in an Expanding Medium

The results of this section are so important that three different versions of the Jeans' instability in an expanding medium are given, each illustrating different aspects of the same basic physical process.

11.4.1 Small Perturbation Analysis

We return first to the full version of (11.24).

$$\frac{d^2\Delta}{dt^2} + 2\left(\frac{\dot{R}}{R} \right) \frac{d\Delta}{dt} = \Delta(4\pi G \varrho - k^2 c_s^2). \tag{11.24}$$

The second term $2(\dot{R}/R)(d\Delta/dt)$ modifies the classical Jeans' analysis in crucial ways. It is apparent from the right-hand side of (11.24) that the

Jeans' instability criterion applies in this case also but the growth rate is significantly modified. Let us work out the growth rate of the instability in the long wavelength limit $\lambda \gg \lambda_J$, in which case we can neglect the pressure term $c_s^2 k^2$. We therefore have to solve the equation

$$\frac{d^2\Delta}{dt^2} + 2\left(\frac{\dot{R}}{R}\right)\frac{d\Delta}{dt} = 4\pi G\varrho_0\Delta. \tag{11.32}$$

Rather than deriving the general solution, let us first consider the special cases $\Omega_0 = 1$ and $\Omega_0 = 0$ for which the scale factor-cosmic time relations are $R = (\frac{3}{2}H_0 t)^{2/3}$ and $R = H_0 t$ respectively.

• *The Einstein–de Sitter Critical Model* $\Omega_0 = 1$. In this case,

$$4\pi G\varrho = \frac{2}{3t^2} \quad \text{and} \quad \frac{\dot{R}}{R} = \frac{2}{3t}. \tag{11.33}$$

Therefore,

$$\frac{d^2\Delta}{dt^2} + \frac{4}{3t}\frac{d\Delta}{dt} - \frac{2}{3t^2}\Delta = 0. \tag{11.34}$$

By inspection, it can be seen that there must exist power-law solutions of (11.34) and so we seek solutions of the form $\Delta = at^n$. Substituting into (11.34), we find

$$n(n-1) + \tfrac{4}{3}n - \tfrac{2}{3} = 0,$$

which has solutions $n = 2/3$ and $n = -1$. The latter solution corresponds to a decaying mode. The $n = 2/3$ solution corresponds to the growing mode we are seeking, $\Delta \propto t^{2/3} \propto R = (1+z)^{-1}$. This is the key result

$$\Delta = \frac{\delta\varrho}{\varrho} \propto (1+z)^{-1}. \tag{11.35}$$

In contrast to the *exponential* growth found in the static case, the growth of the perturbation in the case of the critical Einstein–de Sitter universe is *algebraic*. This is the origin of the problems of forming galaxies by gravitational collapse in the expanding Universe.

• *The Empty, Milne Model* $\Omega_0 = 0$ In this case,

$$\varrho = 0 \quad \text{and} \quad \frac{\dot{R}}{R} = \frac{1}{t}, \tag{11.36}$$

and hence

$$\frac{d^2\Delta}{dt^2} + \frac{2}{t}\frac{d\Delta}{dt} = 0. \tag{11.37}$$

Again, seeking power-law solutions of the form $\Delta = at^n$, we find $n = 0$ and $n = -1$, that is, there is a decaying mode and one of constant amplitude $\Delta = \text{constant}$.

These simple results describe the evolution of small amplitude perturbations, $\Delta = \delta\varrho/\varrho \ll 1$. In the early stages of the matter-dominated phase, the dynamics of the world models approximate to those of the Einstein–de Sitter model, $R \propto t^{2/3}$, and so the amplitude of the density contrast grows linearly with R. In the late stages, when the Universe may approximate to the $\Omega_0 = 0$ model, the amplitudes of the perturbations grow very slowly and, in the limit $\Omega_0 = 0$, do not grow at all. This last result is not particularly surprising since, if $\Omega_0 = 0$, there is no gravitational driving force to make the perturbation grow.

11.4.2 Perturbing the Friedman Solutions

In our second approach, we investigate the behaviour of density perturbations from the perspective of the dynamics of the Friedman world models. We demonstrated in Sect. 7.2 how the dynamics of these models could be understood in terms of a simple Newtonian model. The development of a spherical perturbation in the expanding Universe can be modelled by embedding a spherical region of density $\varrho + \delta\varrho$ in an otherwise uniform Universe of density ϱ (Fig. 11.1). Using the same approach as in Sect. 7.2, the spherical region behaves dynamically like a Universe of slightly higher density. It is simplest to begin with the parametric solutions (7.24a) for the dynamics of the world models

$$\left.\begin{array}{cc} R = a(1 - \cos\theta) & t = b(\theta - \sin\theta); \\[2mm] a = \dfrac{\Omega_0}{2(\Omega_0 - 1)} & b = \dfrac{\Omega_0}{2H_0(\Omega_0 - 1)^{3/2}}. \end{array}\right\} \qquad (7.24a)$$

First, we find the solutions for small values of θ, corresponding to early stages of the matter-dominated era. Expanding to third order in θ, $\cos\theta = 1 - \frac{1}{2}\theta^2$, $\sin\theta = \theta - \frac{1}{6}\theta^3$, we find the solution

$$R = \Omega_0^{1/3}\left(\frac{3H_0 t}{2}\right)^{2/3}. \qquad (11.38)$$

This solution is identical to (7.25) and shows that, in the early stages, the dynamics of all world models tend towards those of the Einstein–de Sitter model, $\Omega_0 = 1$, that is, $R = (3H_0 t/2)^{2/3}$, but with a different constant of proportionality.

Now, let us look at a region of slightly greater density embedded within the background model. To derive this behaviour, we expand the expressions for R and t to fifth order in θ, $\cos\theta = 1 - \frac{1}{2}\theta^2 + \frac{1}{24}\theta^4 \ldots$, $\sin\theta = \theta - \frac{1}{6}\theta^3 + \frac{1}{120}\theta^5 \ldots$. The solution follows in exactly the same manner as (11.38)

$$R = \Omega_0^{1/3}\left(\frac{3H_0 t}{2}\right)^{2/3}\left[1 - \frac{1}{20}\left(\frac{6t}{b}\right)^{2/3}\right]. \qquad (11.39)$$

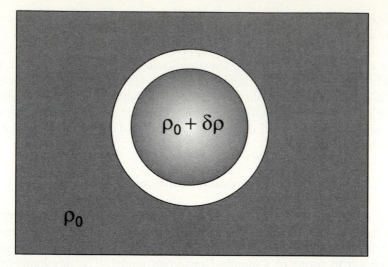

Fig. 11.1. Illustrating a spherical perturbation with slightly greater density than the average in a uniformly expanding Universe. The region with slightly greater density behaves dynamically exactly like a model Universe with density $\varrho_0 + \delta\varrho$.

We can now write down an expression for the change of density of the spherical perturbation with cosmic epoch

$$\varrho(R) = \varrho_0 R^{-3}\left[1 + \tfrac{3}{5}\frac{(\Omega_0 - 1)}{\Omega_0}R\right]. \tag{11.39}$$

Notice that, if $\Omega_0 = 1$, there is no growth of the perturbation. The density perturbation may be considered to be a mini-Universe of slightly higher density than $\Omega_0 = 1$ embedded in an $\Omega_0 = 1$ model. Therefore, the density contrast changes with scale factor as

$$\Delta = \frac{\delta\varrho}{\varrho} = \frac{\varrho(R) - \varrho_0(R)}{\varrho_0(R)} = \tfrac{3}{5}\frac{(\Omega_0 - 1)}{\Omega_0}R. \tag{11.40}$$

This result indicates why density perturbations grow only linearly with cosmic epoch. The instability corresponds to the slow divergence between the variation of the scale factors with cosmic epoch of the model with $\Omega_0 = 1$ and one with slightly greater density. This behaviour is illustrated in Fig. 11.2. This is the essence of the argument developed by Tolman and Lemaître in the 1930s and developed more generally by Lifshitz in 1946 to the effect that, because the instability develops only algebraically, galaxies could not form by gravitational collapse. We will amplify this problem in Sect. 11.7.

This model has another great merit in that it demonstrates that the law of growth of the perturbations applies to fluctuations on any physical scale, including those of wavelength greater than the scale of the horizon, $r > ct$. This follows from the same reasoning which we used in discussing the

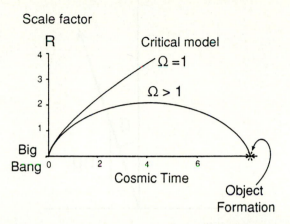

Fig. 11.2. Illustrating the growth of a spherical perturbation in the expanding Universe as the divergence between two Friedman models with slightly different densities.

global dynamics of the Universe in Sect. 7.2. If a perturbation is set up on a scale greater than the horizon, it behaves just like a closed Universe and the amplitude of the fluctuation grows according to $\delta\varrho/\varrho \propto R$. Again, the physics is local physics and the growth is coherent because the perturbation was set up in that way in the first place.

11.4.3 Falling Poles

The third argument contains exactly the same physics. Consider what happens when we attempt to balance a very long, thin pole of length l and mass m on one end. We all know that the situation is unstable and that the pole falls over. This is no more than a gravitational instability in which there is no restoring force to prevent collapse. We can easily work out the growth rate of the instability by conservation of energy in a gravitational field. In Fig. 11.3, the pole is shown at an angle θ to the vertical and then, by conservation of energy, the loss of gravitational potential energy $(gml/2)(1 - \cos\theta)$ must equal the increase in rotational energy $(1/2)I\omega^2$ about the bottom end of the pole O where I is moment of inertia of the pole about O.

$$\frac{gml}{2}(1 - \cos\theta) = \tfrac{1}{2}I\omega^2. \tag{11.41}$$

Since $I = \tfrac{1}{3}ml^2$ and $\omega = \dot{\theta}$, it follows that

$$\dot{\theta}^2 = 3\frac{g}{l}(1 - \cos\theta). \tag{11.42}$$

There is an exact solution for this expression into the non-linear regime but let us only deal with the small angle approximation in which $\cos\theta = (1 -$

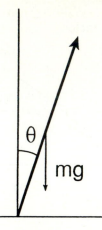

Fig. 11.3. Illustrating a falling pole.

$\theta^2/2 + \ldots$). Then we obtain a simple exponential equation for the collapse of the pole

$$\dot{\theta} = \left(\frac{3g}{2l}\right)^{1/2} \theta. \tag{11.43}$$

The solution $\theta = \theta_0 \exp(\Gamma t)$ with $\Gamma = (3g/2l)^{1/2}$ is the exact analogue of the equation for the growth of the Jeans' instability in the absence of pressure forces in a static medium.

To modify this result for the case of the expanding Universe, we recall that, in the absence of pressure gradients, the differential equation (11.24) for the growth rate of the instability is

$$\frac{\mathrm{d}^2\Delta}{\mathrm{d}t^2} + 2\left(\frac{\dot{R}}{R}\right)\frac{\mathrm{d}\Delta}{\mathrm{d}t} = 4\pi G \varrho_0 \Delta. \tag{11.32}$$

Notice that the force driving the instability on the right-hand side of this expression depends upon the product of the gravitational constant G and the density ϱ_0. Now, in the expanding Universe, $\varrho_0 \propto R^{-3}$ and, in the case of the critical model $\Omega_0 = 1$, $R \propto t^{2/3}$. Therefore, the driving force is proportional $G\varrho_0 \propto t^{-2}$. To simulate this case for our collapsing pole, we can assume that the gravitational acceleration is proportional to t^{-2}, in which case the equation of motion of the pole (11.43) becomes

$$\dot{\theta} = \left(\frac{A}{t^2}\right)^{1/2} \theta. \tag{11.44}$$

Inspection of (11.44) shows that the solutions are of power-law form $\theta \propto t$ rather than exponentially growing solutions. This simple calculation illustrates the origin of the linear algebraic growth of the Jeans' instability in the expanding Universe. The gravitational driving force diminishes with time because the mean density of the Universe decreases as it expands.

11.4.4 The General Solution

The analyses of Sects. 11.4.1 to 11.4.3 give insight into the general solutions of (11.32). Following Heath (1977), Carroll *et al.* (1992) provide a general solution for the growth of the density contrast with scale-factor for all pressure-free Friedman world models. Equation (11.32) can be rewritten in terms of the density parameter Ω_0:

$$\frac{d^2\Delta}{dt^2} + 2\left(\frac{\dot{R}}{R}\right)\frac{d\Delta}{dt} = \frac{3\Omega_0 H_0^2}{2}R^{-3}\Delta, \tag{11.45}$$

where, in general,

$$\dot{R} = H_0\left[\Omega_0\left(\frac{1}{R} - 1\right) + \Omega_\Lambda(R^2 - 1) + 1\right]^{1/2}. \tag{11.46}$$

Following Carroll *et al.* (1992), the solution for the growing mode can be written as follows:

$$\Delta(R) = \frac{5\Omega_0}{2}\left(\frac{1}{R}\frac{dR}{dt}\right)\int_0^R \frac{dR'}{\left(\frac{dR'}{dt}\right)^3}, \tag{11.47}$$

where the constants have been chosen so that the density contrast for the standard critical world model with $\Omega_0 = 1$ and $\Omega_\Lambda = 0$ has unit amplitude at the present epoch, $R = 1$. With this scaling, the density contrasts for all the examples we will consider correspond to $\Delta = 10^{-3}$ at $R = 10^{-3}$. Solutions of this integral can be found in terms of elliptic functions, but it is simplest to carry out the calculations numerically for a representative sample of world models.

In Fig. 11.4, the development of density fluctuations from a scale factor $R = 1/1000$ to $R = 1$ are shown for a range of world models with $\Omega_\Lambda = 0$. These results are consistent with the calculations carried out in Sect. 11.4.1, in which it was argued that the amplitudes of the density perturbations vary as $\Delta \propto R$ so long as $\Omega_0 z \gg 1$, but the growth essentially stops at smaller redshifts.

Since $R = 1/1000$, corresponds to the epoch of recombination, or the last scattering surface of the Cosmic Microwave Background Radiation, the density perturbations developed by relatively modest factors over an interval of cosmic time from about 300,000 to 10^{10} years after the origin of the Big Bang. For example, if $\Omega_0 = 1$, the increase is a factor of 10^3, as expected from (11.35). In the case $\Omega_0 = 0.1$, the amplitudes of the fluctuations grow as $\Delta \propto R$ over the range of scale factors from $R = 10^{-3}$ to 10^{-1}, but grow

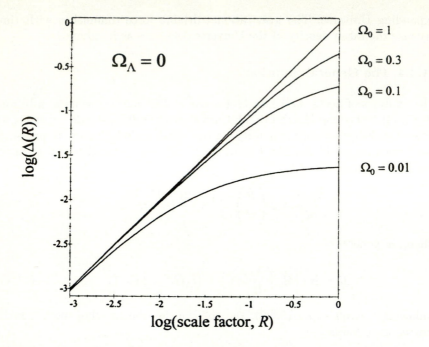

Fig. 11.4. The growth of density perturbations over the range of scale-factors $R = 10^{-3}$ to 1 for world models with $\Omega_\Lambda = 0$ and density parameters $\Omega_0 = 0.01$, 0.1, 0.3 and 1.

only modestly from $R = 10^{-1}$ to 1. In this case, which is of considerable astrophysical interest, the growth of the density contrast is only by a factor of 190 from the epoch of recombination to the present epoch. If Ω_0 were 0.01, the growth of the density contrast would be even smaller, a factor of only 24.

Similar calculations can be carried out for the cases in which $\Omega_\Lambda \neq 0$. Those which are of the greatest interest are the flat models for which ($\Omega_0 + \Omega_\Lambda$) = 1. Fig. 11.5 shows the development of the fluctuations over the range of scale-factors from $R = 1/30$ to the present epoch $R = 1$, in all cases, the fluctuations having amplitude $\Delta = 10^{-3}$ at $R = 10^{-3}$. The growth of the density contrast is somewhat greater in the cases $\Omega_0 = 0.1$ and 0.3 as compared with the corresponding cases with $\Omega_\Lambda = 0$. For example, in the case $\Omega_0 = 0.1$, the growth of the fluctuation from $R = 1/1000$ to 1 is 610. Inspection of Fig. 11.5 shows that the fluctuations continue to grow to greater values of the scale-factor R, corresponding to smaller redshifts, as compared with the models with $\Omega_\Lambda = 0$. The redshift at which the growth rate of the instability slows down can be deduced from (11.46). If $\Omega_0 + \Omega_\Lambda = 1$, the expression for $\dot{R}$ becomes

$$\dot{R} = H_0 \left(\frac{\Omega_0}{R} + \Omega_\Lambda R^2 \right)^{1/2} = H_0 \left[\frac{\Omega_0}{R} + (1 - \Omega_0) R^2 \right]^{1/2} . \qquad (11.48)$$

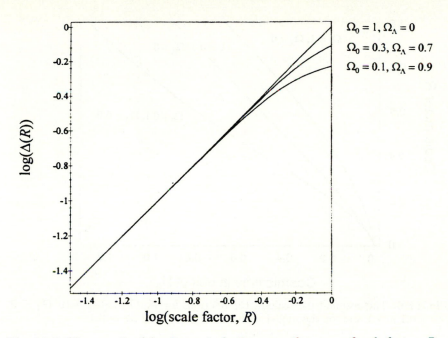

Fig. 11.5. The growth of density perturbations over the range of scale-factors $R = 1/30$ to 1 for world models with $\Omega_0 + \Omega_\Lambda = 0$ and density parameters $\Omega_0 = 0.1$, 0.3 and 1.

Thus, when the first term in square brackets is greater than the second, the same dynamics as in the case of the critical model is found, corresponding to growth as $\Delta \propto R$. When the second term dominates, the universe begins to accelerate under the influence of the vacuum fields, or the repulsive effect of the Λ-term, and then growth is suppressed – recall that even if the expansion is unaccelerated, there is no growth of the perturbations. Thus, the instability can grow to scale factors such that $\Omega_0/R = (1 - \Omega_0)R^2$, that is,

$$R \approx \left(\frac{\Omega_0}{1 - \Omega_0} \right)^{1/3} \quad \text{or} \quad (1+z) \approx \Omega_0^{-1/3} \quad \text{if } \Omega_0 \ll 1. \qquad (11.49)$$

Notice that, expressing the development of the fluctuations in terms of scale-factor rather than cosmic time disguises the fact that the time dependence of the models with and without Ω_Λ are significantly different. Fig. 11.6 shows the cosmic time-scale factor relation for three of the world models discussed above. It can be seen that the effect of the cosmological constant is to stretch out the cosmic time-scale allowing more time for the density perturbation to grow.

We have considered only the cases of Friedman models with $\Omega_\Lambda = 0$ and flat cosmological models with $\Omega_0 + \Omega_\Lambda = 1$, but the space of conceivable world models is much greater than these cases. Generally, the results are

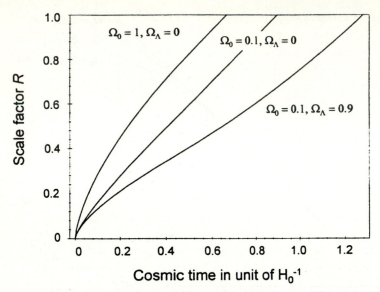

Fig. 11.6. The cosmic-time scale-factor relations for world models with $\Omega_A = 0$, $\Omega_0 = 0.1$ and 1 and for the model with $\Omega_0 + \Omega_A = 1$ and $\Omega_0 = 0.1$.

not too different from those described above, except in cases in which the expansion almost becomes stationary, that is, models close to the 'loitering' line in Fig. 7.7(a). In these cases, much greater growth of the perturbations is possible because the instability behaves like the standard Jeans' instability in a static medium (see Carroll *et al.* 1992). We have argued, however, that these 'loitering' models are unlikely to be applicable to our Universe.

11.5 The Evolution of Peculiar Velocities in the Expanding Universe

The development of velocity perturbations in the expanding Universe can be derived from (11.20). Let us investigate the case in which we can neglect pressure gradients so that the velocity perturbations are only driven by the potential gradient $\delta\phi$.

$$\frac{d\mathbf{u}}{dt} + 2\left(\frac{\dot{R}}{R}\right)\mathbf{u} = -\frac{1}{R^2}\nabla_c\delta\phi. \tag{11.50}$$

We recall that $\mathbf{u}$ is the perturbed *comoving* velocity. Let us split the velocity vector into components parallel and perpendicular to the gravitational potential gradient, $\mathbf{u} = \mathbf{u}_\parallel + \mathbf{u}_\perp$, where $\mathbf{u}_\parallel$ is parallel to $\nabla_c\delta\phi$. The velocity associated with $\mathbf{u}_\parallel$ is often referred to as *potential motion* since it is driven by the potential gradient. On the other hand, the perpendicular velocity component $\mathbf{u}_\perp$ is not driven by potential gradients and corresponds to *vortex* or *rotational motions*. We consider the growth of the velocity perturbations as the gravitational instability develops.

Rotational Velocities. Consider first the rotational component $\mathbf{u}_\perp$. Equation (11.50) reduces to

$$\frac{\mathrm{d}\mathbf{u}_\perp}{\mathrm{d}t} + 2\left(\frac{\dot{R}}{R}\right)\mathbf{u}_\perp = 0. \tag{11.51}$$

The solution of this equation is straightforward $\mathbf{u}_\perp \propto R^{-2}$. Since $\mathbf{u}_\perp$ is a comoving perturbed velocity, the proper velocity is $\delta\mathbf{v}_\perp = R\mathbf{u}_\perp \propto R^{-1}$. Thus, the rotational velocities decay as the Universe expands. This is no more than the conservation of angular momentum in an expanding medium, $mvr = $ constant. This poses a grave problem for models of galaxy formation involving primordial turbulence. Rotational turbulent velocities decay and there must be sources of turbulent energy, if the rotational velocities are to be maintained.

Potential Motions. The development of potential motions is most directly derived from (11.13)

$$\frac{\mathrm{d}\Delta}{\mathrm{d}t} = -\nabla \cdot \delta\mathbf{v}, \tag{11.13}$$

that is, the divergence of the peculiar velocity is proportional to minus the rate of growth of the density contrast. The peculiar velocity $\delta\mathbf{v}_\parallel$ is parallel to the wave vector of the perturbation $\Delta = \Delta_0 \exp \mathrm{i}(\mathbf{k} \cdot \mathbf{x} - \omega t) = \Delta_0 \exp \mathrm{i}(\mathbf{k}_\mathrm{c} \cdot \mathbf{r} - \omega t)$ and so, using comoving derivatives, (11.13) can be rewritten

$$\frac{\mathrm{d}\Delta}{\mathrm{d}t} = -\frac{1}{R}\nabla_\mathrm{c} \cdot (R\mathbf{u}) = -\mathrm{i}k_\mathrm{c} \cdot \mathbf{u}, \tag{11.52}$$

that is,

$$|\delta v_\parallel| = \frac{R}{k_\mathrm{c}}\frac{\mathrm{d}\Delta}{\mathrm{d}t}. \tag{11.53}$$

Notice that we have written this expression in terms of the comoving wave vector k_c which means that this expression describes how the peculiar velocity associated with a particular perturbation changes with cosmic epoch. Let us consider separately the cases $\Omega_0 = 1$ and $\Omega_0 = 0$.

- $\Omega_0 = 1$. As shown above, $\Delta = \Delta_0(t/t_0)^{2/3}$ and $R = (3H_0t/2)^{2/3}$. Therefore,

$$|\delta v_\parallel| = |Ru| = \frac{H_0 R^{1/2}}{k}\left(\frac{\delta\varrho}{\varrho}\right)_0 = \frac{H_0}{k}\left(\frac{\delta\varrho}{\varrho}\right)_0 (1+z)^{-1/2}, \tag{11.54}$$

where $(\delta\varrho/\varrho)_0$ is the density contrast at the present epoch. This calculation shows how potential motions grow with cosmic time in the critical

model, $\delta v_\parallel \propto t^{1/3}$. In addition, it can be seen that the peculiar velocities are driven by both the amplitude of the perturbation and its scale.
Equation (11.54) shows that, if $\delta\varrho/\varrho$ is the same on all scales, the peculiar
velocities are driven by the smallest values of k, that is, by the perturbations on the largest physical scales. Thus, local peculiar velocities can
be driven by density perturbations on the very largest scales, which is an
important result for understanding the origin of the peculiar motion of
the Galaxy with respect to the frame of reference in which the Microwave
Background Radiation is 100% isotropic and of large-scale streaming velocities.

- $\Omega_0 = 0$. In this case, it is simplest to proceed from (11.20) in which there
  is no driving term in the equation

$$\frac{d\mathbf{u}}{dt} + 2\left(\frac{\dot{R}}{R}\right)\mathbf{u} = 0. \tag{11.55}$$

The solution is the same as that for $u_\perp$ given above, that is, $\delta v_\parallel \propto R^{-1}$
– the peculiar velocities decay with time. This is the same result deduced
in Sect. 7.1.

- In the general case, we need the expression for $d\Delta/dt$ which is most simply derived from the numerical solutions of (11.45). There is, however,
  a useful approximation which is often used in the context of the cosmic
  virial theorem for the magnitude of the peculiar velocities induced by density perturbations, such as superclusters and other large scale structures.
  Suppose we write $\Delta = \Delta_0 f(t)$, where Δ_0 is the amplitude of the density
  perturbation at the present epoch $t = t_0$. Then, we can write (11.53) as
  follows:

$$|\delta v_\parallel| = \frac{R}{k_c}\Delta_0\frac{d\Delta}{dt} = \frac{R}{k_c}\Delta_0\frac{df}{dR}\frac{dR}{dt}. \tag{11.56}$$

At the present epoch, $t = t_0$, $R = 1$, $dR/dt = H_0$ and so the peculiar
velocity is

$$|\delta v_\parallel| = \frac{\Delta_0 H_0}{k_c}\left(\frac{df}{dR}\right)_0. \tag{11.57}$$

For the critical model $\Omega_0 = 1$, $f = R$ and we immediately recover (11.54)
at $z = 0$. Thus, the amplitude of the peculiar velocities at the present
epoch depends upon df/dR. To a good approximation, this function can
be written $df/dR = \Omega_0^{0.6}$, the result being exact for the case $\Omega_0 = 1$
(Gunn 1978, Peebles 1980). This is the origin of (8.20).

These solutions provide the general rules for the evolution of peculiar
velocities in the expanding Universe. So long as $\Omega_0 z \gg 1$, velocities driven
by potential gradients grow as $t^{1/3}$, but at redshifts $\Omega_0 z \ll 1$, the velocities
decrease. For a given value of Ω_0, there is a redshift at which the peculiar velocities of galaxies selected randomly from the general field have a maximum
value and, if this could be measured, an estimate of Ω_0 could be obtained.

11.6 The Relativistic Case

We investigate next the case of a relativistic gas because, in the radiation-dominated phase of the Big Bang, the primordial perturbations are in a radiation-dominated plasma, for which the relativistic equation of state $p = \frac{1}{3}\varepsilon$ is appropriate. We therefore require the relativistic generalisations of (11.5), (11.6) and (11.7). Equation (11.5), the equation of continuity, becomes an equation describing the conservation of energy. There is no simple way of demonstrating this except by using the general energy-momentum tensor for a fully relativistic gas (see, for example, Weinberg 1972, d'Inverno 1995). The equation of energy conservation becomes

$$\frac{\partial \varrho}{\partial t} = -\nabla \cdot \left(\varrho + \frac{p}{c^2}\right)\mathbf{v}; \tag{11.58}$$

$$\frac{\partial}{\partial t}\left(\varrho + \frac{p}{c^2}\right) = \frac{\dot{p}}{c^2} - \left(\varrho + \frac{p}{c^2}\right)(\nabla \cdot \mathbf{v}). \tag{11.59}$$

Substituting $p = \frac{1}{3}\varrho c^2$ into (11.58) and (11.59), the relativistic continuity equation is obtained:

$$\frac{d\varrho}{dt} = -\frac{4}{3}\varrho(\nabla \cdot \mathbf{v}). \tag{11.60}$$

Euler's equation for the acceleration of an element of the fluid in the gravitational potential ϕ becomes

$$\left(\varrho + \frac{p}{c^2}\right)\left[\frac{\partial \mathbf{v}}{\partial t} + (\mathbf{v} \cdot \nabla)\mathbf{v}\right] = -\nabla p - \left(\varrho + \frac{p}{c^2}\right)\nabla\phi. \tag{11.61}$$

If we neglect the pressure gradient term, (11.61) reduces to the familiar equation

$$\frac{d\mathbf{v}}{dt} = -\nabla\phi. \tag{11.62}$$

Finally, the differential equation for the gravitational potential ϕ becomes

$$\nabla^2\phi = 4\pi G\left(\varrho + \frac{3p}{c^2}\right). \tag{11.63}$$

For a fully relativistic gas, $p = \frac{1}{3}\varrho c^2$ and so

$$\nabla^2\phi = 8\pi G\varrho. \tag{11.64}$$

The net result is that the equations for the evolution of the perturbations in a relativistic gas are of similar mathematical form to the non-relativistic case. As shown by Coles and Lucchin (1995), the same type of analysis which was carried out in Sect. 11.2 leads to the following equation

$$\frac{d^2\Delta}{dt^2} + 2\left(\frac{\dot{R}}{R}\right)\frac{d\Delta}{dt} = \Delta\left(\frac{32\pi G\varrho}{3} - k^2 c_s^2\right). \tag{11.65}$$

The relativistic expression for the Jeans' length is found by setting the right-hand side equal to zero,

$$\lambda_J = \frac{2\pi}{k_J} = c_s \left(\frac{3\pi}{8G\varrho} \right)^{1/2},$$

(11.66)

where $c_s = c/\sqrt{3}$ is the relativistic sound speed. The result is similar to the standard expression (11.27) for the Jeans' length.

Neglecting the pressure gradient terms in (11.65), the following differential equation for the growth of the instability is obtained

$$\frac{d^2\Delta}{dt^2} + 2 \left(\frac{\dot{R}}{R} \right) \frac{d\Delta}{dt} - \frac{32\pi G\varrho}{3}\Delta = 0.$$

(11.67)

This equation is the same as (11.32) but with a different constant in the last term. Using the same approach as in Sect. 11.4.1, we seek solutions of the form $\Delta = at^n$, recalling that in the radiation-dominated phases, the scale factor-cosmic time relation is given by (9.7), $R \propto t^{1/2}$. Going through precisely the same procedure as before, we find solutions $n = \pm 1$. Hence, for wavelengths $\lambda \gg \lambda_J$, the growing solution corresponds to

$$\Delta \propto t \propto R^2 \propto (1+z)^{-2}.$$

(11.68)

Thus, once again, the unstable mode grows algebraically with cosmic time. It will be noted again that nowhere does the analysis describe the scale of the perturbation relative to the horizon scale.

11.7 The Basic Problem

We will use all the above results in the analyses of structure formation which follow. We have concentrated upon deriving a number of exact results which will prove useful in understanding the more complete analyses found in the literature. Padmanabhan (1993) and Coles and Lucchin (1995) provide many useful solutions of the basic equations for the development of the density perturbations in the expanding Universe and their analyses can be thoroughly recommended.

For the moment, let us note the implications of one of the key results derived above. At redshifts $\Omega_0 z > 1$ throughout the radiation-dominated era, the dynamics of the Universe are approximately those of the critical model $R \propto t^{2/3}$ and the growth rate of perturbations on physical scales much greater than the Jeans' length is

$$\Delta = \frac{\delta\varrho}{\varrho} \propto R = \frac{1}{1+z}.$$

(11.35)

At redshifts less than $1/\Omega_0$, the instability grows much more slowly and, in the limit $\Omega_0 = 0$, does not grow at all (Fig. 11.4).

Since galaxies and astronomers certainly exist at the present day $z = 0$, it follows that $\Delta \geq 1$ at $z = 0$ and so, at the last scattering surface, $z \sim 1,000$, fluctuations must have been present with amplitude at least $\Delta = \delta\varrho/\varrho \geq 10^{-3}$. The same result is found for models with $\Omega_\Lambda \neq 0$, if $\Omega_0 + \Omega_\Lambda = 1$. The key question is whether or not these density perturbations are observable. This is a far from trivial question and the answers depend upon the nature of the density perturbations and upon the processes of decoupling of the matter and the radiation. We will take up these issues in the next three chapters.

In the mean time, we can look at the results of these calculations in two ways.

- On the one hand, the slow growth of density perturbations is the source of a fundamental problem in understanding the origin of galaxies – large-scale structures did not condense out of the primordial plasma by exponential collapse, as may well occur in the formation of stars in dense interstellar clouds. It must be a very much slower process.
- On the other hand, because of the slow development of the density perturbations, we have the real possibility of learning about many aspects of the early Universe which would otherwise have been excluded. We have the opportunity of studying the formation of structure on the last scattering surface at a redshift $z \sim 1,000$ and, even more important, we can obtain crucial information about the spectrum of fluctuations which must have been present in the very early Universe. Thanks to the slow growth of the fluctuations, we have a direct probe of the very early Universe.

12 The Simplest Picture of Galaxy Formation and Why It Fails

We are now in a position to make a first attempt at understanding how galaxies and the large-scale structure of the Universe came about. In this chapter, we examine the case in which it is assumed that all the matter in the Universe, both 'visible' and dark, is in baryonic form. This was the natural starting point for the first plausible models for galaxy formation in the 1960s and 1970s, but they are doomed to failure. Despite this, there are two good reasons for studying this case carefully. The first is that many important physical processes will be introduced which will be needed in the construction of acceptable models. The second aspect is that the failure of these models strongly suggests that the dominant dark matter present in the Universe must be in some non-baryonic form. We need to convince ourselves of the reasoning behind this dramatic conclusion – the nature of the non-baryonic dark matter is of fundamental importance for physics and cosmology.

12.1 Horizons and the Horizon Problem

One of the key concepts in the development of the theory of structure formation in the expanding Universe is that of *particle horizons*. At any epoch t, the particle horizon is defined to be the maximum distance over which causal communication could have taken place by that epoch. In other words, this distance describes how far a light signal could have travelled from the origin of the Big Bang at $t = 0$ by the epoch t. The complication is the fact that the Universe is expanding at a rate which varies with cosmic epoch. We get round this problem by the same reasoning which led to the definition of comoving radial distance coordinates (Sect. 5.4). There, the problem was to define a distance at a single cosmic epoch, despite the fact that observations are made along a past light cone, which spans a wide range of cosmic epochs. In that case, proper distances were projected forward to a reference epoch t_0, taken to be the present epoch, and the same procedure is carried out in this case. The projection takes account of the fact that the fundamental observers are separating as the Universe expands. We can then scale the comoving distance scale back to the epoch t simply by multiplying by the scale factor of the Universe at the time t, according to the definition (5.30).

The comoving radial distance coordinate r corresponding to the distance travelled by a light signal from the origin of the Big Bang to the epoch t is

$$r = \int_0^t \frac{c\,dt}{R(t)} = \int_\infty^z (1+z)c\,dt. \tag{12.1}$$

This integral is similar to that which led to standard result (7.37), the difference being that the comoving radial distance coordinates in that case were integrated from redshifts 0 to z, whereas in the present case, the integral is from $z = \infty$ to z. To find the horizon scale at the epoch corresponding to redshift z, we simply scale r by the scale factor $R(t) = (1+z)^{-1}$. Thus, the definition of the particle horizon $r_H(t)$ at the cosmic epoch t, corresponding to the redshift z is

$$r_H(t) = R(t) \int_0^t \frac{c\,dt}{R(t)} = \frac{1}{1+z} \int_\infty^z (1+z)c\,dt. \tag{12.2}$$

In the case of the standard Friedman models with $\Omega_\Lambda = 0$, we use (7.28) for dt so that, if $\Omega_0 > 1$, the integral is

$$\frac{c}{H_0} \int \frac{dz}{(1+z)(\Omega_0 z + 1)^{1/2}} = \frac{2c}{H_0(\Omega_0 - 1)^{1/2}} \tan^{-1} \left(\frac{\Omega_0 z + 1}{\Omega_0 - 1} \right)^{1/2}. \tag{12.3}$$

Putting in the limits to the integral and carrying out a little algebra, we find

$$r_H(t) = \frac{c}{H_0(\Omega_0 - 1)^{1/2}} R \cos^{-1} \left[1 - \frac{2(\Omega_0 - 1)}{\Omega_0} R \right]. \tag{12.4}$$

The corresponding result for $\Omega_0 < 1$ is:

$$r_H(t) = \frac{c}{H_0(1 - \Omega_0)^{1/2}} R \cosh^{-1} \left[1 + \frac{2(1 - \Omega_0)}{\Omega_0} R \right]. \tag{12.5}$$

In the critical Einstein–de Sitter case, the particle horizon is

$$r_H(t) = \frac{2c}{H_0} R^{3/2}(t) = \frac{2c}{H_0} \frac{t}{t_0} = 3ct. \tag{12.6}$$

Notice that, at early times, corresponding to small values of R, both (12.4) and (12.5) reduce to

$$r_H(t) = \frac{2c}{H_0 \Omega_0^{1/2}} R^{3/2}, \tag{12.7}$$

similar to (12.6), but with a different constant. At early times, all the Friedman models tend toward the dynamics of the critical model

$$R = \Omega_0^{1/3} \left(\frac{3H_0 t}{2} \right)^{2/3}, \tag{7.25}$$

and so the particle horizon becomes $r_{\mathrm{H}}(t) = 3ct$, the same as (12.6). These results make physical sense since one might expect that the typical distance which light could travel by the epoch t would be of order ct. The factor 3 takes account of the fact that fundamental observers were closer together at early epochs and so greater distances could be causally connected than ct.

A similar calculation can be carried out for the radiation-dominated era at redshifts $z \gg 2.4 \times 10^4 \Omega_0 h^2$, during which the dynamics of the expansion were $R \propto t^{1/2}$. Performing the integral (12.2) for this case, we find $r_{\mathrm{H}}(t) = 2ct$. The factor of 2 reflects the difference between the early dynamics of the radiation-dominated and matter-dominated universes.

We can now use these results to illustrate the origin of the *horizon problem* for the standard Friedman models with $\Omega_\Lambda = 0$. Let us work out the angle θ_{H} which the particle horizon on the last scattering surface subtends, according to an observer at the present epoch. At a redshift $z = 1000$, we can safely use the standard matter-dominated solutions of Friedman's equation in the limit $\Omega_0 z \gg 1$, which can be found from (7.38), $D = 2c/H_0\Omega_0$. Therefore, from (5.57) and (12.7), we find

$$\theta_{\mathrm{H}} = \frac{r_{\mathrm{H}}(t)(1+z)}{D} = \frac{\Omega_0^{1/2}}{(1+z)^{1/2}} = 1.8\Omega_0^{1/2} \quad \text{degrees.} \tag{12.8}$$

This result means that, according to the standard Friedman picture, regions of the Universe separated by an angle of more than $1.8\Omega_0^{1/2}$ degrees on the sky could not have been in causal contact on the last scattering surface. Why then is the Cosmic Microwave Background Radiation so uniform over the whole sky to a precision of about one part in 10^5? In the standard picture, it has to be assumed that the remarkable isotropy of the Universe was part of its initial conditions.

This problem is circumvented in the inflationary model of the very early Universe because of the exponential expansion of the scale factor which ensures that opposite directions on the sky were in causal contact. To illustrate this, consider the de Sitter model described by (7.71). Normalising $R(t)$ to the value unity at the present epoch, we find

$$R(t) = \exp\left[\Omega_\Lambda^{1/2} H_0(t - t_0)\right],$$

$$r_{\mathrm{H}}(t) = \frac{c}{H_0} R(t) \frac{\left[\exp\left(\Omega_\Lambda^{1/2} H_0 t\right) - 1\right]}{\Omega_\Lambda^{1/2}}. \tag{12.9}$$

In the inflationary picture described in Sect. 7.3.1, the value of $\Omega_\Lambda^{1/2}$ is enormous and so causal communication can extend far beyond the scale ct when $\Omega_\Lambda^{1/2} H_0 t \gg 1$.

To complete this discussion of horizons, we note that the term *event horizon*, introduced by Rindler (1956), also appears in the cosmological literature. The event horizon is the greatest distance an object can have at a particular cosmic epoch, if it is ever to be observable, however long the observer waits. Consider a light ray emitted at time t_1 which arrives at the observer at the time t. Then, the comoving radial distance coordinate traversed by the light ray is

$$\int_{t_1}^{t} \frac{c\,dt}{R(t)}. \tag{12.10}$$

The question is whether or not this integral converges as $t \to \infty$ in the open models, or as $t \to t_{\max}$ for the collapsing closed models. The definition of the event horizon is therefore

$$r_{\mathrm{E}} = \int_{t_1}^{t_{\max}} \frac{c\,dt}{R(t)}. \tag{12.11}$$

Evidently, in the standard world models with $\Omega_\Lambda = 0$, if $\Omega_0 \leq 1$, the integral (12.11) diverges, since $R \propto t^{2/3}$ for the $\Omega_0 = 1$ model and $R \propto t$ for $\Omega_0 = 0$. Therefore, it is eventually possible, in principle, to observe every particle there is in the Universe in these models. If $\Omega_0 > 1$, the integral (12.11) converges to a finite comoving radial distance coordinate as $t \to t_{\max} = \pi \Omega_0 / H_0 (\Omega_0 - 1)^{3/2}$ (7.22). The integral is the same as (12.3) and, inserting the limits, we find

$$r_{\mathrm{E}} = \frac{c}{H_0(\Omega_0 - 1)^{1/2}} \left\{ 2\pi - \cos^{-1}\left[1 - \frac{2(\Omega_0 - 1)}{\Omega_0} R(t_1) \right] \right\}. \tag{12.12}$$

As Weinberg (1972) pointed out, if $\Omega_0 = 2$, the most distant object we would ever be able to observe before we encounter the Big Crunch has comoving radial distance coordinate 14,000 h^{-1} Mpc at the present epoch.

12.2 Adiabatic Fluctuations in the Standard Big Bang

We now have much of the information we need to discuss the simplest case, the evolution of adiabatic perturbations in the standard Big Bang model. We need the following information:

- *The Jeans' length* is the maximum scale for stable fluctuations at any epoch

$$\lambda_{\mathrm{J}} = \frac{2\pi}{k_{\mathrm{J}}} = c_{\mathrm{s}} \left(\frac{\pi}{G\varrho} \right)^{1/2}. \tag{11.27}$$

- For wavelengths smaller than the Jeans' length, the perturbations are sound waves which have angular frequency

$$\omega = \left[4\pi G \varrho_0 \left(\frac{\lambda_{\mathrm{J}}^2}{\lambda^2} \right) - 1 \right]^{1/2}. \tag{12.13}$$

For wavelengths greater than the Jeans' wavelength, the perturbations are unstable and, for wavelengths $\lambda \gg \lambda_J$, the *growth rates* of the unstable modes are algebraic with epoch. In the matter-dominated phase, the perturbation grows as

$$\Delta = \frac{\delta\varrho}{\varrho} \propto R = (1+z)^{-1}, \qquad (11.35)$$

so long as $\Omega_0 z \gg 1$. The growth is much slower at smaller redshifts and becomes zero in the limit $\Omega_0 = 0$. In the radiation-dominated phases of the standard model, when all the inertial mass and pressure are associated with relativistic matter or radiation, the growth rate is algebraic with $\Delta \propto R^2$.

- *The horizon scale* is the maximum distance over which information can be communicated at cosmic epoch t and, as shown in the last Section, is $r_H(t) = 3ct$.

There are many more complications to be added later, but these rules are adequate to begin our analysis. Let us use these rules to study the evolution of perturbations of different masses in the standard model. There is some uncertainty about how to relate the wavelength λ_J to the mass of the object M_J which ultimately forms from it. The expectation is that the ultimate mass of the object which forms from a perturbation of wavelength λ_J will comprise all the mass within a sphere of diameter λ_J. For illustrative purposes, we adopt the definition of the *Jeans' mass* as the mass contained within a region of diameter λ_J,

$$M_J = (\pi\lambda_J^3/6)\varrho_B, \qquad (12.14)$$

recalling that, in the present chapter, all the mass in the Universe is in baryonic form.

12.2.1 The Radiation-Dominated Era

Let us consider first of all the radiation-dominated phases, when the inertial mass density in the radiation and neutrinos was much greater than that in the baryons, $\varepsilon_{\text{rad}} \gg \varepsilon_B$. During this era, essentially all the inertial mass of the perturbations is in the radiation, but the plasma is strongly coupled to it by Compton scattering, as discussed in Sect. 9.4. Although the Jeans' length is determined by the energy density of radiation, we are interested in the mass of baryons within this scale, since this is the mass which eventually forms bound objects in the matter-dominated era. The mass density in baryons varies with redshift as $\varrho_B = 1.88 \times 10^{-26}\Omega_B h^2(1+z)^3$ kg m^{-3}, where Ω_B is the density parameter in baryons at the present epoch. The Jeans' length in the radiation-dominated phase is

$$\lambda_J = \frac{c}{\sqrt{3}}\left(\frac{3\pi}{8G\varrho}\right)^{1/2}, \qquad (11.66)$$

where ϱ is the total mass density including both photons and neutrinos, that is, $\varrho = 4.7 \times 10^{-31}\chi(1 + z)^4$ kg m^{-3}, recalling that $\chi = 1.7$ when the neutrinos are taken into account. Therefore, the mass within the Jeans' length in baryons in the early stages of the radiation dominated phase, $z \gg 2.4 \times 10^4\Omega_0 h^2$, is

$$M_J = 8.5 \times 10^{28}(1+z)^{-3}\Omega_B h^2 \; M_\odot = 8.5 \times 10^{28}R^3\Omega_B h^2 \; M_\odot. \quad (12.15)$$

Several important conclusions can be drawn from this result. The first is that the baryon mass within λ_J grows as $M_B \propto R^3$ during the radiation-dominated phases. Thus, M_B is one solar mass at a redshift $z \sim 3 \times 10^9$ and increases to the mass of a large galaxy $M = 10^{11}M_\odot$ at redshift $z \sim 10^6$. The second conclusion follows from a comparison of the Jeans' length with the horizon scale $r_H = 2ct$. Using (9.7), the horizon scale can be written

$$r_H = 2ct = c\left(\frac{3}{8\pi G\varrho}\right)^{1/2} \quad \text{compared with} \quad \lambda_J = c\left(\frac{3\pi}{24G\varrho}\right)^{1/2}. \quad (12.16)$$

It is apparent that, during the radiation-dominated phases, the Jeans' length is of the same order as the horizon scale. This is a key result. Consider a perturbation containing a galactic mass of baryons, say $M = 10^{11}M_\odot$. In the early stages of the radiation-dominated phase, its scale far exceeds the horizon scale and so the amplitude of the perturbations grows as $\Delta \propto (1 + z)^{-2}$. At a redshift $z \sim 10^6$, the perturbation enters the horizon and, at more or less the same time, the Jeans' length becomes greater than the scale of the perturbation. The perturbation is therefore stable against gravitational collapse at later times and becomes a sound wave which oscillates at constant amplitude. As long as the Jeans' length remains greater than the scale of the perturbation, the perturbation does not grow in amplitude. The variations with redshift of the Jeans' mass and the baryonic mass within the horizon are shown schematically in Fig. 12.1.

12.2.2 The Matter-Dominated Era

The above results hold good during the radiation-dominated era at redshifts $z \gg 2.4 \times 10^4\Omega_0 h^2$ (see Sects. 9.2 and 9.5). At redshifts $z \ll 2.4 \times 10^4\Omega_0 h^2$, the dynamics of the expansion become matter-dominated, but the matter and the radiation remain strongly thermally coupled so long as the diffuse cosmic plasma remains ionised. As discussed in Sects. 9.3 and 9.4, at a redshift $z \approx 1,500$, the plasma is only 50% ionised and, at a redshift $z \approx 550h^{2/5}\Omega_0^{1/5}$, the thermal coupling between the matter and the radiation ceases. These changes profoundly influence the variation of the Jeans' mass with cosmic epoch.

Let us first work out the variation of the baryonic mass within the particle horizon as a function of redshift, or scale factor, during the matter-dominated

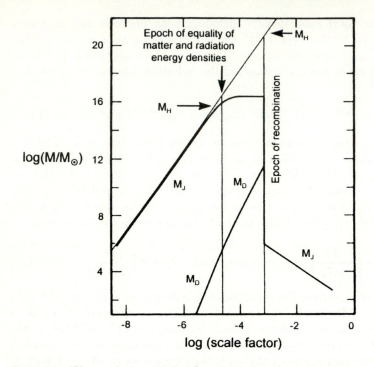

Fig. 12.1. The evolution of the Jeans' mass and the baryonic mass within the particle horizon with scale factor. Also shown is the evolution of the mass scales which are damped by photon diffusion.

era. Using (12.7) for the variation of the particle horizon with scale factor, the baryonic mass within the horizon is

$$M_{\mathrm{H}} = \left(\frac{\pi r_{\mathrm{H}}^3}{6}\right) \varrho_{\mathrm{B}} = \frac{3.0 \times 10^{22}}{(\Omega_0 h^2)^{1/2}} R^{3/2} \, M_{\odot}, \tag{12.17}$$

that is, $M_{\mathrm{H}} \propto R^{3/2} \propto t$, so long as $\Omega_0 z \gg 1$. This relation is shown in Fig. 12.1.

To determine the Jeans' length, we need to know the variation of the speed of sound with redshift. As the epoch of equality of the rest mass energies in matter and radiation is approached, the sound speed becomes less than the relativistic sound speed $c/\sqrt{3}$ and is given by (9.32).

$$c_{\mathrm{s}}^2 = \frac{c^2}{3} \frac{4\varrho_{\mathrm{rad}}}{4\varrho_{\mathrm{rad}} + 3\varrho_{\mathrm{m}}}. \tag{9.32}$$

During this phase, the pressure within the perturbations is provided by the radiation, but the inertial mass by the matter and so the appropriate sound speed is

$$c_s = c \left(\frac{4\varrho_{rad}}{9\varrho_B} \right)^{1/2} = \frac{10^6 (1+z)^{1/2}}{(\Omega_B h^2)^{1/2}} \text{ m s}^{-1}. \tag{9.33}$$

Inserting this result into the expression for the Jeans' mass, we find

$$M_J = \frac{3.75 \times 10^{15}}{(\Omega_B h^2)^2} M_\odot, \tag{12.18}$$

that is, the Jeans' mass is independent of scale factor (or redshift) during this era. Notice the important result that adiabatic perturbations with masses greater than the mass given by (12.18) grow according to the standard result for the matter-dominated era $\Delta \propto R$ from the time they come through the horizon. Even if $\Omega_0 h^2 = 1$, this mass corresponds to the mass of the richest clusters of galaxies – all structures on larger scales continue to grow when they come through the particle horizon.

It is an interesting question whether or not this regime existed prior to the epoch of recombination. According to the analysis of Sect. 9.5, the epoch of equality of the inertial mass densities in matter and radiation occurred at $z = 2.4 \times 10^4 \Omega_0 h^2$. If Ω_0 were as low as 0.1 and $h = 0.5$, the epoch of equality of the matter and radiation energy densities would occur about a redshift of 600, after the epoch of recombination, and close to the epoch when the coupling between matter and radiation ceases, $z = 550 h^{2/5} \Omega_0^{1/5} = 260$. In this case, there would be a precipitous drop in the sound speed from $c/\sqrt{3}$ to the thermal sound speed of a gas at $z = 260$, which would be roughly that of a gas at the temperature of the cosmic background radiation at that redshift, $T \approx 700$ K. It is likely that Ω_0 is greater than 0.1 and enthusiasts for the standard inflation picture will demand $\Omega_0 = 1$, but we should not neglect this possibility.

It will be noted that we have not treated in detail the development of amplitude of the oscillating perturbations on scales less than the Jeans' mass. We have rationalised from the case of the Jeans' instability in a stationary medium that perturbations on scales less than the Jeans' length are sound waves and this is indeed the case. What we ought to have done is to have gone back to (11.24)

$$\frac{d^2 \Delta}{dt^2} + 2 \left(\frac{\dot{R}}{R} \right) \frac{d\Delta}{dt} = \Delta (4\pi G \varrho_0 - k^2 c_s^2), \tag{11.24}$$

and found oscillating solutions when the pressure term is dominant, that is $k^2 c_s^2 \gg 4\pi G \varrho_0$. These calculations have been carried out by Padmanabhan (1993) and Coles and Lucchin (1995). The important results are as follows:

- Throughout the radiation-dominated era, the sound speed is $c/\sqrt{3}$ and the sound waves in the relativistic gas propagate with constant amplitude, $\Delta = $ constant.

- When the Universe becomes matter-dominated, the sound speed is given by (9.33) and then the amplitudes of the oscillating perturbations decrease as $\Delta = \delta\varrho/\varrho \propto t^{-1/6}$. As the above authors show, this result can be interpreted in terms of the adiabatic expansion of the energy in the acoustic waves.

The next crucial epoch is the epoch of recombination when the primordial plasma begins to recombines and soon after the matter and radiation become decoupled thermally. The pressure within the perturbation is no longer provided by the radiation but by the thermal pressure of the baryonic matter. Because of the close coupling between the matter and radiation, the matter and radiation temperatures are more or less the same until a redshift $z = 550h^{2/5}\Omega_0^{1/5}$. Therefore, the appropriate sound speed to include in the expression for the Jeans' mass is the adiabatic sound speed for a gas at temperature 3000 K at a redshift of 1000. The sound speed is $c_s = (5kT/3m_H)^{1/2}$ and so the Jeans' mass at the epoch of recombination is

$$M_J = \left(\frac{\pi\lambda_J^3}{6}\right)\varrho_B = 1.6 \times 10^5 (\Omega_0 h^2)^{-1/2} M_\odot. \tag{12.19}$$

Thus, the Jeans' mass decreases abruptly to masses much less than typical galactic masses. The consequence is that all perturbations with masses greater than about $10^5 M_\odot$ grow according to the standard growth rate $\Delta \propto R$ until $\Omega_0 z \sim 1$. It is intriguing that the Jeans' mass immediately following recombination corresponds roughly to the masses of globular clusters, which are the oldest stellar systems in our Galaxy.

The evolution of the Jeans' mass following recombination depends upon the subsequent thermal history of the gas. If the gas continued to cool adiabatically as the Universe expands, $T \propto R^{-2}$, it is straightforward to show that the Jeans' mass would decrease as $M_J \propto R^{-1.5}$. This is, however, unlikely to be the whole story. As we will discuss in Chap. 19, the intergalactic gas is known to be very highly ionised at epochs corresponding to $z \sim 3 - 4$ from the absence of Lyman-α absorption troughs in the spectra of distant quasars and so it must have been reheated prior to these epochs.

12.3 Dissipation Processes in the Pre-Recombination Era

To complete our discussion of the physics of adiabatic baryonic fluctuations, we need to consider dissipative processes in the pre-recombination era. Although the matter and radiation are closely coupled throughout this era, the coupling is not perfect and radiation can diffuse out of the density perturbations. Since the radiation provides the restoring force for support of the perturbation, the perturbation is damped out if the radiation has time to diffuse out of it. This process was first described by Silk (1968) and is often referred to as *Silk damping*.

Just as in the case of sound waves in an imperfect gas, the damping of the perturbations is associated with the finite viscosity and thermal conductivity of the medium through which the waves are propagating. In the present instance, the transport of energy and momentum is carried by the photons and appropriate expressions for the coefficients of thermal conduction and shear viscosity are given by Weinberg (1972), whose analysis of this problem can be strongly recommended. We can obtain the essential results by somewhat cruder means by realising that the process which impedes the escape of radiation from the perturbations is Thomson scattering by free electrons in the plasma. An order-of-magnitude diffusion calculation illustrates the essence of the physics.

At any epoch, the mean free path for scattering of photons by electrons is $\lambda = (N_e \sigma_T)^{-1}$, where $\sigma_T = 6.665 \times 10^{-29}$ m^2 is the Thomson cross-section. As was shown in Sect. 9.4, the photons and electrons are in close thermal contact throughout the pre-recombination era. In addition, in a fully ionised plasma, the protons and electrons are closely coupled electrostatically and so the photons are closely coupled to the protons as well. To determine how far the photons can diffuse in the cosmic time scale t, it is simplest to work out the diffusion coefficient D for photons which, according to kinetic theory, is related to their mean free path λ by $D = \frac{1}{3}\lambda c$, where c is the speed of light. Therefore, the radial distance over which the photons can diffuse is

$$r_D \approx (Dt)^{1/2} = \left(\tfrac{1}{3}\lambda ct\right)^{1/2}, \tag{12.20}$$

where t is cosmic time. The baryonic mass within this radius, $M_D = (4\pi/3)r_D^3 \varrho_B$, can now be evaluated for the pre-recombination era.

In the early pre-recombination phase, $z > 2.4 \times 10^4 \Omega_0 h^2$, the Universe is radiation-dominated and, from (10.18), the cosmic-time redshift relation is

$$t = \left(\frac{3c^2}{32\pi G\varepsilon}\right)^{1/2} = \left(\frac{3c^2}{32\pi G\chi aT_0^4}\right)^{1/2} \frac{1}{(1+z)^2} = \frac{2.4 \times 10^{19}}{(1+z)^2} \text{ s}, \tag{12.21}$$

where we have adopted $\chi = 1.68$ and taken the present temperature of the Cosmic Microwave Background Radiation to be $T_0 = 2.725$ K. The number density of electrons varies as

$$N_e = \frac{\Omega_B \varrho_{\text{crit}}(1+z)^3}{m_p} = 11\Omega_B h^2 (1+z)^3 \text{ m}^{-3}, \tag{12.22}$$

where m_p is the mass of the proton. Thus, the damping mass, sometimes referred to as the *Silk mass*, is

$$M_D = \frac{4\pi}{3}r_D^3 \varrho_B = 2.4 \times 10^{26} (\Omega_B h^2)^{-1/2}(1+z)^{-9/2} \; M_\odot. \tag{12.23}$$

After the epoch of equality of the matter and radiation energy densities, the cosmic time-redshift relation is given by the matter-dominated relation

$$t = \frac{2}{3H_0\Omega_0^{1/2}} \frac{1}{(1+z)^{3/2}} = \frac{2.06 \times 10^{17}}{(\Omega_0h^2)^{1/2}(1+z)^{3/2}} \text{ s.} \tag{12.24}$$

In this case, we find

$$M_D = 2.0 \times 10^{23}(\Omega_B h^2)^{-5/4}(1+z)^{-15/4} \, M_\odot. \tag{12.25}$$

These variations of the damping mass M_D with scale factor are shown in Fig. 12.1. Damping of the perturbations continues until the epoch of recombination, when the electrons and protons begin to recombine. By the redshift of the last scattering surface at $z = 1000$, the damping mass has attained the value $M_D = 10^{12}(\Omega_B h^2)^{-5/4} \, M_\odot$. Much more detailed calculations were carried out by Peebles (1981), who took into account the details of the damping as the primordial plasma recombined and found the damping mass to be $M_D = 1.3 \times 10^{12}(\Omega_0 h^2)^{-3/2} \, M_\odot$. Notice that, if we accept the constraints of primordial nucleosynthesis, that $\Omega_B h^2 \leq 0.036$, all masses less than about $10^{14} \, M_\odot$ would be damped out.

The upshot of these calculations is that all perturbations with masses $M \leq 10^{12} M_\odot$ are damped out because of the diffusion of photons out of the perturbations. According to the theory of adiabatic baryonic perturbations, only perturbations on the scales of massive galaxies and greater survive into the post-recombination eras. The perturbations which could have resulted in stars, star clusters and even normal galaxies such as our own are damped to exponentially small amplitudes. In this picture, it has to be assumed that these structures were formed by the process of fragmentation of the large scale structures which survive to $z < 1000$.

12.4 Isothermal Perturbations

In the analysis of *adiabatic perturbations*, which has dominated the story so far, it has been assumed that the density perturbations are associated with corresponding perturbations in the pressure by the standard adiabatic relation

$$\frac{\delta p}{p} = \gamma\frac{\delta\varrho}{\varrho}, \tag{12.26}$$

where γ, the ratio of specific heats, is 5/3 for a monatomic non-relativistic gas and 4/3 for a relativistic gas. At the other extreme is the case of *isothermal perturbations*. In the radiation-dominated phase of the standard Big Bang, these are fluctuations in the baryon density which take place against the uniform cosmic background radiation. The perturbations are isothermal in the sense that they cause no fluctuations in the background radiation temperature during the radiation-dominated phases. Their internal temperature is the same as that of the uniform radiation background and they are frozen into the radiation-dominated plasma. In the case of perfect gases, any pressure and density distribution in the radiation-dominated phases can be represented as the superposition of a distribution of adiabatic and isothermal perturbations.

The evolution of isothermal perturbations in the early pre-recombination phases, before the epoch of equality of mass densities in the matter and radiation, is similar to that which will reappear in a different guise latter – the evolution of perturbations of non-relativistic matter in the background of a smooth relativistic gas. This case is pleasantly treated by Coles and Lucchin (1995) (see also Efstathiou 1990). For simplicity, consider a spherical baryonic density perturbation in a smooth expanding background, similar to the little closed Universe discussed in Sect. 11.4.2. If the background were not expanding, the characteristic time-scale for collapse would be given by (11.30), $\tau_g \approx (G\varrho_B)^{-1/2}$. This time-scale can be compared with the characteristic time-scale τ for expansion of the Universe during the radiation-dominated phase, which is readily derived from (9.7)

$$\tau = \frac{R}{\mathrm{d}R/\mathrm{d}t} = \left(\frac{3c^2}{8\pi G\varepsilon_0}R^4\right)^{1/2} = \left(\frac{3}{8\pi G\varrho_{\mathrm{rad}}}\right)^{1/2} \approx (G\varrho_{\mathrm{rad}})^{-1/2}. \quad (12.27)$$

Since $\varrho_{\mathrm{rad}} \gg \varrho_B$ during the radiation-dominated phases, the expansion time-scale is very much shorter than the collapse time-scale and so the perturbation expands with the expanding substratum.

To illustrate this more quantitatively, we can adapt (11.24) for the case in which the growth of the perturbation is driven by the matter density in the perturbation ϱ_B:

$$\frac{\mathrm{d}^2\Delta}{\mathrm{d}t^2} + 2\left(\frac{\dot{R}}{R}\right)\frac{\mathrm{d}\Delta}{\mathrm{d}t} = 4\pi G\varrho_B\Delta. \quad (12.28)$$

Following Coles and Lucchin (1995), we introduce the parameter $y = \varrho_B/\varrho_{\mathrm{rad}} = R/R_{\mathrm{eq}}$, where R_{eq} is the scale factor at which the matter and radiation energy densities were the same. For illustrative purposes, we assume that, in the matter-dominated phases, the dynamics are described by the critical Einstein–de Sitter model, $R = (3H_0t/2)^{2/3}$. Then, the dynamics of the expansion through the radiation and matter-dominated phases can be written

$$\dot{R} = \frac{H_0}{R_{\mathrm{eq}}^{1/2}}\frac{(1+y)^{1/2}}{y}. \quad (12.29)$$

Changing variables from t to y, (12.28) reduces to

$$\frac{\mathrm{d}^2\Delta}{\mathrm{d}y^2} + \frac{2+3y}{2y(1+y)}\frac{\mathrm{d}\Delta}{\mathrm{d}y} - \frac{3\Delta}{2y(1+y)} = 0. \quad (12.30)$$

The growing solution is

$$\Delta \propto 1 + \frac{3y}{2}. \tag{12.31}$$

Thus, throughout the entire radiation-dominated era from $y = 0$ to $y = 1$, the initial perturbation grows by only a factor of 2.5 in amplitude. This important dynamical result is known as the *Meszaros effect* (Meszaros 1974) and will reappear in a different guise later. The key point is that the baryonic perturbations are in a fully ionised plasma and so they are strongly coupled to the cosmic background radiation, which determines the dynamics of the expansion and growth rate of the perturbations. The radiation drag force acting on the plasma retards the growth of the perturbation.

We can illustrate the essential physics by the following order of magnitude calculation. The relation between the rate of growth of the perturbation and the velocity field $\mathbf{v}$ induced by gravitational collapse is given by (11.13),

$$\frac{d}{dt}\left(\frac{\delta\varrho}{\varrho}\right) = \frac{d\Delta}{dt} = -\nabla \cdot \mathbf{v}. \tag{11.13}$$

Let us assume that we can ignore the photon field and estimate the velocity of collapse of the perturbation, if it were to grow as $\Delta \propto R = (1 + z)^{-1}$. Writing div $\mathbf{v} = v/\lambda$, we find

$$v \sim \frac{\lambda}{t}\left(\frac{\delta\varrho}{\varrho}\right). \tag{12.32}$$

The radiation force acting on an electron of the plasma moving at speed v through the isotropic background radiation is given by the first-order Compton scattering formula

$$f_{\mathrm{rad}} = \frac{4}{3}\sigma_{\mathrm{T}}U_{\mathrm{rad}}\frac{v}{c} = \frac{m_{\mathrm{e}}v}{\tau}, \tag{12.33}$$

where τ is given by (9.20). This force is communicated to the protons through the strong electrostatic coupling between electrons and protons. We can compare this force with the typical gravitational force acting on a proton in the perturbation

$$f_{\mathrm{g}} \approx \frac{Gm_{\mathrm{p}}M_{\mathrm{B}}}{\lambda^2} \sim Gm_{\mathrm{p}}\lambda\varrho_{\mathrm{B}}. \tag{12.34}$$

Therefore,

$$\frac{f_{\mathrm{rad}}}{f_{\mathrm{g}}} = \frac{m_{\mathrm{e}}}{m_{\mathrm{p}}}\frac{1}{Gt\tau\varrho_{\mathrm{B}}}\left(\frac{\delta\varrho}{\varrho}\right)$$

$$= 2.6 \times 10^{-4}(\Omega_0 h^2)^{5/2}\left(\frac{\delta\varrho}{\varrho}\right)(1 + z)^{5/2}. \tag{12.35}$$

Because $\Delta = \delta\varrho/\varrho$ can at best grow as $(1 + z)^{-1}$, the value of Δ must be at least 10^{-3} at the last scattering surface at $z = 1000$. Thus, even at the epoch of recombination, the radiation drag exceeded the gravitational force by a factor of at least 10^2. The radiation drag therefore slows up the collapse of the perturbation drastically and the net result is that the isothermal perturbations scarcely grow at all until after the epoch of recombination. After that epoch, they grow according to the standard formula, $\Delta \propto R = (1+z)^{-1}$.

Thus, isothermal perturbations behave in a similar fashion to the adiabatic perturbations, but there is one key difference. There is no Silk damping since the background radiation is uniform and so perturbations on all scales survive to the epoch of recombination. As a result, when the perturbation spectrum passes through the recombination era, the Jeans' mass is $M = 1.6 \times 10^5 (\Omega_0 h^2)^{-1/2} M_\odot$ as in the adiabatic case, but now perturbations on this scale have survived and can begin to collapse to form bound structures.

12.5 Baryonic Theories of Galaxy Formation

We can now put these ideas together to create baryonic theories of the origin of galaxies. This story brings back fond memories for me since I had the privilege of working with Zeldovich, Sunyaev, Doroshkevich and their colleagues in Moscow in 1968–69 while these ideas were being hammered out – it was an extraordinary experience. Zeldovich and his team studied both adiabatic and isothermal perturbations, but put most effort into the adiabatic picture. In contrast, Peebles and his colleagues at Princeton favoured the isothermal picture. These baryonic models resulted in quite different pictures for the formation of galaxies and the large-scale structure of the Universe.

12.5.1 The Adiabatic Scenario

In the adiabatic picture developed by Zeldovich and his colleagues, it was assumed that a spectrum of small adiabatic perturbations was set up in the very early Universe and their evolution was then followed, according to the physical rules developed above. Only large scale perturbations with masses $M \geq M_D = 10^{12} (\Omega_B h^2)^{-5/4} M_\odot$ survived to the epoch of recombination, all fluctuations on smaller mass scales being damped out by photon diffusion, as discussed in Sect. 12.3. During the pre-recombination era, after the perturbations came through their particle horizons, those with masses less than $M_J = 3.75 \times 10^{15} / (\Omega_B h^2)^2 M_\odot$ were sound waves which oscillated until the epoch of recombination, when their internal pressure support vanished and the Jeans' mass dropped to $M_J = 1.6 \times 10^5 (\Omega_0 h^2)^{-1/2} M_\odot$.

Following the remarkable insight of Sakharov (1963), who studied the evolution of perturbation in a Cold Universe, Zeldovich and his colleagues realised that there would be structure in the power spectrum of fluctuations which survived to the epoch of recombination, as illustrated in Fig. 12.2a

(Sunyaev and Zeldovich 1970). Fluctuations of a given mass scale, which develop into bound structures at late epochs, are those with large amplitudes when they came through the horizon. Fig. 12.2a shows perturbations on two different mass scales coming through their particle horizons with the same amplitude. The amplitude of the oscillations at the epoch of recombination depends upon the phase of oscillation of the sound waves at that time. Those oscillations which complete an integral number of oscillations will be observed with maximum amplitude as they begin to collapse freely after recombination, whereas those which reach zero amplitude will not form objects at all. The mass spectrum of objects formed is shown in Fig. 12.2b. This spectrum of oscillations as a function of mass at the epoch of recombination will reappear later in the story – appropriately, they are often referred to as *Sakharov oscillations*. Perturbations with masses $M \geq M_J = 3.75 \times 10^{15}/(\Omega_B h^2)^2 \, M_\odot$ never went through a period of oscillation and continued to grow from the moment they came through the horizon to the present epoch.

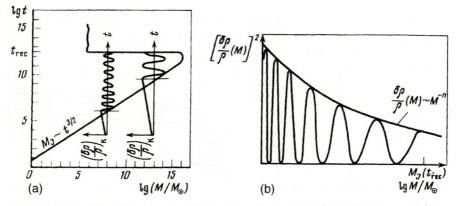

Fig. 12.2a-b. What Sunyaev and Zeldovich refer to as the 'stability diagram' (Sunyaev and Zeldovich 1970). (a) The region of instability is to the right of the solid line. The two additional graphs illustrate the evolution of density perturbations of different masses as they come through the horizon up to the epoch of recombination. (b) Perturbations corresponding to different masses arrive at the epoch of recombination with different phases, resulting in a periodic dependence of the amplitude of the perturbations upon mass.

Following recombination, all the surviving perturbations grew in amplitude as $\Delta \propto (1+z)^{-1}$ until the epoch at which $\Omega_0 z \sim 1$. In the early 1970s, the density parameter in baryons Ω_B was known to be less than about $0.05h^{-2}$ from the constraints provided by primordial nucleosynthesis (Sect. 10.4) and so, even if $h = 0.5$, the perturbations would grow very slowly at redshifts $z \leq 5$. In this case, the amplitudes of the perturbations must have attained $\Delta = 1$ by that epoch, in order to ensure the formation of galaxies and larger scale structures. This was a satisfactory result, since quasars were known to exist at redshifts greater than 2 and the number counts of quasars and radio sources indicated that these objects had flourished at these early epochs (see Sect. 17.2). Zeldovich and his colleagues inferred that galaxies and the large scale structure of the Universe began to form at relatively late epochs, $z \sim 3 - 5$. Since the fluctuations had attained amplitude $\Delta \sim 1$ at $z \sim 5$ and $\Delta \propto (1 + z)^{-1}$, the amplitude of the density perturbations at the epoch of recombination must have been at least $\Delta \geq 3 \times 10^{-3}$.

In the baryonic adiabatic scenario, only large-scale structures survived to $z \sim 5$ and then their evolution became non-linear. We will deal with this topic in more detail in Chap. 16. For the moment, we note that the structures which survived on the scale of clusters and superclusters of galaxies were unlikely to be perfectly spherical and, in a simple approximation, can be described by ellipsoids with three unequal axes. In 1970, Zeldovich derived an analytic solution for the non-linear collapse of these structures and showed that such ellipsoids collapse most rapidly along their shortest axis with the result that flattened structures, which Zeldovich called 'pancakes', are formed. The density becomes large in the plane of the pancake and the infalling matter is heated to a high temperature as the matter collapses into the pancake, a process sometimes called the 'burning of the pancakes'. Galaxies were assumed to form by fragmentation or thermal instabilities within the pancakes. In this picture, all galaxies formed late in the Universe, once the large-scale structures had collapsed. This baryonic pancake theory was developed in some detail by Zeldovich and his colleagues in the 1970s and can be considered a 'top-down' scenario for galaxy formation (Sunyaev and Zeldovich 1972 and the selected papers of Zeldovich 1993). Among the successes of the theory was the fact that it accounted naturally for the large-scale structure in the distribution of galaxies. In three-dimensions, the pancakes formed interconnected flattened, stringy structures, not unlike the great holes and sheets of galaxies observed in the local Universe.

12.5.2 The Isothermal Scenario

In contrast, Peebles and his Princeton colleagues favoured the isothermal perturbations which began to collapse on all scales greater than $M = M_{\rm J} = 1.6 \times 10^5 (\Omega_0 h^2)^{-1/2} \, M_\odot$ immediately after the epoch of recombination. This scenario had the attractive feature that the first objects to form would have masses similar to those of globular clusters, which are the oldest known objects in our Galaxy. The process of galaxy and structure formation was ascribed to the hierarchical clustering of these small-scale structures under the influence of the power spectrum of perturbations, which extended up to the largest scales. One of the attractive features of this picture was that there would be early enrichment of the chemical abundances of the elements as a result of nucleosynthesis in early generations of massive stars. This process could account for the fact that, even in the largest redshift quasars, the abundances of the elements were not so different from those observed locally. Many of these ideas were developed by Peebles in his important monograph *The Large-Scale Structure of the Universe* (Peebles 1980, see also Peebles 1993).

The process of structure formation by hierarchical clustering was put on a formal basis by Press and Schechter in 1974 in a remarkable paper which we will study in Sect. 16.3. Throughout the 1970s, Peebles and his colleagues devoted an enormous effort to determining the correlation functions which describe the statistical properties of the clustering of galaxies (Sects. 2.2 and 14.1). As the power of digital computers increased, it became possible to carry out numerical simulations of the process of structure formation by hierarchical clustering and one of the successes of that programme has been the ability to account for the observed correlation functions for galaxies. We will return to this crucial topic in more detail in Chap. 14. Thus, in contrast to the adiabatic picture, the isothermal scenario is a 'bottom-up' picture, in which galaxies are built up out of smaller objects by clustering and coalescence.

12.6 What Went Wrong?

Despite these successes, there are major problems with both of these scenarios. First of all, the dominant form of matter in the Universe is unlikely to be baryonic. The constraints from primordial nucleosynthesis of the light elements strongly suggest that the mean baryonic mass density of the Universe is about an order of magnitude less than the mean total mass density $\Omega_0 \approx$ 0.2–0.3 and possibly even smaller. In addition, in 1981, the concept of the inflationary expansion of the early Universe, pioneered by Alan Guth and his colleagues, caught the imagination of theorists. One of the consequences of that picture, which can resolve a number of the fundamental cosmological problems, was that the Universe should have flat spatial geometry and so, if $\Omega_A = 0$, it follows that $\Omega_0 = 1$. In this case, there is no question but that most of the mass in the Universe must be in some non-baryonic form.

These are strong theoretical arguments, but there was also observational conflict with the expected amplitude of the temperature fluctuations in the Cosmic Microwave Background Radiation. As we have discussed above, after the epoch of recombination, both adiabatic and isothermal perturbations began to collapse and, for masses on the scales of clusters of galaxies and greater, their behaviour is similar. The problem is that, in the pure baryonic theory, these fluctuations are expected to have large amplitude $\Delta \geq 3 \times 10^{-3}$ on the last scattering surface and these cause observable fluctuations in the radiation temperature of the Cosmic Microwave Background Radiation. We will discuss the theory of temperature fluctuations in the Microwave Background Radiation in some detail in Chap. 15 and so we will simply summarise some of the key points here.

Silk (1968) pointed out that adiabatic density perturbations would result in temperature fluctuations of the matter on the last scattering surface according to the adiabatic relation

$$\left(\frac{\delta T}{T}\right) = \tfrac{1}{3}\left(\frac{\delta \varrho}{\varrho}\right), \tag{12.36}$$

recalling that the matter and the radiation are strongly coupled through the recombination epoch. This calculation suggested that large temperature fluctuations would be expected in the cosmic background radiation $\delta T/T \geq 10^{-3}$, which could be excluded from the observational upper limits to the temperature fluctuations in the background radiation.

As we will show in Chap. 15, the processes by which fluctuations are imprinted on the background radiation are actually much more complex than this. The theory of these processes for both adiabatic and isothermal perturbations was worked out by Sunyaev and Zeldovich (1970). They found the important result that, for both adiabatic and isothermal perturbations, the root-mean-squared temperature fluctuations were predicted to be

$$\left\langle \left(\frac{\delta T}{T}\right)^2 \right\rangle^{1/2} = 2 \times 10^{-5} \left(\frac{M\Omega_0^{1/2}}{10^{15}\, M_\odot}\right)^{1/2} (1 + z_0), \tag{12.37}$$

for masses $M \geq 10^{15}\Omega_0^{-1/2}\, M_\odot$, where z_0 is the redshift at which $\delta \varrho/\varrho = 1$. By the early 1980s, the upper limits to the intensity fluctuations in the Cosmic Background Radiation were in conflict with purely baryonic theories.

It may seem remarkable that I have spent so much time developing a theory which ends up in serious conflict with the observations. There are two important reasons for this. The first is that we have developed many of the useful tools and concepts needed in the formulation of any theory of galaxy formation and we will use these repeatedly in subsequent Sections. The second is that we have presented the best that can be achieved assuming that all the matter in the Universe is in baryonic form. The fact that there are important conflicts with observation is a strong reason for taking seriously the proposition that the Universe is dominated by some unknown form of non-baryonic dark matter.

In some ways, it is a disappointment that the simplest 'dull person's' view of the origin of structure in the Universe runs into these difficulties, since it is entirely based upon known forms of matter and radiation and how they interact. Now, we have to include an additional dominant constituent of the Universe, the nature of which is unknown. The other side of the coin is that there is the exciting prospect of gaining new insights into fundamental physical processes. This would be yet one more example of astronomical problems leading to the understanding of fundamental physical processes

which cannot be studied as yet in the laboratory. This is the story we take
up now.

13 Dark Matter and Galaxy Formation

13.1 Introduction

Let us recall the reasons for taking dark matter, and in particular non-baryonic dark matter, very seriously in the context of galaxy formation.

- Dark matter provides the dominant mass in galaxies and systems of galaxies on the large scale. The mass-to-luminosity ratios of galaxies and clusters of galaxies and the application of different variants of the Cosmic Virial Theorem on larger scales all indicate that most of the mass in the Universe on the large-scale is in some dark form (Sect. 8.5).
- The above evidence suggests that the overall density parameter found from these dynamical techniques amounts $\Omega_0 \sim 0.2 - 0.3$, some of the approaches suggesting values which approach $\Omega_0 = 1$.
- Primordial nucleosynthesis of the light elements provides a firm upper limit to the value of the baryonic density parameter of $\Omega_B \leq 3.6 \times 10^{-2}h^{-2}$, with a best-fitting value of $\Omega_B = 6.2 \times 10^{-3}h^{-2}$ (Sect. 10.4). Thus, even adopting a value of $h = 0.5$, the upper limit to the baryonic mass density would barely be sufficient to account for values of the overall density parameter of the order $\Omega_0 \sim 0.2 - 0.3$, and certainly could not account for $\Omega_0 = 1$.
- The theorist's preferred model of the Universe has flat spatial geometry $\kappa = 0$, in which case, if $\Omega_\Lambda = 0$, $\Omega_0 = 1$. In this case non-baryonic dark matter is essential. There is a loophole in this line of reasoning, in that, if $\Omega_\Lambda \neq 0$, we could adopt $\Omega_0 = 0.1$, which could be baryonic, and $\Omega_\Lambda = 0.9$. This value of Ω_Λ is, however, greater than the upper limit found by Kochanek (1995), discussed in Sect. 8.6.
- The purely baryonic models of galaxy formation result in predicted intensity fluctuations in the Cosmic Microwave Background Radiation which exceed the upper limits on small angular scales.

The upshot of these considerations is that, although it might just be possible to wriggle out of all these constraints and formulate a purely baryonic model for galaxy formation, most cosmologists take the point of view that we should take non-baryonic dark matter very seriously as a candidate for the dominant form of matter in the Universe, despite the fact that we do not understand its nature.

13.2 Forms of Non-Baryonic Dark Matter

We have already discussed the three most popular suggestions for non-baryonic dark matter in Sect. 4.4.2. To recapitulate these are:

1. *Axions.* The discussion by Kolb and Turner (1990) is recommended for the enthusiast who wishes to obtain insight into the motivation for the introduction of these particles into theories of elementary particles. If these particles exist, they would have important astrophysical consequences which are discussed by Kolb and Turner. They show that, if the axions were produced in thermal equilibrium, they would have unacceptably large masses, which would result in the conflict with observations of the Sun and the supernova SN1987A. Specifically, if the mass of the axion were greater than 1 eV, the rate of loss of energy by the emission of axions would exceed the rate at which energy is generated by nuclear reactions in the Sun and so its centre would need to be hotter, resulting in a shorter age than is acceptable and greater emission of high energy neutrinos. In the case of SN1987A, the key observation was the duration of the neutrino pulse of about 12 seconds, consistent with the standard picture of neutrino emission. If the axions had masses in the range 10^{-3} to 2 eV, the cooling of the neutron star would be so rapid that a much shorter pulse of neutrinos would be expected. There is, however, another non-equilibrium route by which the axions could be created in the early Universe. In this variant, described by Kolb and Turner (1990), the axions never acquired thermal velocities, as they were never in equilibrium. The acceptable range of rest mass energies of the axions is $10^{-2} - 10^{-5}$ eV. The axions remain 'cold' and, during the epochs we are interested in, behave in bulk like the massive particles discussed below.

2. *Neutrinos.* The three known types of neutrino may have finite rest masses. This theory was developed by Zeldovich and the Moscow group when Lyubimov *et al.* (1980) reported that they had measured a mass of 30 eV for the electron antineutrino. Although subsequent experiments have not confirmed this result, the possibility that the rest mass energy of the neutrino corresponds to $\varepsilon \sim 10$ eV is of the greatest cosmological interest. Since the neutrinos decoupled from the matter whilst they were fully relativistic, the number density of neutrinos is of the same order as the number density of photons, as discussed in Chap. 10. As was shown in Sect. 4.4.2, if all the relic neutrinos had this mass, they alone would be sufficient to close the Universe, $\Omega_0 = 1$. A rest mass of 10 eV is an intriguing value for the neutrinos since they would have been highly relativistic at the time they decoupled from the electrons at $t \approx 1$ s, and so none of the predictions of standard primordial nucleosynthesis would be affected.

3. *WIMPs.* A third possibility is that the dark matter is in some form of weakly interacting massive particle, or WIMP, with rest mass $\varepsilon \geq 1 - 10$

GeV. These particles might be the gravitino, the supersymmetric partner of the graviton, or the photino, the supersymmetric partner of the photon, or some other type of massive neutrino-like particle as yet unknown.

We discussed a number of limits to the masses of these different candidates in Sect. 4.4.2. An important conclusion from these estimates was that, if the dark matter particles were as common as the neutrinos and photons, their rest mass energies could not be much greater than 10 eV, or the present density of the Universe would exceed the critical mass density $\Omega_0 = 1$. Therefore, there would have to be some suppression mechanism to ensure that hypothetical massive particles are much less common than the photons and electron neutrinos at the present day.

Fortunately, there is a natural way in which this can occur, if the decoupling of the WIMPs from thermal equilibrium took place after they had become non-relativistic, that is, the decoupling took place after the epoch at which $k_B T \sim m_X c^2$, where m_X is the mass of the WIMP. This argument bears a close resemblance to that used in the discussion of the decoupling of neutrinos from the reactions which maintain protons and neutrons in their equilibrium abundances in the early Universe (Sect. 10.1) and the physics of baryon-symmetric Universes discussed in Sect. 10.6. To summarise the calculation carried out in Sect. 10.6, in the relativistic limit, in which all types of particle are maintained in thermal equilibrium, the numbers densities of electrons and neutrinos are given by

$$g = 2 \quad N = \bar{N} = 0.183 \left(\frac{2\pi k_B T}{hc} \right)^3 \mathrm{m}^{-3} \quad \varepsilon = \frac{7}{8} a T^4, \qquad (13.1)$$

$$g = 1 \quad N = \bar{N} = 0.091 \left(\frac{2\pi k_B T}{hc} \right)^3 \mathrm{m}^{-3} \quad \varepsilon = \frac{7}{16} a T^4. \qquad (13.2)$$

When the particles become non-relativistic, $k_B T \ll mc^2$, but the species are maintained in thermal equilibrium by interactions between the particles, the non-relativistic limit of the integral (10.1) gives the equilibrium number density

$$N_X = g_X \left(\frac{m_X k_B T}{2\pi \hbar^2} \right)^{3/2} \exp\left(-\frac{m_X c^2}{k_B T} \right). \qquad (13.3)$$

Thus, once they become non-relativistic, the number densities decrease exponentially until the time-scale of the interactions, which maintain the species in equilibrium, exceeds the expansion age of the Universe. At that point, provided they are stable, the abundances of the massive particles freeze out. Taking the ratio of the (13.3) and (13.1), it can be seen that the suppression, or 'freeze-out', of the species relative to the relativistic values is

$$\frac{N_X}{N_{\mathrm{rel}}} \approx \left(\frac{m_X c^2}{k_B T} \right)^{3/2} \exp\left(-\frac{m_X c^2}{k_B T} \right). \qquad (13.4)$$

The decoupling takes place when the time-scale for the interaction of the particles becomes of the order of the cosmological time-scale, $t \sim \tau_X$, where $\tau_X = (\sigma_X N v)^{-1}$, v now being the non-relativistic thermal speed of the particles. Suitable cross-sections to be used in these calculations are described by Kolb and Turner (1990). We will not enter into these complexities here but simply note that these calculations suggest that WIMP masses greater than about 2–5 GeV would be consistent with the constraint that the density parameter should not exceed $\Omega_X h^2 = 1$. Turner (1997) notes that neutral weakly interacting particles with masses in the range 10 to 1000 GeV are predicted by supersymmetric theories of elementary particles. It may also be suggestive that the range of acceptable masses includes the $W^{\pm}$ and the Z^0 mass scales and the predicted mass of the Higgs boson.

This discussion is a gross simplification of the types of calculation which need to be carried out to develop a convincing physical picture, but it does provide some insight into the origin of the two most popular theories of the nature of non-baryonic dark matter in the context of galaxy formation. The case of standard neutrinos with finite rest masses ~ 10 eV is known as the *Hot Dark Matter (HDM)* picture, because the neutrinos were relativistic when they decoupled from thermal equilibrium. In contrast, WIMPs are examples of what is termed *Cold Dark Matter (CDM)* since they decoupled from the thermal background after they had become non-relativistic or 'cold'. As we will see, the two types of non-baryonic dark matter lead to quite different pictures for the process of galaxy formation.

Experiments are underway to search for these different types of dark matter particles. One important class of experiments involves searching for weakly interacting particles with masses $m \geq 1$ GeV, which could make up the dark halo of our Galaxy. In order to create a dark halo about our Galaxy, the particles would have to have velocity dispersion $\langle v^2 \rangle^{1/2} \sim 250$ km s^{-1} and their total mass is known. Therefore, the number of WIMPs passing through a terrestrial laboratory each day is a straightforward calculation. When these massive particles interact with the sensitive volume of the detector, the collision results in the transfer of momentum to atoms of the material of the detector and it is this recoil which can be measured. In the Boulby Dark Matter Experiment in the UK, a scintillating crystal detector is used and has produced the best upper limits to the properties of galactic halo massive particles (Smith *et al.* 1996). Another programme currently underway involves detecting the tiny temperature increase in a pure germanium or silicon crystal due to the collision of the WIMP with an atom of the crystal – this is observable if the crystal is cooled to about 20 mK. The big problem is that these events are expected to be very rare and so stringent precautions have to be taken to shield the detectors and discriminate against spurious background events. Numerous detector systems are now being developed to search for these rare events, but so far only upper limits are available. At this stage, there is no inconsistency with the preferred range of particle masses discussed above.

13.3 Free Streaming and the Damping of Non-Baryonic Perturbations

One of the key considerations for theories of structure formation, in which non-baryonic dark matter is dominant, is the damping of density perturbations by free-streaming. So long as the dark matter particles are strongly coupled, they behave no differently from ordinary relativistic or non-relativistic particles. At later epochs, however, the dark matter particles no longer interact with other particles. If the particles were relativistic at the epoch when they 'froze-out', they would continue to travel in 'straight lines' at the speed of light. Thus, if the particles belonged to some density perturbation, when it came through the particle horizon, the relativistic dark matter particles continue to stream freely, thus damping out the density perturbation. It is evident that this process is devastating for any density perturbation, so long as the non-baryonic particles are relativistic when the perturbation comes through the horizon. This process is similar to the phenomenon of Landau damping in a collisionless plasma, where it is referred to as 'phase-mixing'. The same occurs in this case – the unrestrained motion of the particles destroys the phase-coherence of the density perturbation.

The masses which are damped out depend upon how far the free-streaming particles can travel at a given epoch. Let us follow the simple treatment of Kolb and Turner (1990). The comoving distance which a free-streaming particle can travel by the epoch t is just

$$r_{FS} = \int_0^t \frac{v(t')}{R(t)} \, dt'. \tag{13.5}$$

It will prove convenient to work in terms of comoving scales at the epoch of equality of the radiation and matter energy densities t_{eq}. Let us split the integral into two parts, the first from time $t = 0$ to the epoch when the particles become non-relativistic, $t = t_{NR}$ and the second from $t = t_{NR}$ to t_{eq}.

$$r_{FS} = \int_0^{t_{NR}} \frac{v(t')}{R(t)} \, dt' + \int_{t_{NR}}^{t_{eq}} \frac{v(t')}{R(t)} \, dt'. \tag{13.6}$$

We recognise that the first part of the integral is just the horizon scale during the radiation-dominated phases while the particle remains relativistic, expressed as a comoving coordinate distance (Sect. 12.1). During the second time interval, the streaming velocities of the particles decrease adiabatically as $v = (R_{NR}/R)c$, as shown in Sect. 7.1. Therefore,

$$r_{FS} = 2 \frac{ct_{NR}}{R_{RN}} + \int_{t_{NR}}^{t_{eq}} \frac{v(t')}{R(t)} \, dt'. \tag{13.7}$$

During the radiation-dominated era, the dynamics of the expansion were described by $R \propto t^{1/2}$ and so we can write $t = t_{NR}(R/R_{NR})^2$. Therefore,

$$r_{FS} = \left(\frac{ct_{NR}}{R_{RN}} \right) \left[2 + \ln \left(\frac{t_{eq}}{t_{NR}} \right) \right]. \tag{13.8}$$

This damping process is not important for cold dark matter perturbations since they were non-relativistic when they decoupled in the early Universe and the mass within the horizon was very small indeed. The process is, however, very important in the Hot Dark Matter scenario.

For the sake of definiteness, let us consider the case of neutrinos with masses $m_\nu c^2 = 10$ eV in the standard critical model with $\Omega_0 = 1$. Then, the epoch of equality of matter and radiation occurred at a redshift $z = 4 \times 10^4$. For neutrinos of rest mass 10 eV, the particles become non-relativistic when $3k_B T_\nu \approx m_\nu c^2$. We recall that, in the standard Big Bang scenario, the temperatures of the neutrinos and the photons are related by $T_\nu = (4/11)^{1/3} T_r$ and so $z_{NR} = T_r/T_0 = 2 \times 10^4$. Thus, the neutrinos became non-relativistic at about the same time that the matter and radiation energy densities were the same. In the neutrino picture, this is not a coincidence. The reason is that, according to the canonical Big Bang, the energy density in the neutrinos is more or less the same as the energy density in the photons during the radiation dominated era. At the epoch when the neutrinos become non-relativistic, their inertial mass no longer decreases as the Universe expands, in contrast to the case of the photons which continue to decrease in energy as R^{-1}. The neutrinos then become the dominant non-baryonic particles which contribute to Ω_0 at the present epoch. As a result, in the logarithmic term in (13.8), t_{eq}/t_{NR} is of order unity. For convenience, we will take the free-streaming scale r_{FS} to be the comoving particle horizon, $r_H = 2ct/R_{eq}$, at z_{NR}. The mass within this scale is

$$M_{FS} \sim \frac{\pi}{6} r_{FS}^3 \varrho_0 = 2 \times 10^{15} \, M_\odot. \tag{13.9}$$

More detailed calculations show that the free-streaming damping mass is

$$M_{FS} = 4 \times 10^{15} \left(\frac{m_\nu}{30 \text{ eV}} \right)^{-2} M_\odot. \tag{13.10}$$

The key result is that all density perturbations on mass-scales less than these very large masses are damped out as soon as they come through the horizon. Note that these masses are of the order of the most massive clusters of galaxies and so, in this picture, only structures on these scales and larger can survive after the epoch of equality of matter and radiation energy densities. This is a key feature of the Hot Dark Matter scenario. It bears some resemblance to the baryonic adiabatic picture in which small masses are damped out by photon diffusion.

13.4 Instabilities in the Presence of Dark Matter

First of all, we should reconsider the concepts of the Jeans' length and Jeans' mass for dark matter. Prior to the decoupling of the dark matter, it behaves like a normal relativistic or non-relativistic gas and the gas dynamical results which we derived in Chap. 11 can be taken over in their entirety. We need to consider the concept of the Jeans' mass a little more carefully when the particles become collisionless. Evidently, the equations of fluid dynamics need to be replaced by the collisionless Boltzmann equation. This topic is dealt with by Coles and Lucchin (1995) who show that the Jeans' stability criterion also applies in the case of a collisionless gas, provided the sound speed c_s is replaced by the velocity v_* where

$$ v_*^{-2} = \frac{\int v^{-2} f(v)\, \mathrm{d}^3 \mathbf{v}}{\int f(v)\, \mathrm{d}^3 \mathbf{v}}. \tag{13.11} $$

$f(v)$ is the velocity distribution of the dark matter particles, which is assumed to be isotropic. In the case of a Maxwellian distribution of velocities v_* is just the root-mean-squared velocity dispersion of the particles.

The physics of this result is straightforward. The issue is whether or not the gravitational pull of the dark matter particles of the perturbation is sufficient to prevent them escaping from the perturbation. Just as in the case of the standard Jeans' analysis, on large enough scales, the mass of the perturbation becomes more than enough to bind the system and collapse ensues, according to the standard results for a relativistic and non-relativistic gas in the absence of pressure support. The one big difference from the case of baryonic matter is that, after decoupling, the dark matter perturbations are subject to damping by free-streaming of the particles out of the perturbation, as described in Sect. 13.3.

Now, we need to study how the dark matter perturbations evolve in the presence of the background radiation and baryonic matter – the problem now involves a three-component fluid. The key result concerns the coupling of density perturbations in the dark matter to the baryonic matter and radiation fields. The important point is that the ordinary matter and radiation are completely decoupled from the dark matter, except through their mutual gravitational influence. Let us study first the development of the gravitational instability when the internal pressure of the fluctuations can be neglected. Equations (11.24) and (11.67) can be written

$$ \ddot{\Delta} + 2 \left(\frac{\dot{R}}{R} \right) \dot{\Delta} = A \varrho \Delta, \tag{13.12} $$

where $A = 4\pi G$ in the matter-dominated case and $A = 32\pi G/3$ in the radiation-dominated case. The first case applies to the post-recombination eras when the baryonic matter and the radiation are decoupled; the second applies to fluctuations in the radiation-dominated era on scales greater than the horizon scale. During the radiation-dominated era, the dynamical role of the dark matter is much less than that of the radiation and thus the dominant gravitational perturbations are associated with fluctuations in the closely coupled radiation-dominated plasma. After the epoch of equality of the matter and radiation energy densities, most of the inertial mass is in the dark matter and the evolution of these perturbations dominates the development of the baryonic perturbations.

Let us write the density contrast in the baryons and the dark matter as Δ_B and Δ_D respectively. We have to solve the coupled equations

$$\ddot{\Delta}_B + 2\left(\frac{\dot{R}}{R}\right)\dot{\Delta}_B = A\varrho_B\Delta_B + A\varrho_D\Delta_D, \tag{13.13}$$

$$\ddot{\Delta}_D + 2\left(\frac{\dot{R}}{R}\right)\dot{\Delta}_D = A\varrho_B\Delta_B + A\varrho_D\Delta_D. \tag{13.14}$$

Rather than find the general solution, let us find the solution for the case in which the dark matter has $\Omega_0 = 1$ and the baryon density is negligible compared with that of the dark matter. Then (13.14) reduces to (13.12) for which we have already found the solution $\Delta_D = BR$ where B is a constant. Therefore, the equation for the evolution of the baryon perturbations becomes

$$\ddot{\Delta}_B + 2\left(\frac{\dot{R}}{R}\right)\dot{\Delta}_B = 4\pi G\varrho_D BR. \tag{13.15}$$

Since the background model is the critical model for which $R = (3H_0t/2)^{2/3}$ and $3H_0^2 = 8\pi G\varrho_D$, this equation simplifies to

$$R^{3/2}\frac{\mathrm{d}}{\mathrm{d}R}\left(R^{-1/2}\frac{\mathrm{d}\Delta}{\mathrm{d}R}\right) + 2\frac{\mathrm{d}\Delta}{\mathrm{d}R} = \frac{3}{2}B. \tag{13.16}$$

We find that the solution, $\Delta = B(R - R_0)$, satisfies (13.16). This is a rather pleasant solution because it has the property that at the epoch corresponding to $R = R_0$, the amplitude of the baryon perturbations is zero.

This result has the following significance. Suppose that, at some redshift z_0, the amplitude of the baryon fluctuations is very small, that is, very much less than that of the perturbations in the dark matter. The above result shows how the amplitude of the baryon perturbation develops subsequently under the influence of the dark matter perturbations. In terms of redshift we can write

$$\Delta_B = \Delta_D\left(1 - \frac{z}{z_0}\right). \tag{13.17}$$

Thus, it can be seen that the amplitude of the perturbations in the baryons grows rapidly to the same amplitude as that of the dark matter perturbations.

To put it crudely, the baryons fall into the dark matter perturbations and, within a factor of two in redshift, have amplitudes half that of the dark matter perturbations.

Before the epoch of equality of matter and radiation energy densities, the roles of the dark matter and the radiation-dominated plasma were reversed. The same result is found in the early development of the perturbations when the dark matter and baryonic perturbations have scales greater than the horizon. Most of the inertial mass is in the radiation and so the development of the perturbations in the dark matter is closely tied to those in the radiation-dominated plasma.

There are, however, important differences when the perturbations come through the horizon. Let us consider the case of adiabatic perturbations. When they come through the horizon, the amplitudes of the perturbations in the three-component fluid are

$$\frac{\Delta \varrho_B}{\varrho_B} = \frac{3}{4} \frac{\Delta \varrho_{\text{rad}}}{\varrho_{\text{rad}}} = \frac{\Delta \varrho_D}{\varrho_D}. \qquad (13.18)$$

The baryonic perturbations are stabilised because the Jeans' length is of the same order as the horizon scale and the radiation-dominated plasma can provide pressure support for the perturbation. The baryonic perturbations become sound waves which oscillate with more or less the same amplitude up to the epoch of recombination, when the decoupling of the matter and radiation takes place. After the epoch of equality of the energy densities in the dark matter and the radiation, the dark matter perturbations grow independently of those in the plasma. We see now why the above calculation is of considerable importance. The baryon perturbations are stabilised from the redshift at which they enter the horizon to the epoch of recombination, but the amplitude of the perturbations in the dark matter grows from z_{eq} to the epoch of recombination as $\Delta_D \propto (1+z)^{-1}$. Therefore, the relative amplitudes of the fluctuations in the dark matter and the baryons is roughly $\Delta_B/\Delta_D \approx 1500/z_{\text{eq}}$, that is, the baryon perturbations are of much smaller amplitude than those in the dark matter at the epoch of recombination, typically by a factor of about 10.

Perturbations on scales larger than those which come through the horizon at redshift z_{eq} have relatively smaller differences between Δ_D and Δ_B at the epoch of recombination. In the limit in which the perturbations come through the horizon at the epoch of recombination, the amplitudes of the fluctuations are of the same order of magnitude. As soon as the matter and radiation decouple, the amplitude of the perturbations in the baryonic matter rapidly grows to the same amplitude as that in the dark matter as demonstrated by (13.17). As shown above, the amplitude of the perturbations in the baryons has grown to values close to that in the dark matter by a redshift a few times smaller than the recombination redshift. Thus, even if the baryonic perturbations were completely washed out, the presence of fluctuations in the dark matter ensures that baryon fluctuations are regenerated after recombination.

13.5 The Evolution of Hot and Cold Dark Matter Perturbations

We need to introduce a little more of the terminology commonly found in the literature. In the purely baryonic picture, we noted that the distribution of perturbations in the early Universe could be decomposed into isothermal and adiabatic modes. In the three-component case, the decomposition can be made into similar modes, but the names 'isothermal' and 'adiabatic' are scarcely appropriate for fluids containing collisionless dark matter particles. The corresponding modes are referred to as *curvature* and *isocurvature* modes.

- The *curvature modes* are the equivalent of the adiabatic modes in that, during the radiation-dominated era, the amplitudes of the perturbations in the radiation, the baryonic matter and the dark matter were all more or less the same and given by (13.18). As a result, there were variations in the local mass-energy density from point to point in the Universe, resulting in local perturbations to the curvature of space.
- In the *isocurvature modes*, the mass-energy density is constant throughout space and so there are no perturbations to the spatial curvature of the background world model, despite the fact there may be fluctuations in the mass-energy density of each of the three components from point to point in the Universe.

The two models which have received most attention are those involving the two types of dark matter discussed in some detail in Sect. 13.2, which are commonly referred to as the Hot and Cold Dark Matter scenarios for the origin of the large scale structure of the Universe. These examples illustrate the basic physics involved in any picture of the formation of galaxies and the reader is invited to use these tools to develop the many possible variants of Dark Matter cosmologies.

13.5.1 Hot Dark Matter Scenario

In this scenario, it is assumed that the dark matter is in the form of standard neutrinos with rest masses of the order 10 eV. For the sake of definiteness, let us assume that the rest mass of the neutrino is 10 eV. Then, these particles became non-relativistic at an energy $m_\nu c^2 = 3k_B T$, which corresponds to an redshift $z = 2 \times 10^4$, assuming $\Omega_0 h^2 = 1$. As we have shown, the fact that the particles were highly relativistic when they decoupled, means that this epoch also corresponds closely to the epoch at which the Universe changed from being radiation to matter-dominated.

It is assumed that all perturbations of astrophysical interest were set up on a very wide range of scales, which far exceeded their particle horizons in the very early Universe. These perturbations then grew according to the rules established in Sect. 13.4, until they came through their particle horizons. As we have shown in Sect. 13.3, so long as the neutrinos remained relativistic when they came through the horizon, the perturbations on these scales were rapidly damped out by free streaming of the neutrinos. This process continued until the epoch of equality of the matter and radiation energy densities at $z = 2 \times 10^4$ and so wiped out all perturbations in the dark matter with masses less than $M_{\mathrm{FS}} = 4 \times 10^{15} (m_\nu/30\,\mathrm{eV})^{-2}\, M_\odot$.

At this epoch, the perturbations in the three components had more or less the same amplitude, but now the perturbations in the dark matter became dynamically dominant and, since they were not coupled to the radiation-dominated plasma, they continued to grow in amplitude as $\Delta_{\mathrm{D}} \propto (1+z)^{-1}$. In contrast, when the perturbations in the plasma came through the horizon, they were very quickly stabilised, as the Jeans' mass for the radiation-dominated plasma remained roughly the same as the mass within the particle horizon, as discussed in Sect. 12.2.1. Therefore, the perturbations in the plasma oscillated with more or less constant amplitude between the epoch when they came through the horizon to the epoch of recombination. As discussed in the last section, the result is that the amplitudes of the perturbations in the dark matter became progressively greater than those in the plasma, and, by the epoch of recombination, the perturbations in the dark matter with masses $M \sim 4 \times 10^{15} (m_\nu/30\,\mathrm{eV})^{-2}\, M_\odot$ were greater than those in the plasma by a factor $\Delta_{\mathrm{D}}/\Delta_{\mathrm{B}} \approx (1 + z_{\mathrm{eq}})/(1 + z_{\mathrm{rec}}) \approx 10$.

The perturbations in the plasma were subject to the dissipation processes discussed in Sect. 12.3 and so baryonic perturbations with masses up to about $10^{12}\, M_\odot$ were damped out, but this is not such an important feature of the Hot Dark Matter scenario – the perturbations which determine the structures which form in the Universe now are those in the dark matter which survived to the epoch of equality of matter and radiation energy densities, that is, $M \geq 4 \times 10^{15} (m_\nu/30\,\mathrm{eV})^{-2}\, M_\odot$. The key point is that, at the epoch of recombination, the matter and radiation were decoupled, and then the baryonic matter could collapse into the dark matter perturbations, which had much greater amplitudes than those in the plasma. As discussed in Sect. 13.4, the density perturbations in the matter grew rapidly to the same amplitude as those in the dark matter, which continued to grow uninterrupted as $\Delta_{\mathrm{D}} \propto (1 + z)^{-1}$, until they became non-linear when $\Delta_{\mathrm{D}} \sim 1$. At this point, the perturbations separated out as discrete entities and began to form the large-scale structures we observe today.

The Hot Dark Matter scenario successfully avoids creating excessively large fluctuations in the Cosmic Background Radiation because, at the last scattering surface, the amplitude of the perturbations in the plasma were significantly less than those in the Dark Matter. We will take this important theme up in more detail in Chap. 15.

The Hot Dark Matter scenario bears a close resemblance to the adiabatic baryonic model, with all its attendant successes and problems. The key prediction is that the first structures to form are those on the largest scales in the Universe. It is assumed that smaller scale structures such as galaxies and their contents formed by fragmentation and instabilities, once the structures on the scales of clusters of galaxies and greater began to form. This picture was developed by Zeldovich and his colleagues in a remarkable series of papers in 1980 (see the selected works of Zeldovich 1993). The great success of this picture is that it accounts naturally for the large-scale 'cellular' structure in the distribution of galaxies. Indeed, as we will show in the next chapter, the theory is too successful in producing structure on the very largest scales – it is difficult to account for the form of the correlation functions for galaxies on scales less than those of clusters of galaxies. The big problem with this theory has been that galaxies can only form once the large-scale structures have collapsed, and so it is inevitable that galaxies form rather late in the Universe. This may pose problems for topics such as the early heating and ionisation of the intergalactic gas and for the early enrichment of the intergalactic gas.

13.5.2 Cold Dark Matter Scenario

In the standard Cold Dark Matter scenario, the cold dark matter particles decoupled early in the Universe, after they had already become non-relativistic. As discussed in Sect. 13.2, since the cold dark matter particles decoupled not too long after they became non-relativistic, the mass within the horizon at these times was very small, $M \ll M_\odot$. Free-streaming was unimportant as soon as the particles became non-relativistic and so the Cold Dark Matter scenario begins with the big advantage that dark matter perturbations on all scales of astrophysical interest can survive from the early Universe. This picture became popular in the early 1980s, the term Cold Dark Matter being coined by Peebles (1982) (see also Peebles 1993).

The rules we have developed above apply equally to the cold dark matter perturbations. We require the cold dark matter to be the dominant form of mass at the present epoch and, assuming $\Omega_0 h^2 \sim 1$, it follows that the epoch of equality of matter and radiation energy densities occurred about $z \approx 2 \times 10^4$. Up till this epoch, the perturbations in the dark matter, either hardly grew at all, or were dynamically coupled to the perturbations in the radiation-dominated plasma. In either case, after this epoch, the dark matter became dynamically dominant and the perturbations in the dark matter grew independently of the behaviour of the perturbations in the plasma. As in the standard baryonic picture, adiabatic baryonic fluctuations came through the horizon and were then stabilised by the pressure of the radiation-dominated plasma. The diffusion of photons from these perturbations leads to Silk damping of masses up to about $10^{11} M_\odot$ by the epoch of recombination. Perturbations with greater masses survive as oscillating sound waves up to the epoch of recombination, when they imprint their presence on the last scattering surface, just as in the case of the hot dark matter and baryonic scenarios. Once again, the amplitude of the perturbations associated with a particular

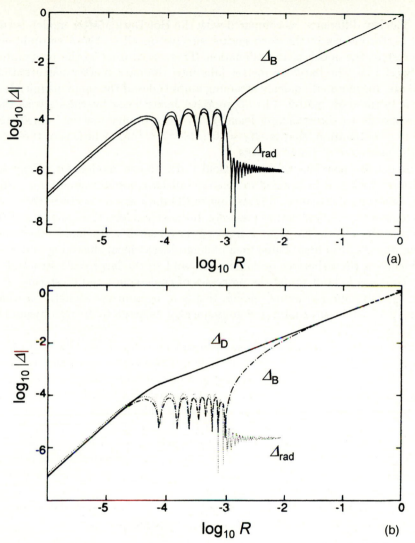

Fig. 13.1a-b. Illustrating the evolution of density perturbations in (**a**) baryonic matter and radiation in the standard baryonic adiabatic model, and (**b**) the baryonic matter, the radiation and the dark matter according to the Cold Dark Matter scenario. In both cases, the mass of the perturbation is $M \sim 10^{15} \, M_{\odot}$ (Coles and Lucchin 1990).

baryonic mass are reduced as compared to the corresponding fluctuations in the dark matter by factors of the order 10 and this helps reduce the amplitude of the predicted fluctuations in the Cosmic Background Radiation. The evolution of the density perturbations in the three components for a mass $M = 10^{15} \, M_{\odot}$ is contrasted with the case of the adiabatic baryonic model in Fig. 13.1, which is taken from Coles and Lucchin (1990).

The key difference, as compared with the Hot Dark Matter model, is that the perturbations in the dark matter survive on all scales of astrophysical interest to the epoch of recombination. Then, according to the prescription of Sect. 13.4, the baryonic matter falls into the dark matter perturbations and are 'regenerated', quickly attaining amplitudes of the same magnitude as those in the dark matter. The appropriate Jeans' mass for the baryons then corresponds to the standard Jeans' mass for ordinary matter immediately after recombination, that is, $M_J = 10^5 (\Omega_B h^2)^{-1/2} M_\odot$, which is of the order of the masses of globular clusters.

It is now clear how to develop Cold Dark Matter models for galaxy formation which can be studied in some detail by computer simulation. Once the spectrum of the initial fluctuations in the dark matter is given, their evolution can be followed rather precisely. In most models, the spectrum of the fluctuations is such that there is most power on small scales and so the lowest mass objects form first. These then undergo hierarchical clustering under the influence of perturbations on large scales and so the large scale structure of the Universe is built up by the processes of coalescence and clustering. Just like the baryonic isothermal model, this is a 'bottom-up' scenario in which small scale structures form and subsequently coalesce to create objects like galaxies and clusters.

The model has a number of successes and some problems. One of the most attractive features is that it is amenable to detailed computation and these have shown that the resulting correlation functions for galaxies can be rather beautifully explained (see Chap. 14 and Efstathiou 1990). Furthermore, the formation of stars and the other contents of galaxies can begin soon after the epoch of recombination and so there is no problem in accounting for the early heating and ionisation of the intergalactic gas and the early chemical enrichment of the gas by the products of stellar nucleosythesis. The problem with this picture is that it is difficult to match the observed power spectrum of galaxies and the spectrum of fluctuations in the Cosmic Microwave Background Radiation over all scales. We need the more detailed studies discussed in Chap. 14 to understand the nature of this problem.

13.6 Where We Go from Here?

We have now made considerable progress towards the development of consistent scenarios for galaxy formation. The broad outlines of the two most appealing theories, the Hot and Cold Dark Matter pictures, have been outlined above, but what is now needed is to convert these stories into quantitative models for galaxy formation. To do this, we need to look into the development of the spectra of density fluctuations and this is the subject of Chap. 14. Then, in Chap. 15, we need to make the crucial confrontation of these theories with the observed properties of the temperature fluctuations in the Cosmic Microwave Background Radiation.

It is worthwhile stressing a crucial aspect of the discussion of this chapter. We have demonstrated how it is possible to understand, in principle, how to overcome the problems inherent in the baryonic picture of galaxy formation, provided we introduce non-baryonic dark matter. This is a really momentous step, because the nature of the dark matter is an unresolved problem. The enormous interest of the particle physicists in these endeavours is all the more understandable if it really is correct that we need some form of non-baryonic dark matter to account for the formation of the large scale structure of the Universe. Furthermore, the types of dark matter particles needed in this scenario may well be closely related to those predicted by theories of elementary particles, but which have not yet been created in terrestrial accelerators. In my view, the incorporation of non-baryonic particles into the models is essential – the problems of forming galaxies force us to take these ideas very seriously.

14 Correlation Functions and the Spectrum of the Initial Fluctuations

To make quantitative comparisons between the theories of galaxy formation described in Chap. 13 and the observed distribution of galaxies, we need to make assumptions about the spectrum of the fluctuations from which galaxies and larger scale structures formed. The natural way of describing the distribution of galaxies on the large scale is in terms of *correlation functions* and their associated *power spectra*. The objective is to relate these properties of the distribution of galaxies at the present day to the spectrum of initial fluctuations which must have been present in the very early Universe. Let us first review the properties of the correlation function of galaxies at the present day, extending our introduction of Sect. 2.2.

14.1 The Two-point Correlation Function for Galaxies

The simplest description of the distribution of galaxies on a large scale is the *two-point correlation function*, which describes the excess probability of finding a galaxy at distance r from a galaxy selected at random over that expected in a uniform, random distribution. The two-point correlation function $\xi(r)$ was introduced in Sect. 2.2 and describes the number of galaxies in the volume element dV at distance r in the form

$$dN(r) = N_0[1 + \xi(r)]\,dV, \tag{14.1}$$

where N_0 is a suitably-defined average background number density of galaxies. The function $\xi(r)$ can also be written in terms of the probability of finding pairs of galaxies separated by distance r, the normal way in which correlation functions are defined:

$$dN_{\text{pair}} = N_0^2[1 + \xi(r)]\,dV_1\,dV_2. \tag{14.2}$$

The two-point correlation function can be directly related to the *density contrast* $\Delta(x) = \delta\varrho/\varrho$. We can write $\varrho = \varrho_0[1 + \Delta(x)]$ and so the pairways numbers of galaxies separated by distance r, in the sense of (14.2), is

$$dN_{\text{pair}}(\mathbf{r}) = \varrho(\mathbf{x})\,dV_1\,\varrho(\mathbf{x} + \mathbf{r})\,dV_2. \tag{14.3}$$

Therefore

$$dN_{\text{pair}}(\mathbf{r}) = \varrho_0^2[1 + \Delta(\mathbf{x})][1 + \Delta(\mathbf{x} + \mathbf{r})]\, dV_1\, dV_2. \tag{14.4}$$

When we take averages over a large volume, the average value of Δ is zero by definition and therefore the two-point correlation function is just

$$dN_{\text{pair}}(r) = \varrho_0^2[1 + \langle \Delta(\mathbf{x})\Delta(\mathbf{x} + \mathbf{r}) \rangle]\, dV_1 dV_2. \tag{14.5}$$

This shows explicitly the relation between the density contrast on different scales r and the two point correlation function which can be derived from the distribution of galaxies in space,

$$\xi(r) = \langle \Delta(\mathbf{x})\Delta(\mathbf{x} + \mathbf{r}) \rangle. \tag{14.6}$$

It will be appreciated that, in order for this procedure to work, it has to be assumed that the galaxies are tracers of the underlying distribution of dark matter. Notice also that the two-point correlation function is a rather sweeping, broad-brush description of the spatial distribution of galaxies, in that it is assumed that the distribution is spherically symmetric about a randomly selected galaxy. Inspection of Fig. 2.5 suggests that this is at best a crude approximation on large physical scales. Higher order correlation functions, such as the three- and four-point correlation functions can be defined to take more account of the real three-dimensional structure seen in the large scale distribution of galaxies. These topics have been treated by Peebles (1980, 1993), to whose works the reader is referred for further details.

Despite its broad-brush character, the two-point correlation function has the big advantage that it can be derived with excellent statistics from large galaxy surveys such as the Cambridge Southern Galaxy Survey (Fig. 2.4). There would be advantages in determining the spatial two-point correlation functions $\xi(r)$ directly from projects such as the Harvard–Smithsonian Center for Astrophysics Galaxy Survey (Fig. 2.5) and the Las Campanas Redshift Survey (Fig. 2.6), for both of which large numbers of redshifts have been measured. The problem is that the numbers cannot compete with the 2 million galaxies observed in the Cambridge Southern Galaxy Survey. Redshifts are not available for these galaxies, but the two-point spatial correlation function can be derived from the angular two-point correlation function, assuming that the clustering is a stationary, random process. The angular two-point correlation function is defined by

$$N(\theta)\, d\Omega = n_{\text{g}}[1 + w(\theta)]\, d\Omega, \tag{2.4}$$

and so, if $\xi(r)$ were of power-law form $\xi(r) \propto r^{-\gamma}$, then $w(\theta)$ would be proportional to $\theta^{-(\gamma-1)}$. The success of the scaling procedures described by (2.7), as applied to the Cambridge Southern Galaxy Survey and illustrated in Fig. 2.3, suggests that the derived form of $\xi(r)$ indeed represents a stationary, random process.

To recapitulate the results described in Sect. 2.2, the function $\xi(r)$ can be well represented by a power-law of the form

$$\xi(r) = \left(\frac{r}{r_0}\right)^{-\gamma}, \tag{2.6}$$

on physical scales from about 100 h^{-1} kpc to 10 h^{-1} Mpc with the scale $r_0 = 5h^{-1}$ Mpc and the exponent $\gamma = 1.8$. On scales greater than about $10h^{-1}$ Mpc the two-point correlation function decreases more rapidly than the power-law (2.6).

There are several points to be made about this correlation function.

1. Fig. 2.3 shows that the correlation function for galaxies is quite smooth. There are no obvious preferred scales, say, on the scale of the rich clusters of galaxies or superclusters. Whether or not any real structure on these scales would have been washed out in the averaging process is an interesting question. Be that as it may, the evidence of Fig. 2.3 is that fluctuations on a very wide range of scales must have been been present in the initial perturbation spectrum.

2. There is a characteristic scale $r_0 = 5h^{-1}$ Mpc which defines the scale at which the density of galaxies is greater than that of the background by a factor of two. This may be interpreted roughly as a measure of the scale on which the perturbations have become non-linear, in the sense that all structures on smaller scales have $\xi(r) > 1$. Notice that this means that structures on the scales of groups and clusters of galaxies have had time to become strongly non-linear by the present epoch, entirely consistent with the formation of these virialised structures by the present epoch. This cannot be the whole story, however, since the cellular structure of the large scale distribution of galaxies extends to scales very much greater than $5h^{-1}$ Mpc.

3. Of particular interest is the behaviour of the two-point correlation function on large physical scales. In addition to falling off more rapidly than a power-law on large scales, in some determinations, the function $\xi(r)$ becomes negative on scales greater than $20h^{-1}$ Mpc. Notice that, on these very large scales, the amplitude of the two-point correlation function for galaxies, in general, is very much less than one. There is, however, clustering on very large scales. Bahcall and her colleagues have found that Abell clusters, the richest clusters of galaxies, are correlated with a characteristic clustering scale $r_0 \approx (15-25)h^{-1}$ Mpc (see e.g. Bahcall 1988, 1997). We described their clustering properties in Sect. 4.1.2, as well as a schematic picture of what would be required to account for Bahcall's observations. There are other phenomena which may have to be explained on these very large scales. There is now good evidence for the clustering of quasars on scales $\leq 10\,h^{-1}$ Mpc with $r_0 = 6h^{-1}$ Mpc (Shanks et al. 1987, Boyle et al. 1991, Iovino et al. 1991) and we have yet to discuss the evidence for the streaming velocities of galaxies on large scales.

14.2 The Perturbation Spectrum

So far we have dealt with the growth of density perturbations in terms of the quantity $\Delta = \delta\varrho/\varrho$, but now we need to treat the evolution of the spectrum of fluctuations in more detail. The natural way of developing the theory is in terms of the spatial Fourier transforms of $\Delta(\mathbf{r})$, so that the amplitude of the perturbations with different wavelengths λ, or wavevectors $\mathbf{k} = (2\pi/\lambda)\mathbf{i}_k$, can be found. Since we are dealing with a three-dimensional distribution of galaxies, we have to take a three-dimensional Fourier transform. A number of simplifications are, however, needed since one of our aims is to relate the spectrum of the fluctuations to the two-point correlation function, which is, by definition, spherically symmetric about each point.

14.2.1 The Relation Between $\xi(r)$ and the Power Spectrum of the Fluctuations

First of all, we define the Fourier transform pair for $\Delta(\mathbf{r})$

$$\Delta(\mathbf{r}) = \frac{V}{(2\pi)^3} \int \Delta_{\mathbf{k}} e^{-i\mathbf{k}\cdot\mathbf{r}} \, d^3k; \tag{14.7}$$

$$\Delta_{\mathbf{k}} = \frac{1}{V} \int \Delta(\mathbf{r}) e^{i\mathbf{k}\cdot\mathbf{r}} \, d^3x. \tag{14.8}$$

We now use Parseval's theorem to relate the integrals of the squares of $\Delta(\mathbf{r})$ and its Fourier transform $\Delta_{\mathbf{k}}$

$$\frac{1}{V} \int \Delta^2(\mathbf{r}) \, d^3x = \frac{V}{(2\pi)^3} \int |\Delta_{\mathbf{k}}|^2 \, d^3k. \tag{14.9}$$

The quantity on the left-hand side of (14.9) is the mean square amplitude of the fluctuations per unit volume, and $|\Delta_{\mathbf{k}}|^2$ is the *power spectrum* of the fluctuations, which is often written as $P(k)$. Therefore, we can write

$$\langle \Delta^2 \rangle = \frac{V}{(2\pi)^3} \int |\Delta_{\mathbf{k}}|^2 \, d^3k = \frac{V}{(2\pi)^3} \int P(k) \, d^3k. \tag{14.10}$$

Since the two-point correlation function is spherically symmetric, the element of $\mathbf{k}$-space can be written $d^3k = 4\pi k^2 \, dk$ and so

$$\langle \Delta^2 \rangle = \frac{V}{2\pi^2} \int |\Delta_{\mathbf{k}}|^2 k^2 \, dk = \frac{V}{2\pi^2} \int P(k) k^2 \, dk. \tag{14.11}$$

The final step is to relate $\langle \Delta^2 \rangle$ to the two-point correlation function through (14.6). It is simplest to begin with a Fourier series and then transform the series summation into a Fourier integral. We begin by writing $\Delta(\mathbf{x})$ as

$$\Delta(\mathbf{x}) = \sum_{\mathbf{k}} \Delta_{\mathbf{k}} \, e^{-i\mathbf{k}\cdot\mathbf{x}}. \tag{14.12}$$

$\Delta(\mathbf{x})$ is a real function and therefore we can find $|\Delta(\mathbf{r})|^2$ by writing $|\Delta(\mathbf{r})|^2 = |\Delta(\mathbf{r})\Delta^*(\mathbf{r})|$ where $\Delta^*(\mathbf{r})$ is the complex conjugate of $\Delta(\mathbf{r})$. Taking the average value of the product of $\Delta(\mathbf{x})$ and $\Delta(\mathbf{x}+\mathbf{r})$ in the same way, in order to find $\xi(\mathbf{r})$ from (14.6), we find

$$\xi(r) = \left\langle \sum_{\mathbf{k}} \sum_{\mathbf{k}'} \Delta_{\mathbf{k}} \Delta_{\mathbf{k}'}^* \, e^{-i(\mathbf{k}-\mathbf{k}')\cdot\mathbf{x}} \, e^{i\mathbf{k}'\cdot\mathbf{r}} \right\rangle. \tag{14.13}$$

When we multiply out the cross terms in this summation, they all vanish except for those for which $\mathbf{k} = \mathbf{k}'$. Therefore,

$$\xi(r) = \sum_{\mathbf{k}} |\Delta_{\mathbf{k}}|^2 \, e^{i\mathbf{k}\cdot\mathbf{r}}. \tag{14.14}$$

We now convert this Fourier summation into a Fourier integral

$$\xi(r) = \frac{V}{(2\pi)^3} \int |\Delta_{\mathbf{k}}|^2 \, e^{i\mathbf{k}\cdot\mathbf{r}} \, d^3k. \tag{14.15}$$

Finally, we note that $\xi(r)$ is a real function, and so we are only interested in the integral of the real part of $e^{i\mathbf{k}\cdot\mathbf{r}}$, that is, the integral over $\cos(\mathbf{k}\cdot\mathbf{r}) = \cos(kr\cos\theta)$. Because of the spherical symmetry of the two-point correlation function, we integrate over an isotropic distribution of angles θ, that is, we integrate $\cos(kr\cos\theta)$ over $\frac{1}{2}\sin\theta\,d\theta$. Performing this integral, we obtain the final answer

$$\xi(r) = \frac{V}{2\pi^2} \int |\Delta_k|^2 \frac{\sin kr}{kr} k^2 \, dk = \frac{V}{2\pi^2} \int P(k) \frac{\sin kr}{kr} k^2 \, dk. \tag{14.16}$$

This is the relation between the two-point correlation function $\xi(r)$ and the power spectrum of the fluctuations $|\Delta_k|^2$ we have been seeking. Notice what this procedure has achieved. The function $\sin kr / kr$ acts as a 'window function' which allows only wavenumbers $k \leq r^{-1}$ to contribute to the amplitude of the fluctuations on the scale r. Fluctuations with larger wavenumbers, corresponding to smaller scales, average out to zero on the scale r.

14.2.2 The Initial Power-Spectrum

The observations described in Sect. 14.1 suggest that the spectrum initial fluctuations must have been very broad with no preferred scales and it is natural therefore to begin with power spectra of power-law form

$$P(k) = |\Delta_k|^2 \propto k^n. \tag{14.17}$$

According to (14.16), the correlation function $\xi(r)$ should have the form

$$\xi(r) \propto \int \frac{\sin kr}{kr} k^{(n+2)} \, dk. \tag{14.18}$$

Because the function $\sin kr/kr$ has value unity for $kr \ll 1$ and decreases rapidly to zero when $kr \gg 1$, we can integrate $k^{(n+2)}$ from 0 to $k_{max} \approx 1/r$ to estimate the amplitude of the correlation function on the scale r.

$$\xi(r) \propto r^{-(n+3)}. \tag{14.19}$$

Since the mass of the fluctuation is proportional to r^3, this result can also be written in terms of the mass within the fluctuations on the scale r, $M \sim \varrho r^3$.

$$\xi(M) \propto M^{-(n+3)/3}. \tag{14.20}$$

Finally, to relate ξ to the root-mean-square density fluctuation on the mass scale M, $\Delta(M)$, we take the square root of ξ, that is,

$$\Delta(M) = \frac{\delta \varrho}{\varrho}(M) = \langle \Delta^2 \rangle^{1/2} \propto M^{-(n+3)/6}. \tag{14.21}$$

The above analysis illustrates the relations between the various ways of describing the density perturbations. To summarise, a power spectrum $P(k) = |\Delta_k|^2$ of power-law form $P(k) \propto k^n$ corresponds to a two-point correlation function $\xi(r) \propto r^{-(n+3)} \propto M^{-(n+3)/3}$ and to a spectrum of density perturbations $\Delta(M) \propto M^{-(n+3)/6}$. These results merit a number of comments.

Large-scale Perturbations. So long as $n > -3$, the mass spectrum of fluctuations decreases to large mass scales and so the Universe is isotropic and homogeneous on the very largest scales.

Poisson Noise. A natural model to consider is one in which the mass fluctuations are associated with random statistical fluctuations in the numbers of particles N on the scale r. According to Poisson statistics, the fluctuations in the numbers of particles is $\delta N/N = 1/N^{1/2}$, or, in terms of mass fluctuations, $\Delta(M) = \delta M/M = 1/M^{1/2}$. It follows from the relations (14.21) that $n = 0$, in other words, a 'white-noise' power spectrum with equal power on all scales. The corresponding correlation functions are $\xi(r) \propto r^{-3}$ and $\xi(M) \propto M^{-1}$.

The Harrison–Zeldovich Spectrum. The case $n = 1$ is of special interest and results in a power-spectra of the form

$$\Delta(M) \propto M^{-2/3} \quad \text{and} \quad \xi \propto r^{-4} \propto M^{-4/3}. \tag{14.22}$$

This spectrum has the property that the density contrast $\Delta(M)$ had the same amplitude on all scales when the perturbations came through the horizon. Let us illustrate this important result. We consider the early development of the perturbations before they came through their particle horizons and before the epoch of equality of matter and radiation energy densities. In both variants of the Dark Matter picture described in Sect. 13.5, the perturbations in the dark matter were driven by the perturbations in the radiation-dominated plasma and these grew as $\Delta(M) \propto R^2$. Therefore, we can write the development of the spectrum of perturbations as

$$\Delta(M) \propto R^2 \, M^{-(n+3)/6}. \tag{14.23}$$

A perturbation of scale r came through the horizon when $r \approx ct$, and so the mass of dark matter within it was $M_{\mathrm{D}} \approx \varrho_{\mathrm{D}}(ct)^3$. During the radiation dominated phases, $R \propto t^{1/2}$ and the number density of dark matter particles, which will eventually form bound structures at $z \sim 0$, varied as $N_{\mathrm{D}} \propto R^{-3}$. Therefore, the horizon mass increased as $M_{\mathrm{H}} \propto R^3$, or, $R \propto M_{\mathrm{H}}^{1/3}$. We can insert this result into (14.23) in order to find the mass spectrum when the fluctuations came through the horizon at different cosmic epochs

$$\Delta(M) \propto M^{2/3} \, M^{-(n+3)/6}. \tag{14.24}$$

Thus, if $n = 1$, the amplitudes of the density perturbations were all the same when they came though their particle horizons during the radiation dominated era.

This rather special value, $n = 1$, is known as the *Harrison–Zeldovich spectrum* (Harrison 1970, Zeldovich 1972). These authors realised that this form of spectrum has a number of appealing features. If the amplitudes of the fluctuations were $\sim 10^{-4}$ on all scales when they came through the horizon, this would result in the formation of structure on a wide range of scales by the present epoch. An additional attraction of this form of spectrum is that it avoids the creation of excessive numbers of low-mass black holes in the early Universe. If the initial power-spectrum were of such a form that $\Delta\varrho/\varrho \gg 1$ when these perturbations came through the horizon, these would inevitably lead to the formation of black holes (see, for example, Zeldovich and Novikov 1983). Furthermore, the spectrum does not diverge on large angular scales either, and so is consistent with the observed large-scale isotropy of the Universe. Sunyaev and Zeldovich (1970) used a variety of arguments of this type to constrain the form of the power-spectrum of density perturbations as they came through the horizon and this must have played a part in Zeldovich's thinking. Thus, the Harrison–Zeldovich spectrum is strongly motivated by observational constraints, rather than being simply a theoretical construct. At that time, the calculations were carried out for baryonic scenarios, but the same result is found for the Dark Matter scenarios as well. The epoch of equality of the inertial mass densities in the matter and radiation occurred at a redshift $2.4 \times 10^4 h^2$ if $\Omega_0 = 1$. In the subsequent matter-dominated era, the dark matter perturbations grew as $(1 + z)^{-1}$, from that epoch to the present day, so long as they remained linear. We know that galaxies and clusters exist now with $\Delta \gg 1$ and therefore the amplitude of the perturbations when they came through the horizon must have been $\Delta \sim 10^{-4}$.

It is intriguing that, if $n = 1$, the Universe is *fractal*, in the sense that every perturbation came through the horizon with the same amplitude – as the Universe expands, we always find perturbations of the same amplitude coming through the horizon. Proponents of the inflationary picture of the early Universe find that fluctuations with the Harrison–Zeldovich spectrum occur rather naturally in that picture.

The Observed Two-point Correlation Function. For the sake of interest, we might ask what value of n would give the observed exponent of the two-point correlation function for galaxies, $\gamma = 1.8$. Since $\xi \propto r^{-(n+3)}$, the preferred value would appear to be $n = -1.2$. Unfortunately, this is very much an academic exercise since it is assumed that the observed two-point correlation function represents the initial fluctuation spectrum. This cannot be the case, since non-linear effects must have dramatically modified its initial form on scales less than $r_0 \sim 5h^{-1}$ Mpc. This point is illustrated rather dramatically by N-body simulations of the growth of structure in Dark Matter cosmologies, which we will discuss below. To put it another way, on the scales over which the two-point correlation function has been well measured, relaxation under gravity has been well under way and there has been time for systems on the scales of, say, 1 Mpc to have become virialised.

14.3 The Evolution of the Initial Perturbation Spectrum

In Chaps. 12 and 13, we described many of the effects, which modify significantly the initial perturbation spectrum. We now need to put all of these together to understand how the initial perturbation spectrum evolves to form the structures we observe in the Universe today. Again, we devote most attention to the Hot and Cold Dark matter scenarios and follow the pleasant presentation of Kolb and Turner (1990). Fig. 14.1 shows the spectrum of fluctuations in the dark matter at some epoch after the epoch of equality of energy densities in the cold matter and radiation, at $z \approx 2.4 \times 10^4 \Omega_0 h^2$. This diagram repays some study.

First of all, we note that the scale on the abscissa is the wavelength associated with the perturbation of mass M, corresponding to $k = 2\pi/\lambda$. The ordinate shows the amplitude of the density perturbation associated with each wavelength. This is derived from (14.21), substituting $n = 1$ for the Harrison–Zeldovich spectrum and writing $M \propto r^3 \propto \lambda^3$. Thus, for the 'unprocessed' Harrison–Zeldovich spectrum, we expect

$$\Delta(M) = \frac{\delta\varrho}{\varrho}(\lambda) \propto M^{-(n+3)/6} \propto \lambda^{-2}. \tag{14.25}$$

It can be seen that, for large scales, corresponding to large masses, all three models follow this spectrum. For smaller masses, however, the predictions are somewhat different. In the case of the *Hot Dark Matter* model, small-scale

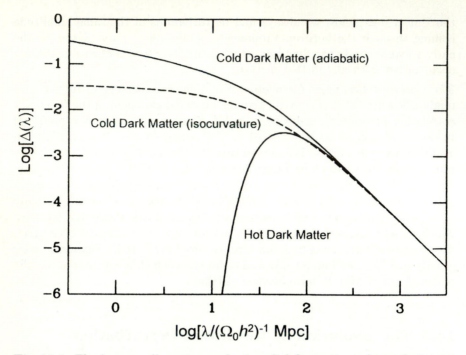

Fig. 14.1. The 'processed' spectrum of primordial fluctuations observed at some epoch after that corresponding to the epoch of equality of matter and radiation energy densities (Kolb and Turner 1990). Rather than mass, the abscissa is plotted in terms of the wavelength associated with the perturbation. The ordinate shows the amplitude of the fluctuations on the scale λ. The three models show the processed spectra for Hot and Cold Dark Matter, assuming the fluctuations are adiabatic and for an isocurvature Cold Dark Matter model. In all three cases, it is assumed that the input spectrum is of Harrison–Zeldovich form, $P(k) \propto k$. The vertical normalisation is arbitrary.

perturbations are damped by the free-streaming of neutrinos as soon as the perturbations come through the horizon during the radiation-dominated era. The spectrum cuts off exponentially below the critical mass given by (13.10). An analytic expression for the distortion of the input spectrum is quoted by Kolb and Turner (1990), following Bond and Szalay (1983). The function describing changes to the input spectrum of the perturbations is called the *transfer function*.

In the case of the *Cold Dark Matter* picture, we have to look in a little more detail at the evolution of the spectrum of perturbations, prior to the epoch of equality of matter and radiation energy densities. On scales greater than the horizon, the mass spectrum preserved its initial form, $\Delta \propto M^{-(n+3)/6}$, and masses on all scales grew as R^2, until they entered their particle horizons. In other words, during these early phases, the spectrum evolved as $\Delta \propto M^{-(n+3)/6} R^2$. When a particular mass-scale came through the horizon during the radiation-dominated phases, the dark matter perturbations were dynamically coupled to the dominant mass, which is in the radiation-dominated plasma. The perturbations in the radiation-dominated plasma oscillated as sound waves and so the amplitude of the perturbations in the cold dark matter were stabilised until the end of the radiation-dominated era, when they began to grow as $(1 + z)^{-1}$. As shown in Sect. 14.2.2, the scale factor R at which a perturbation of mass M entered the horizon is proportional to $M^{1/3}$, and so the density perturbation spectrum for masses less than the horizon mass became

$$\Delta \propto M^{-(n+3)/6} \times M^{2/3} = M^{-(n-1)/6}. \tag{14.26}$$

Thus, for small masses, the 'processed' spectrum is flatter than the input spectrum of perturbations. Rewriting the spectrum in terms of wavelengths λ rather than masses, we find

$$\Delta(M) = \frac{\delta \varrho}{\varrho}(\lambda) \propto M^{-(n-1)/6} \propto \lambda^{-(n-1)/2}. \tag{14.27}$$

For the Harrison–Zeldovich spectrum with $n = 1$, the spectrum becomes flat at small wavelengths. The detailed calculations shown in Fig. 14.1 indicate that the curvature of the spectrum is very gradual. Nonetheless, the key point is that perturbations survive to very small wavelengths and masses in the Cold Dark Matter picture. Analytic expressions for the transfer function are given by Peebles (1983) and Davis et al. (1985).

Isocurvature Cold Dark Matter modes behave like the isothermal modes discussed in Sect. 12.4. On scales greater than the horizon, they behave like the Hot and Cold Dark matter modes described above. On small scales, however, after they have come through the horizon, they evolve like the isothermal perturbations discussed in Sect. 12.4 and are subject to the Meszaros effect. This drastically slows their growth and results in the flattened spectrum seen in Fig. 14.1. Efstathiou and Bond (1986) have presented analytic forms for these transfer functions.

Notice that the evolution, which ultimately leads to the formation of the large scale structure of the Universe as we know it, is in the growth and non-linear development of the dark matter perturbations after the epoch of equality of matter and radiation energy densities. The amplitudes of the spectra shown in Fig. 14.1 grow as $(1+z)^{-1}$ throughout the matter-dominated era, until they become non-linear when $\Delta_{\mathrm{D}} \sim 1$. We have already described the general features of the subsequent evolution of the dark matter perturbations in Sect. 13.5 and we will look at various aspects of the non-linear development of the perturbations in Chap. 16, but, for the moment, let us outline some of the key features, which will appear in the remarkable computer simulations of their subsequent evolution.

The perturbations with the largest amplitudes attain $\Delta_{\mathrm{D}} \sim 1$ first and begin to collapse to form bound systems. A key point is that the dark matter perturbations consist of collisionless particles and so dissipative processes, such as the release of energy by radiative processes or by friction and viscosity, are not relevant. In order to form bound structures satisfying the virial theorem (3.19), the collapsing dark matter clouds have to lose internal kinetic energy and this is achieved primarily through the process of *violent relaxation*, which was first described by Lynden-Bell (1967). As the collapse of the cloud of non-interacting particles gets underway, large gravitational potential gradients are set up, since the collapse is unlikely to be precisely spherically symmetric. The system relaxes under the influence of these large potential gradients and Lynden-Bell showed that the particles of the system rapidly attain an equilibrium configuration in which all the masses attain the same velocity distribution. In the process, the system gets rid of half of its kinetic energy and so ends up as a bound system satisfying the virial theorem. Subsequently, energy exchange between the dark matter particles can take place by *dynamical friction*, the exchange of energy in gravitational encounters between particles, but this is a slow process (see, for example, Longair 1997).

In the Cold Dark Matter picture, the first objects to form are low mass systems and these then begin to be assembled into larger scale systems by the process of hierarchical clustering. We can think of this as being the process by which the dark matter haloes of galaxies, groups and clusters are formed. The baryonic matter falls into these structures and, since it can lose energy by dissipative processes, stars and gas clouds begin to form within the dark matter haloes. In contrast, in the Hot Dark Matter picture, the lowest mass objects to form are those on the scale of clusters of galaxies. As discussed in Sect. 13.5.1, the asymmetric collapse of these structures is expected to lead to flattened 'pancake' structures, into which the baryonic matter collapses.

These models have been the subject of detailed computer simulation to determine how well they can reproduce the observed large scale structure in the Universe. Fig. 14.2 is a sample of the results of computer simulations of the Hot and Cold Dark Matter models carried out by Frenk (1986). In the *Hot Dark Matter* picture, flattened structures like pancakes are produced very effectively. The baryonic matter forms pancakes within the large neutrino haloes and their evolution is similar to that of the adiabatic baryonic picture from that point on. The model is, in fact, too effective in producing flattened, stringy structures. Essentially everything collapses into thin pancakes and filaments, and the observed Universe is not as highly structured as this. As we have remarked, galaxies must form rather late in this picture because only the most massive structures survive to the epoch of recombination. This means that it is difficult to produce stars and galaxies which are younger than the structures on the scale of $\sim 4 \times 10^{15} M_{\odot}$.

In the *Cold Dark Matter* picture, masses on all scales can begin to collapse soon after recombination and star clusters and the first generations of stars can be old in this picture. The large-scale structure is formed by gravita-

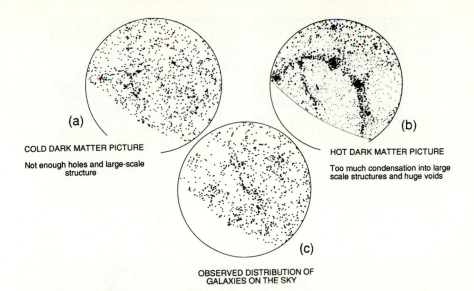

Fig. 14.2a-c. Simulations of the expectations of (a) the Cold Dark Matter and (b) the Hot Dark Matter models of the origin of the large scale structure of the Universe compared with the observations (c) (Frenk 1986). The unbiased cold dark matter model does not produce sufficient large scale structure in the form of voids and filaments of galaxies whereas the unbiased hot dark matter model produces too much structure.

tional clustering under the influence of the 'processed' power spectrum of the perturbations. Large scale systems such as galaxies and clusters of galaxies are assembled from their component parts by the dynamical processes discussed above. Fig. 14.2 shows that structure indeed develops but it is not as pronounced on the large scale as is observed in the local Universe. This is because it is difficult to produce elongated structures by gravitational clustering alone, which tends to make more symmetrical structures than the sheets and filaments of galaxies found in the Universe on the largest scales.

One of the important successes of the Cold Dark Matter picture is that it is can account for the observed two-point correlation function of galaxies over a remarkably wide range of scales. The review by Efstathiou (1990) can be thoroughly recommended for those who wish to enter much more deeply into the details of these remarkable simulations. We will give only one example of the simplest case in which the initial spectrum is of standard Harrison–Zeldovich form in an $\Omega_0 = 1$ universe, the phases of the waves which make up Δ_k are random, and the transfer function is that appropriate for the adiabatic Cold Dark Matter model shown in Fig. 14.1. Fig. 14.3 shows a two-dimensional projection of the evolution of the particle positions in such a Cold Dark Matter simulation. The initial conditions are shown in the top-left hand

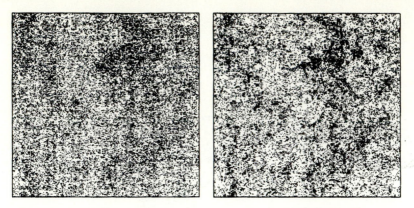

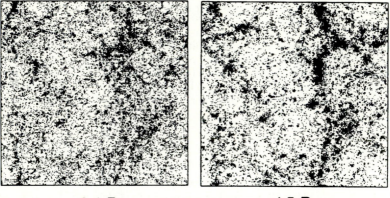

Fig. 14.3. Computer simulations of the development of structure in a standard Cold Dark Matter model with $\Omega_0 = 1$ (Efstathiou 1990, see also Davis *et al.* 1985). The pictures are two-dimensional projections of the positions of 32768 particles within a box the sides of which have comoving length $32.5h^{-1}$ Mpc. The initial conditions are shown in the top-left box with scale factor R_0 and the other boxes show the development of structure at scale factors $1.8R_0$, $2.4R_0$ and $4.5R_0$.

picture which has scale factor R_0. The other boxes show the development of structure at scale factors $1.8R_0$, $2.4R_0$ and $4.5R_0$.

The corresponding forms of the two-point correlation function $\xi(r)$ are shown in Fig. 14.4. It is evident that the process of gravitational clustering converts the 'processed' input power-spectrum into one which much more closely resembles the power-law form of the two-point correlation function. Specifically, at scale factor $1.8R_0$, $\xi(r)$ has the observed slope of $\gamma = 1.8$ over a wide range of scales. As the model evolves, the correlation function

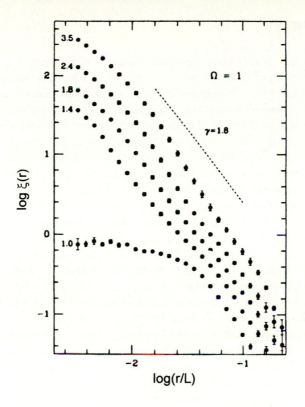

Fig. 14.4. The two-point correlation functions $\xi(r)$ at different scale-factors for the simulations shown in Fig. 14.3 (Efstathiou 1990, see also Davis *et al.* 1985). The error bars show the uncertainties in the correlation functions from 5 independent runs of the computational model. The dashed line shows the slope $\gamma = 1.8$ of the observed two-point correlation function for galaxies.

becomes steeper. As explained by Efstathiou, this is very far from the end of the story. In particular, in this realisation of the model, the velocity dispersion of galaxies chosen at random from the field would be too large. As he shows, the match to observation can be improved if it is assumed that the galaxies provide a biased view of the large-scale distribution of matter.

From our present perspective, the important point is that these models show how it is possible, in principle, to account for the power-law form of the two-point correlation functions for galaxies, within the context of the Cold Dark Matter scenario, and this must be considered a remarkable achievement.

14.4 Biasing

In the story so far, it has been assumed that the visible parts of galaxies trace the distribution of the dark matter, but one can imagine many reasons why this should not be so. The generic term for this phenomenon is *biasing*, meaning the preferential formation of galaxies in certain regions of space rather than in others. Part of the motivation behind the introduction of biasing was to improve the agreement between the predictions of the Cold Dark Matter scenario and the observed distribution of galaxies. In the Hot Dark Matter picture, *anti-biasing* is needed so that the formation of galaxies is not so highly concentrated into sheets and filaments.

Is there any direct evidence for biasing in the Universe? A good example is provided by the rich clusters of galaxies, such as the Coma Cluster. As described in Sect. 4.3.1, this cluster has been studied kinematically in great detail and estimates of its mass and mass-to-light ratio are well established. The mass of the dark matter amounts to about a factor of 10 greater than that in the visible matter, but this factor is only about a quarter or one third of the value necessary to achieve a density parameter $\Omega_0 = 1$. A similar conclusion has been reached by Bahcall (1997), who has found that estimates of the overall density parameter for the largest systems, for which reasonable dynamical estimates can be made, cluster about values of $\Omega_0 \sim 0.2 - 0.3$. If the Universe really has the critical density, $\Omega_0 = 1$, there must be biasing by a factor of three towards the formation of galaxies on the scale of clusters and superclusters as opposed to the general field.

Many possible biasing and anti-biasing mechanisms have been described by Dekel and Rees (1987) and by Dekel (1987) who discuss how these can be tested by observations of the nearby Universe. As they emphasise, perhaps the most important aspects of this story concern reliable observational estimates of how much biasing, or anti-biasing, actually occurs in the Universe.

Some of the biasing mechanisms come about rather naturally in the theory of the formation of galaxies, and we restrict attention to one or two of the more important of these. Kaiser (1984) realised that, inherent in the notion of the power-spectrum of the perturbations, as formulated in Sect. 14.2.1, is the fact that the perturbations have a Gaussian distribution of amplitudes about the root mean squared value $\bar{\Delta}$ with variance $\bar{\Delta}^2$, so that the probability of encountering a density contrast Δ at some point in space is proportional to $\exp(-\Delta^2/\bar{\Delta}^2)$. Kaiser argued that galaxies are most likely to form in the highest peaks of the density distribution. Thus, if we require the density perturbation to exceed some value Δ_{crit} in order that structures form, galaxy formation would be biased towards the highest density perturbations over the mean background density. This picture has a number of distinct attractions. For example, it can account for the fact that the clusters of galaxies are more strongly clustered than galaxies in general (see Sect. 4.1.2). If structure only forms if the density contrast exceeds a certain value Δ_{crit}, then galaxy formation within a large-scale density perturbation, which will eventually form a cluster of galaxies, is strongly favoured. As Kolb and Turner (1990) remark, the reason that all the highest mountains in the world are in the Himalayas is that they are superimposed upon the large-scale plateau, or long wavelength perturbation, caused by the plate supporting the Indian subcontinent crashing into the Asian plate. This concept has been worked out in detail by Peacock and Heavens (1985) and by Bardeen *et al.* (1986). The numerical simulations described by Efstathiou (1990) illustrate very clearly how the density peaks of a Gaussian random field result in a much more highly structured distribution of galaxies as compared with the underlying mass distribution (Fig. 14.5).

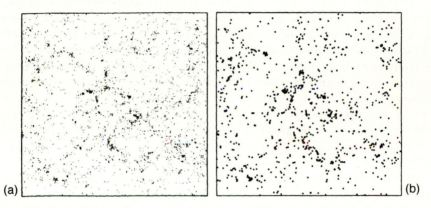

Fig. 14.5a-b. Comparison of the density field in a Cold Dark Matter simulation in an $\Omega_0 = 1$ universe showing, (**a**) the projected particle distribution for all particles, and (**b**) the distribution of 'galaxies', defined as peaks in the density distribution with $b \geq 2.5$, where b is the bias factor. The high density peaks in (**b**) are much more prominent than those in the underlying density field (Efstathiou 1990, Davis *et al.* 1985).

In a simple Gaussian model, Coles and Lucchin (1995) showed that the two-point correlation functions in the underlying dark matter and that of the galaxies are related by

$$\xi_{\text{gal}}(r) = b^2 \xi_{\text{D}}(r), \tag{14.28}$$

where b is defined to be the *bias factor*. Notice that there are a number of more-or-less equivalent definitions of the bias factor b. From the relation between the two-point correlation function and the density contrast (14.14), we see that

$$\left(\frac{\delta\varrho}{\varrho}\right)_{\text{B}} = b \left(\frac{\delta\varrho}{\varrho}\right)_{\text{D}}. \tag{14.29}$$

Another essentially equivalent definition often found in the literature is in terms of the variance in the counts of galaxies within a sphere of radius $8h^{-1}$

Mpc relative to the variance in the underlying mass distribution in the same sphere

$$b^2 = \frac{\sigma_8^2(\text{galaxies})}{\sigma_8^2(\text{mass})}.$$ (14.30)

The value $8h^{-1}$ Mpc has been chosen since that is the scale on which the two-point correlation function for galaxies has unit amplitude, $\xi(r) \sim 1$. The introduction of the bias factor b brings with it a number of important advantages for the Cold Dark matter model. Since galaxies form preferentially in peaks of the density distribution, their random velocities are smaller than those associated with the underlying dark matter distribution. Efstathiou (1990) has shown how a Cold Dark Matter model with $b = 2.5$ can be reconciled with a large number of independent aspects of the large scale distribution of galaxies, including the amplitude and slope of the two-point correlation function and the mean velocity dispersion of galaxies in the general field. Although these must be considered very considerable achievements, notice that it is an entirely empirical procedure. We need more astrophysical understanding of the non-linear stages of galaxy formation to develop a concrete astrophysical picture of biasing.

Another example of the type of astrophysics, which might be involved, concerns galactic explosions which may sweep the gas away from the vicinity of a galaxy and this could result in positive or negative biasing. A violent explosion can remove the gas from the vicinity of the galaxy and make it too hot for further galaxy formation to occur in its vicinity. On the other hand, the swept-up gas may be greatly enhanced in density at the interface between the hot expanding sphere and the intergalactic gas. By analogy with the case of galactic supernova remnants, in which star formation can be stimulated by the passage of a shock wave, the same process on a galactic scale might stimulate the formation of new galaxies (Ostriker and Cowie 1981). This mechanism could thicken pancakes in the Hot Dark Matter picture. Another possibility is that the gas in the voids between superclusters may be so hot that galaxies cannot form in these regions. It is evident from these examples that the understanding of biasing is an astrophysical problem. The role of star formation in the early development of galaxies is of central importance in understanding how biasing could influence the process of galaxy formation.

14.5 Reconstructing the Initial Power Spectrum

Granted that biasing must play a role in determining the amplitudes of the correlation functions on different scales, is it possible to produce self-consistent models for the formation of structure from a single initial power spectrum? There have been many studies of this problem but we will follow the impressive analysis of Peacock and Dodds (1994) as the basis for the analysis of viable models.

The analysis of Peacock and Dodds is based upon the assumption that, despite the different amplitudes of the correlation and cross-correlation functions for galaxies and clusters of galaxies, these can be derived from a single smooth initial power spectrum which has been subject to different degrees of bias, depending upon which samples of galaxies are selected. An important insight is that it is possible to relate the observed spectrum of perturbations in the non-linear regime, $\xi(r) \geq 1$, to the initial spectrum in the linear regime, following the pioneering analysis of Hamilton *et al.* (1992) (see also Padmanabhan 1997). Peacock and Dodds have shown how it is possible to formulate analytically the relation between the observed non-linear perturbations on large scales to the initial spectrum and then to determine the biasing factors for different samples of galaxies and clusters which will generate a smooth power-spectrum. We will not go into the details of how this is achieved but illustrate the results of going through this procedure and its consequences for structure formation.

Peacock and Dodds use eight separate data sets in their reconstructions. The most straightforward of these is the two-point correlation function derived from the large APM survey of galaxies in the Southern hemisphere (Figs. 2.3 and 2.4). The others are derived from large redshift surveys of galaxies and clusters of galaxies. The large redshift surveys have the advantage of providing much more information about the three-dimensional space distribution of galaxies and clusters, but their numbers are much more limited than the 2 million galaxies available from the APM survey. In the simplest picture, the redshifts provide estimates of the distances of the galaxies according to Hubble's law $r = cz/H_0$, but these distance estimates have to be corrected for the effects of *redshift bias*. This arises is two ways. First of all, the 'fingers of God' seen in large redshift surveys, such as the Harvard–CfA survey (Fig. 2.5), significantly distort the cosmological redshifts of the galaxies in such a way that the galaxies would be shifted from their true spatial positions, reducing the amplitude of the correlation function. The second effect is associated with the fact that the density perturbation itself induces potential motions according to the expression derived in Sect. 11.5

$$\delta v = H_0 r \Omega_0^{0.6} \left(\frac{\delta \varrho}{\varrho} \right).$$ (14.31)

As a result, galaxies are observed to be 'falling into' the perturbation and so the projected velocity component along the line of sight is less than that associated with its true cosmological redshift. For large scales, on which the linear relation (14.31) is valid, Kaiser (1987) has shown how this redshift bias can be eliminated from the inferred two-point correlation functions.

The results of these analyses are illustrated in Fig. 14.6. Relative bias parameters were derived which produced the best fit to a smooth power spectrum and the following values were found

$$b_\mathrm{A} : b_\mathrm{R} : b_\mathrm{O} : b_\mathrm{I} = 4.5 : 1.9 : 1.3 : 1,$$ (14.32)

where the subscripts refer to the bias factors for Abell clusters (A), radio galaxies (R), optically selected galaxies (O) and IRAS galaxies (I). We recall that the amplitudes of the correlation functions are related by the square of the bias factor (14.28) and so these values reflect well-known results about the relative clustering tendencies of these classes of objects. Thus, the clusters of galaxies are much more strongly correlated than galaxies in general (Sect. 4.1.2) and radio galaxies tend to belong to groups and clusters of galaxies. The IRAS galaxies are strong dust emitters and so are preferentially associated with spiral and starburst galaxies. Since the spiral galaxies tend to avoid groups and clusters of galaxies, the IRAS sample displays the weakest clustering.

The results of this analysis are shown in Fig. 14.6. The quantity plotted on the ordinate is $\Delta^2(k)$ which is related to the power spectrum of the density perturbations (14.16) by

$$\Delta^2(k) = \frac{V}{(2\pi)^2} 4\pi k^3 |\Delta_k|^2 = \frac{V}{(2\pi)^2} 4\pi k^3 P(k). \qquad (14.33)$$

Inspection of (14.16) shows that $\Delta^2(k)$ is the contribution of the power spectrum per unit logarithmic interval in k to the two-point correlation function. Notice that often the power spectrum $P(k)$ is plotted which differs from $\Delta^2(k)$ by three powers of k. Fig. 14.6a shows that there is excellent agreement between the eight independent determinations of the power spectrum. Peacock and Dodds find that there must be a significant amount of redshift bias present such that $\Omega_0^{0.6}/b_{\rm I} = 1.0 \pm 0.2$

These estimates of $\Delta^2(k)$ are averaged in bins of width 0.1 in $\log_{10} k$ in Fig. 14.6b and compared with the predictions of various Cold Dark Matter models in Fig. 14.6c. The shapes of the curves are derived by assuming that the initial power spectrum is of scale invariant form $P(k) \propto k^n$ and this is modified by the transfer function for different values of the parameter $\Omega_0 h$. The best-fitting value of $\Omega_0 h$ is

$$\Omega_0 h = 0.255 \pm 0.017 + 0.32(n^{-1} - 1). \qquad (14.34)$$

Thus, for the standard Harrison–Zeldovich spectrum, $n = 1$, this analysis suggests that the 'standard' Cold Dark Matter model with $\Omega_0 = 1$, is a poor fit to the data. Even if $h = 0.5$, the best fitting models would have Ω_0 significantly less than 1. This result has been found by a number of workers, namely, that the simplest Cold Dark Matter models predict too much power on scales corresponding to $0.1 < k/h < 1$ Mpc^{-1}, if the large scale power-spectrum in the linear region is to be accommodated. The other key constraint, which we have not yet built into this picture, is the observation of intensity fluctuations in the Cosmic Microwave Background Radiation, which we take up in Chap. 15. The predicted power spectra shown in Fig. 14.6c can be extrapolated to the scale of the COBE observations and excellent agreement is found with the observed fluctuation spectrum.

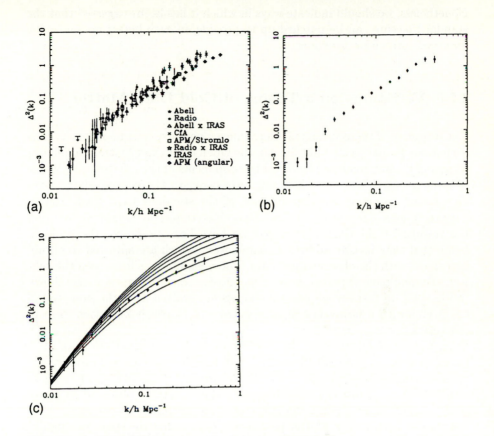

Fig. 14.6a-c. The power spectrum of perturbations on large physical scales, linearised assuming $\Omega_0 = b_{\mathrm{I}} = 1$ (Peacock and Dodds 1994). **(a)** The data derived from eight independent surveys. Notice that all the surveys are consistent with the detection of a break in the power spectrum about $k = 0.03h$. **(b)** The linearised data in (a) averaged over bins of width 0.1 in $\log_{10}k$ (see text for explanation of the notation used). **(c)** Comparison of the data of (b) with different variants of the Cold Dark Matter model. The models have scale-invariant input spectra which have been modified by the Cold Dark Matter transfer functions. Different values of the fitting parameter $\Omega_0 h$ are shown, $\Omega_0 h = 0.5, 0.45, \ldots 0.25, 0.2$, in decreasing order of power at short wavelengths.

We have spent some effort in understanding the 'standard' Cold Dark Matter picture. This is the picture in which the Universe has $\Omega_0 = 1$, $\Omega_\Lambda = 0$ and $n = 1$. In many ways, the model is remarkably successful in providing more or less the observed structures we observe in the Universe on large-scales, but the consensus of opinion among cosmologists is that it is probably not good enough. It remains to be seen how serious this problem is. One might argue that, in view of the empirical way in which biasing has been included into the models, there must be scope for fine-tuning the predictions.

Nonetheless, we should indicate ways in which it has been suggested that the standard picture can be patched up to provide an even better match to the observed correlation functions.

14.6 Variations on a Theme of Cold Dark Matter

The objective of these variants is to produce a power spectrum of density perturbations similar to that obtained in the Open Cold Dark Matter model, but retaining the flat geometry favoured by proponents of the inflationary picture of the very early Universe. It is simplest to regard the various alternatives as ways of distorting the expectations of the standard Cold Dark Matter picture to look like the Open Cold Dark Matter picture. Many variants on the standard Cold Dark Matter picture have been proposed, the problem being that they involve additional parameters, which are adjusted to obtain agreement with the observational data. These possibilities have been the subject of remarkable supercomputer simulations, some examples being shown in Fig. 14.7. In these examples, the models are evolved from the same initial conditions with a power-law input spectrum of Harrison–Zeldovich form in the very early Universe.

Open Cold Dark Matter (OCDM). As Fig. 14.6c shows, Open Cold Dark Matter models can account satisfactorily for all the observations and produce the type of sponge-like structure seen in the lower right panel of Fig. 14.7. These structures bear a distinct resemblance to the structures observed in large-scale redshift surveys (Figs. 2.5 and 2.6). The differences as compared with the standard Cold Dark Matter picture are, firstly, that the epoch of equality of matter and radiation occurs rather later and, secondly, that the growth of structure proceeds over a somewhat smaller range of redshifts, only until the epoch at which $\Omega_0 z \approx 1$. Consequently, the break in the power spectrum takes place at greater masses, resulting in less power at short wavelengths. This is a perfectly viable picture and would be consistent with the dynamical evidence that the overall density parameter is about $\Omega_0 = 0.3$, the value used in the simulation of Fig. 14.7. This model has been advocated by Gott (1997). The concern with this picture is that proponents of the inflationary Universe would much prefer the geometry of the Universe to be flat.

Cold Dark Matter with a finite cosmological constant (ΛCDM). A way of preserving the flat geometry of space is to include the cosmological constant into the model so that $\Omega_0 + \Omega_\Lambda = 1$. There is not a great deal of difference in the dynamics of the underlying model as compared with the Open Cold Dark Matter model (see the top-left and bottom right panels of Fig. 14.7). This can be understood from the analysis of Sect. 11.4.4 which showed that the dynamics only differ from the $\Omega_\Lambda = 0$ case at redshifts $(1 + z) \leq \Omega_0^{1/3}$. Thus, the differences occur in the late stages of evolution when the effect of the cosmological constant is to stretch out the time-scale of the model,

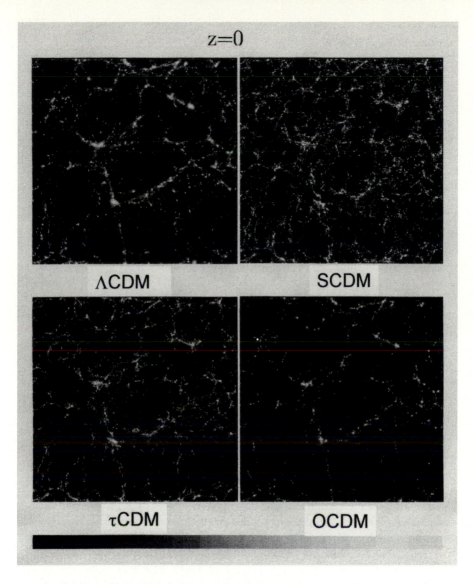

Fig. 14.7 Some examples of the predicted large-scale structure in the distribution of galaxies from supercomputer simulations by the Virgo consortium. Each panel has side $240h^{-1}$ Mpc and involved following the gravitational interactions of $256^3 = 1.7 \times 10^7$ particles. In the standard CDM picture, there is a great deal of small-scale structure, more than is found in the distribution of galaxies. The other models have been designed to create less structure on the small scale while preserving the power on large scales (Courtesy of Prof. Carlos Frenk and his colleagues).

allowing some further development of the perturbations, as may be seen by comparing Figs. 11.4 and 11.5.

Mixed Dark Matter (HCDM) In this variant, $\Omega_0 = 1$ and the dark matter is made up of a mixture of Hot and Cold Dark Matter. This model attempts to capitalise upon the advantages of the Hot and Cold Dark Matter pictures discussed in Sect. 13.5. By adding in the Hot Dark Matter, the power on large physical scales can be boosted. In terms of the transfer functions, inspection of Fig. 14.1 shows that a suitable weighting of the hot and cold components would enable more power to be included on large scales, relative to the standard Cold Dark Matter picture. It might seem remarkable that the two quite distinct types of dark matter end up making similar contributions to the overall dark matter density, but this cannot be excluded.

Cold Dark Matter with Decaying Neutrinos (τCDM) In this scenario, the objective is to enhance the radiation to matter energy densities so that the epoch of equality of matter and radiation energy densities is shifted to lower redshifts, as in the case of the Open Cold Dark Matter picture. It is essential to ensure that the predictions of primordial nucleosynthesis are not violated in that, if there were additional relativistic components present in the Universe during the epoch of nucleosynthesis, the expansion rate would be increased and excessive amounts of helium are produced. The trick is to suppose that there exist particles which decay after the epoch of nucleosynthesis, thus enhancing the radiation relative to the matter energy densities and so delaying the epoch of equality of matter and radiation energy densities. The lower left-hand panel of Fig. 14.7 shows the results of this type of simulation. Whether or not such particles with the appropriate half-lives exist is a question for the particle physicists.

Tilted Cold Dark Matter In this variant, the power spectrum of the initial fluctuations departs from the standard value $n = 1$. In one variant on this theme, there might be additional power on large physical scales, associated with primordial gravitational waves which would boost the fluctuations in the Cosmic Microwave Radiation observed by COBE relative to the fluctuations associated with the galaxies on smaller physical scales (see Sect. 15.2).

Broken Scale Invariant Cold Dark Matter In this picture, the initial power spectrum changes slope at the appropriate wave-number.

At this stage, it is largely a matter of taste which of these the reader prefers. A number of groups have made supercomputer simulations to find acceptable models for many of these variants (see Turok 1997). The models which seem to provide the best fit to the observations are those involving Open Cold Dark Matter (OCDM), Hot and Cold Dark Matter (HCDM), the Cold Dark Matter with a finite cosmological constant (ΛCDM) and the Cold Dark Matter with decaying particles (τCDM). It remains to be seen which, if any, of these models are relevant to the real Universe, but one route which will undoubtedly cast significant light on these issues is the study of the fluctuations in the Cosmic Microwave Background Radiation. This is the subject of the next chapter.

15 Fluctuations in the Cosmic Microwave Background Radiation

The study of the power-spectrum of spatial fluctuations in the Cosmic Microwave Background Radiation provides one of the most important means of confronting theories of the origin of the large-scale structure of the Universe with observation. The problem addressed in this chapter is how to relate the fluctuations in the dark and baryonic matter on the last scattering surface at the epoch of recombination to the intensity fluctuations which they imprint upon the background radiation. Of crucial importance in this calculation is the ionisation state of the intergalactic gas during the epochs when it changed from being a fully ionised plasma to a neutral gas. We have dealt with a number of important aspects of the physics of this process in Sect. 9.3, but now we need to understand these in a little more detail.

15.1 The Ionisation of the Intergalactic Gas Through the Epoch of Recombination

An expression for the optical depth of the fully ionised intergalactic gas due to Thomson scattering was derived in Sect. 9.3. Expression (9.16) shows how rapidly the optical depth increases with redshift once the gas becomes fully ionised. As a result, temperature fluctuations which originate at redshifts greater than the redshift of recombination are damped out by Thomson scattering and the fluctuations we observe originate in a rather narrow redshift range about that at which the optical depth of the intergalactic gas is unity. Hence, the ionisation history of the intergalactic gas through the epoch of recombination needs to be determined in some detail.

The problem, first discussed by Zeldovich, Kurt and Sunyaev (1968) and Peebles (1968), is well-known. At the epoch of recombination, the plasma was 50% ionised when the temperature of the background radiation was about 4,000 K. Photons emitted in the recombination of hydrogen atoms must have energies $h\nu \geq h\nu_\alpha$, where ν_α is the frequency of the Lyman-α transition which has wavelength 121.6 nm. Therefore, these photons can either reionise other hydrogen atoms directly, or else raise them to an excited state H^*, from which the electron can be ejected by the much more plentiful soft photons in the black-body spectrum. Thus, there is no direct way of destroying the energetic photons liberated in the recombination process. The way in which Lyman-α photons are destroyed is through the *two-photon process* in which two photons are liberated from the 2s state of hydrogen in a rare quadrupole transition. The spontaneous transition probability for this process is very small, $w = 8.23$ sec^{-1}, but it turns out to be the dominant process which determines the rate at which the intergalactic gas recombines.

Detailed calculations of the degree of ionisation through the critical red-shift range have been carried out by Jones and Wyse (1985) who find a very strong dependence of the fractional ionisation x upon redshift at re-combination. They provide a convenient analytic expression for the degree of ionisation through the critical redshift range:

$$x = 2.4 \times 10^{-3} \frac{(\Omega_0 h^2)^{1/2}}{\Omega_B h^2} \left(\frac{z}{1000}\right)^{12.75}. \tag{15.1}$$

Ω_0 is the density parameter for the Universe as a whole and Ω_B the density parameter of baryons. Using the same formalism which led to (9.16), we can find a simple expression for the optical depth of the intergalactic gas at redshifts $z \sim 1000$

$$\tau = 0.37 \left(\frac{z}{1000}\right)^{14.25}. \tag{15.2}$$

Because of the enormously strong dependence upon redshift, the optical depth of the intergalactic gas is always unity very close to a redshift of 1070, inde-pendent of the exact values of Ω_0, Ω_B and h. We can now work out the range of redshifts from which the photons of the background radiation we observe today were last scattered. This probability distribution is given by

$$p(z) \, \mathrm{d}z = e^{-\tau} \frac{\mathrm{d}\tau}{\mathrm{d}z} \, \mathrm{d}z, \tag{15.3}$$

which can be closely approximated by a Gaussian distribution with mean red-shift 1070 and standard deviation $\sigma = 80$ in redshift. This result formalises the statement that the photons of the Microwave Background Radiation were not last scattered at a single redshift. Rather, half of those we observe today were last scattered between redshifts of 1010 and 1130. For more exact calcu-lations, improved analytic formulae for $p(z)$ are given by Hu and Sugiyama (1995).

15.2 The Physical and Angular Scales of the Fluctuations

First of all, let us work out the physical scale at the present epoch corre-sponding to the thickness of the last scattering layer. The element of radial comoving distance at redshift z is given by (8.25)

$$dr = \frac{c\,dz}{H_0[(1+z)^2(\Omega_0 z + 1) - \Omega_\Lambda z(z+2)]^{1/2}}. \qquad (8.25)$$

Because of the use of comoving coordinates, this is the radial distance element projected to the present epoch. At large redshifts, $\Omega_0 z \gg 1$, the term involving the cosmological constant can be neglected and so we find

$$dr = \frac{c}{H_0} \frac{dz}{z^{3/2}\Omega_0^{1/2}}. \qquad (15.4)$$

If we take the thickness of the last scattering layer to correspond to a redshift interval $\Delta z = 120$ at $z = 1070$, this is equivalent to a physical scale of $10(\Omega_0 h^2)^{-1/2}$ Mpc at the present epoch. The mass contained within this scale is $M \approx 3 \times 10^{14}(\Omega_0 h^2)^{1/2} M_\odot$, corresponding roughly to the mass of a cluster of galaxies. The comoving scale of $d = 10(\Omega_0 h^2)^{-1/2}$ Mpc corresponds to a proper distance $d/(1+z)$ at redshift z and hence to an angular size

$$\theta = \frac{d(1+z)}{D} = \frac{10(\Omega_0 h^2)^{-1/2}}{D_{\text{Mpc}}} = 6\Omega_0^{1/2} \text{ arcmin}, \qquad (15.5)$$

since $D = 2c/H_0\Omega_0$, if $\Omega_0 z \gg 1$.

On physical scales smaller than $10(\Omega_0 h^2)^{-1/2}$ Mpc at the present epoch, we expect a number of independent fluctuations to be present along the line of sight through the last scattering layer. Consequently, the random superposition of these perturbations leads to a statistical reduction in the amplitude of the observed intensity fluctuations by a factor of roughly $N^{-1/2}$, where N is the number of fluctuations along the line of sight.

These dimensions and angular scales can be compared with the horizon scales derived in Sect. 12.1. From (12.8), we observe that the horizon scale corresponds to an angular scale of $1.8\,\Omega_0^{1/2}$ degrees at a redshift of 1000. Perturbations on this scale are the largest which can be in causal contact at that epoch. They correspond to physical scales about 20 times greater than the thickness of the last scattering surface and so to comoving scales of about $200(\Omega_0 h^2)^{-1/2}$ Mpc at the present day.

Thus, we have to consider the influence of fluctuations on a very wide range of physical scales relative to the thickness of the last scattering layer, as illustrated schematically in Fig. 15.1. For reference, it is useful to remember that the thickness of the last scattering layer corresponds to the masses of clusters of galaxies, the horizon scale to the very largest voids, and the smallest angular scales observed by the COBE satellite, $\theta = 10°$, to about 5 to 10 times the size of the largest voids. The problem is now to convert the density perturbations and their associated velocities into intensity fluctuations on the last scattering layer.

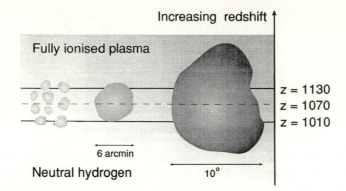

Fig. 15.1. A schematic diagram illustrating perturbations on different physical scales at the epoch of recombination, relative to the thickness of the last scattering layer. The largest scale perturbations, such as those observed by the COBE satellite, are much larger than the thickness of the last scattering layer from which most of the photons of the Cosmic Microwave Background Radiation were last scattered. On small scales, a number of perturbations is observed along the line of sight through the last scattering layer.

15.3 Large Angular Scales

On the very largest scales, the dominant source of intensity fluctuations results from the fact that the photons we observe have to climb out of the gravitational potential wells associated with perturbations which are very much greater in size than the thickness of the last scattering layer (Fig. 15.1). During their subsequent propagation to the Earth, although the photons pass through gravitational potential fluctuations, what they gain by falling into them is exactly compensated by the gravitational redshift coming out, so long as the perturbations continue to grow linearly with redshift, as we show below. Thus, it is the escape from density perturbations at the epoch of recombination which provides a first-order effect in the gravitational redshift. This phenomenon was first analysed by Sachs and Wolfe (1967) and is known as the *Sachs–Wolfe* effect. It is expected to be one of the principal sources of intensity fluctuations in the Cosmic Microwave Background Radiation observed by the COBE satellite on angular scales of 10° and greater. There might also be a contribution from primordial gravitational waves on these scales. Thus, these large scale perturbations provide direct evidence about processes in the very early Universe.

15.3.1 The Sachs–Wolfe Effect – Physical Arguments

We can illustrate the origin of the amplitude and spectrum of these fluctuations by simple calculations. First, we note that, in the simplest approximation, the frequency shift and the corresponding change in the thermodynamic temperature of the background radiation can be related to the gravitational redshift z_{grav} by the Newtonian expression

$$\frac{\Delta \nu}{\nu} = \frac{\Delta T}{T} \approx z_{\mathrm{grav}} = \frac{\Delta \phi}{c^2} \approx \frac{G \Delta M}{dc^2}, \tag{15.6}$$

where ΔM is the mass excess in the perturbation and d its physical scale at the epoch of recombination. Expressions (15.6) are far from trivial relations. On the scales of interest, the fluctuations at the epoch of recombination far exceed the horizon scale and so the perturbations would represent a change of the gravitational potential of everything within the horizon. More properly, we should describe these perturbations as *metric perturbations*. These 'super-horizon' perturbations raise the thorny question of the choice of gauge to be used in relativistic perturbation theory, the issues involved being clearly described by Hu (1996). A general relativistic treatment, first performed by Sachs and Wolfe (1967), is needed. Padmanabhan (1993) gives a straightforward analysis of how the relevant result can be derived by perturbing the Friedman metric and relating the temperature fluctuation to the perturbation in the Newtonian gravitational potential $\Delta \phi$ (see also his even better solution in Padmanabhan 1996). The result is $\Delta T/T = (1/3)\Delta\phi/c^2$, recalling that $\Delta \phi$ is a negative quantity. Coles and Lucchin (1995) rationalise how the Sachs–Wolfe answer can be found. In addition to the Newtonian gravitational redshift, because of the perturbation of the metric, the cosmic time, and hence the scale factor R, at which the fluctuations are observed, are shifted to slightly earlier cosmic times. Temperature and scale factor change as $\Delta T/T = -\Delta R/R$. For all the standard models in the matter-dominated phase $R \propto t^{2/3}$ and so the increment of cosmic time changes as $\Delta R/R = (2/3)\Delta t/t$. But $\Delta \nu/\nu = -\Delta t/t$ is just the Newtonian gravitational redshift, with net result that there is a positive contribution to $\Delta T/T$ of $-(2/3)\Delta\phi/c^2$. The net temperature fluctuation is $\Delta T/T = \frac{1}{3}\Delta\phi/c^2$.

We can derive a number of important results using (14.10) and (14.11), which relate the power spectrum of the initial fluctuations to the mass spectrum, by order of magnitude calculations. If the power-spectrum of the fluctuations is taken to have the form $P(k) = |\Delta_k|^2 \propto k^n$, the mass of an object of dimension $d \sim k^{-1}$ is proportional to k^{-3} and so, taking the integral (14.16) over wavenumber to be

$$\langle \Delta^2 \rangle \propto |\Delta_k^2| k^3, \tag{15.7}$$

we find that

$$\langle \Delta^2 \rangle^{1/2} = \frac{\delta\varrho}{\varrho} = \frac{\delta M}{M} \propto M^{-(n+3)/6}. \tag{15.8}$$

Let us work out the fluctuation in the gravitational potential $\Delta\phi$ at the epoch of recombination in terms of the properties of the density perturbations observed at the present epoch. Throughout this calculation, we assume that the perturbations remain small until the present epoch and that they developed linearly as $\Delta \propto (1+z)^{-1}$. The physical size of the perturbation d at redshift z corresponds to a physical size d_0 at the present epoch, such that $d(1+z) = d_0$. Assuming that $\Omega_0 = 1$, the density perturbation grows as $\Delta\varrho/\varrho \propto (1+z)^{-1}$ and so the density fluctuation $\Delta\varrho$ at redshift z was

$$\Delta\varrho = \frac{\Delta\varrho_0}{(1+z)}\frac{\varrho}{\varrho_0} = \Delta\varrho_0(1+z)^2. \tag{15.9}$$

Since $\Delta M \approx \Delta\varrho\, d^3$ and $d = d_0/(1+z)$, it follows that

$$\Delta\phi \approx \frac{G\Delta M}{d} \approx G\Delta\varrho_0 d_0^2. \tag{15.10}$$

This is the first important result of this analysis – the perturbation of the gravitational potential is independent of cosmic epoch, since all dependences upon redshift z have disappeared from (15.10). Thus, $\Delta\phi$ at any redshift is the same as that which would be estimated for the same perturbation once it had evolved linearly to the present epoch.

We can now incorporate the mass spectrum into the calculation. Since $\delta\varrho_0 \propto \varrho_0 M^{-(n+3)/6}$, and $M \approx \varrho_0 d_0^3$, we find $\delta\varrho_0 \propto d_0^{-(n+3)/2}$ and hence

$$\Delta\phi \approx G\Delta\varrho_0 d_0^2 \propto d_0^{(1-n)/2}. \tag{15.11}$$

It follows immediately from (15.6) and the expression relating the physical size d_0 of the perturbation to the angular size it would subtend at a large redshift, $d_0 = \theta D$, where $D = 2c/H_0\Omega_0$, that

$$\frac{\Delta T}{T} \approx \frac{1}{3}\frac{\Delta\phi}{c^2} \propto \theta^{(1-n)/2}. \tag{15.12}$$

This is the important result we have been seeking. The amplitude of the temperature fluctuations as a function of angular scale depends only upon the spectral index n of the initial power spectrum of the fluctuations. In particular, the amplitude is independent of angular scale if $n = 1$. This is another aspect of the fact that the perturbation spectrum with $n = 1$ is known as a *scale-invariant* spectrum. The results of a typical computation of the expected amplitude of intensity fluctuations on a wide range of angular scales is shown in Fig. 15.2 for the case in which $n = 0.86$ (Bersanelli *et al.* 1996). The reason for this choice of n will become apparent from the considerations of Sect. 15.3.3. We will discuss the notation used in Fig. 15.2 in the next section. The predicted spectrum of density perturbations due to the Sachs–Wolfe effect on large angular scales, that is, at small multipoles $l < 30$, can be seen to be more or less independent of angular scale, in agreement with the simple arguments given above. Thus, the observed spectrum of intensity fluctuations on large angular scales provides a direct estimate of the power-spectrum of the initial perturbations, assuming there is no contribution from gravitational waves (see Sect. 15.3.3). According to Partridge (1998), the power-spectrum determined from the 4-year data set from the COBE experiment provides an estimate of $n = 1.1 \pm 0.3$ (see also, Bennett *et al.* 1996). Combining the COBE data with the results of the Tenerife ground-based experiment, which had angular resolution $4°$, Hancock and his colleagues (1997) find a similar result, $n = 1.1 \pm 0.1$. It is striking that this value is in close agreement with the expectations of the scale-free Harrison–Zeldovich spectrum, $n = 1$.

Fig. 14.1 shows that the spectrum of density perturbations deviates from the primordial Harrison-Zeldovich spectrum with $n = 1$ at comoving scales $\sim 20(\Omega_0 h^2)^{-1}$ Mpc and so it is expected that the spectrum of perturbations associated with the Sachs-Wolfe effect will diminish on the corresponding angular scales, $\theta \leq 0.2° \times h(\Omega_0 h^2)^{-2}$. Recalling the arguments presented in Sect. 14.3, this break in the power-spectrum is associated with the scale of perturbations which come through the horizon at the epoch of equality of matter and radiation energy densities. Thus, if this break in the spectrum were observable, it would provide direct information about the epoch of equality of matter and radiation energy densities. As we will show in Sect. 15.4, however, it is likely that this key feature of the temperature fluctuation spectrum will be masked by the greater fluctuations associated with acoustic waves at the last scattering layer. Nonetheless, the important point of principle is that features are imprinted on the spectrum of temperature fluctuations which are related directly to physical processes in the early Universe.

15.3.2 The Statistical Description of the Temperature Fluctuations

The tools used in deriving (15.12) were somewhat rough and ready. Let us introduce briefly the nomenclature used in making precise comparisons between theories and observations of temperature fluctuations in the Cosmic Microwave Background Radiation. The temperature fluctuations are distributed over the surface of a sphere and so we require the spherical polar equivalent of the relation between the density distribution and its power-spectrum, which was given by (14.14) for the case of Fourier series in three Cartesian dimensions. For the surface of a sphere, the appropriate complete sets of orthonormal functions are the spherical harmonics. The first step is to make a spherical harmonic expansion of the temperature distribution over the whole sky.

$$\frac{\Delta T}{T}(\theta, \phi) = \frac{T(\theta, \phi) - T_0}{T_0} = \sum_{l=0}^{\infty} \sum_{m=-l}^{m=l} a_{lm} Y_{lm}(\theta, \phi), \tag{15.13}$$

where the normalised functions Y_{lm} are given by the expression

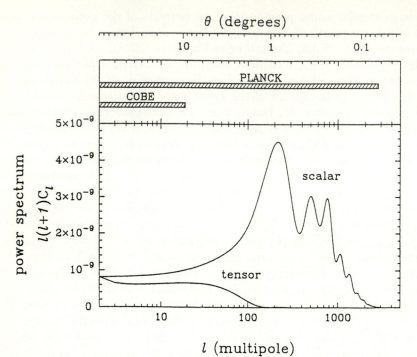

Fig. 15.2. The predicted power spectrum of temperature fluctuations in the Cosmic Microwave Background Radiation plotted as a function of the multipole l for an inflationary cold dark matter cosmology (Bersanelli *et al.* 1996). The quantity plotted on the ordinate is $l(l+1)C_l$, where C_l is the power spectrum defined by (15.17). The multipole l corresponds to an angular scale $\theta = \pi/l$. The curve labelled *scalar* shows the contribution of small density perturbations and that labelled *tensor* the contribution of gravitational waves. As discussed in Sect. 15.3.3, the relative amplitudes of these contributions depend upon the specific inflationary model. In the case of the scale-invariant spectrum, $n = 1$, there is expected to be no contribution of gravitational waves and the spectrum of density perturbations would be flat, corresponding to $n = 1$, at multipoles $l \leq 30$. The estimates shown in the diagram are for $n_s = 0.86$, for which the contributions of scalar and tensor components are equal for $l = 2$. The bars along the top of the diagram show the ranges of multipoles probed by COBE and projected to be observed by the *Planck* mission of the European Space Agency.

$$Y_{lm}(\theta, \phi) = \left[\frac{2l+1}{4\pi} \frac{(l - |m|)!}{(l + |m|)!} \right]^{1/2} P_{lm}(\cos \theta)\, e^{im\phi}$$

$$\times \begin{cases} (-1)^m & \text{for } m \geq 0 \\ 1 & \text{for } m < 0 \end{cases}, \qquad (15.14)$$

and $P_{lm}(\cos \theta)$ are the associated Legendre polynomials of order l. With this normalisation, the orthogonality condition for the spherical harmonics becomes

$$\int_{4\pi} Y_{lm}^* Y_{l'm'} \, d\Omega = \delta_{ll'} \delta_{mm'}, \tag{15.15}$$

where the asterisk means the complex conjugate of the spherical harmonic and the integral is taken over the whole sky, that is, over $d\Omega = \sin\theta \, d\theta \, d\phi$, the element of solid angle. The δs are delta functions which take the value unity if $l = l'$ and $m = m'$, and are zero otherwise. From the orthogonality condition (15.15), the values of a_{lm} are found by multiplying the temperature distribution over the sphere by Y_{lm}^* and integrating over the sphere.

$$a_{lm} = \int_{4\pi} \frac{\Delta T}{T}(\theta, \phi) \, Y_{lm}^* \, d\Omega. \tag{15.16}$$

Peebles (1980) provides a physical interpretation of how the power on angular scale θ is related to the power in the spherical harmonic l. The zeros of the real and imaginary parts of Y_{lm} divide the sky into roughly rectangular cells, the minimum dimension of each cell at low latitude being close to π/l. Near the poles, the zeros of the azimuthal functions $\sin m\phi$ and $\cos m\phi$ are crowded together, but there the values of P_{lm} are close to zero. The net result is that each spherical harmonic has a well-defined angular resolution $\theta \approx \pi/l$.

In general, the temperature distribution over the sky need not be Gaussian. By this term, we mean that the phases of the waves which make up the spherical harmonic decomposition over the sky are random. As expressed by Kogut *et al.* (1996), the distribution of the coefficients a_{lm} should follow a Gaussian probability distribution with phases which are uniformly distributed between 0 and 2π. The temperature fluctuations on the sky might display non-Gaussian features such as abrupt temperature discontinuities, intense hot spots, linear structures and so on. These types of features are predicted by theories in which large scale structures are seeded by topological defects, cosmic strings, or by cosmic textures. This in itself provides an important test of this class of theory. The non-Gaussian features would result in strongly correlated values of the coefficients a_{lm}.

In contrast, many theories predict that the fluctuations should obey Gaussian statistics. For example, in the inflation scenario of the early Universe, density perturbations arise from random quantum fluctuations in the pre-inflationary era and it is expected that these perturbations will have random phases. In what follows, we will assume that the perturbations are Gaussian, an assumption which results in a number of simplifications in interpreting the angular power-spectrum. Specifically, if the fluctuations can be represented by a superposition of waves of random phase, each of the $(2l+1)$ coefficients a_{lm} associated with the multipole l provides an independent estimate of the amplitude of the temperature fluctuations associated with that multipole. The power spectrum is assumed to be circularly symmetric about each point and so the mean value of $a_{lm}a_{lm}^*$, averaged over the whole sky, provides an estimate of the power associated with the multipole l.

$$C_l = \frac{1}{2l+1} \sum_m a_{lm} a_{lm}^* = \langle |a_{lm}|^2 \rangle. \tag{15.17}$$

Furthermore, because of the assumption that the fluctuations are Gaussian, the power spectrum C_l provides a *complete* statistical description of the temperature fluctuations. The analyses which have been made of the four-year COBE data have found no evidence for non-Gaussian features over regions of sky away from the Galactic plane (Kogut *et al.* 1997).

Various ways of presenting the results of these statistical analyses are found in the literature. One approach is to derive the autocorelation function for the distribution of temperature over the sky in angular coordinates, the analogue of determining the two-point correlation function for galaxies (Sect. 14.2.1).

$$C(\theta) = \left\langle \frac{\Delta T(\mathbf{i}_1)}{T} \frac{\Delta T(\mathbf{i}_2)}{T} \right\rangle. \tag{15.18}$$

where $\mathbf{i}_1$ and $\mathbf{i}_2$ are unit vectors in the directions 1 and 2 and the average is taken over the sky with a fixed angular separation θ. Because of the orthogonality properties of the spherical harmonics, it follows that

$$C(\theta) = \frac{1}{4\pi} \sum_l (2l+1) C_l P_l(\cos\theta), \tag{15.19}$$

where again $C_l = \langle |a_{lm}|^2| \rangle$ is the power spectrum of the temperature fluctuations on the sky as defined by (15.17). In deriving this result, we have used the addition theorem for spherical harmonics, namely, that

$$\sum_{lm} Y_{lm}^*(\mathbf{i}_1) Y_{lm}(\mathbf{i}_2) = \sum_l \frac{2l+1}{4\pi} P_l(\cos\theta), \tag{15.20}$$

where $P_l(\cos\theta)$ is the Legendre polynomial of order l (see, for example, Mathews and Walker 1970). The expression (15.19) bears a close resemblance to the autocorrelation theorem in Fourier transform theory, namely, that if $f(x)$ has Fourier transform $F(s)$, then its autocorrelation function has Fourier transform $|F(s)|^2$ (Bracewell 1986). The exponential has been replaced by the Legendre polynomial associated with multipole l. Note also the pleasant point emphasised by Bracewell that the autocorrelation theorem in Fourier theory wipes out all phase information. The same happens in (15.17) and (15.19), but we do not worry since the phases are assumed to be random and hence perfectly well-defined for our purposes. As Peebles (1993) points out, it is a matter of convenience whether one works with the angular two-point correlation function $C(\theta)$ or the power-spectrum C_l.

The use of the quantity $l(l+1)C_l$ plotted on the ordinate of Fig. 15.2 is a consequence of working out the power in each multipole for a Harrison–Zeldovich spectrum of density perturbations. As shown by Peebles (1993) and White, Scott and Silk (1994), if the power-spectrum of the initial perturbations is of standard power-law form, $P(k) = Ak^n$, the values of C_l are given by

$$C_l \propto A 2^n \pi^2 \frac{\Gamma \left(3 - n\right) \Gamma \left(l + \frac{n-1}{2}\right)}{\Gamma^2 \left(\frac{4-n}{2}\right) \Gamma \left(l + \frac{5-n}{2}\right)}, \tag{15.21}$$

where the Γs are Gamma-functions. In the case of the Harrison–Zeldovich spectrum, $n = 1$ and (15.21) reduces to

$$C_l \propto \frac{A}{l(l+1)}. \tag{15.22}$$

Thus, the quantity $l(l + 1)C_l$ is expected to be independent of multipole l for the standard Harrison–Zeldovich spectrum, exactly as inferred from the analysis which led to (15.12).

Another way of expressing the amplitude of the temperature fluctuations has been used by the COBE investigators. They assume the perturbations have a flat spectrum and then work out the rms temperature fluctuation Q_{rms-PS}, which can be related to the best estimate of C_2, the quadrupole component of the temperature power-spectrum (White, Scott and Silk 1994).

One important aspect of the above analysis is that it brings out clearly the problem of *cosmic variance*, which ultimately limits the precision with which the amplitude of the temperature fluctuations on any scale can be estimated. The spherical harmonic analysis shows that we obtain $(2l + 1)$ independent estimates of the value of C_l for a given multipole l. Thus, for small values of l, we obtain few independent samplings of the correlation function. Evidently, the precision with which value of C_l is known is proportional to $1/N^{1/2}$, where N is the number of individual estimates. Bennett *et al.* (1996) quote an uncertainty in the values of C_l of

$$\frac{\sigma(C_l)}{C_l} \approx \left[\frac{2}{2l + 1}\right]^{1/2}. \tag{15.23}$$

According to their analysis of the 4-year COBE data-set, the observations are cosmic-variance limited, rather than noise-limited, for multipoles $l <$ 20. Thus, there is little prospect of improving the estimates of the power-spectrum on large angular scales.

After this long digression into the techniques of spherical harmonic analysis of the COBE data, let us show how the power spectrum of the fluctuations on large angular scales can be related to the spectrum of density perturbations. Perturbations on an angular scale of $10°$ correspond to physical dimensions of about $1,000(\Omega_0 h)^{-1}$ Mpc at the present epoch, at least an order of magnitude greater than the large holes seen in the distribution of galaxies in Fig. 2.2. Because we expect the temperature fluctuations to be independent of angular scale on large angular scales as a result of the analysis which led to (15.12), let us extrapolate from scales on which the amplitude of density perturbations is known to the COBE scales, assuming a scale-invariant spectrum. The problem is discussed in detail by Efstathiou (1990), Peebles (1993) and Peacock and Dodds (1994). For the purposes of illustration, let us use Peacock and Dodds' estimate for the amplitude of the density fluctuations on the largest scales for which they are defined with reasonable statistics, that is, at wavenumber $k = 0.028h$ Mpc^{-1}. The power-spectrum is shown in Fig. 14.6 in the notation explained in Sect. 14.5. The value of $\delta\varrho_0/\varrho_0 = [\Delta^2(k)]^{1/2} = (0.0087)^{1/2} = 0.093$ on the scale $d_0 = (2\pi/k) = 244/h$ Mpc. Adopting the critical density, $\varrho_0 = 2 \times 10^{-26}h^2$ kg m^{-2}, we find from (15.11) that $\Delta T/T \sim 2 \times 10^{-5}$. This temperature fluctuation is roughly the same as that observed by the COBE experiment. Obviously, the quality of the COBE data merit a much more detailed analysis than this crude estimate, but it serves to illustrate the orders of magnitude of the quantities involved in the calculation. Detailed comparison of the COBE data with the power-spectrum of galaxy clustering have been carried out by many authors, for example by Peacock and Dodds (1994) and by Turner (1997), and these confirm the excellent agreement between the COBE temperature fluctuations and the power spectrum of galaxy clustering on large angular scales. Indeed, many analyses nowadays use the COBE temperature fluctuations in order to normalise the spectrum of density perturbations which is used to construct models of structure formation. There remain the problems of reconciling the details of the galaxy power-spectrum with the expectations of the standard Harrison-Zeldovich spectrum which we discussed in detail in Sect. 14.6.

15.3.3 Primordial Gravitational Waves

It will have been observed that Fig. 15.2 includes a contribution from primordial gravitational waves on large angular scales. The treatment we have given of perturbations in the standard Big Bang cosmologies has been essentially Newtonian, with modifications to adapt these equations to the relativistic case where appropriate. The complete treatment according to General Relativity automatically includes small perturbation modes corresponding to the propagation of gravitational waves. The mathematical analysis is somewhat lengthy and complex and would take us far beyond the scope of this introductory text. As explained by Coles and Lucchin (1995), problems arise from the fact that, when the metric is perturbed, it is not trivial to distinguish betweeen physically meanful modes from those arising purely from the choice of coordinate system (see also Kolb and Turner 1990). These problems have, however, been solved and we simply quote the results of these analyses.

It is not unexpected that the analysis results in three types of perturbation, *tensor modes* corresponding to propagating gravitational waves, *vector modes*, corresponding to the rotational modes discussed in Sect. 11.5, and *scalar modes* corresponding to the growing and decaying modes of density perturbations which we analysed in detail in Sects. 11.4 and 11.6. The new feature is the presence of gravitational waves, which might well have been generated in the very early Universe. On small physical scales, these behave like massless particles with equation of state $p = \frac{1}{3}\varepsilon$, corresponding to a ratio of specific heats $\gamma = 4/3$. As a result, after they come through the horizon, $\lambda \sim r_{\mathrm{H}}$, their energy density decreases as R^{-4}. Thus, unlike the scalar modes in the relativistic plasma which are stabilised, the amplitudes of the tensor modes are damped adiabatically.

This is not true of the perturbations which have scale $\lambda \gg r_{\mathrm{H}}$. In both the matter-dominated and radiation-dominated eras, there are tensor modes which have constant amplitude as the Universe expands. In these cases, the perturbations correspond to redshifting everything within the horizon and, as a result, they produce qualitatively the same type of Sachs–Wolfe temperature perturbations on scales greater than the horizon scale at the last scattering layer. Thus, we can immediately set limits to the possible energy density of gravitational waves at the present epoch from the observed power in the COBE fluctuations. As pointed out by Padmanabhan (1993), gravitational waves at the present epoch with $\lambda \sim c/H_0$ would produce a quadrupole anisotropy in the Cosmic Microwave Background Radiation through the Sachs–Wolfe effect and so the upper limit to the amplitude of these waves must be of the order of the quadrupole anisotropy $h \sim \Delta T/T \approx 10^{-5}$. Therefore, the energy density of these gravitational waves is

$$\varepsilon_{\mathrm{G}} = \varrho_{\mathrm{G}} c^2 \sim (32\pi G)^{-1}(\omega^2 h^2 c^2), \tag{15.24}$$

where ω is the angular frequency of the waves and h their amplitude. Setting $\omega^2 = c^2 k^2 = (4\pi^2 c^2/\lambda^2) = 4\pi^2 H_0^2$, we find that

$$\Omega_{\mathrm{G}} = \frac{8\pi G \varrho}{3H_0^2} = \frac{\pi^2}{3} h^2 \tag{15.25}$$

Thus, the upper limit to gravitational waves on the scale of the horizon corresponds to $\Omega_{\mathrm{G}} \leq 3 \times 10^{-10}$.

According to inflationary theories of the early evolution of the Universe, the quantum fluctuations which are thought to be the source of the density perturbations from which the large scale structure of the Universe developed might well be accompanied by a background of gravitational waves (Starobinsky 1985, Davis et al. 1992, Crittenden et al. 1993). Indeed, it is claimed that this is a general feature of a wide class of inflationary models for the formation of structure. According to one variant of the inflationary picture, to a good approximation, the spectral indices of the tensor (gravitational wave) and scalar (density perturbation) modes are related by $n_{\mathrm{t}} \approx 1 - n_{\mathrm{s}}$ and the ratio of their quadrupole power spectra $r = C_2^{\mathrm{t}}/C_2^{\mathrm{s}}$ depends upon the spectral index of the scalar perturbations as

$$r = 7(1 - n_{\mathrm{s}}). \tag{15.26}$$

This is the origin of the power spectrum of gravitational waves shown in Fig. 15.2, in which the value of r has been set equal to unity. It can be seen that the power spectrum of perturbations due to gravitational waves cuts off at angular scales $\theta \sim 2-3°$, corresponding to the horizon scale at the epoch of recombination. The perturbations on smaller scales are damped adiabatically when they come through the horizon, as explained above.

Thus, the presence of primordial gravitational waves could enhance the amplitude of temperature fluctuations associated with the scalar modes on large angular scales. In particular, it can be seen that the effect of the primordial gravitational waves would be to add power to the long wavelength modes relative to those expected from an extrapolation from the physical scales $\lambda \sim 20-50$ Mpc which have been determined from studies of the correlation function of galaxies. In the analysis of Peacock and Dodds (1995), which led to the primordial fluctuation spectrum shown in Fig. 14.6, it was assumed that there was no gravitational wave contribution to the spectrum of fluctuations in the Microwave Background Radiation. It is noteworthy that the Tenerife observations of Hancock *et al.* (1997) find that the power spectrum of the temperature fluctuations on angular scales greater than $4°$ is $n = 1.1 \pm 0.1$, which they claim leaves little scope for a significant contribution of gravitational waves to the fluctuation spectrum, within the context of the inflationary theory which led to (15.26).

15.4 Intermediate Angular Scales – the Acoustic Peaks

Inspection of Fig. 15.2 shows that the amplitude of the temperature power spectrum is expected to oscillate dramatically on angular scales $\theta \leq 1°$, that is, on physical scales less than the horizon scale at the last scattering layer. Let us disentangle the different physical processes contributing to the temperature fluctuations on these scales. Excellent surveys of the physics of temperature fluctuations in the Cosmic Microwave Background Radiation have been given by Hu and Sugiyama (1995), Hu (1996) and by Hu, Sugiyama and Silk (1997).

First of all, let us make an inventory of the relevant physical scales. From (12.7), the proper *horizon scale* at the epoch of recombination is

$$r_{\mathrm{H}} = 3ct = \frac{2c}{H_0 \Omega_0^{1/2}}(1+z)^{-1.5} = 5.8 \times 10^{21}(\Omega_0 h^2)^{-1/2} \text{ m}, \qquad (15.27)$$

corresponding to a comoving scale of $200(\Omega_0 h^2)^{-1/2}$ Mpc.

The next important scale is the *Jeans' length* of the photon-dominated plasma, which is decoupled from the dark matter. The sound speed is given by (9.32), the mass density now referring only to the baryonic component Ω_{B},

$$c_s^2 = \frac{c^2}{3} \frac{4\varrho_{\text{rad}}}{4\varrho_{\text{rad}} + 3\varrho_B}. \tag{15.28}$$

The speed of sound approaches the relativistic sound speed $c/\sqrt{3}$ if $4\varrho_{\text{rad}} > 3\varrho_B$, that is, at redshifts $z \geq 4 \times 10^4 (1 + z)\Omega_B h^2$. We are interested in the Jeans' length at the last scattering surface at $z \approx 1000$, and so, provided $\Omega_B h^2 \geq 2.5 \times 10^{-2}$, the baryonic Jeans' length is

$$\lambda_J = c_s \left(\frac{\pi}{G\varrho}\right)^{1/2} = \frac{1.4 \times 10^{21}}{(\Omega_B h^2)} \text{ m}. \tag{15.30}$$

This scale corresponds to a comoving length scale of $46(\Omega_B h^2)^{-1}$ Mpc. From the considerations of primordial nucleosynthesis of Sect. 10.4, however, we have argued that $\Omega_B h^2 \leq 0.036$, with a best estimate of $\Omega_B h^2 \approx 10^{-2}$. In the latter case, the appropriate sound speed to use in (15.29) would be $c/\sqrt{3}$, in which case the Jeans' length is of the same order as the horizon scale at the last scattering layer. This is an important result because it means that, as the perturbations in the dark matter come through the horizon, baryonic perturbations on the same scale are stabilised and begin to oscillate as sound waves. In other words, *all* baryonic perturbations with wavelengths on the horizon scale and less are *acoustic waves*. As a result, they give rise to Sakharov oscillations at the last scattering layer (Sect. 12.5.1) and this is the origin of the oscillatory behaviour observed on angular scales less than the horizon scale at the epoch of recombination as observed in Fig. 15.2.

The next key scale is the *sound horizon* at the last scattering layer. This scale, which we take to be $\lambda_s = c_s t$, where t is the age of the Universe, is the distance which sound waves could travel at the epoch of recombination. This scale sets an upper limit to the wavelengths which acoustic waves could have at the epoch of recombination, once they have come through the horizon. From the considerations of the last paragraph, the value of c_s is likely to be close to the relativistic sound speed, $c_s = c/\sqrt{3}$. Adopting this value as an upper limit to c_s, the sound horizon has physical scale

$$\lambda_s = \frac{c}{\sqrt{3}} t = \frac{10^{21}}{(\Omega_0 h^2)} \text{ m}, \tag{15.30}$$

corresponding to a comoving distance scale of $32(\Omega_0 h^2)^{-1}$ Mpc at the present epoch. If $\Omega_B h^2 > 2.5 \times 10^{-2}$, the speed of sound becomes

$$c_s = \frac{c}{\sqrt{3}} \left(\frac{4\varrho_{\text{rad}}}{3\varrho_B}\right)^{1/2} = \frac{10^6 z^{1/2}}{\Omega_B h^2} \text{ m s}^{-1}, \tag{15.31}$$

and then the sound horizon is

$$\lambda_s = \frac{2.1 \times 10^{20}}{(\Omega_B \Omega_0 h^4)^{1/2}} \text{ m}. \tag{15.32}$$

Because we have argued that $\Omega_B h^2 \leq 0.036$, the sound horizon cannot be too different from (15.30). Notice a key point concerning the comparison of the Jeans' length with the sound horizon. In purely baryonic cosmologies, these would be essentially the same thing, but in dark matter cosmologies they are quite distinct because, whilst the overall dynamics and time-scales of the Universe are determined by the dark matter, the dynamical stability of the baryonic matter is determined by the baryonic sound speed, which is decoupled from the overall mass density of the Universe. The fact that $\lambda_s \ll \lambda_J$ also means that we can use the short wavelength approximation for the dispersion relation of the acoustic waves, that is, we can write (11.25) in the form

$$\omega^2 = c_s^2 k^2 - 4\pi G \varrho_B = c_s^2 (k^2 - k_J^2) \approx c_s^2 k^2, \qquad (15.33)$$

in the short wavelength limit $k \gg k_J$. In other words, the waves really are pure acoustic waves, the internal pressure being provided by the photons of the Cosmic Background Radiation, which are strongly coupled to the plasma.

Next, we recall that the thickness of the last scattering layer corresponds to a comoving physical scale of

$$\lambda_{LSL} = 10 (\Omega_0 h^2)^{-1/2} \text{ Mpc}. \qquad (15.34)$$

Finally, we can evaluate the damping scale for baryonic perturbations at the epoch of recombination using (12.20) and (12.22), for the case in which the dynamics of the expansion are determined by the dark matter $t = (2/3H_0)\Omega_0^{-1/2}(1+z)^{-1.5}$,

$$\lambda_D = \left(\tfrac{1}{3}\lambda ct\right)^{-1/2} = \frac{3 \times 10^{19}}{(\Omega_B h^2)^{1/2} (\Omega_0 h^2)^{1/4}} \text{ m}. \qquad (15.35)$$

Again, since we have argued that $\Omega_B h^2 \leq 0.036$, the damping scale at the epoch of recombination is at least $1.6 \times 10^{20} (\Omega_0 h^2)^{-1/4}$ m, corresponding to a comoving scale of $5(\Omega_0 h^2)^{-1/2}$ Mpc at the present epoch. Notice that the treatment of the damping scale through the last scattering layer needs some care, because, as the plasma recombines, the mean free path of the photons increases dramatically (see Hu and Sugiyama 1995).

We can therefore put the various scales in order, using comoving scales for convenience

$$r_H : \lambda_J : \lambda_s : \lambda_{LSL} : \lambda_D$$
$$= 200(\Omega_0 h^2)^{-1/2} : 46(\Omega_B h^2)^{-1} : 32(\Omega_0 h^2)^{-1}$$
$$: 10(\Omega_0 h^2)^{-1/2} : 5(\Omega_0 h^2)^{-1/4} \text{ Mpc}. \qquad (15.36)$$

To develop a theory of the predicted temperature fluctuation spectrum, the evolution of the power spectrum of the density perturbations on all scales needs to be followed through the epoch of recombination in some detail. This involves using the collisional Boltzmann equation to follow the evolution of the independent Fourier modes of the perturbations in the dark matter, the baryonic matter and the radiation field. The computations are described by Peebles and Yu (1970) for baryonic perturbations and by Efstathiou (1990) and Hu and Sugiyama (1995) for dark matter cosmologies. Nonetheless, we can understand the basic features of Fig. 15.2 using physical arguments.

First of all, let us ignore the presence of the dark matter perturbations and consider only the acoustic waves. The first acoustic peak is associated with perturbations on the scale of the sound horizon at the epoch of recombination. As discussed in Sect. 12.5.1, when density perturbations come through the horizon, those which eventually form bound objects are those with maximum density constrasts on the horizon scale at that time, their amplitudes being given by the power spectrum of the initial perturbations. The dark matter perturbations on that scale continue to grow in amplitude while perturbations in the radiation-dominated plasma become acoustic waves (Fig. 13.1b). The amplitudes of the acoustic waves at the last scattering layer therefore depend upon the phase difference from the time they came through the horizon to last scattering layer, that is, they depend upon

$$\int \mathrm{d}\phi = \int c_{\mathrm{s}} \, \mathrm{d}t \qquad (15.37)$$

The first peak in the temperature spectrum corresponds to waves with wavelength equal to the sound horizon at the last scattering layer. The angular scale of these perturbations is

$$\theta \approx \frac{\lambda_{\mathrm{s}}(\text{comoving})}{D} = 0.3 \, \Omega_0^{1/2} \, {}^{\circ} \qquad (15.38)$$

where we have assumed that the sound speed is the relativistic sound speed $c_{\mathrm{s}} = c/\sqrt{3}$ and that the distance measure $D = (2c/H_0\Omega_0)$, the approximation for large redshifts. This is the origin of the first acoustic peak in the temperature power-spectrum and it can be seen that it provides a means of determining the value of Ω_0. The detailed solutions of the coupled Boltzmann equations for the baryonic and dark matter and the radiation indicate that the first acoustic peak should occur at $l = 200$ if $\Omega_0 = 1$ and at $l = 430$ if $\Omega_0 = 0.2$, in good agreement with the $\Omega_0^{1/2}$ dependence of (15.38) (Bersanelli et al. 1996).

Let us label the wavenumber of the first acoustic peak k_1. Then, oscillations which are $n\pi$ out of phase with the first acoustic peak also correspond to maxima in the temperature power spectrum at the epoch of recombination. There is, however, an important difference between the even and odd harmonics of k_1. The odd harmonics correspond to the maximum compression of the waves and so to increases in the temperature, whereas the even harmonics correspond to rarifactions of the acoustic waves and so to temperature minima. The perturbations with phase differences $\pi(n+\frac{1}{2})$ relative to that of the first acoustic peak have zero amplitude at the last scattering layer and correspond to the minima in the power spectra, as discussed in

Sect. 12.5.1. Thus, to find the acoustic peaks, we need to find the wavelengths corresponding to frequencies

$$\omega t_{\mathrm{rec}} = n\pi. \tag{15.38}$$

Adopting the short wavelength dispersion relation (15.33), the condition becomes

$$c_{\mathrm{s}} k_n t_{\mathrm{rec}} = n\pi \qquad k_n = \frac{n\pi}{\lambda_{\mathrm{s}}} = n k_1. \tag{15.39}$$

Thus, the acoustic peaks are expected to be evenly spaced in wavenumber. The separation between the acoustic peaks thus provides us with further information about various combinations of cosmological parameters, specifically, about $c_{\mathrm{s}} t_{\mathrm{rec}}/(c/H_0)\Omega_0$.

The next task is to determine the amplitudes of the acoustic peaks in the power spectrum. The complication is that the acoustic oscillations take place in the presence of growing density perturbations in the dark matter, which have greater amplitude than those in the acoustic oscillations. Therefore, in dark matter scenarios, the acoustic waves are driven by the larger density perturbations in the dark matter with the same wavelength, that is, the perturbations are forced oscillations.

Let us develop the equation which describes these oscillations, neglecting the slow growth of the density perturbations in the dark matter and the time variation of the ratio of energy densities in the radiation and the matter. We begin with (11.24) and (11.65), but with the difference that the driving potential on the right-hand side is associated with the dark matter, and neglect the expansion term associated with $\mathrm{d}\Delta/\mathrm{d}t$,

$$\frac{\mathrm{d}^2 \Delta_{\mathrm{B}}}{\mathrm{d}t^2} = \Delta_{\mathrm{D}} 4\pi G \rho_D - \Delta_{\mathrm{B}} k^2 c_{\mathrm{s}}^2. \tag{15.40}$$

This procedure is similar to that which led to (13.15), but now we are interested in wave solutions rather than in growing solutions. The temperature fluctuations are related to the density perturbations by the standard adiabatic relation

$$\Theta_0 = \frac{\delta T}{T} = \tfrac{1}{3}\frac{\delta\rho}{\rho} = \tfrac{1}{3}\Delta_{\mathrm{B}}, \tag{15.41}$$

and so,

$$\frac{\mathrm{d}^2 \Theta_0}{\mathrm{d}t^2} = \frac{4\pi G \Delta\rho_{\mathrm{D}}}{3} - k^2 c_{\mathrm{s}}^2 \Theta_0. \tag{15.42}$$

This equation can be compared with that presented by Hu (1996)

$$\frac{\mathrm{d}^2 \Theta_0}{\mathrm{d}t^2} = -\tfrac{1}{3} k^2 \Psi - k^2 c_{\mathrm{s}}^2 \Theta_0, \tag{15.43}$$

where Ψ is the gravitational potential of the dark matter perturbations of the same wavenumber k, which in our notation is the same as $\Delta\phi$ in (15.6). The equivalence of the two expressions is evident if we write

$$-\Psi = -\Delta\phi \sim \frac{G\Delta M}{d} \sim \frac{4\pi G d^2 \Delta\rho_\mathrm{D}}{3} \sim \frac{4\pi G \Delta\rho_\mathrm{D}}{k^2} \qquad (15.44)$$

where $d \sim k^{-1}$ is the scale of the perturbation, and recalling that Ψ and $\Delta\phi$ are negative quantities. In fact, (15.44) is an approximation to the full equations for the evolution of the spectrum of acoustic oscillations which were derived by Peebles and Yu (1970) and Hu and Sugiyama (1995). In their approach, the analysis is carried out in terms of the evolution of the Fourier components of the acoustic oscillations, resulting in the wave equation (15.43). The sound speed is given by (15.28) which we write

$$c_\mathrm{s} = \frac{c}{\sqrt{3}} \left(\frac{4\varrho_\mathrm{rad}}{4\varrho_\mathrm{rad} + 3\varrho_\mathrm{B}} \right)^{1/2} = \frac{c}{\sqrt{3(1+\mathcal{R})}}, \qquad (15.45)$$

where $\mathcal{R} = 3\varrho_\mathrm{B}/4\varrho_\mathrm{rad}$.

The solution of the forced-oscillator equation (15.43) is

$$\Theta_0(t) = \left[\Theta_0(0) + \frac{(1+\mathcal{R})}{c^2}\Psi \right] \cos\omega t + \frac{1}{kc_\mathrm{s}}\dot{\Theta}_0(0) \sin\omega t - \frac{(1+\mathcal{R})}{c^2}\Psi. \quad (15.46)$$

where $\Theta_0(0)$ and $\dot{\Theta}_0(0)$ are the amplitudes of the initial temperature fluctuation and its time derivative respectively. In the more complete version of this analysis, these would be the values appropriate to the epoch at which the perturbations came through the horizon. We require this solution at the epoch of recombination, $t = t_\mathrm{rec}$ and then, setting $\omega t = kc_\mathrm{s}t_\mathrm{rec} = k\lambda_\mathrm{s}$, we obtain the result

$$\Theta_0(t) = \left[\Theta_0(0) + \frac{(1+\mathcal{R})}{c^2}\Psi \right] \cos k\lambda_\mathrm{s} + \frac{1}{kc_\mathrm{s}}\dot{\Theta}_0(0) \sin k\lambda_\mathrm{s} - \frac{(1+\mathcal{R})}{c^2}\Psi. \quad (15.47)$$

The expression (15.47) describes adiabatic temperature oscillations of the acoustic waves in the potential wells of the dark matter. In addition to the purely adiabatic effect, the matter of the perturbation must be in motion because of the continuity equation (11.13), which can be written

$$\frac{\mathrm{d}}{\mathrm{d}t}\left(\frac{\delta\rho}{\rho_0} \right) = -\nabla \cdot \delta\mathbf{v} = -\mathbf{k} \cdot \delta\mathbf{v}. \qquad (15.48)$$

The motions of the perturbations also give rise to temperature fluctuations because of the Doppler effect such that $\Theta_1 = \delta T/T_0 = \delta v \cos\theta/c$. Therefore, (15.48) can be written in terms of the derivative of Θ_0 using (15.41)

$$3\dot{\Theta}_0(t) = -kc\Theta_1(t). \qquad (15.49)$$

The temperature perturbations due to the Doppler motions at the last scattering surface are then found by differentiating (15.46) and evaluating it at $\omega t = k\lambda_\mathrm{s}$:

$$\Theta_1(t) = \frac{3c_s}{c} \left[\Theta_0(0) + \frac{(1+\mathcal{R})}{c^2}\Psi\right] \sin k\lambda_s - \frac{3\dot{\Theta}_0(0)}{kc} \cos k\lambda_s. \qquad (15.50)$$

The temperature perturbation $\Theta_0(t)$ is referred to as the 'monopole' contribution to the temperature fluctuations spectrum and $\Theta_1(t)$ as the 'dipole' term.

The expressions (15.47) and (15.50) are illuminating forms for the temperature fluctuation spectrum as a function of wavenumber k. The cosine term in (15.47) can be identified with the adiabatic modes, since they had finite amplitudes when they came through the horizon. In contrast, isocurvature modes had zero amplitude when they crossed the horizon, corresponding to the sine term in (15.47). As a result, the maxima of the two modes are $\pi/2$ out of phase.

Let us concentrate first upon the adiabatic modes. The pleasant aspect of (15.47) is that it enables us to compare the amplitude of the perturbations associated with the acoustic waves with the amplitude of the Sachs-Wolfe effect on larger angular scales. We have not yet taken account of the gravitational redshifting of the temperature perturbations since those described by (15.47) take place within the gravitational potential well of amplitude Ψ. Therefore, the temperature fluctuation observed by the distant observer is $\Theta_0(t) + \Psi/c^2$, recalling that Ψ is a negative quantity. As a result, what is observed is the oscillating part of (15.47), that is,

$$\Theta_0(t) + \frac{\Psi}{c^2} = \left[\Theta_0(0) + \frac{(1+\mathcal{R})}{c^2}\Psi\right] \cos \omega t - \frac{\mathcal{R}}{c^2}\Psi. \qquad (15.51)$$

$\Theta_0(t) - \Psi/c^2$ is what Hu *et al.* (1997) refer to as the *effective temperature* of the fluctuations. An important aspect of this analysis is that the potential perturbation $\Psi = \Delta\phi$ is independent of cosmic epoch (see 15.10). We can therefore relate these temperature perturbations to the amplitude of the Sachs–Wolfe perturbations by noting that, at $t = 0$, the fluctuations result in the Sachs-Wolfe temperature perturbation, that is, the effective temperature should have the value $(1/3)\Psi/c^2$. This enables us to relate the value of $\Theta_0(0)$ to Ψ/c^2. We then find that

$$\left(\frac{\Delta T}{T}\right)_{\text{eff}} = \frac{\Psi}{3c^2}(1 + 3\mathcal{R}) \cos k\lambda_s - \frac{\mathcal{R}}{c^2}\Psi. \qquad (15.52)$$

This solution illuminates features of the complete solutions. First of all, in the limit $\mathcal{R} \to 0$, (15.52) reduces to

$$\left(\frac{\Delta T}{T}\right)_{\text{eff}} = \frac{\Psi}{3c^2} \cos k\lambda_s, \qquad (15.53)$$

corresponding to acoustic oscillations in the photon-dominated plasma in the relativistic limit. The monopole oscillations are accompanied by dipole oscillations, the amplitude of these being given by the sine term in (15.50).

In fact, we only observe a dipole temperature fluctuation of $\Theta_1/\sqrt{3}$ when we average the radial components of randomly oriented velocity vectors over the last scattering layer and so, using (15.49), we find

$$\left(\frac{\Delta T}{T}\right)_{\text{eff}} = \frac{\Psi}{3c^2} \sin k\lambda_{\text{s}}, \tag{15.54}$$

Thus, in the limit $\mathcal{R} \to 0$, the monopole and dipole temperature fluctuations are of the same amplitude. However, when the inertia of the baryons can no longer be neglected, the monopole contribution becomes significantly greater than the dipole term. At maximum compression, $k\lambda_{\text{s}} = \pi$, the amplitude of the observed temperature fluctuation is $(1 + 6\mathcal{R})$ times that of the Sachs–Wolfe effect. Furthermore, the amplitudes of the oscillations are asymmetric if $\mathcal{R} \neq 0$, the temperature excursions varying between $-(\Psi/c^2)(1+6\mathcal{R})$ for $k\lambda_{\text{s}} = (2n + 1)\pi$ and (Ψ/c^2) for $k\lambda_{\text{s}} = 2n\pi$. These results can account for the some of the prominent features of the temperature fluctuation spectrum displayed in Fig. 15.2. The temperature perturbations associated with the acoustic peaks are much larger than the Sachs–Wolfe fluctuations. The asymmetry between the even and odd peaks in the fluctuation spectrum is associated with the extra compression at the bottom of the gravitational potential wells when account is taken of the inertia of the perturbations associated with the baryonic matter.

This simplified discussion, which has closely followed the exposition of Hu *et al.* (1997), is intended to give some flavour for the various physical phenomena which have to be taken into account in understanding the origin of the temperature perturbation spectrum on angular scales $\theta \leq 1°$. In the full analysis, many other feature have to be taken into account. Specifically,

1. The predicted spectrum has to be statistically averaged over a random distribution of acoustic waves and integrated over all wavenumbers;
2. The Doppler (or dipole) contributions to the fluctuation spectrum have to be included as well as the monopole contributions;
3. The acoustic waves evolve in an expanding substratum in which the speed of sound and the density perturbations in the dark matter vary with time.

These effects tend to reduce the amplitude of the acoustic peaks seen in Fig. 15.2, but the expectation is that these features should be readily detectable in high sensitivity observation of the background radiation on angular scales $\theta \leq 1°$. Some examples of the detailed predictions by Hu (1996) are shown in Fig. 15.3, indicating how the different features of the power-spectrum can be used to estimate cosmological parameters, within the context of the scale-invariant adiabatic model.

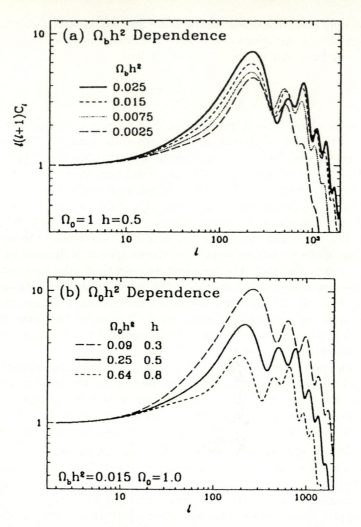

Fig. 15.3. Examples of the predicted temperature power-spectrum of fluctuations in the Microwave Background Radiation (Hu 1996). A scale-invariant adiabatic model ($n = 1$) with $\Omega_0 = 1$ has been assumed and the specta evaluated for different values of the baryonic density parameter Ω_B and h. Increasing Ω_B increases the asymmetry between odd and even maxima in the power-spectrum. Decreasing h while keeping Ω_0 fixed brings the epoch of equality of matter and radiation energy densities closer to the recombination epoch.

15.5 Small Angular Scales

On small angular scales, the fluctuations are damped by two effects. The first is the statistical reduction in the amplitude of the predicted perturbations when the dimension of the perturbation becomes less than the thickness of the last scattering layer. As indicated in Sect. 15.2, this leads to a reduction in the amplitude of the predicted fluctuation spectrum by a factor of order $N^{-1/2}$ where N is the average number of wavelengths through the last scattering layer.

The second is the effect of photon diffusion, or Silk damping, discussed in Sects. 12.3 and 15.4. A convenient way of relating the temperature perturbations through the last scattering layer to what is observed on the sky is through a modification of (15.3) to take account of the effects of photon diffusion. This *suppression factor* for waves of wavenumber k can be written

$$\int \frac{\mathrm{d}\tau}{\mathrm{d}z} \mathrm{e}^{-\tau(z)} \mathrm{e}^{-k/k_D(z)} \, \mathrm{d}z, \qquad (15.55)$$

where $k_D = 2\pi/\lambda_D$ is given by (15.35). The predicted spectra in Figs. 15.2 and 15.3 show clearly the damping expected at large wavenumbers, or short wavelengths. The effects of damping and the random superposition of the perturbations lead to a strong damping of all primaeval temperature fluctuations with wavenumbers greater than about 2000, that is, on the scale of a few arcminutes.

Despite the effects of damping, Figs. 15.2 and 15.3 show that there should be a wealth of structure in the temperature fluctuation spectrum on angular scales $1° > \theta > 1$ arcmin. It is apparent from the above discussion that, if that spectrum could be measured with high precision, a vast amount of information about cosmological parameters and the physics of the early Universe could be obtained.

15.6 Other Sources of Primordial Fluctuations

The main thrust of the arguments presented above is based upon the assumption that the spectrum of primordial fluctuations is of power-law form. According to proponents of the inflationary scenario of the early Universe, this is expected in a wide range of plausible models. The standard theory of adiabatic, Cold Dark Matter perturbations also provides a simple, robust quantitative theory which can be compared with observations. It is worth cautioning, however, that there might well be much more to the story than this. For example, an alternative picture for the origin of primordial density perturbations is that they are associated with phase transitions which occur as symmetry-breaking in Grand Unified Theories takes place during the inflationary era. As these symmetries are broken, disorderly phase transitions can take place, giving rise to topological defects such as cosmic strings, monopoles and textures (Kibble 1976, Turok 1989). These could be the fluctuations which seed the process of galaxy formation.

The big difference between this class of theory and the standard Cold Dark Matter picture is that the fluctuations are expected to have a non-Gaussian

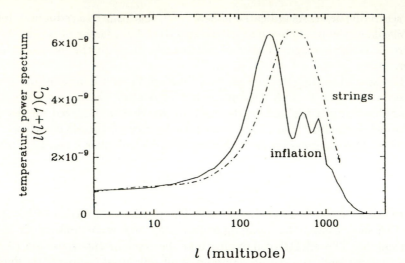

l (multipole)

Fig. 15.4 Comparison of the predicted temperature power-spectrum for a cosmic string model with $\Omega_B = 0.05$ and $h = 0.5$ (dot-dashed line) with a scale-invariant Cold Dark Matter model (solid line), designed to have the same amplitude as the string model (Bersanelli *et al.* 1996).

character. For example, cosmic strings would produce line-like discontinuities on the sky, and cosmic textures would produce distinct non-Gaussian hot and cold spots. A key test of this class of theory is therefore the presence of non-Gaussian features in the distribution of the radiation temperature over the sky. If the temperature fluctuations are Gaussian, the higher moments of the temperature distribution, such as its skewness and kurtosis, should be zero. If the fluctuations are non-Gaussian, these moments will be significantly different from zero. A second test is the power-spectrum of temperature fluctuations, which is expected to be significantly different from the predictions of the standard Cold Dark Matter model (Fig. 15.4). The differences arise from the fact that topological defects give rise to isocurvature perturbations with non-random phases. The result is that the structure in the acoustic peaks expected in the standard adiabatic picture are not present and the peak of the temperature power-spectrum is shifted to higher multipoles. Taken at face value, this form of spectrum seems to be in poor agreement with the observations discussed in Section 15.8, which suggest that the first acoustic peak is indeed present at multipoles $l \sim 300$.

Although this class of theory appears to be a poor fit to the observations, there remains the concern that, although topological defects may not be the origin of the overall power-spectrum of density perturbations, they are a rather general feature of phase transitions during symmetry-breaking in Grand Unified Theories, and so there might well be a significant contribution from these to the temperature fluctuation spectrum, in addition to primordial perturbations, which would complicate the interpretation of the observations.

15.7 Other Sources of Fluctuations

In addition to effects associated with temperature fluctuations at the last scattering layer at the epoch of recombination, there are a number of other effects which could cause fluctuations in the background radiation at about the same level.

15.7.1 The Reheating of the Intergalactic Gas

At some stage the intergalactic gas must have been heated and re-ionised, a topic which we deal with in much more detail in Chap. 19. It is not yet entirely clear how this came about, although the most plausible mechanism is photoionisation at redshifts less than 10, and quite possibly at redshifts as low as 4. The optical depth of the intergalactic gas to Thomson scattering was derived in Sect. 9.3 in the limit of large redshifts $\Omega_0 z \gg 1$

$$\tau_T \approx 0.035 \frac{\Omega_B}{\Omega_0^{1/2}} h \, z^{3/2}, \tag{15.56}$$

which is adequate for our present purposes. Adopting $\Omega_0 h^2 = 1$, fully ionised gas with $\Omega_B h^2 = 0.036$ would have optical depth $\tau_T = 1$, if it extended out to a redshift $z = 86$. It is not inconceivable that this could have occurred, but it would have required a substantial amount of star formation, not long after the epoch of recombination. Within the Cold Dark Matter picture, perturbations with masses $\sim 10^6 \, M_\odot$ can form soon after recombination and it might be that the formation of large numbers of massive stars could photoionise the intergalactic gas at these large redshifts.

If the intergalactic gas were reionised at these early epochs, Thomson scattering of the background radiation would decrease the amplitude of all temperature fluctuations by the factor $e^{-\tau}$, smoothing out the primordial temperature fluctuations. The formation of these early generations of objects would, however, necessarily give rise to significant velocity perturbations during the reionisation epoch, resulting in Doppler shifts of the background radiation, in exactly the same way that the dipole temperature perturbations are generated at the last scattering layer. The problem is that these effects are expected to be small, partly because the optical depth of the perturbations is small, and also because the random superposition of these perturbations leads to a statistical decrease of their amplitude. There is, however, a second order effect discussed by Vishniac (1987) which does not lead to the statistical cancellation of the velocity perturbations. These perturbations are associated with second-order terms in $\delta n_e v$, where δn_e is the perturbation in the electron density and v the peculiar velocity associated with the collapse of the perturbation. This effect only becomes significant on small angular scales at which the density perturbations are large. According to Vishniac's calculations, in the standard Cold Dark Matter picture, these temperature fluctuations might amount to $\Delta T/T \sim 10^{-5}$ on the scale of 1 arcmin. Similar calculations have been carried out by Efstathiou (1988) with essentially the same result. These are important conclusions since it is expected that all primordial perturbations on these scales would have been damped out by the processes discussed in Sect. 15.5.

Temperature fluctuations on these angular scales have been searched for in very long integrations in deep survey fields by the Very Large Array. The analyses of Partridge *et al.* (1997) have found no evidence for such perturbations at a level $\Delta T/T \approx 10^{-5}$ on angular scales $\theta \sim 1$ arcmin.

15.7.2 The Sunyaev–Zeldovich Effect in Clusters of Galaxies

One effect which is known to cause temperature fluctuations on angular scales $\theta \sim 1$ arcmin is the Sunyaev-Zeldovich effect in the directions of clusters of galaxies which contain large amounts of hot gas. As shown in Sect 4.3.3, the temperature fluctuation due to the passage of the Cosmic Microwave Background Radiation through a region of hot gas such as that contained within a cluster of galaxies is

$$\frac{\Delta T}{T} = -2y = -2 \int \left(\frac{kT_e}{m_e c^2} \right) \sigma_T N_e dl, \tag{15.57}$$

in the Rayleigh–Jeans region of the spectrum. This effect has now been observed in at least 12 rich clusters of galaxies which are known to be strong X-ray sources (R. Saunders 1997, personal communication). For these clusters of galaxies, the temperature decrements amount of $\Delta T/T \sim 10^{-4}$. Some indication of the likely importance of this effect can be estimated from the number counts of rich clusters of galaxies. As a generous upper limit to the number of such decrements due to clusters of galaxies, we can use the statistics of the numbers of clusters of richness class 1 and greater given in Sect. 4.1.1. Within a distance $r = c/H_0$, the surface density of such clusters is about 30 per square degree and this is very much an upper limit to the likely number of clusters which could cause these perturbations because only a fraction of the richest clusters are strong X-rays sources. These temperature perturbations are, however, of the greatest astrophysical interest since the combination of the X-ray observations with the corresponding decrements in the background radiation enable the evolution of the gaseous content of clusters of galaxies with cosmic epoch to be studied in detail. Notice that, whereas the Sunyaev-Zeldovich decrement is independent of the redshift of the hot gas cloud, its X-ray flux density is strongly redshift dependent, because the high-energy X-ray bremsstrahlung spectrum cuts off exponentially.

In addition to the above effect, known as the *thermal* Sunyaev-Zeldovich effect, there is also an effect associated with the peculiar motion of hot gas clouds with respect to the frame of reference in which the Microwave Background Radiation is isotropic. For such a gas cloud, the background radiation appears anisotropic in its rest frame and those photons scattered by the hot electrons are redistributed isotropically in the moving frame. Sunyaev and Zeldovich (1980) showed that first order Thomson scattering of the background results in temperature perturbations due to these peculiar motions

$$\frac{\Delta T}{T} = \sigma_T \int N_e \frac{v_{\parallel}}{c} \, dl \qquad (15.58)$$

This effect, known as the *kinematic* Sunyaev-Zeldovich effect, is independent of frequency, similar to the spectrum of primordal perturbations. Inserting reasonable values for a hot gas cloud in a rich cluster of galaxies, $N_e = 3 \times 10^3$ m^{-3}, $v_{\parallel} = 500$ km s^{-1} and taking the core radius of the cluster to be 0.4 Mpc, a temperature perturbation of about 30 μK is expected. This effect provides an important means of estimating the radial component of the peculiar velocities of clusters of galaxies. Sunyaev and Zeldovich also pointed out that the peculiar motion perpendicular to the line of sight could be measured from the polarisation of the temperature perturbation associated with the first-order Thomson scattering process.

15.7.3 Confusion due to Discrete Sources

Finally, there has always been the worry that discrete sources would eventually be the limiting factor in measuring the power-spectrum of the primordial temperature perturbations. At centimetre wavelengths this is certainly an important limitation. It has only recently become possible to make the first realistic estimates of the deep number counts of discrete sub-millimetre sources, thanks to the development of the sub-millimetre bolometer array receiver, SCUBA, for the James Clerk Maxwell Telescope in Hawaii. Smail, Ivison and Blain (1997) have measured a surface density of sub-millimetre sources of about 3000 sources degree^{-2} at a flux density of 3 mJy at 850 μm. These sources are almost certainly dust-emitting star-forming galaxies at large redshifts. By fitting the number counts, they have estimated the level at which source confusion is the dominant source of temperature fluctuations over the sky (see Sect. 17.2.3). The results depend sensitively upon the angular resolution of the telescope and the frequency at which the observations are made. For the *Planck* mission (see below), Blain *et al.* (1998) find that the temperature fluctuations due to sub-millimetre sources are comparable to those of the specifications of the instrument itself at frequencies greater than 300 GHz.

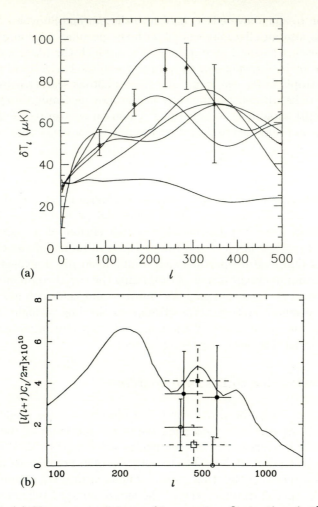

Fig. 15.5. (a) The power-spectrum of temperature fluctuations in the Cosmic Microwave Background Radiation measured in the Princeton Saskatoon experiment (Netterfield *et al.* 1997). The lines show the predictions of some of the different variants on the standard Cold Dark Matter model discussed in Sect. 14.6. (b) The power-spectrum determined on angular scales beyond the predicted first acoustic peak by the Cambridge CAT experiment (Scott *et al.* 1996).

15.8 Present and Future Observations

The whole subject of temperature fluctuations in the Cosmic Microwave Background Radiation has undergone a revolution following the outstanding success of the COBE mission. Many of the pre-COBE observations were described by Readhead and Lawrence (1992) and the results of the final analysis of the COBE observations are contained in the paper by Bennett *et al.*

(1996) and the immediately following papers in that issue of the *Astrophysical Journal*. In parallel with these observations, there has been a concerted effort by ground-based astronomers to measure the temperature perturbations in the vicinity of the predicted acoustic peaks of their power spectra. These are technically very challenging experiments, the objective being to measure with precision a noise signal at the level of about one part in 10^5 of the total intensity. Ground-based observations have to cope with a myriad of technical problems such as ground emission, the variable water vapour content of the atmosphere, and so on. This plethora of data has been judiciously scrutinised by Partridge (1998).

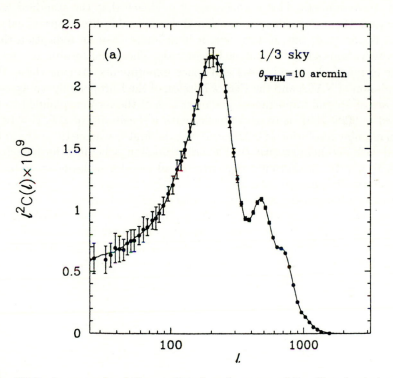

Fig. 15.6. An example of the predicted performance of the *Planck* mission of the European Space Agency. The error bars correspond to the total cosmic and experimental variances expected from a survey of one third of the sky for a period of one year (Bersanelli *et al.* 1996).

Of the many experiments, particular interest is attached to the Princeton Saskatoon experiment, which has made systematic observations with the same instrument in the vicinity of the first acoustic peak. Clear evidence has been found for a significant excess as compared with the COBE power-spectrum (Netterfield *et al.* 1997) (Fig. 15.5a). A second important set of observations has been made by the Cambridge group with the Cosmic

Anisotropy Telescope (CAT), which has made sensitive observations in the region beyond the first acoustic peak (Scott *et al.* 1997)(Fig. 15.5b). In these observations, the amplitude of the temperature fluctuations is significantly less than the maxima found in the Saskatoon experiment. On small angular scales, $6 < \theta < 80$ arcsec, observations with the Very Large Array (VLA) have found only upper limits to the amplitude of temperature fluctuations, at the level $\Delta T/T \leq 2 \times 10^{-5}$, consistent with a cut-off to the perturbation spectrum at large values of l (Partridge *et al.* 1997).

Partridge concluded that there is evidence that the first acoustic peak is present in the temperature power-spectrum and that the spectrum falls off at large wavenumbers. This is encouraging evidence that the standard cold dark matter theory of galaxy formation is a reasonable first approximation to what is observed, but, in my view, it is no more than an indication that these observations should be pursued vigorously. These are the aims of a large number of planned ground-based and space experiments. Among these, the *MAP* mission of NASA and the *Planck* Mission of the European Space Agency (ESA) are of special importance. The objective of the former, planned to be launched in 2000-2001, is to map the sky with a sensitivity of $\Delta T/T = 10^{-5}$ with an angular resolution of $0.22°$, enabling the first few acoustic peaks to be mapped with excellent sensitivity. The latter mission, scheduled for launch in 2005, will map the background radiation on all angular scales down to $0.12°$ with a sensitivity per pixel of $\Delta T/T = 2 \times 10^{-6}$, enabling the detailed power spectrum well beyond the first acoustic peak to be determined with very high sensitivity. An indication of the expected performance of the *Planck* mission is shown in Fig. 15.6. It can be seen that the error bars are expected to be so small at large multipoles l, that very precise discrimination can be made between models.

Part IV

The Post-Recombination Universe

16 The Post-Recombination Universe – the Dark Ages

In the final part of this volume, we consider the post-recombination Universe and the physical processes responsible for the formation of galaxies and clusters of galaxies as we know them at the present day. The post-recombination era spans the redshift range from about 1000 to zero and it is convenient to divide it into two parts. The more recent of these, spanning the redshift range $0 < z < 5$, may be termed the *observable Universe of galaxies*, in the sense that this is the range of redshifts over which galaxies and quasars have now been observed. Although there have been remarkable developments in detector and telescope technology over recent years, it is still the case that only the most luminous objects can be readily detected at large redshifts, $z \geq 1$. As we will show in Chap. 17, there is plentiful evidence that the populations of galaxies and quasars have evolved dramatically over the redshift range $0 < z < 5$. There is now good evidence that much of the star formation activity in galaxies and the synthesis of the heavy elements took place during these late phases of the post-recombination era, as we discuss in Chap. 18. Many of the most important physical processes which led to the formation of galaxies as we know them took place during this era, which, fortunately, is now accessible to astronomical observation. Many of these phenomena will be discussed in Chap. 20.

The earlier phase of the post-recombination era, the redshift interval $1000 > z > 5$, is often referred to as the *Dark Ages*. At the beginning of this redshift interval, we can learn a considerable amount about the early development of the perturbations from which galaxies and larger scale structures formed from observations of the Cosmic Background Radiation. Until galaxies became visible as bound, star-forming systems at redshifts $z \leq 5$, however, there are few observational tools which can be used to study precisely what took place during the immediate post-recombination era. The perturbations were still in the linear regime at $z \sim 1000$ but, as they collapsed to form bound systems, their evolution became non-linear. Many of the processes which led to the variety of structures we observe today must have taken place during these epochs. In this chapter we investigate some aspects of the physical processes which are likely to be important during the post-recombination epochs, before the galaxies became visible at redshifts $z < 5$.

16.1 The Non-linear Collapse of Density Perturbations

This is a large and important topic which has been the subject of a great deal of detailed theoretical analysis and computer simulation. We have already discussed some of the results of these computations in Chap. 14, for example, the formation of the two-point correlation function of galaxies and the reconstruction of the initial power-spectrum of the perturbations. It is useful to study some of the analytic solutions for the non-linear development of density perturbations to illuminate the results of the numerical computations.

One calculation, which can be carried out exactly, is the collapse of a uniform spherical density perturbation in an otherwise uniform Universe, a model sometimes referred to as 'spherical top-hat collapse'. As discussed in Sect. 11.4.2, the dynamics of such a region are precisely the same as those of a closed Universe with $\Omega_0 > 1$. The variation of the scale factor of the perturbation R_p is cycloidal and given by the parametric solution (7.24a)

$$\left. \begin{array}{c} R_p = a(1 - \cos\theta) \qquad t = b(\theta - \sin\theta), \\[2mm] a = \dfrac{\Omega_0}{2(\Omega_0 - 1)} \quad \text{and} \quad b = \dfrac{\Omega_0}{2H_0(\Omega_0 - 1)^{3/2}}. \end{array} \right\} \qquad (16.1)$$

The perturbation reached maximum radius at $\theta = \pi$ and then collapsed to infinite density at $\theta = 2\pi$, as illustrated in Fig. 11.2. To express this in another way, the perturbation stopped expanding, $\dot{R}_p = 0$, and separated out of the expanding background at $\theta = \pi$. This occurred when the scale factor of the perturbation was $R = R_{max}$, where

$$R_{max} = 2a = \frac{\Omega_0}{\Omega_0 - 1} \quad \text{at time} \quad t_{max} = \pi b = \frac{\pi \Omega_0}{2H_0(\Omega_0 - 1)^{3/2}}. \qquad (16.2)$$

It is now straightforward to work out the density of the perturbation at maximum scale factor ρ_{max} relative to that of the background ρ_0, which, for illustrative purposes, we take to be the critical model, $\Omega_0 = 1$. Recalling that the density within the perturbation was Ω_0 times that of the background model to begin with,

$$\frac{\rho_{max}}{\rho_0} = \Omega_0 \left(\frac{R}{R_{max}} \right)^3 = 9\pi^2/16 = 5.55, \qquad (16.3)$$

where the scale factor of the background model has been evaluated at cosmic time t_{max}. Thus, by the time the perturbed sphere had stopped expanding, its density was already 5.55 times greater than that of the background density.

In the spherical top-hat model, the perturbation has no internal pressure and so it collapsed to infinite density at time $t = 2\pi b$, twice the time it took to reach maximum expansion. Since $R \propto t^{2/3}$, it follows that the relation between the redshift of maximum expansion z_{max} and the redshift of collapse z_c is

$$1 + z_c = \frac{1 + z_{max}}{2^{2/3}}. \tag{16.4}$$

This means that the subsequent collapse occurred very rapidly, within less than a factor of two in redshift, once the perturbation had separated out from the background. For example, if $z_{max} = 20$, $z_c = 12$, if $z_{max} = 10$, $z_c = 6$, and so on.

Interpreted literally, the spherical perturbed region collapsed to a black hole. In practice, however, it is much more likely to form some sort of bound object. In one scenario, as the gas cloud collapsed, its temperature increased until internal pressure gradients became sufficient to balance the attractive force of gravitation. Another possibility is that, during collapse, the cloud fragmented into sub-units and then, through the process of *violent relaxation*, these regions came to a dynamical equilibrium under the influence of large scale gravitational potential gradients (Lynden-Bell 1967). In either case, the end result is a system which satisfies the Virial Theorem. In the case of a self-gravitating system of masses, the internal kinetic energy of the system would be just half its (negative) gravitational potential energy (see Sect. 3.4.1); if the system were held up by the internal pressure of hot gas, the Virial Theorem would require the internal thermal energy to be half the gravitational potential energy.

In either case, we can work out the final dimensions of the virialised system, or gas cloud, by the following argument. At z_{max}, the sphere is stationary and all the energy of the system is in the form of gravitational potential energy. For a uniform sphere of radius r_{max}, the gravitational potential energy is $-3GM^2/5r_{max}$. If the system does not lose mass and collapses to half this radius, its gravitational potential energy becomes $-3GM^2/(5r_{max}/2)$ and, by conservation of energy, the kinetic energy, or internal thermal energy, acquired is

$$\text{Kinetic Energy} = \frac{3GM^2}{5(r_{max}/2)} - \frac{3GM^2}{5r_{max}} = \frac{3GM^2}{5r_{max}} \tag{16.5}$$

Thus, by collapsing by a factor of two in radius from its maximum radius of expansion, the kinetic energy, or internal thermal energy, becomes half the negative gravitational potential energy, the condition for dynamical equilibrium according to the Virial Theorem. Therefore, the density of the perturbation increased by a further factor of 8, while the background density continues to decrease. From (16.1), the scale factor of the perturbation reached the value $R_{max}/2$ at time $t = (1.5 + \pi^{-1})t_{max} = 1.81t_{max}$, when the background density was a further factor of $(t/t_{max})^2 = 3.3$ less than at maximum. The net result of these simple calculations is that, when the collapsing cloud became a bound virialised object, its density was $5.55 \times 8 \times 3.3 \approx 150$ times the background density at that time.

Although this is a highly idealised picture, similar results are found from N-body computer simulations of the process of formation of bound structures in the expanding Universe. According to Coles and Lucchin (1995), these simulations suggest that the systems can be considered virialised after a few crossing times, $t \approx 3t_{max}$, when the density contrast would be closer to 400. It is this type of reasoning which has allowed the initial power-spectrum of the density perturbations to be reconstructed well into the non-linear regime, as discussed in Sect. 14.5 (Hamilton *et al.* 1991, Peacock and Dodds 1994). The general rule which comes out of these considerations is that discrete objects such as galaxies and clusters of galaxies only separated out as distinct gravitationally bound objects when their densities were at least 100 times the background density.

For illustration, let us see what this means for the redshifts at which galaxies and other large scale systems could have separated out of the expanding background. The factor of order 100 derived above is the minimum enhancement in the density of the bound object relative to the background. The density contrast might be even greater if dissipative processes played a rôle in determining the final stable configuration, as discussed in Sect. 16.2. We can therefore state with some confidence that the density of the virialised object should be at least 100 times the background density, that is,

$$\rho_{vir} \geq 100 \times \frac{3\Omega_0 H_0^2}{8\pi G}(1 + z_{vir})^3, \tag{16.6}$$

where z_{vir} is the redshift at which the system became virialised. We can make an estimate of ρ_{vir} from the Virial Theorem. If M is the mass of the system and v^2 its velocity dispersion, the condition that the kinetic energy be half the gravitational potential energy is

$$\frac{1}{2}Mv^2 = \frac{1}{2}\frac{GM^2}{R}, \tag{16.7}$$

where R is some suitably defined radius. Therefore,

$$\rho_{vir} \approx \frac{v^6}{(4\pi/3)G^3M^2} \tag{16.8}$$

Inserting this value into (16.6), we can find a limit to the redshift at which the object became virialised:

$$(1 + z_{vir}) \leq 0.47 \left(\frac{v}{100\,\text{km s}^{-1}}\right)^2 \left(\frac{M}{10^{12}\,M_\odot}\right)^{-2/3} (\Omega_0 h^2)^{-1/3} \tag{16.9}$$

What this calculation amounts to is an improved version of the simple estimates presented at the beginning of Chap. 11 concerning the mean densities of objects relative to the background density and when they could have formed. Let us put in some representative figures. For *galaxies* having $v \sim 300$ km s^{-1} and $M \sim 10^{12}M_\odot$, the redshift of formation must be less than about 10. For *clusters of galaxies* for which $v \sim 1000$ km s^{-1} and $M \sim 10^{15}M_\odot$, the redshift of formation cannot be much greater than 1. This conclusion follows immediately from the rule that the density contrast of virialised objects should be at least 100, which is not so different from the ratio of the mean density of clusters of galaxies to typical cosmological densities at the present day. This means that, typically, the clusters of galaxies must have formed in the relatively recent past.

One of the obvious deficiencies of the top-hat model is that the perturbations are unlikely to be precisely spherically symmetric. The next best approximation is to assume that they are ellipsoidal with three unequal principal axes. Peacock and Heavens (1985) have discussed the expected ellipticity distributions for perturbations arising from random Gaussian fields. One of the general rules which comes out of the study of the collapse of ellipsoidal distributions of matter is that collapse occurs most rapidly along the shortest axis (Lin, Mestel and Shu 1965). For the case of primordial density fluctuations, Zeldovich (1970) showed in a remarkable paper how the collapse could be followed into the non-linear regime in the more general case.

In the *Zeldovich approximation*, the development of perturbations into the non-linear regime is followed in Lagrangian coordinates, the same procedure described in Sect. 11.2. In other words, rather than working out the development of the perturbation in some external Eulerian reference frame, the motion of particles in a comoving coordinate frame is followed. If $\mathbf{x}$ and $\mathbf{r}$ are the proper and comoving position vectors of the particles of the fluid, the Zeldovich approximation can be written

$$\mathbf{x} = R(t)\mathbf{r} + b(t)\mathbf{p}(\mathbf{r}). \qquad (16.10)$$

The first term on the right-hand side describes the uniform expansion of the background model and the second term the perturbations of the particles' positions about the Lagrangian (or comoving) coordinate $\mathbf{r}$. Zeldovich showed that, in the coordinate system of the principal axes of the ellipsoid, the motion of the particles in comoving coordinates is described by a 'deformation tensor' D

$$D = \begin{bmatrix} R(t) - \alpha b(t) & 0 & 0 \\ 0 & R(t) - \beta b(t) & 0 \\ 0 & 0 & R(t) - \gamma b(t) \end{bmatrix}. \qquad (16.11)$$

Because of conservation of mass, the density ϱ in the vicinity of any particle is

$$\varrho[R(t) - \alpha b(t)][R(t) - \beta b(t)][R(t) - \gamma b(t)] = \bar{\varrho} R^3(t) \qquad (16.12)$$

where $\bar{\varrho}$ is the mean density of matter in the Universe. The clever aspect of the Zeldovich solution is that, although the constants α, β and γ vary from point to point in space depending upon the spectrum of the perturbations, the functions $R(t)$ and $b(t)$ are the same for all particles. In the case of the critical model, $\Omega_0 = 1$,

$$R(t) = \frac{1}{1+z} = \left(\frac{t}{t_0}\right)^{2/3} \quad \text{and} \quad b(t) = \tfrac{2}{5}\frac{1}{(1+z)^2} = \tfrac{2}{5}\left(\frac{t}{t_0}\right)^{4/3} \qquad (16.13)$$

where $t_0 = 2/3H_0$. The function $b(t)$ has exactly the same dependence upon scale factor (or cosmic time) as was derived from perturbing the Friedman solutions in Sect. 11.4.2. This can be demonstrated by expanding (16.12) for small values of $b(t)$ for the case $\alpha = \beta = \gamma$ and comparing it with (11.39).

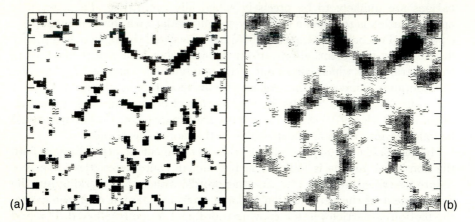

Fig. 16.1a-b. A comparison between the formation of large-scale structure according to (a) N-body simulations, and (b) the Zeldovich approximation, which began with the same initial conditions, which were assumed to be Gaussian with power-spectrum $P(k) \propto k^{-1}$ (Coles *et al.* 1993, Coles and Lucchin 1995). The agreement between the two approaches is good. In these simulations, the Zeldovich approximation was continued beyond the point at which caustics were formed which accounts for the somewhat more diffuse structures formed.

If we consider the case in which $\alpha > \beta > \gamma$, collapse occurs most rapidly along the x-axis and the density becomes infinite when $R(t) - \alpha b(t) = 0$. At this point, the ellipsoid has collapsed to a 'pancake' and the solution breaks down for later times. Although the density becomes formally infinite in the pancake, the surface density remains finite, and so the solution still gives the correct result for the gravitational potential at points away from the caustic surface. It is clear that the Zeldovich approximation cannot deal with the more realisitic situation in which collapse of the gas cloud into the pancake gives rise to strong shock waves, which heat the matter falling into the pancake – in the Zeldovich approximation, the particles move purely under gravity and have no pressure.

The results of numerical N-body simulations have shown that the Zeldovich approximation is quite remarkably effective in describing the evolution of the non-linear stages of the collapse of large scale structures up to the point at which caustics are formed. An comparison between the results of adopting the Zeldovich approximation and those of N-body simulations is shown in Fig. 16.1, which is taken from Coles *et al.* (1993) (see also, Coles and Lucchin 1995). The two simulations begin with the same Gaussian power-spectrum with $P(k) \propto k^{-1}$. In the case of the Zeldovich approximation, the coefficients of the deformation tensor have to be evaluated at each point in space. There are ways of continuing the evolution of the Zeldovich approximation beyond the formation of caustics, among these being the *adhesion model* in which the particles are assumed to stick together once they enter the region of the caustic. Details of these developments and the techniques of N-body simulations are clearly described by Coles and Lucchin (1995).

16.2 The Role of Dissipation

So far we have mostly considered the development of perturbations under the influence of gravity alone. In addition, we need to consider the rôle of dissipation, by which we mean energy loss by radiation, resulting in the loss of thermal energy from the system. In a number of circumstances, once the gas within the system is stabilised by thermal pressure, loss of energy by radiation can be an effective way of decreasing the internal pressure, allowing the region to contract in order to preserve pressure equilibrium. If the radiation process is effective in removing pressure support from the system, this can result in a runaway situation, known as a *thermal instability*. This is the process which may be responsible for the cooling flows which are present in the hot gas in the central regions of rich clusters of galaxies (Sect. 4.3.2).

There are at least two ways in which dissipation plays an important rôle in galaxy formation. The first is that star formation is an essential feature of the formation of protogalaxies. Evidence on the evolution of star and metal formation rates in galaxies with cosmic epoch is the subject of Chap. 18. In the case of our own Galaxy, stars are formed within cool, dusty regions, most of the star formation occurring within Giant Molecular Clouds, which pervade the disk of our Galaxy. The likely sequence of events is that a region within a cool dust cloud becomes unstable, either through the standard Jeans' instability, described in Sect. 11.3, or the equivalent instability in a differentially rotating medium (see Sect. 18.3.1 and the expression (18.5)). In addition, the collapse of the gas cloud may be stimulated by an external influence, such as the passage of the gas cloud through a spiral arm, or by compression of the gas by the blast wave of a supernova remnant.

The star can only be formed if the collapsing cloud can get rid of its gravitational binding energy – the most important means of effecting this is by radiation. This process continues until the cloud becomes optically thick to its own radiation. The loss of energy from the protostar is then mediated by the dust grains in the contracting gas cloud. The dust grains are heated to temperatures of about $60 - 100$ K, and consequently the binding energy of the protostar is radiated away at far-infrared wavelengths, at which the collapsing cloud is transparent to radiation. Observations of star-forming regions by the IRAS satellite and by ground-based millimetre and submillimetre telescopes make it wholly convincing that this is the process by which stars are formed at the present day in galaxies. In the case of the very first generations of stars, there is an obvious problem, in that there were no heavy elements present in the primordial gas out which dust grains could be formed. For the very first generation of stars, star formation presumably had to take place in a gas of essentially pure molecular hydrogen. This suggests that the process of formation of the first generation of stars may well be different from the star formation we observe at the present day in our Galaxy. Once the first generation of massive stars has formed, however, the fraction of the heavy elements in the interstellar medium can build up quickly and it is out of this enriched gas that dust and subsequent generations of stars form. The observation of molecular gas and dust emission in some of the most distant galaxies and quasars known at redshifts $z > 4$ shows that there must have been considerable early enrichment of the interstellar media in at least some massive galaxies (Omont 1996, Ohta et al. 1996, Omont et al. 1996).

Dissipative processes clearly play a dominant rôle in the formation of stars and this naturally leads to the question of whether or not similar processes might be important in the formation of larger scale structures. The rôle of dissipative processes in galaxy formation was elegantly described by Rees and Ostriker (1977). We follow the more recent presentation of Silk and Wyse (1993). The key relation is the energy loss rate of an astrophysical plasma by radiation as a function of temperature and is shown in Fig. 16.2 for a wide range of abundances of the heavy elements. The cooling function is presented in a form such that the energy loss rate per unit volume is $dE/dt = -N^2 \Lambda(T)$, where N is the number density of nuclei. In the case of the primordial plasma with no metal heavy elements, the dominant loss mechanism at high temperatures, $T > 10^6$ K, is thermal bremsstrahlung, the energy loss rate being proportional to $N^2 T^{1/2}$. At lower temperatures, the main loss mechanisms are free-bound and bound-bound transitions of hydrogen and ionised helium, corresponding to the two maxima in the cooling curve. Fig. 16.2 also shows how the cooling rate increases when the primordial plasma in enriched by heavy elements – the overall energy loss rate can be more than an order of magnitude greater than that of the primordial plasma at temperatures $T < 10^6$ K by the time the heavy elements have attained their present cosmic abundances.

For the case of a fully ionised plasma, the cooling time is defined to be the time it takes the plasma to radiate away its thermal energy

$$\tau_{\text{cool}} = \frac{E}{|dE/dt|} = \frac{3Nk_\text{B}T}{N^2 \Lambda(T)}. \tag{16.14}$$

This time scale can be compared with the time-scale for gravitational collapse, given by (11.30), $\tau_{\text{dyn}} \approx (G\rho)^{-1/2} \propto N^{-1/2}$. The significance of these timescales is best appreciated by inspecting the locus of the equality $\tau_{\text{cool}} = \tau_{\text{dyn}}$ in a number density–temperature diagram (Fig. 16.3). The locus $\tau_{\text{cool}} = \tau_{\text{dyn}}$ is a mapping of the cooling curve of the plasma energy loss-rate

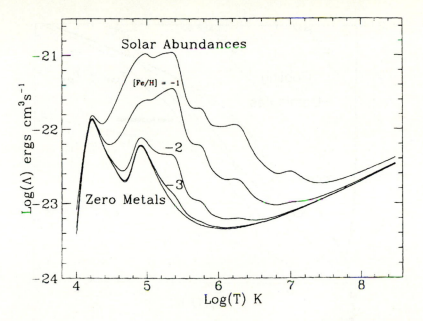

Fig. 16.2. The cooling rate per unit volume $\Lambda(T)$ of an astrophysical plasma of number density 1 nucleus cm^{-3} by radiation for different cosmic abundances of the heavy elements ranging from zero metals to the present abundance of the heavy elements as a function of temperature T (Silk and Wyse 1993, after Sutherland and Dopita 1993). In the zero metal case, the two maxima of the cooling curve are associated with the recombination of hydrogen ions and doubly ionised helium (see also Sect. 19.5 and Fig. 19.3).

into the $N-T$ plane. Inside this locus, in the region indicated in Fig. 16.3, the cooling time is shorter than the collapse time, and so it is expected that dissipative processes are more important than dynamical processes under these physical conditions. Also shown on Fig. 16.3 are lines of constant mass, as well as loci corresponding to the radiation loss time being equal to the age of the Universe, and to the perturbations having such low density that they do not collapse gravitationally in 10^{10} years. It can be seen that, even in the case of zero metals, the range of masses which lie within the critical locus, and which can cool in 10^{10} years, corresponds to $10^{10} \leq M/M_{\odot} \leq 10^{13}$ – this is the key conclusion of this analysis. The fact that the masses lie naturally in the range of observed galaxy masses suggests that the typical masses of galaxies may not only be determined by the initial fluctuation spectrum, but by astrophysical processes as well. As expected, the greater the abundance of the heavy elements, the shorter the time-scale for cooling of a region of a given temperature and density.

This diagram can be used astrophysically in the following way. For any theory of the origin of the large-scale structure of the Universe, the density, temperature and metal abundance of the gas can be worked out at each epoch.

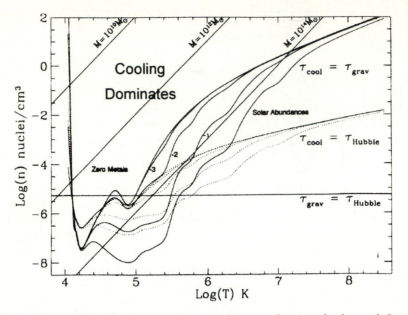

Fig. 16.3. A number density–temperature diagram showing the locus defined by the condition that the collapse time of a region τ_{dyn} should be equal to the cooling time of the plasma by radiation τ_{cool} for different abundances of the heavy elements (after Silk and Wyse 1993). Also shown are lines of constant mass, a cooling time of 10^{10} years (dotted lines), and the density at which the perturbations are of such low density that they do not collapse in the age of the Universe.

Fig. 16.3 can then be used to determine whether or not cooling by radiative losses is important. A good example is found in the various versions of the pancake theory. When the gas cloud collapses to form a pancake, the matter falls onto a singular plane and, as a result, a shock wave passes out through the infalling matter, heating it to a high temperature. In this picture, galaxies can form by thermal instabilities in the heated gas. Inspection of Fig. 16.3 shows that, if the gas is heated above 10^4 K, there is no stable region for masses in the range 10^{10} to $10^{13} M_\odot$.

A second exercise, carried out by Blumenthal *et al.* (1984), is to plot the observed location of galaxies on the number density–temperature diagram. An effective temperature associated with the velocity dispersion of the stars in the galaxy or the galaxies in the cluster, $\frac{1}{2}k_{\mathrm{B}}T_{\mathrm{eff}} \approx \frac{1}{2}mv^2$, can be plotted rather than the thermal temperature of the gas. The irregular galaxies fall well within the cooling locus and the spirals, S0 and elliptical galaxies all appear to lie close to the critical line. On the other hand, the clusters of galaxies lie outside the cooling locus. This suggests that cooling may well have been an important factor in certain scenarios for the formation of galaxies. In the case of the clusters, it is likely that cooling was not important in their formation, but once they are formed, cooling flows can develop in the hot dense gas which can form in their central regions, as may be deduced from Fig. 16.3.

16.3 The Press-Schechter Mass Function

According to the Cold Dark Matter scenario for galaxy formation, galaxies and larger scale structures are built up by the process of hierarchical clustering. An elegant description of this process was developed in a remarkable paper by Press and Schechter (1974). Their objective was to provide an analytic formalism for the process of structure formation once the density perturbations had reached such an amplitude that they could be considered to have formed bound objects. Press and Schechter were well aware of the limitations of their approach, but it has turned out that their mass function and its evolution with cosmic epoch are in good agreement with more detailed analyses and with the results of supercomputer simulations.

The analysis begins with the assumption that the primordial density perturbations are *Gaussian fluctuations*. Thus, the phases of the waves which make up the density distribution are random and the distribution of the amplitudes of the perturbations of a given mass M can be described by a Gaussian function

$$p(\delta) = \frac{1}{\sqrt{2\pi}\sigma(M)} \exp\left[-\frac{\delta^2}{2\sigma^2(M)}\right], \tag{16.15}$$

where $\delta = \delta\varrho/\varrho$ is the density contrast associated with perturbations of mass M. Being a Gaussian distribution, the mean value of the distribution is zero and its variance is $\sigma^2(M)$, that is, the mean-squared fluctuation is

$$\langle\delta^2\rangle = \left\langle\left(\frac{\delta\rho}{\rho}\right)^2\right\rangle = \sigma^2(M). \tag{16.16}$$

This is exactly the statistical description of the perturbations implicit in the analysis of Sect. 14.2, in particular, the statistical average (14.13).

The Press-Schechter analysis begins with the assumption that, when the perturbations have developed to amplitude greater than some critical value δ_c, they develop rapidly into bound objects with mass M. This assumption sweeps an enormous amount of astrophysics under the carpet, but let us first see where it leads. The problem is now completely defined, since we can assume that the perturbations have a power-law power-spectrum $P(k) = k^n$ and we know the rules which describe the growth of the perturbations with cosmic epoch. For illustrative purposes, let us assume that the background world model is the critical Einstein-de Sitter model, $\Omega_0 = 1, \Omega_\Lambda = 0$, so that the perturbations develop as $\delta \propto R \propto t^{2/3}$ right up to the present epoch.

For fluctuations of a given mass M, the fraction $F(M)$ of those which become bound at a particular epoch are those with amplitudes greater than δ_c

$$F(M) = \frac{1}{\sqrt{2\pi}\sigma(M)} \int_{\delta_c}^{\infty} \exp\left[-\frac{\delta^2}{2\sigma^2(M)}\right] d\delta = \tfrac{1}{2}\left[1 - \Phi(t_c)\right] \qquad (16.17)$$

where $t_c = \delta_c/\sqrt{2}\sigma$ and $\Phi(x)$ is the probability integral defined by

$$\Phi(x) = \frac{2}{\sqrt{\pi}} \int_0^x e^{-t^2}\, dt \qquad (16.18)$$

The expression (14.21) can be used to relate the mean square density perturbation to the power spectrum of the perturbations.

$$\sigma^2(M) = \left\langle \left(\frac{\delta\varrho}{\varrho}\right)^2 \right\rangle = \langle \delta^2 \rangle = A M^{-(3+n)/3} \qquad (16.19)$$

where A is a constant. We can now express t_c in terms of the mass distribution

$$t_c = \frac{\delta_c}{\sqrt{2}\sigma(M)} = \frac{\delta_c}{\sqrt{2}A^{1/2}} M^{(3+n)/6} = \left(\frac{M}{M^*}\right)^{(3+n)/6} \qquad (16.20)$$

where we have introduced a reference mass $M^* = (2A/\delta_c^2)^{3/(3+n)}$.

Since the amplitude of the perturbation $\delta(M)$ grows as $\delta(M) \propto R \propto t^{2/3}$, it follows that $\sigma^2(M) = \delta^2(M) \propto t^{4/3}$, that is, $A \propto t^{4/3}$. Therefore,

$$M^* \propto A^{3/(3+n)} \propto t^{4/(3+n)} \qquad (16.21)$$

which can be rewritten

$$M^* = M_0^* \left(\frac{t}{t_0}\right)^{4/(3+n)} \qquad (16.22)$$

where M_0^* is the value of M^* at the present epoch t_0.

The fraction of perturbations with masses in the range M to $M + dM$ is $dF = (\partial F/\partial M)\,dM$. In the linear regime, the mass of the perturbation is $M = \bar{\varrho}V$ where $\bar{\varrho}$ is the mean density of the background model. Once the perturbation becomes non-linear, collapse ensues and ultimately a bound object of mass M is formed. The space density $N(M)\,dM$ of these masses is V^{-1}, that is,

$$N(M)\,dM = \frac{1}{V} = -\frac{\bar{\varrho}}{M}\frac{\partial F}{\partial M}\,dM, \qquad (16.23)$$

the minus sign appearing because F is a decreasing function of increasing M.

We now have everything we need to determine the mass distribution and how it evolves with time from (16.17), (16.22) and (16.23). Noting that

$$\frac{d\Phi}{dx} = \frac{2}{\sqrt{\pi}} e^{-x^2}, \qquad (16.24)$$

we find

$$N(M) = \frac{1}{2\sqrt{\pi}} \left(1 + \frac{n}{3}\right) \frac{\bar{\varrho}}{M^2} \left(\frac{M}{M^*}\right)^{(3+n)/6} \exp\left[-\left(\frac{M}{M^*}\right)^{(3+n)/3}\right], \quad (16.25)$$

in which all the time dependence of $N(M)$ has been absorbed into the variation of M^* with cosmic epoch (16.22).

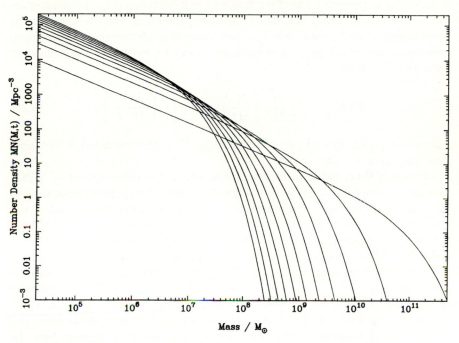

Fig. 16.4. Illustrating the variation of the form of the Press-Schechter mass function as a function of cosmic time (Courtesy of Dr. Andrew Blain).

It can be seen that this formalism results in only half of the total mass density being condensed into bound objects, entirely because of the fact that, according to this simple development, only the positive density fluctuations of the symmetrical Gaussian distribution (16.15) develop into bound systems. The underlying cause of this factor of two discrepancy is the fact that the above analysis is based upon the linear theory of the growth of the perturbations. Once the perturbations develop to large amplitude, mass is accreted from the vicinity of the perturbation and the N-body simulations show that most of the mass is indeed condensed into discrete structures. Press and Schechter were well aware of this problem and argued that the mass spectrum (16.25) should be multiplied by a factor of 2 to take account of the accretion of mass during the non-linear stages.

A further problem concerns 'peaks-within-peaks' in the sense that a full treatment would take account of the fact that, in general, any perturbation of a particular wavenumber k is superimposed upon other longer wavelength perturbations. These types of issue have been the subject of intensive study (see, for example, Bardeen *et al.* 1986, Efstathiou 1990, Kauffman and White 1993). The results of the N-body simulations for the standard Cold Dark Matter scenario show that the Press-Schechter mass function is an excellent description of the evolution of the mass function of dark matter halos with cosmic epoch, provided the extra factor of two is included in the mass spectrum. Efstathiou (1990) has shown how the problem of the evolution of the mass function can be formulated in such a way that all the mass is condensed into bound objects. Thus, the mass function can be written in the somewhat more compact form

$$N(M) = \frac{\bar{\varrho}}{\sqrt{\pi}} \frac{\gamma}{M^2} \left(\frac{M}{M^*} \right)^{\gamma/2} \exp\left[-\left(\frac{M}{M^*} \right)^{\gamma} \right], \qquad (16.26)$$

where $\gamma = 1 + (n/3)$ and $M^* = M^*(t_0)(t/t_0)^{4/3\gamma}$. The variation of the function with time is shown in Fig. 16.4.

This is a very useful formalism for studying the development of galaxies and clusters of galaxies in hierarchical scenarios for galaxy formation. As an example of the use of the function, Efstathiou (1995) has matched the Press-Schechter mass function to the results of N-body simulations of the development of galaxies and clusters within the context of the standard $\Omega_0 = 1$ Cold Dark Matter model. Since the hierarchical clustering is entirely associated with the dynamical behaviour of the dark matter, the structures formed are referred to as *dark matter haloes*, rather than as galaxies. Visible galaxies as we know them are presumed to form from the baryonic material which was present in the cold dark matter perturbations. Following Efstathiou and Rees (1988), Efstathiou (1995) presented a useful diagram showing how the comoving number density of dark matter haloes with masses greater than a given value $N(> M, z)$ changes with redshift (Fig. 16.5). The present number density of L^* galaxies is shown as a dotted line.

This diagram illustrates a number of important aspects of hierarchical clustering models. First of all, it demonstrates vividly how the most massive systems form rather late in the Universe according to the hierarchical scenario. This reinforces the point that it is not unexpected that rich clusters of galaxies should display strong evolutionary effects at small redshifts.

Secondly, massive galaxies only formed relatively late in the Universe. Fig. 16.5 shows that galaxies only built up to masses $M \sim 10^{12} \, M_\odot$ in substantial numbers at redshifts $z \leq 4$. Efstathiou and Rees (1988) used this result to show that there would necessarily be a rather dramatic cut-off to the distribution of quasars at redshifts $z \geq 4$ according to the Cold Dark Matter picture. The basis of the argument was that supermassive black holes with $M \sim 10^9 \, M_\odot$ in the nuclei of galaxies are necessary to power the quasars

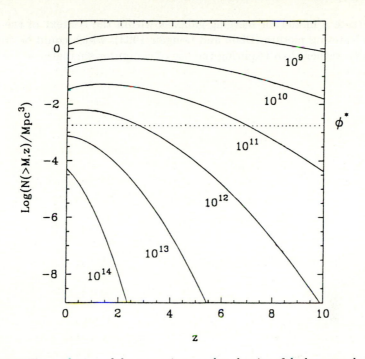

Fig. 16.5 The evolution of the comoving number density of dark matter haloes with masses greater than M as a function of redshift for a standard Cold Dark Matter model with $\Omega_0 = 1$. The curves have been derived using the Press-Schechter form of evolution of the mass spectrum which is a good fit to the results of N-body simulations. The dotted line labelled ϕ^* shows the present number density of L^* galaxies. (after Efstathiou 1995).

observed at redshifts $z \sim 4$ and these can only form from the baryonic component of the galaxy. Since the baryonic matter probably amounts to only about 10% of the mass of the dark matter halo, it follows that, to create a $10^9 \, M_\odot$ black hole, 1% of the baryonic mass would have to form a black hole in the nucleus of the galaxy. Among nearby galaxies at the present epoch, supermassive black hole masses are typically about 0.3% of their halo masses (Kormendy and Richstone 1995) and so it would just be feasible for $10^{12} \, M_\odot$ galaxies to contain massive enough black holes to power the most luminous quasars. As pointed out by Efstathiou and Rees, inspection of Fig. 16.5 shows that the comoving number density of galaxies with masses $M = 10^{12} \, M_\odot$ decreases rapidly with increasing redshift beyond $z = 4$ and so the likelihood of observing quasars as luminous as those at $z \sim 3 - 4$ decreases dramatically.

Thirdly, notice the large number density of low mass objects present throughout the redshift interval $0 < z < 10$. It is a challenge to find the evidence for the formation of these low mass haloes at large redshifts.

The Press-Schechter formalism is very convenient for evaluating the predictions of hierarchical models of galaxy formation. As an example, we were able to create a model for merging galaxies within the context of the hierarchical clustering picture (Blain and Longair 1993), which could be treated analytically, rather than requiring the use of computer simulations.

17 The Evolution of Galaxies and Active Galaxies with Cosmic Epoch

17.1 Introduction

Over recent years, there has been a deluge of data concerning changes with cosmic epoch of essentially all classes of extragalactic object. Evidence for strong evolutionary changes with cosmic epoch was first found in the 1950s and 1960s as a result of surveys of extragalactic radio sources and quasars. The key evidence was the excess of faint sources found in radio source and quasar surveys, as compared with the expectations of all uniform world models. The inference was that there were many more of these classes of objects at early cosmic epochs as compared with their number at the present epoch. During the 1980s, as the first deep counts of galaxies became available, it was discovered that there is a large excess of blue galaxies at faint apparent magnitudes. These studies culminated in the remarkable observations of the Hubble Deep Field by the Hubble Space Telescope, which will be discussed in some detail in Chap. 18.

In the 1990s, the first deep surveys of the X-ray sky were carried out by the ROSAT X-ray observatory and evidence for an excess of faint X-ray sources was found, similar in many ways to the evolution inferred from studies of extragalactic radio sources and quasars. Finally, the IRAS survey, although not extending to as large redshifts as the surveys mentioned above, has also provided evidence for an excess of faint sources, which appear to be evolving in a manner similar to the active galaxies.

The evidence for evolutionary changes came from simply counting the numbers of objects in well-defined complete samples. There is, however, vast amounts of new data on many different aspects of galaxy formation and evolution – counts of particular classes of galaxies, absorption line systems, abundances of the elements in large redshift absorption systems and so on. In this chapter, we will restrict attention rather narrowly to the discussion of the type of information which can be derived from counts of different classes of object and their redshift distributions and how they contribute to the background radiation.

17.2 Counts of Galaxies and Active Galaxies

In his earliest studies of galaxies as extragalactic systems, Hubble realised that the number counts of galaxies potentially contain information about the large-scale structure of the Universe. In his famous monograph *The Realm of the Nebulae* (1936), he used counts of galaxies to the limit of the Mount Wilson 100-inch telescope to demonstrate that, overall, the distribution of galaxies is homogeneous on the large scale. He also found deviations from the local Euclidean counts, which he interpreted as evidence for the non-Euclidean nature of space-time when observed over large distances. Hubble's argument concerning the homogeneity of the distribution of galaxies is a powerful one and applies to objects studied in any waveband. It is worth repeating that argument as a preliminary to what follows.

17.2.1 Euclidean Source Counts

Suppose the sources have a luminosity function $N(L)\,dL$ and that they are uniformly distributed in Euclidean space. The numbers of sources with flux densities greater than different limiting values S in a particular solid angle Ω on the sky is denoted $N(\geq S)$. Consider first sources with luminosities in the range L to $L + dL$. In a survey to a limiting flux density S, these sources can be observed out to some limiting distance r, given by the inverse square law, $r = (L/4\pi S)^{1/2}$. The number of sources brighter than S is therefore the number of sources within distance r in the solid angle Ω

$$N(\geq S, L)\,dL = \frac{\Omega}{3} r^3 N(L)\,dL. \qquad (17.1)$$

Therefore, substituting for r, the number of sources brighter than S is

$$N(\geq S, L)\,dL = \frac{\Omega}{3} \left(\frac{L}{4\pi S}\right)^{3/2} N(L)\,dL. \qquad (17.2)$$

Integrating over the luminosity function of the sources,

$$N(\geq S) = \frac{\Omega}{3(4\pi)^{3/2}} S^{-3/2} \int L^{3/2} N(L)\,dL, \qquad (17.3)$$

that is, $N(\geq S) \propto S^{-3/2}$, independent of the luminosity function $N(L)$. The result $N(\geq S) \propto S^{-3/2}$ is known as the *Euclidean source counts* for any class of extragalactic object. In terms of apparent magnitudes, $m = \text{constant} - 2.5 \log_{10} S$, the Euclidean source counts become

$$N(\leq m) \propto 10^{0.6m}. \qquad (17.4)$$

This was the homogeneity test carried out by Hubble with the results shown in Fig. 17.1.

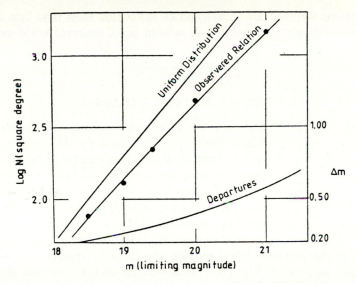

Fig. 17.1. The counts of galaxies determined by Hubble (1936). The counts follow the Euclidean prediction at bright apparent magnitudes but depart from the prediction $N(\leq m) \propto 10^{0.6m}$ towards the limit of the survey. Hubble interpreted this as evidence for the non-Euclidean nature of the space-time geometry of the Universe at large redshifts.

It will be noted that the integral counts $N(\geq S)$ suffer from the disadvantage that the numbers of sources counted to different limiting flux densities are not independent since bright objects contribute to the counts at all lower flux densities. It is therefore statistically preferable to work in terms of *differential source counts* rather than integral counts, so that the numbers of sources counted in each flux density interval are independent. In this case,

$$dN(S) = N(S)\, dS \propto S^{-5/2}\, dS. \tag{17.5}$$

The corresponding expression in terms of apparent magnitudes is

$$dN(m) = N(m)\, dm \propto 10^{0.6m}\, dm. \tag{17.6}$$

These are useful reference relations and it is convenient to compare the observed counts and the expectations of various world models with them, as we illustrate below.

17.2.2 Source Counts for the Standard World Models

We have developed all the tools necessary for predicting number counts and redshift distributions in Chap. 5 and given explicit formulae for the distance measure D for world models with $\Omega_\Lambda = 0$ (7.38) and $\Omega_\Lambda \neq 0$ (7.83). The two

key relations we need are (5.67) and (5.75) which show how flux densities
and number densities are related to redshift in all isotropic world models.

$$S(\nu_0) = \frac{L(\nu_1)}{4\pi D^2 (1+z)}; \tag{17.7}$$

$$dN(z) = N(z)\,dz = \Omega D^2 N_0\,dr, \tag{17.8}$$

where $\nu_0 = \nu_1/(1+z)$ and N_0 is the local number density of sources.

A number of important differences are immediately obvious as compared
with the Euclidean formulae. First of all, the flux density of the source now
depends upon its spectrum, because the radiation emitted at frequency ν_1
is observed at the redshifted frequency $\nu_0 = \nu_1/(1+z)$. As discussed in
Sect. 5.5.4 , the differences between an inverse square law in luminosity distance and the predictions of the standard world models is often expressed in
terms of K-corrections for observations made in the optical waveband (5.73).
Second, in the standard world models, the distance measure D tends to a
finite limit as $z \to \infty$ (Fig. 7.3b). However, unless the spectrum is strongly
inverted, as is the case for dust spectra observed in the millimetre and submillimetre wavebands, the effects of observing the source spectrum at a redshifted frequency generally result in the sources becoming fainter and fainter
with increasing redshift. To put it another way, the effects of the K-correction
compensate for the fact that D tends to a finite value as $z \to \infty$ Third, the
volume element per unit redshift interval changes from $dN(z) \propto z^2\,dz$ at
small redshifts, $z \ll 1$, to $dN(z) \propto z^{-3/2}\,dz$ at redshifts $\Omega_0 z \gg 1$ – the volume elements become smaller and smaller with increasing redshift, resulting
in a 'cut-off' to the source distribution at redshifts $\Omega_0 z \gg 1$. Thus, it becomes progressively more and more difficult to discover large redshift objects
since they are discriminated against both in flux density and observable volume. For these reasons, the number counts of sources in the standard world
models normally result in fewer sources as compared with the expectations
of the Euclidean source counts when the source distribution extends to large
redshifts. Nonetheless, one of the great triumphs of recent years has been that
it has been possible to overcome these disadvantages and the study of the
distant Universe is unquestionably one of the most exciting areas of modern
astrophysical cosmology. Part of the reason for this success has been that
many classes of distant object are more luminous, or populous, than they are
nearby, as we discuss in more detail below.

For illustrative purposes, let us begin with the simplest case of a population of sources which have power-law spectra, $L(\nu) \propto \nu^{-\alpha}$, which is a good
approximation for the spectra of extragalactic radio sources, X-ray sources
and quasars; α is known as the spectral index. The flux density-redshift relation then becomes

$$S(\nu_0) = \frac{L(\nu_0)}{4\pi D^2 (1+z)^{1+\alpha}}. \tag{17.9}$$

For the case of a uniform population of sources of local space density $N_0(L)$, (17.8) becomes

$$dN = N(z)\,dz = \frac{c}{H_0}\frac{\Omega N_0(L)\,D^2}{(1+z)(\Omega_0 z + 1)^{1/2}}\,dz, \qquad (17.10)$$

and, differentiating (17.7), we find

$$\frac{dS}{dz} = -\frac{L(\nu_0)}{4\pi D^2(1+z)^{1+\alpha}}\left[\frac{(1+\alpha)}{(1+z)} + \frac{2\left(\dfrac{dD}{dz}\right)}{D}\right]. \qquad (17.11)$$

Now, for a locally Euclidean population of sources, we expect

$$N(\geq S) = \frac{\Omega}{3}N_0(L)\left[\frac{L(\nu_0)}{4\pi S}\right]^{3/2}; \qquad (17.12)$$

$$dN_0 = -\frac{\Omega}{2}N_0(L)\left[\frac{L(\nu_0)}{4\pi}\right]^{3/2}S^{-5/2}\,dS. \qquad (17.13)$$

Hence

$$\frac{dN}{dN_0} = \frac{(dN/dS)}{(dN/dS)_0} = \frac{2c(1+z)^{-(3/2)(1+\alpha)}}{H_0(\Omega_0 z + 1)^{1/2}\left[D(1+\alpha) + 2(1+z)\left(\dfrac{dD}{dz}\right)\right]}. \qquad (17.14)$$

In general, this result for dN/dN_0 is cumbersome, but there is a simple solution for the case of the Einstein–de Sitter model, $\Omega_0 = 1$, $\Omega_\Lambda = 0$.

$$\frac{dN}{dN_0} = \frac{(1+z)^{-1.5(1+\alpha)}}{\left[(1+\alpha)(1+z)^{1/2} - \alpha\right]}. \qquad (17.15)$$

To illustrate the expected behaviour of dN/dN_0, (17.14) and (17.15) have been evaluated for a range of world models as a function of flux density and redshift (Fig. 17.2a and b). The Euclidean prediction $dN/dN_0 = \text{constant}$ is represented by the abscissa, $\log_{10}(dN/dN_0) = 0$. It can be seen that the predicted differential counts depart rapidly from the Euclidean prediction even at relatively small redshifts. For example, for the case $\alpha = 0.75$ and $\Omega_0 = 1$, the source counts at a redshift of 0.5 have differential slope -2.08 rather than -2.5, corresponding to a slope for the integral source counts of -1.08 rather than -1.5. Thus, the effects of redshift set in at much smaller redshifts than might be expected.

In practice, the source populations cannot be represented by a single luminosity, but rather, the counts shown in Fig. 17.2 must be convolved with the luminosity function $N_0(L)$ of the sources. It is evident, however, that because all the relations shown in Fig. 17.2 are monotonically decreasing functions of decreasing flux density, convolution with *any* function must also

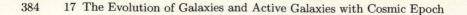

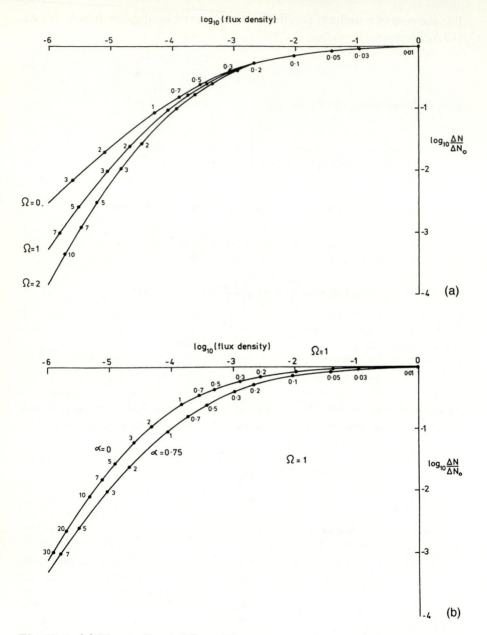

Fig. 17.2. (a) The predicted differential source counts for a single luminosity class of source having spectral index $\alpha = 0.75$ for different values of the density parameter. The numbers opposite each point on the curves are the redshifts at which the sources are observed. (b) The same as (a) but for the model with $\Omega_0 = 1$ and spectral indices $\alpha = 0$ and 0.75 (Longair 1978).

produce a monotonically decreasing function of decreasing flux density. In other words, in all realistic world models, the slope of the differential and integral source counts must be smaller than the Euclidean prediction, that is, if $N(\geq S) \propto S^{-\beta}$, $\beta < 1.5$.

Historically, this was an important result because the counts of radio sources at high flux densities have integral slope $\beta = 1.8$, much steeper than expected in the uniform world models. In the 1960s, this was strong evidence that there must have been many more radio sources at large redshifts than were predicted by the uniform, isotropic models, indicating that the source population must have evolved strongly with cosmic epoch. The choice of spectral indices $\alpha = 0.75$ and 0 in Fig. 17.2 betrays their radio astronomical origin. These are the typical spectral indices of the radio sources appearing in radio source catalogues at metre and centimetre wavelengths. It can be seen that there is not a particularly strong dependence upon spectral index and the argument holds good for all models in which the flux densities of the sources decrease with increasing frequency, as is the case for galaxies observed in the optical waveband.

The following points should be noted:

- The predicted counts depend upon knowledge of the spectra of the sources counted. If they are known to have power-law spectra, there is no problem. In the case of galaxies, however, the predicted number counts depend upon knowledge of their spectra in the ultraviolet waveband, if their distribution extends to large redshifts. The K-corrections are strongly dependent upon the rates of star formation in galaxies of different types, and so are somewhat more uncertain than those in, say, the near infrared waveband. We will find that star formation activity dominates the deep counts of galaxies in the optical waveband.

- The predictions illustrated in Fig. 17.2 are strongly modified in the millimetre and submillimetre wavebands, in which the spectra of galaxies are dominated by the emission of heated dust grains. The spectrum of the emission of heated dust grains takes the form of a modified blackbody spectrum, with a maximum intensity at about 100 μm. To the long-wavelength side of this peak, the dust clouds are optically thin and the spectral index of their thermal emission is typically $\alpha \approx -3$ to -4. Fig. 17.3a shows the typical far-infrared spectrum of a star-burst galaxy. The star-burst galaxies can have luminosities in the far-infrared waveband as great as, and in some cases greater than, those in the optical waveband and are the dominant contributors to the counts of sources in the millimetre and submillimetre wavebands. This form of 'inverted' spectrum strongly modifies the predicted flux density-redshift relation and the predicted number counts of sources in these wavebands. The flux density-redshift relations at millimetre wavelengths are shown in Fig. 17.3b, in which it can be seen that the relation flattens, or even inverts, because of the very large negative K-corrections in these wavebands (Blain and

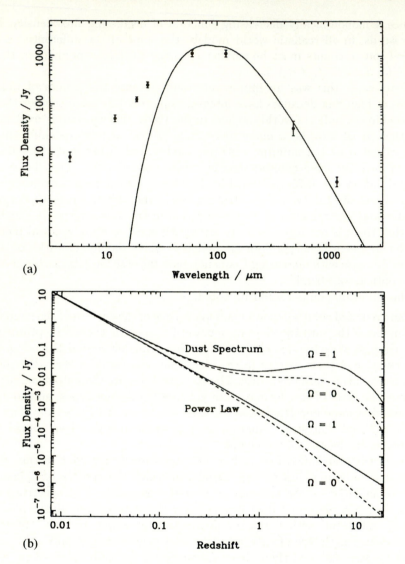

Fig. 17.3. (a) The typical far-infrared spectrum of a star-burst galaxy, the irregular star-forming galaxy M82. The solid line shows a fit of the observed millimetre and submillimetre spectrum by a standard dust continuum spectrum at a dust temperature of 45 K. (b) The flux density-redshift relation for a star-forming galaxy with the spectrum shown in (a), if the source has intrinsic far-infrared luminosity 10^{13} $L_\odot$ for world models with $\Omega_0 = 1$ and 0. For comparison, the flux density-redshift relations for a population of sources with power-law spectra $S_\nu \propto \nu^{-1}$ are also shown (Blain and Longair 1993).

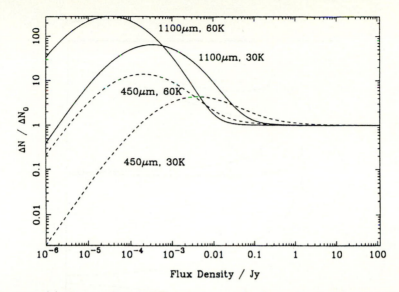

Fig. 17.3 (c) The predicted differential normalised counts of sources at 450 and 1100 μm assuming the galaxies have spectra similar to those of Fig. 17.3a and that they have a far-infrared luminosity function given by that of IRAS galaxies at 60μm (Saunders *et al.* 1990). The predictions for dust temperatures of 30 and 60 K are shown (Blain and Longair 1993).

Longair 1993, 1996). As a result, the differential source counts are quite different from those shown in Fig. 17.2. In Fig. 17.3c, the predicted differential source counts are shown for a population of far-infrared sources with the far-infrared luminosity function of IRAS galaxies, which has been determined at a wavelength of 60 μm. It can be seen that, when the source distribution extends to redshifts $z \geq 1$, the source counts are inverted with $\beta > 1.5$, even if there is no cosmological evolution of the source population.

- Similar analyses can be carried out for the world models with $\Omega_\Lambda \neq 0$. The cases of most interest are the Lemaître models in which the Universe almost reaches a stationary Eddington–Lemaître state at some redshift z_c. As discussed in Sect. 7.3.3, the flux-density redshift relation is strongly modified for those Lemaître models which differ only slightly from the static Eddington–Lemaître models (Fig. 7.9). It can be seen that the strictly homogeneous models have pronounced maxima at those redshifts corresponding to the poles and antipoles in the relation between comoving radial distance coordinate and redshift. It takes a little more care to work out the source counts in these cases (Longair and Scheuer 1970) and the example shown in Fig. 17.4 corresponds to the Lemaître model shown in Fig. 7.9. The counts are displayed as normalised integral source counts for a single luminosity class of source with spectral index

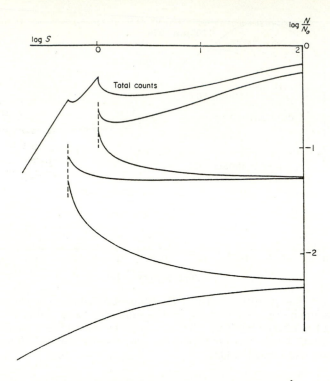

Fig. 17.4. The normalised integral source counts for the Lemaître model shown in Fig. 7.9. The value of the cosmological constant is $\Lambda = \Lambda_0(1 + \epsilon)$, where Λ_0 is the value of the cosmological constant which would result in a static universe at redshift $z = 2$. In this model, $\epsilon = 0.02$ and the sources are assumed to have power-law spectra $S \propto \nu^{-1}$ (Longair and Scheuer 1970). The vertial dashed lines indicate the flux densities corresponding to the pole and antipole seen in Fig. 7.9. The solid line labelled 'Total counts' is the integral source count found by summing the contributions from the ranges of $r/\Re$ from 0 to $\pi/2$, $\pi/2$ to π and so on.

$\alpha = 1$. It can be seen that the source counts are flatter than the models with $\Omega_\Lambda = 0$ at high flux densities, but once observations extend beyond the coasting phase at $z \approx z_{\rm c}$, the counts converge dramatically. In practice, to work out a more realistic source count, the function displayed in Fig. 17.4 would have to be convolved with the luminosity function of the sources and this would wipe out the jagged structure in the counts, resulting in a monotonically decreasing function of decreasing flux density. In reality, it is unlikely that the sharp maxima in the flux density-redshift relation would actually be observed. The existence of the poles and antipoles in the flux density-redshift relation depends upon the Universe being perfectly isotropic and homogeneous, acting as a perfect lens for the focussing of the light rays from the distant source to the observer. As pointed out by Petrosian and Salpeter (1968), if galaxies were present at the epochs corresponding to the poles and antipoles, they would cause gravitational deflections of the light rays, resulting in the splitting up of background point sources into a number of components, qualitatively similar to those shown in Fig. 4.10. It is therefore unlikely that the predictions of the perfectly uniform model would be observed in the real Universe. The gravitational deflections would have the effect of blurring the peaks in the predicted source counts.

17.2.3 Fluctuations in the Background Radiation due to Discrete Sources

A topic of importance directly related to the number counts of sources is the amplitude of *fluctuations* in the background radiation due to discrete sources. This problem, which was first solved for the case of observations made with a radio interferometer by Scheuer (1957), may be stated as follows. Suppose the sky is observed with a telescope of finite beamwidth θ and the integral number counts of sources are $N(\geq S) \propto S^{-\beta}$. If the survey extends faint enough, a flux density is reached at which there is one source per beam area and then fainter, more numerous sources cannot be detected individually. In this circumstance, the noise level of the survey is due to the random superposition of faint sources within the beam of the telescope, as illustrated schematically in Fig. 17.5. The problem of making observations of radio sources when the 'noise' is due to the presence of faint unresolved sources in the beam is referred to as *confusion*. This problem afflicted the early radio surveys and is the source of fluctuations in the X-ray background emission when observed at low angular resolution. These types of fluctuations may ultimately limit the detection of fluctuations in the Cosmic Microwave Background Radiation.

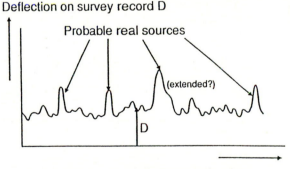

Fig. 17.5. Illustrating the fluctuations in the intensity of the background radiation due to the superposition of faint sources.

Scheuer (1974) also solved the problem of interpreting the background fluctuations in a survey made with a single beam telescopes. In this approach, we forget about the detection of individual sources, and deal directly with the probability distribution of the intensity fluctuations on the map of the sky. If the map is sampled at the information rate, that is, at twice per beam-width in a one-dimensional scan, a probability distribution $P(D)$ is found for the deflections D on the record measured from some zero level. The term 'deflection' D is used since the original radio astronomy surveys were recorded on strip charts and the 'deflections' really were the deflections of the recording pen. Typical $P(D)$ distributions are illustrated in Fig. 17.6. The distributions are non-Gaussian but, according to the central limit theorem, the noise level is given by the standard deviation of the probability distribution $P(D)$. The very large deflections are identified as discrete sources and the $P(D)$ distribution tends asymptotically to the differential source count $P(D)\,\mathrm{d}D \propto D^{-(\beta+1)}\,\mathrm{d}D$. Normally, some criterion for the detection of discrete sources, such as an intensity 5 times the standard deviation of the confusion noise, or one source per twenty or thirty beam areas, is adopted. In a confusion-limited survey, the flux densities of sources are systematically overestimated because of the random presence of faint sources in each beam.

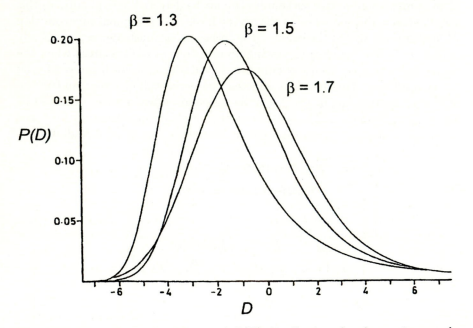

Fig. 17.6. Examples of the theoretical $P(D)$ distributions for observations made with a single-beam telescope for different assumed slopes of the differential source counts $\mathrm{d}N(S) \propto S^{-(\beta+1)}\,\mathrm{d}S$. The zero point of the abscissa is the mean amplitude $\bar{D}$ and the areas under the probability distributions have been normalised to unity. The distributions tend asymptotically to $\mathrm{d}N(D) \propto D^{-(\beta+1)}\,\mathrm{d}D$ at large deflections D (Scheuer 1974).

By carrying out a statistical analysis for the function $P(D)$ for sources selected randomly from a differential source count $dN(S) \propto S^{-(\beta+1)} dS$, Scheuer showed how the slope of the source counts could be found. Fig. 17.6 shows the normalised $P(D)$ distributions for different value of β. It can be seen that the shape of the $P(D)$ distribution provides a means of determining the form of the source counts. To order of magnitude, the most probable value of $P(D)$ corresponds to the flux density of those sources which have a surface density of roughly one source per beam-area. At higher flux densities, the sources are too rare to make a large contribution to the beam-to-beam variation in background signal. At lower flux densities, many faint sources add up statistically and so contribute to the background intensity, but the fluctuations are dominated by the brightest sources present in each beam. At roughly one source per beam area, the fluctuations can be thought of as arising from whether or not a source of that flux density is present by chance within the beam. Thus, whereas the reliable detection of individual sources can only be made to about 5 or 6 times the confusion noise level, statistical information about the source counts can be obtained to about one source per beam area.

Scheuer's analysis was entirely analytic but nowadays it is simpler to use Monte Carlo methods to work out the functions $P(D)$ for the assumed form of source count. The first of these studies using Monte Carlo modelling procedures was carried out by Hewish (1961) in his analysis of the original records of the 4C survey, in which he found the first evidence for the convergence of the radio source counts at low flux densities. These procedures have been used to determine the source counts to the very faintest flux densities in the radio waveband (see, for example, Fomalont et al. 1988 and Fig. 17.8) and in the deep ROSAT surveys by Hasinger et al. (1993) (see Fig. 17.13).

A different approach to the interpretation of fluctuations in the background radiation is to search for a signal in the correlation function of the fluctuations. This has been carried out succesfully in the optical waveband by Shectman (1974), who found a clear signature corresponding to the two-point correlation function for galaxies. The observed fluctuation spectrum is in quite remarkable agreement with the standard correlation function found in studies of large samples of galaxies. Another application of this approach in the ultraviolet waveband was carried out by Martin and Bowyer (1989). In a short rocket flight, they made a survey of a small region of sky and found a significant correlated signal among the spatial distribution of the counts. With a number of reasonable assumptions, they were able to show that they had detected the ultraviolet emission from galaxies. Similar analyses have been carried out for the fluctuations in the X-ray background as observed by the HEAO1 A-2 experiment (Persic et al. 1989) and in five deep *Einstein* fields (Barcons and Fabian 1989), but with negative results. The same type of fluctution analysis has been carried out by Kashlinsky et al. (1996) upon the COBE diffuse background emission in the near infrared region of the spectrum.

17.3 The V/V_{max} or Luminosity-Volume Test

A more direct method of investigating the uniformity of a distribution of objects in space is to use what is known as the V/V_{max} or *luminosity-volume test* (Schmidt 1968, Rowan-Robinson 1968). A sample of objects is selected which is known to be complete within well-defined flux density and apparent magnitude limits and for which the distances, or redshifts, are known for all the objects. The essence of the test is to ask whether or not the distribution of objects is statistically uniform within the accessible region of space defined by the observational selection criteria. The test has been of particular value in the study of the space distribution of quasars and we begin by discussing it in that context.

Suppose a sample of quasars is available which is complete in that all quasars having flux densities greater than some limiting value S_0 at a particular wavelength have been detected, and their redshifts measured. Consider first quasars of a single luminosity L. Then, for each quasar having flux density S, we may evaluate the quantity V/V_{max}, where V is the volume of space enclosed by the redshift z of the quasar and V_{max} is the volume of space within which this source could have been observed and still be included in the complete sample. The volume V_{max} corresponds to the redshift z_{max} at which a source of intrinsic luminosity L would be observed to have flux density S_0. Thus,

$$\frac{V}{V_{max}} = \frac{\int_0^r D^2\,\mathrm{d}r}{\int_0^{r_{max}} D^2\,\mathrm{d}r}, \tag{17.16}$$

where D is the distance measure and r and r_{max} are the comoving radial distance coordinates corresponding to z and z_{max} respectively. The volumes used in the test are comoving-coordinate volumes. Now suppose the distribution of quasars in space is uniform. The mean value of V/V_{max} is then

$$\left\langle \frac{V}{V_{max}} \right\rangle = \frac{\int_0^{r_0} \left(\frac{V}{V_{max}} \right) D^2\,\mathrm{d}r}{\int_0^{r_0} D^2\,\mathrm{d}r}. \tag{17.17}$$

Setting

$$\int_0^r D^2\,\mathrm{d}r = X, \qquad \frac{\mathrm{d}X}{\mathrm{d}r} = D^2, \tag{17.18}$$

and hence

$$\left\langle \frac{V}{V_{max}} \right\rangle = \frac{\int_0^{r_0} X \left(\frac{dX}{dr}\right) dr}{X^2 (r_0)} = \tfrac{1}{2}. \qquad (17.19)$$

As is intuitively expected, for a uniform distribution of quasars, we observe sources on average half-way volume-wise to the limits of their observable volumes, independent of the luminosities of the sources.

Furthermore, the values of V/V_{max} should be uniformly distributed between 0 and 1, if the source distribution is uniform, that is, if they have constant comoving number densities. In the case of samples of extragalactic objects such as quasars, the statistics of complete samples are often limited and so it has been normal practice to evaluate the quantity $\langle V/V_{max} \rangle$ and compare this with the expected mean value of 0.5. The simplest statistical test for the significance of departures from the mean value of 0.5 can be developed as follows. For a uniform distribution of V/V_{max} between 0 and 1, the standard deviation is $\sigma_0 = 1/\sqrt{12} = 0.288$. When the number of sources N is sufficiently large, the central limit theorem may be applied and the probability distribution of $\langle V/V_{max} \rangle$ approaches a Gaussian distribution with standard deviation $N^{-1/2}$ of the original distribution, that is $\sigma = \sigma_0/N^{1/2}$.

Sometimes, the samples of sources can only be considered to be complete within certain radio and optical limits – for example, in Schmidt's original sample of thirty-three quasars, the completeness criteria corresponded to a radio flux density limit at 178 MHz of $S_{178} = 9$ Jy and to a limiting optical visual magnitude of 18.4 (Schmidt 1968). Thus, in some cases the limiting volume was determined by the radio flux density and in others by the optical apparent magnitude. The mean value of V/V_{max} is, however, still expected to be 0.5 for a uniform distribution of sources, provided the larger of $(V/V_{max})_{radio}$ and $(V/V_{max})_{optical}$ is chosen for each source.

Nowadays, thanks to the efforts of many astronomers, the redshifts of complete samples of quasars and radio galaxies are becoming available and an example of the V/V_{max} distributions for 3CR radio galaxies and radio quasars is shown in Fig. 17.7. These objects form complete samples of the brightest radio sources in the northern sky and they have similar redshift distributions, both of them extending to redshifts $z \sim 2$ (Longair 1997). It can be seen that both distributions are biased towards values of V/V_{max} greater than 0.5. The values of $\langle V/V_{max} \rangle$ for the quasars and radio galaxies are $\langle V/V_{max} \rangle = 0.686 \pm 0.042$ and 0.697 ± 0.031 respectively, showing that the objects in these samples were more populous at large redshifts.

The V/V_{max} procedure is a very powerful technique for determining the uniformity and space density of any class of object. For example, in Felten's determination of the luminosity function of galaxies, he took account of the effects of extinction by interstellar dust in our Galaxy upon the luminosities and observable volumes within which galaxies of different intrinsic luminosities could be observed (Felten 1977). The observable volume need not be spherically symmetric about the observer – the key point is that the sam-

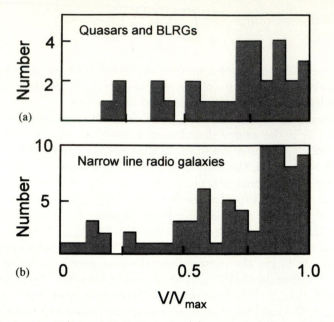

Fig. 17.7a-b. The V/V_{max} distributions for (**a**) quasars and broad-line radio galaxies and (**b**) the narrow line radio galaxies in a complete sample of extragalactic radio sources selected from the 3CR complete sample of Laing, Riley and Longair (1983). Redshifts have been measured for all the objects in the sample, which is complete to a limiting radio flux density of 9.7 Jy at 178 MHz (Longair 1997).

ples of sources studied should be statistically complete within well-defined observational selection criteria.

17.4 The Evolution of Active Galaxies with Cosmic Epoch

It may seem strange to begin the analysis of number counts and redshift distributions with the case of active rather than normal galaxies. The reason is that the analysis is somewhat more straightforward than is the case for the populations of normal galaxies because, for many purposes, the sources may be assumed to have power-law spectra. It may appear even stranger that I have chosen to begin with the counts of extragalactic radio sources – the reason is that their counts are particularly well-defined over a very wide range of flux densities, thanks to the remarkable efforts of the radio astronomers (see, for example, Wall 1996). We will then consider the case of optically selected samples of quasars, X-ray sources and IRAS galaxies. The counts of normal galaxies will be discussed in the Sect. 17.5.

17.4.1 Number Counts and V/V_{max} Tests for Extragalactic Radio Sources

Historically, the number counts of extragalactic radio sources and the V/V_{max} tests for quasars were among the first pieces of direct evidence for the evolution of certain classes of extragalactic object with cosmic epoch. The history of the early controversies over the counts of radio sources has been recounted by Scheuer (1990) and Sullivan (1990) and more briefly, in the overall context of the development of astrophysical cosmology, in my history of 20th century astrophysics and cosmology (Longair 1995a). A recent compilation of the counts of radio sources at a wide range of frequencies throughout the radio waveband is shown in Fig. 17.8.

The cause of the controversies in the 1950s and early 1960s was the steepness of the radio source counts at high flux densities. The early surveys were all carried out at low radio frequencies, the counts at metre wavelengths (0.178 GHz) being the first to show the 'inverted' behaviour of the counts with $\beta = 1.8$ at high flux densities. The radio spectra of the sources have mean spectral index $0.7 - 0.9$ and so the arguments developed in Sect. 7.2, and illustrated by Fig. 17.2, show that there must be many more faint sources than are expected in all uniform world models. By the mid-1960s, it was known that the majority of these sources had to be at large redshifts and this led to the realisation that the evolution of the source population had to be very strong indeed (Longair 1966). When the redshift data for complete samples of the radio quasars in the 3CR samples became available, Schmidt (1968) and Rowan-Robinson (1968) showed that these sources were concentrated towards the limits of their observable volumes, as illustrated in Fig. 17.7a. The radio galaxies in the 3CR sample were much more difficult to identify, but this was achieved thanks to the introduction of CCD cameras in the late 1970s which enabled optical identifications with galaxies to be made to about V = 23.5 (Gunn et al. 1981). Fortunately, many of these radio galaxies have strong narrow emission lines in their spectra and this enabled Spinrad and his colleagues to measure their redshifts. The $\langle V/V_{max} \rangle$ test for the 3CR radio galaxies gave almost exactly the same result as for the radio quasars (Fig. 17.7b).

Since these pioneering days, the radio source counts have been determined over the entire radio waveband from low radio frequencies (0.15 GHz) to short centimetre wavelengths (8.44 GHz) as illustrated in Fig. 17.8. They all show the same overall features – a steep source count at high flux densities, a plateau at intermediate flux densities and convergence at low flux densities. At the very lowest flux densities, $S \le 10^{-3}$ Jy, the source counts flatten out again.

In interpreting the counts of sources, the objective is to find out how the luminosity function of the sources $N(L)$ has changed with cosmic epoch. The problem is that, to achieve this, it is necessary to measure the redshifts for large, complete samples of sources, spanning the range of flux densities shown

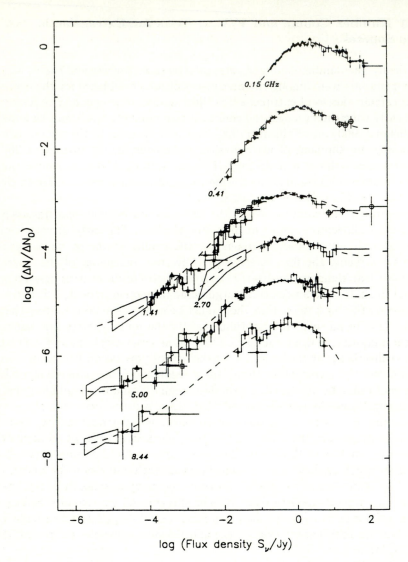

Fig. 17.8. The differential, normalised counts of extragalactic radio sources at a wide range of frequencies throughout the radio waveband. This compilation was kindly provided by Dr. Jasper Wall (see also, Wall 1990, 1996). The points show the number counts derived from surveys of complete samples of radio sources. The boxes indicate extrapolations of the source counts to very low flux densities using the $P(D)$ technique described in Sect. 17.2.3

in Fig. 17.8. Unfortunately, even with the very considerable efforts of many radio and optical astronomers, it remains a difficult and time-consuming task to find the optical identifications for deep samples of radio sources, many of which lie at redshifts $z \geq 1$, and then to measure their redshifts. Optical observations of the galaxies are generally difficult, but observations in the near infrared waveband have proved valuable in identifying distant galaxies and in estimating their redshifts by multicolour photometry (see, for example, Dunlop 1994, 1998).

In analysing the counts of sources, it is conventional to consider separately the radio sources with steep ($\alpha \sim 0.8$) and flat ($\alpha \sim 0$) radio spectra. The former are the extended double radio sources which are the dominant population at low radio frequencies, $\nu < 1$ GHz; the latter are the compact radio sources, often exhibiting variability and superluminal motions observed in high frequency samples, $\nu > 1$ GHz. In order to define how the luminosity function of radio sources with steep and flat radio spectra have changed with cosmic epoch, free-form modelling techniques were developed by Peacock (1985) and extended by Dunlop and Peacock (1990). In their approach to the problem, all the available radio, infrared, optical and redshift data are used in constructing the models. The radio luminosity function is split into two populations, one consisting of intrinsically weak radio sources associated with normal and Seyfert galaxies and the other consisting of the powerful extended and compact radio sources which typify the strong radio source phenomenon. Two examples of the types of model which are consistent with all the available data are shown in Fig. 17.9. Both the steep and flat spectrum radio sources show the same forms of evolutionary behaviour.

There is general agreement that the changes in the radio luminosity function out to redshifts $z \approx 2$ can be very well described by what is termed *luminosity evolution*. In this form of evolution, the radio luminosity function of the powerful radio galaxies is shifted to greater radio luminosities with increasing redshift, whilst the normal galaxy radio luminosity function remains unchanged. There is agreement that these changes can be accounted for by evolution of the form $L(z) = L_0(1+z)^3$. It should be emphasised that this is simply a convenient way of parameterising the necessary changes of the radio luminosity function to account for all the data – it can be seen from Fig. 17.9 that, although the shift in luminosity amounts to only a factor of 27 by a redshift of 2, the number density of sources of a given luminosity increases by a much greater factor. In the functions shown in Fig. 17.9, the comoving number density of sources of luminosity 10^{27} W Hz^{-1} sr^{-1} increases by a factor of about 1000, indicating how strongly the population of extragalactic radio sources has evolved with cosmic epoch.

This form of evolution cannot continue to redshifts $z > 2$ or else the observed convergence of the counts would not be reproduced and the radio background emission would exceed the observed intensity at low radio frequencies, which amounts to a brightness temperature of $T_b = (\lambda^2/2k)I_\nu \approx 23$

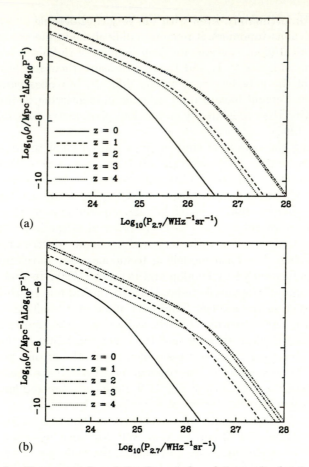

(a)

(b)

Fig. 17.9a-b. Two examples of the forms of evolving radio luminosity function which can account for the radio source counts and available redshift and identification data (Dunlop 1998). In model (**a**), the change in the radio luminosity function is described by pure luminosity evolution. In model (**b**), the changes in the radio luminosity function involve in addition negative density evolution at large redshifts. In both cases, the changes in the form of the radio luminosity function are described in terms of the number densities of sources per unit comoving volume (Dunlop 1998).

K at 0.178 GHz (see, Longair 1995b and Sect. 17.6.3). There has been considerable debate as to whether the evolution function displays a cut-off at redshifts $z > 2$ or whether it remains at a constant enhanced level to large redshifts. As Dunlop (1998) shows, the identification content and redshift distributions of deep radio surveys can give the answer to this question. Suitable deep surveys for these studies are now becoming available and two examples involving the 6C/B2 sample of Eales, Rawlings and their collaborators and the Leiden–Berkeley Deep Survey are shown in Fig. 17.10. The first of these

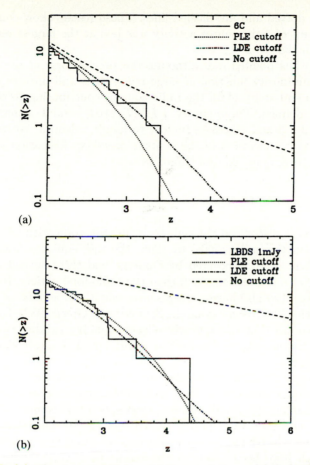

Fig. 17.10. (a) The observed high-redshift cumulative redshift distribution for the complete 6C/B2 samples of Eales, Rawlings and their colleagues (Eales and Rawlings 1996), compared with the large redshift predictions of the two evolution models shown in Fig. 17.9, and a model in which the comoving luminosity function of the radio sources is constant at redshifts $z \geq 2$. (b) The same comparison for the Leiden–Berkeley Deep Survey (Dunlop *et al.* 1995). For more details, see Dunlop (1998).

surveys extends to a flux density about a factor five deeper than the 3CR samples and the second to a factor of about 100 fainter than the 6C/B2 sample. The redshift distributions found in these samples are shown by the solid lines and the predictions of various models by the dashed lines. The important result is that, if the comoving luminosity function of the sources remained constant at redshifts $z \geq 2$, the redshift distribution would be that shown by the dashed lines. It can be seen that significantly fewer large redshift sources are observed in both samples than would be expected according to this model. The expectations of the two models illustrated in Fig. 17.9 are shown by the dotted and dot-dashed lines. These results show convincingly that the amount of radio source activity was less at the largest redshifts as compared with that at $z = 2$.

As Dunlop has pointed out, although the precise form of the evolution of the radio luminosity function at large redshifts is still uncertain, the integrated radio emissivity of all the radio sources contributing to the counts is rather well defined. The quantity $\int LN(L, z)\,\mathrm{d}L$ attains a maximum at redshifts $z \approx 2 - 3$ and decreases rapidly to larger redshifts. We will return to this important conclusion in the context of galaxy formation and metal and star-formation in galaxies in Chap. 18.

17.4.2 Radio Quiet Quasars

The quasars were discovered through their identification with extragalactic radio sources in the early 1960s. In 1965, the radio-quiet counterparts of the radio quasars were discovered by Sandage and these turned out to be about $50 - 100$ times more common than the radio-loud variety. It was soon established that they also have a steep source count and a value of $\langle V/V_{\mathrm{max}} \rangle$ much greater than 0.5 (see for example, Schmidt and Green 1983). Subsequent studies have shown that the optically selected quasars exhibit evolutionary behaviour over cosmological time scales similar to that established for the radio-loud quasars.

The definition of complete samples of optically-selected quasars is somewhat more complex than that of their radio counterparts because the quasars have much more complex and varied optical spectra as compared with the radio quasars and they vary with time. Woltjer (1990) has given a concise summary of the relative merits of the different selection procedures. The approaches which have been taken to the definition of statistically complete quasar samples are as follows:

- One of the most successful techniques for finding radio-quiet quasars has been the search for star-like objects which have ultraviolet excesses, one of the characteristics of the first samples of radio quasars. This technique relies upon the fact that the UV-optical continuum spectrum of quasars is, to a first approximation, a power-law and so they are relatively more intense ultraviolet, and infrared, emitters than normal stars. The pioneering studies of Bracessi *et al.* (1970) first demonstrated convincingly that the UV excess objects have a much steeper number count than that expected in a uniform Euclidean model. This technique is very successful in discovering quasars with redshifts $z \leq 2.2$. At this redshift, the Lyman-α emission line is redshifted into the B filter and so the quasars no longer exhibit ultraviolet excesses. At larger redshifts, absorption lines associated with the Lyman-α forest depress the ultraviolet emission beyond the Lyman-α line so that quasars appear even redder. The UV excess technique was exploited by Schmidt and Green (1983) who derived a complete

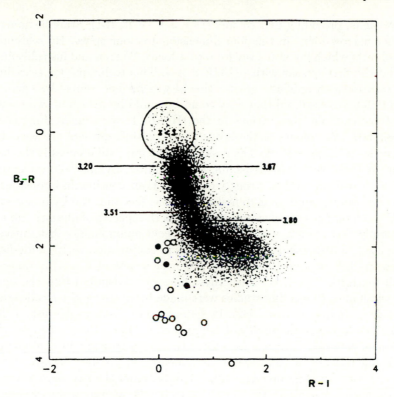

Fig. 17.11. The two-colour (B_J-R, R-I) diagram for stellar objects in a high latitude field observed by the UK Schmidt Telescope (Irwin *et al.* 1991). The region occupied by the majority of quasars with redshifts $z < 3$ is shown by the large ellipse. The redshifts of a number of quasars in the range $3 < z < 4$, discovered by the objective prism technique, are also shown. The quasars discovered in this field with redshifts $z > 4$ are shown by open circles, as well as the three quasars discovered by Warren and his colleagues in the South Galactic Pole (filled circles).

sample of over 114 bright radio-quiet quasars with B, on average, less than 16.16 in a survey which covered about a quarter of the whole sky. The most extensive use of this technique has been by Boyle and his colleagues (1990) who have defined a complete sample of 420 quasars with $B \le 21$ and this has enabled the evolution of the optical luminosity function of quasars out to redshift $z \sim 2$ to be determined.

- The extension of this technique has involved the use of multicolour photometry to discriminate between stars and objects with the typical spectra of large redshift quasars. Koo and Kron (1982) used (U, J, F, N) photometry to find radio quiet quasars with ultraviolet excesses to B = 23. This survey provided the first evidence for the convergence of the counts of radio-quiet quasars. This technique was extended by Warren *et al.* (1987) to four-colour photometry using observations in the U, J, V, R and I wavebands, providing four colours (U-J, J-V, V-R, R-I). Stars lie along a rather narrow locus in this four-dimensional colour space. By searching for objects which lay well away from that locus, Warren and his colleagues found the first quasar with redshift $z > 4$. This technique for searching for quasars with redshifts greater than 4 was further refined by Irwin *et al.* (1991), who realised that they could be found by means of two-colour photometry from observations in the (B_J, R, I) wavebands. Fig. 17.11 shows how the colours of these very large redshift quasars are very different from those of stars. The reason for the large differences is the fact that, at these very large redshifts, the redshifted Lyman-α forest enters the B_J waveband and so strongly depresses their continuum intensities.

- An alternative approach is to make use of the fact that the Lyman-α and CIV emission lines are always very strong in the spectra of quasars and are superimposed upon a roughly power-law continuum energy distribution. The use of a dispersion prism, or grating, in conjunction with a wide-field telescope has proved to be a very powerful means of discoving quasars with redshifts $z > 2$, at which these lines are redshifted into the optical waveband. Pioneering studies were made by Osmer and his colleagues (see, for example, Osmer 1982). Perhaps the most remarkable use of this technique has been the survey of Schmidt *et al.* (1991, 1995) who used the Palomar 200-inch telescope as a fixed transit instrument in conjunction with a grism and a large area CCD camera, which was clocked at the siderial rate. In this way, six narrow bands across the sky were scanned both photometrically in the v and i wavebands, as well as spectroscopically, resulting in a total scanned area of 62 degree^2. Of 1660 candidate emission line objects, 141 were found to be quasars in the redshift interval $2.0 < z < 4.7$ (Schmidt *et al.* 1991, 1995).

- Finally, one of the most important characteristics of the quasars is that they are variable over time-scales which range from days to decades. If a sufficiently long baseline is used, say, of the order of 10 years, all quasars are found to be variable. Following the pioneering efforts of Hawkins (1986), it has been confirmed that this is a successful approach in selecting complete samples of quasars. It is found that the degree of variability of the quasar is correlated with its luminosity, but not with redshift (Hook *et al.* 1991). In one variant of this approach, Majewski *et al.* (1991) used a combination of variability and the lack of proper motions to estimate the completeness of various approaches to defining complete quasar samples. They found the important result that the multicolour surveys miss at most up to 34% of the quasars and probably far less than this percentage.

The upshot of all these studies is that the quasars display a steep number count which converges at faint magnitudes. The changes of the optical luminosity function of optically selected quasars with cosmic epoch has been derived by Boyle and his colleagues out to a redshift $z \sim 2$ and is shown in Fig. 17.12 (Boyle *et al.*, 1991). At zero redshift, the luminosity function of the overall quasar population joins smoothly onto the luminosity function of Seyfert galaxies, an important but natural continuity of the properties of these classes of object. With increasing redshift, there are many more quasars of a given luminosity than expected in a uniform model. Over this redshift range, the changes in the luminosity function can be described by luminosity evolution, of almost exactly the same form as that needed to account for the evolution of the radio luminosity function for extragalactic radio sources, namely, $L(z) \propto (1 + z)^{\beta}$ out to $z = 2$ with $\beta = 3.5$.

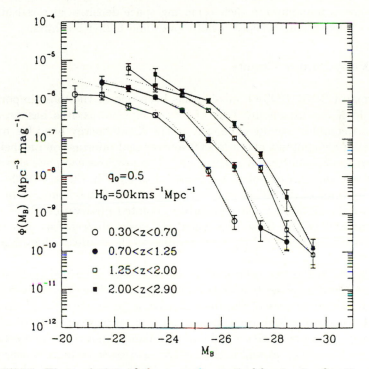

Fig. 17.12. The evolution of the comoving optical luminosity function of radio-quiet quasars in the redshift interval $0.3 < z < 2.9$. The redshift bins have been selected to correspond to equal intervals in $\log(1+z)$. The luminosity functions have been derived from complete samples of quasars and include almost 1000 quasars. The faint dotted lines show the the expectations of a luminosity evolution model in which the luminosities of the quasars change with cosmic epoch as $L(z) \propto (1+z)^{3.5}$ in the redshift interval $0 < z < 2$ and $L(z) = $ constant for redshifts $2 < z < 2.9$ (Boyle *et al.* 1991).

Systematic surveys have been made to determine the large redshift evolution of the radio-quiet quasars by a number of authors (Warren, Hewett and Osmer 1994, Schmidt, Schneider and Gunn, 1995, Kennefick *et al.* 1995). There is good agreement between the results of these surveys, which are similar in many ways to the results from studies of the extragalactic radio sources. First, there are fewer large redshift optically-selected quasars than expected if the comoving optical luminosity function had remained constant at all redshifts $z \geq 2$. The comoving number densities of luminous quasars decreases by a factor of about 5 to 7 over the redshift interval $2 \leq z \leq 4$. A second similarity is that the strength of the evolution may well be a function of luminosity. There is evidence that there are relatively more luminous quasars at large redshifts than lower luminosity objects, similar to the type of differential evolution suggested by Fig. 17.9b for the radio sources. There is general agreement that both the radio sources and the optically selected quasars show a maximum in their comoving space densities at a redshift of about $2 - 3$ and decline steeply at both lower and higher redshifts.

17.4.3 X-ray Source Counts

The German-US-UK ROSAT mission has been one of the most important missions in X-ray astronomy. One of its principal objectives has been to carry out a complete survey of the sky in the X-ray energy band 0.1 to 2.4 keV. The survey contains about 60,000 sources and information on their X-ray spectra is available in 4 X-ray 'colours'. The previous complete survey of the X-ray sky was carried out by the HEAO-1 satellite in the late 1970s. The flux density limit of the ROSAT survey is about 100 times fainter than that of HEAO-1. In addition to the sky survey, pointed observations have been carried out and among these have been very deep observations of a small region of sky to define the X-ray source counts to the faintest achievable flux densities.

The results of the first deep ROSAT surveys were described by Hasinger *et al.* (1993) and involved two separate surveys. The first was derived from 26 fields observed as part of the ROSAT medium deep survey and discrete sources in these fields have been catalogued. The second consisted of a very long, deep exposure of duration 42 hours in a region known as the 'Lockman Hole', a region of sky in which the neutral hydrogen column density has a very low value, $N_H = 5.7 \times 10^{19}$ cm^{-2}. As a result, in this direction, there is minimum photoelectric absorption by the neutral component of the interstellar gas which becomes important at X-ray energies $\epsilon < 1$ keV (see, for example, Sect. 4.2 of Longair 1992).

The X-ray source counts were derived in two ways from these observations. In the case of the medium deep survey, particular care was taken to understand the effects of source confusion at low X-ray flux densities. The very deep survey was analysed using a $P(D)$ analysis described in Sect.17.2.3. The resulting source counts are shown in Fig. 17.13.

The differential X-ray source counts bear a strong resemblance to the differential counts of radio sources (Fig. 17.8). The slopes of the differential source counts quoted by Hasinger *et al.* (1993) are as follows. At high X-ray flux densities, $S > 3 \times 10^{-14}$ erg s^{-1} cm^{-1}, the *differential* counts have

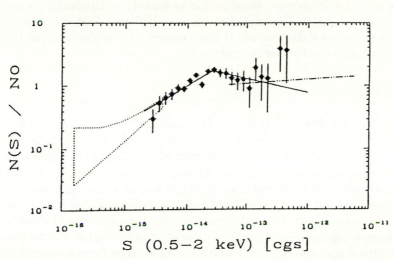

Fig. 17.13. The differential, normalised counts of faint X-ray sources observed by the ROSAT X-ray observatory. The filled circles are sources from the ROSAT Medium Deep Sky Survey, corrected for systematic effects. The dot-dash line is the best-fit source count from the Einstein Observatory surveys. The dotted area at faint flux densities shows the 90% confidence limits from the fluctuation analysis of the deepest ROSAT survey in the Lockman hole (Hasinger *et al.* 1993).

slope $(\beta + 1) = 2.72 \pm 0.27$ and below this flux density the counts from the medium deep survey have slope $(\beta + 1) = 1.94 \pm 0.19$. The best-fit slope of the very deep survey, which extends statistically to flux densities just greater than 10^{-16} erg s^{-1} cm^{-2}, is 1.8. It is a useful exercise, which I leave to the reader, to use the order of magnitude rules described in Sect. 17.2.3 to work out the number density of sources at the confusion limit of the ultra-deep survey in the Lockman Hole. It will be found that the result is within order-of-magnitude agreement with the number densities derived from the Monte Carlo simulations, which are indicated by the dotted area in Figs. 17.13. The identifications of sources in the medium deep survey are consistent with a picture in which the X-ray sources follow the same type of cosmological evolutionary behaviour as the radio galaxies, radio quasars and optically selected quasars.

It is straightforward to work out how much of the X-ray background in the $0.5 - 2$ keV energy band can be attributed to discrete X-ray sources. According to Hasinger *et al.* (1993), the background intensity which can be attributed to discrete sources with flux densities greater than 2×10^{-15} erg cm^{-2} s^{-1} amounts to 59% of the total background intensity in this energy range. Extending the counts to the bottom of the deep survey, about 75% of the background intensity can be accounted for. Thus, not more than 25%

of the background radiation in this waveband is not associated with discrete sources and it is entirely plausible that essentially all the background can be attributed to discrete sources. For example, if the slope of the source counts found in the very deep survey is extrapolated to zero flux density, the total background intensity can be explained.

17.4.4 IRAS Galaxy Counts

The IRAS Satellite carried out the first essentially complete sky survey in the infrared wavebands between 12.5 and 100 μm, which can only be observed from space. Among the many important discoveries of the mission was the realisation that many galaxies are intense far-infrared emitters, the most intense of these being referred to as star-burst galaxies. The catalogues of IRAS galaxies have been very important cosmologically because from them complete samples of galaxies can be selected which are unaffected by obscuration by interstellar dust. As a result of a major effort by many astronomers, the redshifts of complete samples of IRAS galaxies have been measured and the local luminosity function at 60 μm determined.

Counts of IRAS galaxies have been made at 60 μm from the IRAS Point Source Catalogue, the IRAS Faint Source Survey and a survey in the region of the ecliptic poles (for details, see Oliver *et al.* 1992). The normalised differential counts are shown in Fig. 17.14. Since the luminosty function of the IRAS galaxies has been determined at 60 μm, it is straightforward to work out the expected differential source counts, which are shown by the solid line in Fig. 17.14. Again, it is apparent that there are more faint IRAS sources than expected, in the same sense as the counts of radio sources, X-ray sources and quasars. Various parametric fits to the observed counts of IRAS galaxies are shown. At this stage, the counts do not extend deep enough to constrain the large redshift behaviour of the source popultion, but the observed counts are consistent with exactly the same form of evolution found for the active galaxies, in other words, their number counts can be explained by luminosity evolution of the IRAS luminosity function such that $L(z) = L_0(1+z)^3$ out to a redshift $z \sim 2$. Although there is not direct evidence for convergence of the source distribution from the counts themselves, there are constraints from the upper limits to the millimetre and submillimetre background radiation due to discrete sources. These show that the evolution must level off beyond $z \sim 2$, or else the submillimetre background radiation would be exceeded (Blain and Longair 1993).

An interesting consequence of the evolution of the IRAS galaxies results from the strong correlation between the radio emission of normal and starburst galaxies and their far-infrared emission (Helou *et al.* 1985). The proportionality extends over many orders of magnitude and can be written $S(60\,\mu\text{m}) = 90\,S(1.4\,\text{GHz})$, where both flux densities are measured in Jy. As a result, it is a straightforward calculation to predict the counts of starburst

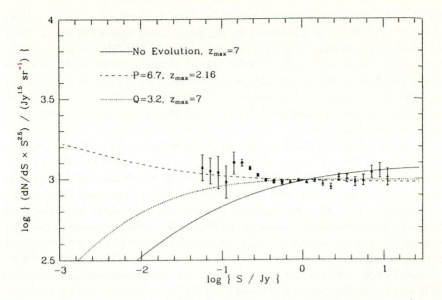

Fig. 17.14. The normalised, differential counts of IRAS galaxies at 60 μm (Oliver *et al.* 1992). The solid line shows the expected differential counts on the basis of the luminosity function of IRAS galaxies measured at 60 μm. The dashed and dotted lines show the counts expected for different possible forms of the cosmological evolution of the population of IRAS galaxies. The observed number counts would be consistent with the forms of strong luminosity evolution which can account for the number counts of active galaxies and quasars described in Sects. 17.4.1 to 17.4.3.

and normal galaxies in the radio waveband. It turns out that it is possible to account for the flattening of the radio source counts at radio flux denisties $S \leq 10^{-3}$ Jy seen in Fig. 17.8 in terms of the evolution of the population of IRAS galaxies (Rowan-Robinson *et al.* 1993). This conclusion is supported by the identification content of the millijansky radio sources, many of which are blue and have spectra similar to those of star-burst galaxies (Windhorst *et al.* 1987, 1995).

17.5 Counts of Galaxies

The determination of precise counts of galaxies has proved to be one of the more difficult areas of observational cosmology. The review of counts of faint galaxies by Ellis (1997) describes vividly the complications in determining them precisely and in interpreting them. The reasons for these complications are multifold. First of all, galaxies are extended objects, often with complex brightness distributions, and great care must be taken to ensure that the same types of object are compared at different magnitude limits and redshifts. Unlike the radio and X-ray sources, which are uniformly distributed over the sky, the distribution of galaxies is far from uniform on scales less than about 50 h^{-1} Mpc, as illustrated by the large voids and walls in the local distribution of galaxies (Figs. 2.5 and 2.6). Even at the faintest magnitudes, this 'cellular' structure in the distribution of galaxies results in fluctuations in the number counts of galaxies which exceed the statistical fluctuations expected in a random distribution. Furthermore, Fig. 3.3 shows that the probability of finding galaxies of different morphological types depends upon the galaxy environment.

A major complication concerns the K-corrections which should be used for galaxies of different types. For example, in order to estimate the number counts of galaxies at large redshifts in the optical waveband, their spectra in the ultraviolet waveband need to be known and these can only be observed from above the Earth's atmosphere. It turns out that there have been remarkably few systematic surveys of the ultraviolet spectra of normal galaxies and only in a few cases have images of galaxies in the ultraviolet waveband been obtained (see, for example, Giavalisco et al. 1996). The problem is exacerbated by the fact that the ultraviolet spectra of galaxies can be dominated by bursts of star formation and these influence very strongly the observability of galaxies at large distances. As Ellis points out, this fact alone makes the comparison of the optical images of galaxies at the present epoch with those at redshifts of one and greater problematic. One important advance has been the determination of the counts of galaxies in the infrared K waveband at 2.2 μm, which has a number of advantages. First of all, the galaxy counts are much less affected by dust extinction. Second, the light of the galaxies in the infrared waveband is dominated by the majority old, stellar populations in the galaxies. Third, the K-corrections are much better known since, even at redshifts $z \sim 4$, the observed light of the galaxies was emitted at 400 nm in the galaxies' rest frames.

These problems are clearly and carefully described in Ellis's excellent review. Granted these problems, Fig. 2.10 presented by Metcalfe et al. (1996) gives a good impression of the current state of play in determining the overall counts of galaxies in the B(440 nm), I(800 nm) and K(2.2 μm) wavebands. These number counts are based upon a number of separate determinations by ground-based optical and infrared telescopes, as well as deep number counts in the Hubble Deep Field. It can be seen that the number counts of galaxies in the Hubble Deep Field join smoothly onto the ground-based counts. Furthermore, they have enabled the counts in the I waveband, in which the background emission from the Earth's atmosphere becomes an increasingly important problem, to be extended by about a factor of 100 fainter than is possible from the ground. The lines labelled 'No evoln.' shown the expectations of uniform world models and include appropriate K-corrections for the types of galaxy observed in bright galaxy smaples. A complication concerns the normalisation of the number counts at bright magnitudes, and a high normalisation, also advocated by Ellis (1997), has been used. The model counts can then give a good account of the total number counts of galaxies, as well as the separate counts for spiral and elliptical galaxies, in the magnitude interval $18 < B < 22.5$.

A number of general features of the galaxy counts are apparent from the diagram. In the infrared K waveband (2.2 μm), the counts follow reasonably closely the expectations of uniform world models with $q_0 \sim 0 - 0.5$. This is perhaps not too unexpected since the old stellar populations of galaxies are the principal contributors to the luminosities of galaxies in these wavebands. There is some evidence that there are more galaxies than might be expected if $q_0 = 0.5$. In contrast, in the B and I wavebands, there is a large excess of faint galaxies, particularly in the B waveband. The departure from the expectations of the uniform models sets in at about $B = 23$ and, at fainter magnitudes, there is a large excess of faint blue galaxies. The lines on the diagram illustrate the results of various modelling exercises to account for the observed counts on the basis of models for the evolution of the stellar populations of spiral and elliptical galaxies.

The nature of the excess of faint blue galaxies is a key cosmological problem. Redshift distributions for complete samples of galaxies at which the excess of blue galaxies is observed are required but, unfortunately, this is just beyond the capabilities of the present generation of 4-metre telescopes. Ellis (1997) notes that the complete redshift surveys with such telescopes are effectively limited to $B \leq 24$, $I \leq 22$ and $K \leq 18$. Examples of what has been possible to these limits is illustrated in Fig. 17.15, which shows the results of complete spectroscopic surveys from the Canada–France Redshift Survey (CFRS), which extends to $I = 22$ (Lilly *et al.* 1995), and the Autofib-LDSS survey carried out with the Anglo-Australian Telescope (Ellis *al.* 1996) which extended to $b_J = 24$. The numbers of galaxies observed in the different redshift intervals are indicated in brackets.

In the analysis of Lilly *et al.*, the change in the form of the luminosity function with increasing redshift is apparent and shows a modest amount of evolution of the luminosity function of galaxies out to redshifts $z \sim 1$ in the sense that the whole luminosity function appears to be shifted to greater luminosities (Fig. 17.15a). An excess of galaxies is also observed in the Autofib-LDSS survey, but it appears to be primarily associated with galaxies which have luminosities $L \leq L^*$ (Fig. 17.15b). In any case, both surveys are in agreement that there is a marked increase in the comoving density of star-forming galaxies with redshift.

The first indications of what will become possible with the coming generation of $8 - 10$ metre optical-infrared telescopes is provided by the first surveys of faint galaxies carried out with the Keck 10-metre telescope (Cowie *et al.* 1995, 1996). Fig. 17.16 shows the redshift distribution for an almost complete sample of galaxies in the magnitude interval $22.5 < B < 24$, taken from the presentation by Metcalfe *et al.* (1996). The important feature of this redshift distribution is the comparison between the 'no evolution' pre-

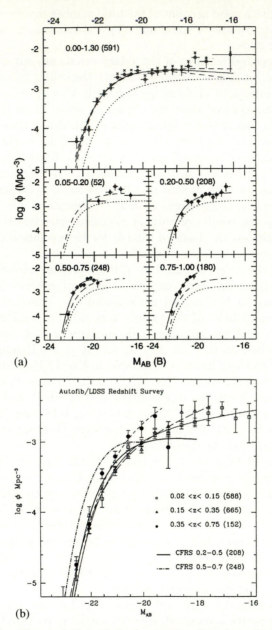

Fig. 17.15a-b. The evolution of the luminosity function of galaxies in the rest-frame B waveband as determined by (**a**) the CFRS survey (Lilly *et al.* 1995) and (**b**) the Autofib/LDSS survey (Ellis *et al.* 1996). In (**a**), the overall luminosity function of galaxies is shown as well as portions of the luminosity function derived in different redshift intervals. The numbers of galaxies included in each determination is shown in brackets in each panel. The dotted line shows the local luminosity function of galaxies determined by Loveday *et al.* (1992). (**b**) Comparison of the luminosity functions determined from the CFRS and Autofib/LDSS surveys (Ellis 1997).

dictions, indicated by the solid and long-dashed lines in the figure, and the histogram showing the observed distribution. It can be seen that there is a significant excess of blue galaxies extending to redshifts $z = 1.7$. According to Cowie *et al.* (1996), the distribution is composed of a mixture of normal galaxies at small redshifts plus galaxies undergoing rapid star-formation from $z = 0.2$ to beyond $z = 1.7$. These results are in accord with the conclusions of Ellis (1997) that there is unquestionably an increase in the numbers of star-forming galaxies with increasing redshift. Eqully intriguing, Cowie *et al.* (1996) find that there is little change in the K-band luminosity function out to redshifts $z \approx 1$, suggesting that most of their stellar populations were already in place by a redshift of 1. We will return to the issue of star-formation rates in galaxies in general when we study the significance of observations of the Hubble Deep Field (Chap. 18).

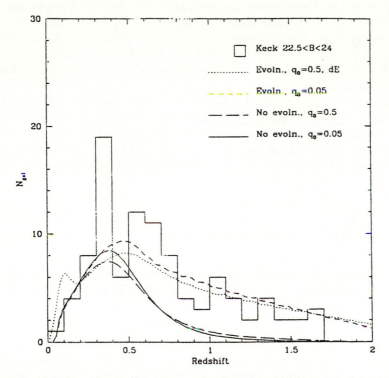

Fig. 17.16. The redshift distribution of galaxies from the deep survey of Cowie *et al.* (1996) in the magnitude interval $22.5 < B < 24$ (from Metcalfe *et al.* 1996). The solid line and the long-dashed line show the expected redshift distributions if there were no evolution of the local luminosity function of galaxies. It can be seen that there is a long high-redshift tail of blue galaxies, which, from the presence of strong [OII] lines, are inferred to be undergoing rapid star formation.

The nature of the excess of blue galaxies has been at least partially elucidated by studies with the Hubble Space Telescope. By combining observations from the Medium Deep Survey, which utilised observations of 13 fields taken in parallel mode with the Wide Field Planetary Camera 2, with the observations of the Hubble Deep Field, number counts have been determined for galaxies of different morphological types. The high resolution images have enabled the morphologies of galaxies to be classified into spheroidal/compact, spiral and irregular/peculiar/merger categories. The results of the analysis of Abraham *et al.* (1996) are shown in Fig. 17.17. It is apparent that the spheroidal and spiral galaxies more or less follow the expectations of the uniform world models, while the objects classified as irregular/peculiar/merger systems show a distinct excess relative to their populations in bright galaxy samples, which amount to only about $1-2\%$ of the galaxy population. These results are consistent with the visual impression of the Hubble Deep Field which suggests that about 25% of the galaxies seem to be irregular/interacting/merging systems. They are also consistent with the imaging results of Cowie *et al* (1995) and Schade *et al.* (1995) which indicate that about the same fraction of the blue galaxies in their surveys have peculiar morphologies.

While the evidence presented in Fig. 17.17 appears conclusive, Ellis (1997) rightly urges caution in interpreting these results for the reasons discussed above. We do not possess a good control sample of ultraviolet images of nearby galaxies with which the objects in the Hubble Deep Field can be compared. Nonetheless, as we will discuss in Chap. 18, a great deal of important astrophysics can be derived from these observations.

17.6 The Background Radiation

It is traditional to refer to the study of the background radiation as the 'oldest problem in cosmology'. What is commonly called *Olbers' paradox* revolves around the the question

Why is the sky dark at night?

Olbers was in fact only one of many distinguished scientists who realised that the darkness of the night sky provides us with general information about the large-scale distribution of matter and radiation in the Universe. Harrison (1987) was written the history of Olbers' paradox in his delightful book *Darkness at Night*. Indeed, the integrated emission from a population of extragalactic objects can provide useful constraints on their spatial distribution and, to complete this chapter, we carry out some simple illustrative calculations.

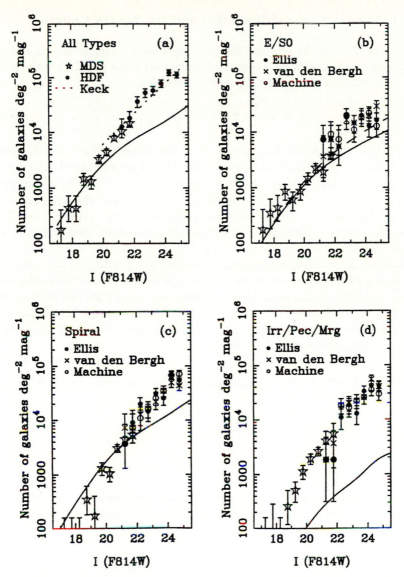

Fig. 17.17. The number-magnitude relation for morphologically segregated samples of galaxies from the Medium Deep Survey (MDS) and the Hubble Deep Field (Abraham *et al.* 1996). The observations from the MDS survey are indicated by stars. The other symbols show the counts based upon morphological classifications carried out by Ellis, van den Bergh and by an automated machine-based classification algorithm. The dotted line in panel (a) shows the total counts of galaxies from a field observed by the Keck Telescope. The solid lines show the expected counts of the different morphological classes assuming their properties do not change with cosmic epoch (Glazebrook *et al.* 1995).

17.6.1 The Background Radiation and the Source Counts

Let us look first at the relation between the observed source counts and the background radiation. The background radiation from a population of sources with differential source count $dN \propto S^{-(\beta+1)} \, dS$ is

$$I \propto \int_{S_{\min}}^{S_{\max}} S \, dN \propto \int_{S_{\min}}^{S_{\max}} S^{-\beta} dS = \frac{1}{1-\beta} \left[S^{(1-\beta)} \right]_{S_{\min}}^{S_{\max}}. \tag{17.20}$$

Notice that, thoughout this chapter, β is the slope of the *integral* counts. Thus, there is a critical value $\beta = 1$ for the slope of the integral source counts. If the slope of the counts is steeper than $\beta = 1$, the background intensity $I_\nu \propto S_{\min}^{(1-\beta)}$. On the other hand, if the slope of the integral source counts is less than $\beta = 1$, the background intensity is proportional to $S_{\max}^{1-\beta}$. Thus, most of the background radiation originates from that region of the counts with slope $\beta = 1$.

Now, for a Euclidean population of sources $\beta = 1.5$. In real world models, the slope is 1.5 at small redshifts but decreases at larger redshifts, as discussed in Sect. 17.2. From the considerations which led to (17.15), we showed that the slope of the integral counts is about 1 by a redshift 0.5, showing that the bulk of the background radiation originates from redshifts $z \ll 1$. Let us carry out some more exact calculations.

17.6.2 Evaluating the Background due to Discrete Sources

For illustrative purposes, we assume that the sources have power-law spectra, $S \propto \nu^{-\alpha}$, and then the flux density-luminosity relation is as usual

$$S(\nu_0) = \frac{L(\nu_0)}{4\pi D^2 (1+z)^{1+\alpha}}. \tag{17.21}$$

The numbers of sources per steradian in the increment of comoving coordinate distance dr in the case of a uniform distribution of sources is

$$dN = N_0 D^2 \, dr. \tag{17.22}$$

Therefore, the background intensity $I(\nu_0)$ due to this uniform distribution of sources is

$$I(\nu_0) = \int S(\nu_0) \, dN = \int_0^\infty \frac{L(\nu_0)}{4\pi D^2 (1+z)^{1+\alpha}} N_0 D^2 \, dr$$
$$= \frac{L(\nu_0) N_0}{4\pi} \int_0^\infty (1+z)^{-(1+\alpha)} \, dr. \tag{17.23}$$

For the Friedman world models with $\Omega_\Lambda = 0$,

$$dr = \frac{c \, dz}{H_0 (\Omega_0 z + 1)^{1/2} (1+z)}, \tag{7.36}$$

and so we obtain the result

$$I(\nu_0) = \frac{c}{H_0}\frac{L(\nu_0)N_0}{4\pi} \int_0^\infty \frac{dz}{(\Omega_0 z + 1)^{1/2}(1+z)^{2+\alpha}}. \qquad (17.24)$$

This result can be compared with the Newtonian version of the same calculation which, from the small redshift limit, $z \to 0$, $r = cz/H_0$, becomes

$$I(\nu_0) = \frac{L(\nu_0)N_0}{4\pi} \int_0^\infty dr. \qquad (17.25)$$

This is the standard exposition of *Olbers' paradox*, namely, that, in a isotropic, infinite, stationary Euclidean Universe, the background radiation diverges. This calculation has not taken account of the finite sizes of the sources, nor does it take account of thermodynamics, since in an infinite static Universe all the matter must come into thermodynamic equilibrium at the same temperature.

Unlike the integral (17.25), the integral (17.24) converges provided $\alpha > -1.5$. Even if the spectral index of the sources were more negative than this value, any realistic spectrum must eventually turn over at a high enough frequency and so a finite integral is always obtained. Let us work out the background intensity for world models with $\Omega_0 = 0$ and $\Omega_0 = 1$.

$$\Omega_0 = 0 \qquad I(\nu_0) = \frac{c}{(1+\alpha)H_0}\frac{L(\nu_0)N_0}{4\pi}; \qquad (17.26a)$$

$$\Omega_0 = 1 \qquad I(\nu_0) = \frac{c}{(1.5+\alpha)H_0}\frac{L(\nu_0)N_0}{4\pi}. \qquad (17.26b)$$

Thus, for typical values of α, to order of magnitude, the background intensity is just that originating within a typical cosmological distance (c/H_0), that is,

$$I(\nu_0) \sim \frac{c}{H_0}\frac{L(\nu_0)N_0}{4\pi}. \qquad (17.27)$$

A combination of factors leads to the convergence of the integral for the background intensity. Inspection of the integral (17.23) shows that part of the convergence is due the factor $(1+z)^{-(1+\alpha)}$ which is associated with the redshifting of the emitted spectrum of the sources. The second is the dependence of r and D upon redshift z. This relation is linear at small redshifts, $z \ll 1$, but, as shown in Figs. 7.3a and b, r and D converge to the values

$$r_{max} = \frac{2c}{H_0(\Omega_0-1)^{1/2}}\tan^{-1}(\Omega_0-1)^{1/2} \qquad D_{max} = 2c/H_0\Omega_0, \qquad (17.28)$$

as $z \to \infty$. This convergence is associated with the fact that the Friedman models have finite ages and consequently there is a finite maximum radial distance from which electromagnetic waves can reach the Earth.

Let us look in a little more detail at the origin of the background radiation in the uniform models. For the critical model $\Omega_0 = 1$ with $\alpha = 1$, the background intensity out to redshift z is

$$I(\nu_0) = \frac{2c}{5H_0} L(\nu_0) N_0 \left[1 - (1+z)^{-5/2} \right]. \qquad (17.29)$$

From this result, it is evident that half of the background intensity originates at redshifts $z \leq 0.31$. A similar calculation for the case of the empty world model, $\Omega_0 = 0$, shows that half the intensity comes from redshifts less than 0.42. Thus, although it might be thought that the background radiation would probe the very distant Universe, in fact, most of the radiation originates at redshifts $z < 1$. Furthermore, since half of the background is expected to originate at redshifts less than about 0.5, the principal contributors to the background radiation are not difficult to identify nowadays, provided their positions are accurately known. If the main sources of the background are associated with galaxies, there should be no difficulty in discovering the principal contributors to the background radiation, provided the sources are uniformly distributed in space. This statement is not correct if the properties of the sources have evolved strongly with cosmic epoch and we take up that topic now.

17.6.3 The Effects of Evolution – the Case of the Radio Background Emission

If the luminosity function of the sources evolves with cosmic epoch, the background intensity is

$$I(\nu_0) = \frac{c}{H_0} \frac{L(\nu_0) N_0}{4\pi} \int_0^\infty \frac{f(L, z, \text{type}, \dots) \, dz}{(\Omega_0 z + 1)^{1/2}(1 + z)^{2+\alpha}}, \qquad (17.30)$$

where the evolution of the comoving luminosity function is described by the function $f(L, z, \text{type}, \dots)$. For simplicity, let us consider one of the forms of evolution of the luminosity function suggested by the analyses described in Sects. 17.4.1 to 17.4.4, that is, 'luminosity evolution' of the form

$$L(z) = L_0(1 + z)^3 \qquad 0 < z_m \leq 2;$$
$$L(z) = 27 L_0 \qquad z_m > 2. \qquad (17.31)$$

The integrated background emission from a population of sources which locally has luminosity L_0 and space density N_0 is then

$$I(\nu_0) = \frac{c}{H_0} \frac{N_0}{4\pi} \int_0^\infty \frac{L(z) \, dz}{(1 + z)^{7/2}}, \qquad (17.32)$$

where we have assumed that the spectral index of the sources α is 1 and that $\Omega_0 = 1$. Performing this integral, we find

$$I(\nu_0) = \frac{c}{H_0} \left(\frac{N_0 L_0}{4\pi} \right) \left[\frac{12}{5}(1 + z_m)^{1/2} - 2 \right]. \tag{17.33}$$

The contributions to the background from redshifts 0 to z_m and from z_m to infinity are in the ratio $5[1 - (1 + z_m)^{-1/2}] : 1$

In the *no evolution* case, $L(z) = L_0$ for all redshifts and the background intensity is

$$I(\nu_0) = \frac{2}{5} \frac{c}{H_0} \left(\frac{N_0 L_0}{4\pi} \right). \tag{17.34}$$

In contrast, in the *evolution* case, adopting the variation of luminosity with redshift given by (17.31), the background intensity is $[6(1 + z_m)^{1/2} - 5]$ times greater than the no evolution background intensity. Taking as an example, the canonical value $z_m = 2$, the effects of cosmological evolution result in a background intensity 5.4 times greater than in the case of no evolution. It is interesting that, without taking account of the effects of cosmological evolution, the background brightness temperature due to strong radio sources would amount to only a few K, but amounts to about $16 - 19$ K when the effects of evolution are taken into account. Notice that, in the canonical model, most of the background originates at redshifts $z \sim z_m$, only 32% of the background originating at $z > 2$.

This is a case in which the background emission *does* indeed originate from the distant Universe, the bulk of the background coming from redshifts of the order 2, but this only occurs because of the *very* strong effects of cosmological evolution. The important point is that the evolution has to be very drastic to make a significant impact upon the intensity of the background emission due to discrete sources.

18 The Evolution of Star and Element Formation Rates with Cosmic Epoch

18.1 Star and Element Formation in Galaxies

Some of the most important clues to the sequence of events which must have taken place as galaxies took up the forms they have today come from observations of the variation of the overall star formation rate in galaxies with cosmic epoch. This story is very closely related to studies of the rate at which the chemical elements were built up in the stars and in the interstellar media of galaxies – the chemical enrichment of the interstellar gas is associated with the formation of heavy elements in short-lived massive stars and their recirculation through the interstellar gas by supernova explosions. One of the most important developments in observational cosmology has been the realisation that star and element formation rates can be derived by a number of independent methods and these help constrain scenarios for the evolution of galaxies.

These constraints come from a variety of different types of argument.

- We can obtain limits to the total amount of element and star formation which could have taken place from the intensity of the isotropic background radiation in the optical, infrared and submillimetre wavebands.
- Spectroscopic studies of absorption line systems in the spectra of quasars provide estimates of the abundances of the elements in these gas clouds over a wide range of redshifts and, as we will show, these provide direct evidence concerning the build up of heavy elements in galaxies.
- The characteristic flat, blue spectra of regions of star formation can be observed over a wide range of redshifts and, these enable estimates to be made of the rates of star and element formation as a function of cosmic epoch. These observations can be influenced by the effects of dust in the star-forming regions and in the galaxy itself, but the absorbed energy is reradiated in the far-infrared and submillimetre wavebands and so the overall production of heavy elements can be constrained by observations in these wavebands.

In this chapter, these approaches to the determination of star and element formation rates at different cosmic epochs are described and then the observations are reviewed in the light of models for the chemical evolution of galaxies. This involves deriving and solving the equations of cosmic chemical evolution.

18.2 The Background Radiation and Element Formation

Heavy elements such as carbon and oxygen are formed in the central nuclear-burning regions of massive stars. In Fig. 18.1, the evolution of the internal chemical structure of a $5M_\odot$ star is shown, illustrating the synthesis of oxygen and carbon in its central regions (Kippenhahn and Weigert 1990). It can be seen that the first step in the synthesis of these elements involves the burning of all the hydrogen into helium in the core of the star. The formation of carbon and oxygen only begins when the central regions have become sufficiently hot and dense for the famous triple-α reaction, $3\,^4\mathrm{He} \rightarrow\,^{12}\mathrm{C}$, to take place after about 6×10^7 years. By 8×10^7 years, the inner 20% by mass of the star has been converted into carbon and oxygen, which is recirculated through the interstellar gas following the demise of the star. During these brief terminal phases, the star evolves in the red giant region of the H-R diagram, as shown in the lower part of Fig. 18.1.

Thus, the essential first step in the formation of the heavy elements during stellar evolution is the conversion of hydrogen into helium within the core of the star. Inspection of a table of the binding energies of the chemical elements shows that by far the most important source of energy generation during the processes of nucleosynthesis is the conversion of hydrogen into helium, in which 0.7% of the rest mass energy of the hydrogen atoms is liberated. Thus, to produce a given mass of heavy elements, the energy release during the main-sequence lifetime of a massive star is just the energy liberated in the hydrogen to helium conversion necessary to create that mass of heavy elements. In other words, to create a mass $M_Z = ZM$ of heavy elements, an energy $0.007ZMc^2$ must be liberated during the steady main-sequence hydrogen burning phase of the star, where Z is the fraction of the total mass in the form of heavy elements created by stellar nucleosynthesis. Therefore, taking averages over the lifetime of the star, if a mass of metals $\dot{M}_Z$ is created per unit time, the average luminosity of the star during its main-sequence phase is $0.007\dot{M}_Z c^2$, that is,

$$L = 0.007\dot{M}_Z c^2. \tag{18.1}$$

Cowie (1988) has argued that this expression should be modified to take account of the fact that not all the heavy elements become available to be recycled through the interstellar gas and so incorporated into the next generation of stars. Some of the mass is locked up in white dwarfs or neutron stars formed at the end of the star's lifetime and, if a black hole is formed,

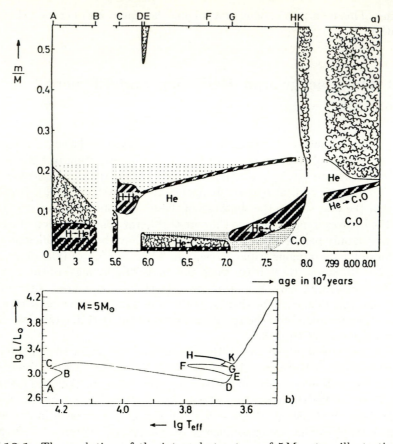

Fig. 18.1. The evolution of the internal structure of $5M_\odot$ star, illustrating the synthesis of carbon and oxygen in its central regions. The abscissa shows the age of the model star after the ignition of hydrogen in its core in units of 10^7 years. The ordinate shows the radial coordinate in terms of the mass m within a given radius relative to the total mass M of the star. The cloudy regions indicate convective zones. The corresponding positions of the star on an H-R diagram at each stage in its evolution are shown in the lower part of the diagram (Kippenhahn and Weigert 1990).

the infalling material is lost to us. If α_m is the fraction of the heavy elements which is recycled to the interstellar medium, which Cowie takes to be 0.67, a greater luminosity is required to create the observed interstellar abundance of the elements by a factor α_m^{-1}. Integrating over the lifetime of the galaxy, we can rewrite (18.1) in terms of the total energy density of radiation ε which is built up in the process of creating a total mass density of heavy elements (or metals) ϱ_m ,

$$\varepsilon = \frac{0.007\varrho_m c^2}{\alpha_m}, \tag{18.2}$$

where $\rho_m = V^{-1} \int \dot{M}_Z \, dt$ and $\varepsilon = V^{-1} \int L \, dt$. If this energy was all liberated at redshift z, (18.2) describes the energy density of radiation and the mean density of metals produced at that redshift. The energy density of radiation decreases with redshift as $(1 + z)^{-4}$, whereas the average density of metals decreases as only $(1 + z)^{-3}$. Writing the energy density of radiation in terms of the intensity, $I = \varepsilon c/4\pi$, we find

$$I = \frac{0.007 \varrho_m c^3}{4\pi \alpha_m (1 + z)}. \tag{18.3}$$

This is the background radiation intensity which should be observed today if a mass density ϱ_m of heavy elements is observed at the present epoch.

Let us insert some plausible values into (18.3) to illustrate how the present abundance of the metals is constrained by the intensity of the background radiation. If the density parameter in baryons is Ω_B, $\varrho_m = Z \varrho_{crit} \Omega_B$ and the intensity of the background radiation would be

$$I = 3.0 \times 10^{-4} \frac{Z\Omega_B h^2}{\alpha_m (1 + z)} \; \mathrm{W\,m^{-2}\,sr^{-1}}. \tag{18.4}$$

This estimate can be compared with various limits to the isotropic cosmic background radiation, which are shown in Fig. 18.2. This figure is plotted in units of $\nu I(\nu)$ and so represents the intensity of the radiation in the same units as (18.4). Taking as representative values $\Omega_B h^2 = 0.01$, $Z = 0.01$ and $\alpha_m = 0.67$, we find $I = 4 \times 10^{-8}/(1 + z) \; \mathrm{W\,m^{-2}\,sr^{-1}}$. If this radiation were due to the emission of stars, we would expect this to be the typical intensity emitted over roughly a decade in frequency. It can be seen from Fig. 18.2 that this value lies close to the upper limit to the optical background radiation due to Toller (1990) (open square box at 440 nm or 7×10^{14} Hz), and is also similar to the lower limit to the background light due to galaxies found by integrating the counts of faint galaxies (Tyson 1990) (boxes with vertical arrows). It is interesting that the simplest picture for the background radiation gives a result not too different from what might be expected from the abundance of heavy elements in the Universe now.

One way in which the background intensity in the optical waveband could be reduced would be if a significant fraction of the optical-ultraviolet emission of galaxies were absorbed by dust. It is not implausible that this should be important at some level since stars are formed in the coolest, dustiest regions of galaxies, and much of the synthesis of the heavy elements may well have taken place when a much larger fraction of the baryonic mass of galaxies was in the form of interstellar gas than it is today. It is also plausible that the very first generations of stars created a significant abundance of heavy elements, which would lead to the formation of interstellar dust. In turn, the presence of dust would facilitate the formation of subsequent generations of stars. The absorbed optical-ultraviolet radiation would be re-radiated in the submillimetre–far-infrared wavebands, in which case the same argument

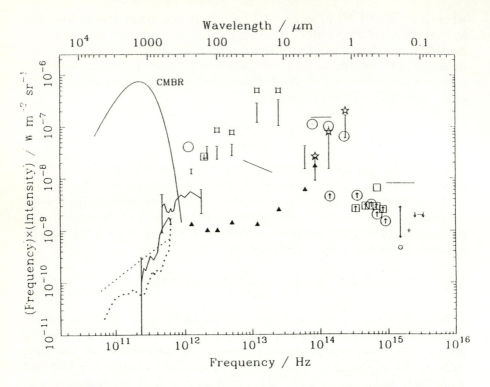

Fig. 18.2. The extragalactic background radiation from the millimetre to ultra-violet wavebands. Background radiation has been detected in the millimetre and submillimetre wavebands, but only upper limits are available in the far-infrared to ultraviolet wavebands. The upper limit to the diffuse optical background radiation, shown as an open box at 7×10^{14} Hz, is taken from Toller (1990) and the integrated light from galaxies, shown by open boxes with vertical arrows, is taken from Tyson (1990). In the submillimetre waveband, the upper limits come from the lack of distortion of the spectrum of the Cosmic Background Radiation observed by COBE (straight dotted line) and the possible background component in the 100 to 1000 μm region of the spectrum extracted by Puget *et al.* (1996) from the COBE data (jagged solid lines between 100 and 1000 μm). The other limits between 10^{12} and 10^{14} Hz are from observations by the IRAS and COBE observatories (Courtesy of Dr. Andrew Blain.)

which led to (18.4) can be used, but now applied to the upper limits to the background radiation intensity in these wavebands provided by COBE (Fig. 18.2). At wavelengths longer than 0.5 mm, strong upper limits to the background intensity are provided by the lack of distortion of the spectrum of the Cosmic Background Radiation (straight dotted line) (see Sect. 2.2.1), while at shorter wavelengths, the background intensity extracted by Puget *et al.* (1996) can be safely used as an upper limit to the background intensity (jagged solid lines between 100 and 1000 μm). It can be seen that these limits to the background intensities are of the same order of magnitude as those in the optical waveband.

These limits are somewhat weaker if the formation of the heavy elements took place at redshift z. Then, according to (18.4), for a fixed metal abundance now, the background radiation observed at the present epoch would be reduced by a factor $(1+z)^{-1}$. We will return to a stronger version of this argument in relation to direct measures of star-formation rates in Sect. 18.4.

Notice that this argument can be used to argue that the limits to the average baryonic density of the Universe from primordial nucleosynthesis, $\Omega_B h^2 \leq 0.035$, are consistent with the present production rate of metals by stellar nucleosynthesis. Values of $\Omega_B h^2 \sim 1$ would result in a background intensity which would far exceed the limits shown in Fig. 18.2.

18.3 The Lyman-α Absorption Clouds

Important clues to the history of metal formation, as well as estimates of the intensity of the far-ultraviolet background radiation, can been derived from the Lyman-α absorption line systems observed in the ultraviolet spectra of quasars. The nature and properties of these absorption line systems are vast subjects and, for many more details, reference should be made to the following volumes, *QSO Absorption Lines: Probing the Universe* (Blades *et al.* 1988), *QSO Absorption Lines* (Meylan 1995) and *The Hubble Space Telescope and the High Redshift Universe* (Tanvir *et al.* 1997).

18.3.1 The Properties of the Lyman-α Absorption Clouds

To summarise briefly, there are two broad classes of absorption line system which are of direct importance for these studies. By far the most common are those belonging to the *Lyman-α forest* which dominate the spectra of large redshift quasars, such as that illustrated in Fig. 18.3. These systems have neutral hydrogen column densities[4] in the range $N_{\rm HI} \sim 10^{16} - 10^{21}$ m^{-2}. In those systems with $N_{\rm HI} \geq 10^{19}$ m^{-2} evidence for low abundances of the heavy elements, corresponding to about 10^{-2} of the solar value, is usually found (Boksenberg 1997). Systems with lower column densities do not possess detectable CIV lines, and so it is uncertain whether or not this is also true for all the weak Lyman-α absorbers. The absorbers responsible for the Lyman-α forest are interpreted as intergalactic clouds, containing largely unprocessed primordial material. There must, however, be some means of mildly enriching these primordial clouds, the chemical abundances of which are similar to those of halo stars in our Galaxy. We will return to the interpretation of these systems in Sect. 19.8.

---

[4] Note that, for consistency, I express column densities in units of m^{-2}, although essentially all the literature uses units of cm^{-2}.

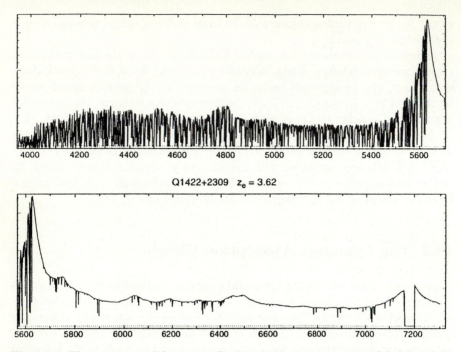

Q1422+2309 $z_e = 3.62$

Fig. 18.3. The spectrum of the quasar Q1422+2309 at an emission redshift $z = 3.62$ showing the remarkable 'Lyman-α forest' to the short wavelength side of the strong redshifted Lyman-α emission line, which has observed wavelength 560 nm (Courtesy of Dr. W.L.W. Sargent, from Boksenberg 1997). To the long wavelength side of the Lyman-α line, the spectrum is very much smoother, only weak metal absorption lines being observed, most of them associated with the CIV absorption line.

In contrast, the rarer *Lyman-limit* and *damped Lyman-α systems* have much larger column densities, $N_{HI} \sim 10^{21} - 10^{26}$ m^{-2}, and have correspondingly larger optical depths for Lyman-α absorption. In the Lyman-limit systems, those with column densities in the range $N_{HI} \sim 10^{21} - 2 \times 10^{24}$ m^{-2}, a continuum break is observed at the redshifted wavelength of the 91.2 nm Lyman limit, and they display absorption lines of the common elements. The Lyman-limit systems can be associated with the extended gaseous haloes of galaxies, similar to those observed about nearby galaxies (Bergeron 1988). The abundances of the heavy elements in the latter systems are less than 10% of their cosmic abundances, consistent with the inference that the halos are very extensive, $\sim 50 - 100$ kpc, and so the gas in these regions is not expected to be as enriched as the gas in the disc of a galaxy.

The *damped Lyman-α* systems have the largest column densities, $N_{HI} \geq 2 \times 10^{24}$ m^{-2}. These systems are so-called because the optical depth of the Lyman-α line is so large that the continuum flux density of the quasar reaches essentially zero intensity at the line centre, and the profile of the absorption line can be fitted by a Voigt profile, characteristic of the natural broadening of an absorption line in the limit $\tau \gg 1$. These systems almost certainly contain most of the mass density of neutral gas in the Universe (Lanzetta *et al.* 1995).

As demonstrated by Wolfe (1988), the damped Lyman-α systems can be convincingly associated with galactic discs and, in his pioneering paper, he identified them with the progenitors of the stellar discs of present-day spiral galaxies. This is a natural assumption since stars form in the coldest regions of the interstellar gas. It is also strongly reinforced by the important analysis of Kennicutt (1989), who made detailed observations of a number of nearby spiral galaxies and showed that active star formation only takes place in their discs if the column density of neutral gas exceeds 2×10^{24} m^{-2}. It is striking that this criterion is identical to the lower limit at which absorbers are identified as damped Lyman-α systems. Kennicutt argued that this criterion is consistent with the stability criterion for rotating thin gaseous discs, namely, that the disc becomes unstable if the surface density exceeds the critical value Σ_{crit}

$$\Sigma_{\mathrm{crit}} = A \frac{\kappa c_{\mathrm{s}}}{3.36G}, \tag{18.5}$$

where κ is the epicyclic frequency, c_{s} is the sound speed in the disc and $A \sim 1$. Thus, the damped Lyman-α systems can provide key information about the chemical evolution of galaxies.

The observed number density distribution of Lyman-α absorbers follows closely a power-law distribution from the weakest detectable lines with $N_{\mathrm{HI}} \sim 2 \times 10^{16}$ m^{-2} to the high density damped Lyman-α systems, and can be described by the relation $N(N_{\mathrm{HI}}) \propto N_{\mathrm{HI}}^{-1.5}$ (Fig. 18.4).

18.3.2 The Evolution of Lyman-α Absorption Clouds with Cosmic Epoch

An important aspect of the distribution of both classes of absorption system is their variation with redshift. It is convenient to parameterise the variation of the number density of absorbers with redshift by a power-law distribution of the form

$$N(z) \, \mathrm{d}z = A(1+z)^{\gamma} \, \mathrm{d}z. \tag{18.6}$$

Typically, it is found that, for the Lyman-α forest systems, $A \approx 10$ and $\gamma = 2 - 3$ whereas, for the Lyman-limit systems $A \approx 1$ and $\gamma \sim 1$. These variations with redshift can be compared with the expected distribution if the properties of the absorbers were unchanged with cosmic epoch, that is, if the absorbers had the same proper cross-sections and constant *comoving* number density. The numbers of absorbers intercepted along any line of sight in the interval of *proper* length $\mathrm{d}r_{\mathrm{prop}}$ is

$$N(z) \, \mathrm{d}z = \sigma_{\mathrm{A}} N_{\mathrm{A}}(z) \, \mathrm{d}r_{\mathrm{prop}}, \tag{18.7}$$

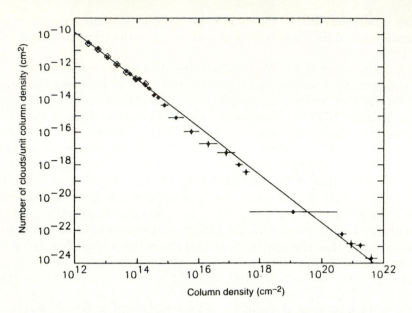

Fig. 18.4. The number density distribution of Lyman-α absorbing clouds as derived by Songaila *et al.* (1995) from Keck Telescope observations of the quasar Q0302-003.

where σ_A is the cross-section of each absorber and $N_A(z)$ is their proper number density at redshift z. Now, $N_A(z) = N_0(1 + z)^3$, where N_0 is the comoving number density of absorbers, that is, the number density they would have at the present epoch, and

$$\mathrm{d}r_{\mathrm{prop}} = c\,\mathrm{d}t = \frac{\mathrm{d}r}{(1 + z)} = \frac{c}{H_0}\frac{\mathrm{d}z}{(\Omega_0 z + 1)^{1/2}(1 + z)^2}. \tag{18.8}$$

Therefore,

$$N(z)\,\mathrm{d}z = \sigma_A N_0 \frac{c}{H_0}\frac{(1 + z)}{(\Omega_0 z + 1)^{1/2}}\,\mathrm{d}z. \tag{18.9}$$

If $\Omega_0 = 1$, $N(z) \propto (1 + z)^{1/2}$ and, if $\Omega_0 = 0$, $N(z) \propto (1 + z)$. Thus, the observed number density of Lyman-α forest absorbers changes more rapidly with increasing redshift than expected according to the uniform absorber model. The sense of the evolution is that there were more Lyman-α forest systems at large redshifts as compared with low redshifts. On the other hand, the Lyman-limit systems seem to show little variation with redshift other than that expected if their cross-sections and comoving number densities remained unchanged with cosmic epoch. The story is much more intriguing, however, when the detailed properties of the clouds are taken into account.

18.3.3 The Abundances of Elements in Lyman-α Absorbers

Fall (1997) has emphasised that one of the great attractions of using absorption lines in Lyman-α absorbers to estimate the relative abundances of any species along the line of sight to distant quasars is that, provided the absorbers are randomly oriented, average relative abundances can be found for the different species, independent of the structures or clumpiness of the clouds. We can readily adapt (18.7) to the case in which the optical depth for absorption has been transformed into a column density N_x for absorption by some species x along an arbitrary line of sight through a cloud. The subscript x might refer to hydrogen atoms (x = HI), metal ions (x = m), or dust grains (x = d). Then, for a line of sight of proper length dr_{prop} through the cloud,

$$N(x)\sigma_x \, dr_{prop} = \sigma_x N_x, \qquad (18.10)$$

where $N(x)$ is the number density of species x in the cloud, σ_x is the absorption, or scattering, cross-section for the species x, which is known from atomic physics, and $N_x = N(x) \, dr_{prop}$ is the column density of species x through the cloud. Thus, knowing the atomic cross-sections, the observations provide direct estimates of the column densities of each species in the clouds intercepted along a particular line of sight.

As illustrated by Fig. 18.3, there will generally be a distribution of column densities $N(N_x) \, dN_x$ for any species, and this is expected to change with cosmic epoch as gas is condensed into stars, and the interstellar gas in galaxies is enriched as a result of nucleosynthesis in stars and the recycling of processed material to the gas. We can determine how the relative abundances of different species change with redshift by the following calculation. Let us relate the total column density of the species x in the redshift interval z to $z + dz$ to the density parameter $\Omega_x(z)$ to which it would correspond at the present epoch. Notice that this will *not* be the density parameter of the species at that epoch – we adopt this procedure simply as a way of comparing abundances at different cosmic epochs in a self-consistent manner.

First of all, we define the function $N(N_x, z) \, dN_x \, dz$ to be the number of absorbers with column densities in the range N_x to $N_x + dN_x$ and in the redshift interval z to $z + dz$. Then, the total column density of the species x in the redshift interval dz is

$$\mathcal{N}_x(z) \, dz = dz \int_0^{\infty} N_x N(N_x, z) \, dN_x \qquad (18.11)$$

This column density can be related to the average *comoving* number density of the species x at redshift z and the interval of proper length dr_{prop} by introducing the density parameter $\Omega_x(z)$ as follows:

$$\Omega_x(z) = \frac{8\pi G m_x N_x(z)}{3H_0^2(1+z)^3} \quad \text{where} \quad \mathcal{N}_x(z) \, dz = N_x(z) \, dr_{prop}, \qquad (18.12)$$

$N_{\rm x}(z)$ is the average number density of the species x at redshift z and $m_{\rm x}$ is the mass of the species x. Therefore, using (18.8), we find

$$\Omega_{\rm x}(z) = \frac{8\pi G m_{\rm x}}{3 H_0 c} \frac{(\Omega_0 z + 1)^{1/2}}{(1 + z)} \mathcal{N}_{\rm x}(z) \qquad (18.13)$$

The beauty of this result is that, if we average over many lines of sight, the linearity of (18.11) and (18.12) mean that we average over all systems in all orientations which contribute to $\Omega_{\rm x}(z)$ at that redshift. Furthermore, we can determine how the global metallicity $Z = \Omega_{\rm m}/\Omega_{\rm g}$ and other relative abundances change with redshift by taking appropriate ratios.

The damped Lyman-α systems are of particular importance in understanding the evolution of neutral gas and the build-up of the heavy elements with cosmic epoch. The studies of Lanzetta *et al.* (1995), Storrie-Lombardi *et al.* (1996) and Pettini *et al.* (1996) are excellent examples of what can now be achieved. It is necessary to take averages over as many different lines of sight as possible and about 80 damped Lyman-α systems are now available for study. The picture which emerges is that, at redshifts $z \sim 3$, the comoving density parameter in neutral hydrogen was $\Omega_{\rm HI} \approx (1 - 2) \times 10^{-3} h^{-1}$ and this decreases with decreasing redshift until, at $z = 0$, its value is only about 2×10^{-4}, a value consistent with independent measures of the amount of neutral hydrogen present in galaxies and their environs at the present epoch (Fig. 18.5a). As Lanzetta *et al.* (1995) remark, although the comoving number density of damped Lyman-α systems does not change markedly with redshift, the column densities themselves are a strong function of redshift. Thus, all the systems at $z \sim 3.5$ have column densities greater than $N_{\rm HI} = 10^{25}$ m^{-2}, whereas only 20% of them exceed this value at $z \approx 1$.

In Fig. 18.5a, the crosses with error bars show the estimates of $\Omega_{\rm HI}$ in damped Lyman-α systems at different redshifts. These estimates have been corrected to take account of the effects of dust extinction (circles) and also for the presence of neutral hydrogen in systems with lower column densities (boxes). It is striking that the density parameter in neutral gas at a redshift $z = 2.5$ is of the same order of magnitude as the density parameter corresponding to the visible mass of galaxies at the present time, which is indicated by the hatched area in Fig. 18.5b. These observations are consistent with a picture in which the discs of spiral galaxies observed today were formed from the damped Lyman-α discs observed at large redshifts.

The big advantage of studying the damped Lyman-α systems is that they have such large column densities that relatively rare species can be used to probe the chemical abundances of the elements. Pettini *et al.* (1996), for example, used observations of singly ionised zinc, Zn^+ or ZnII, which has a number of advantages as a tracer of the overall abundance of the heavy elements. According Pettini *et al.*, zinc shows little affinity for dust and is predominantly in the form of Zn^+ in HI regions. As a result, observations of ZnII are likely to provide robust estimates of the total zinc abundance, independent of the condensation of the heavy elements into dust grains. Although

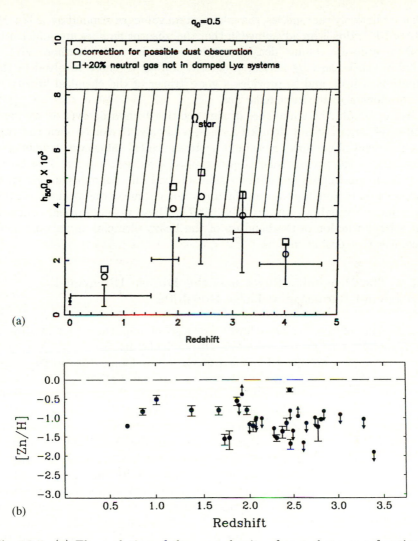

Fig. 18.5. (a) The evolution of the mass density of neutral gas as a function of redshift as determined by the mass density of damped Lyman-α absorbers in the spectra of distant quasars (Storrie-Lombardi *et al* 1996). Note that these authors use h_{50} rather than $h = h_{100}$ used throughout this text. The circles show the estimates of Ω_{HI} corrected for the effects of dust extinction and the boxes corrected values which take account of neutral hydrogen not associated with damped Lyman-α systems. The shaded band indicates estimates of the density parameter in stars, that is, the mass associated with the visible light of galaxies, at the present day. (b) The [Zn/H] abundance in 24 damped Lyman-α systems at large redshifts observed by Pettini *et al.* (1996) relative to the solar value, which is indicated by the dashed line. The usual convention is used of plotting the logarithm of the abundance ratio [Zn/H] on the ordinate.

zinc is a relatively rare species, the solar system value corresponding to [Zn/H] $= 3.8 \times 10^{-8}$, this is an advantage in that the absorption lines are sufficiently weak for accurate column densities to be determined. Furthermore, zinc is probably synthesised by nuclear processes similar to those involved in the formation of iron, and so provides a good tracer of the chemical history of the enrichment of the interstellar gas.

In Fig. 18.5b, the abundance of zinc relative to hydrogen is shown for a number of large redshift damped Lyman-α clouds. It can be seen that the heavy element abundances have built up from typical values which were only about 10% of the present solar abundances at a redshift $z \approx 2$ to $Z \approx Z_\odot$ at the present epoch. Over the same redshift interval, the mean dust-to-gas ratio has increased by roughly the same factor, while the mean dust-to-metals ratio has remained roughly constant (Fall 1997). These results strongly suggest that a large fraction of the build-up of the heavy elements in galaxies took place over the redshift interval $1 < z < 3$.

18.3.4 The Proximity Effect and the Diffuse Ultraviolet Background Radiation at Large Redshifts

Observations of the Lyman-α forest are also of importance for studies of the ultraviolet background radiation. Of particular interest is the *proximity effect*, in which a deficit of Lyman-α absorbers is found close to the emission line redshift of the quasar relative to the expectations of (18.6) (see, for example, Murdoch *et al.* 1986). The lines of the Lyman-α forest are interpreted as clouds which are almost fully ionised by the background of far ultraviolet Lyman continuum radiation. Close to the quasar, in addition to ionisation by intergalactic continuum radiation, the clouds are also ionised by the Lyman continuum radiation of the quasar itself. Since the intensity at any distance from the quasar can be estimated once its luminosity is known, the distance at which the cross-over takes place between ionisation predominantly due to the quasar and to the intergalactic continuum radiation can be found from the proximity effect, and hence estimates of the intensity of the background ultraviolet radiation can be made. If it is assumed that the background emission spectrum has the form $I(\nu) \propto \nu^{-0.5}$, the background intensity at the Lyman limit is roughly 10^{-24} W m^{-2} Hz^{-1} sr^{-1} at redshift $z \sim 2 - 3$, corresponding to $\nu I_\nu \approx 3 \times 10^{-9}$ W m^{-2} sr^{-1}.

The same technique can be used to estimate the flux of intergalactic ionising radiation at small redshifts from observations of low redshift quasars made with the Hubble Space Telescope. Kulkarni and Fall (1993) have tentatively found evidence for the proximity effect in a sample of 13 low redshift quasars observed by Bahcall and his colleagues as part of the Quasar Absorption Line key project. They find a deficit of Lyman-α absorption clouds close to the redshifts of these quasars, their estimate of the intergalactic flux of ionising radiation at a typical redshift $z \sim 0.5$ lying in the range 4×10^{-26}

to 2×10^{-27} W m^{-2} Hz^{-1} sr^{-1} with a best estimate of 6×10^{-27} W m^{-2} Hz^{-1} sr^{-1}.

It is interesting to compare these values with other methods of estimating the local far ultraviolet background radiation. Hα emission has probably been detected from two high velocity neutral hydrogen clouds in the halo of our Galaxy and this provides an upper limit to the local flux of ionising radiation of $I_\nu \leq 2 \times 10^{-25}$ W m^{-2} Hz^{-1} sr^{-1} (Kutyrev and Reynolds 1989, Songaila, Bryant and Cowie 1989). Sunyaev (1968) first proposed using the existence of neutral hydrogen in the peripheries of galaxies to set limits to the flux of intergalactic Lyman continuum radiation. The results depend somewhat upon the assumptions made about the spectrum of the ionising radiation and the thickness of the layer of neutral hydrogen in the galaxies. Bochkarev and Sunyaev (1977), Corbelli and Salpeter (1993) and Maloney (1993) have found values in the range $(1 - 10) \times 10^{-26}$ W m^{-2} Hz^{-1} sr^{-1}. Thus, the flux of ionising radiation at redshifts $z \sim 0.5$ may well be up to two orders of magnitude less than the intensity at redshifts $z \sim 2 - 3$.

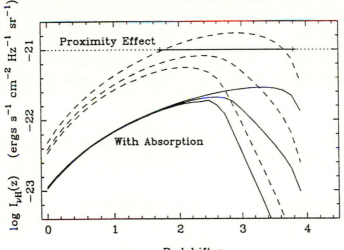

Fig. 18.6. The predicted background radiation at the Lyman limit due to quasars for several different models of their evolution with cosmic epoch (Bajtlik, Duncan and Ostriker 1988). The lower solid curves show the same results but now including attenuation to take account of the absorption by all types of Lyman-α forest and Lyman-limit systems. The background intensity derived from the proximity effect is also shown (Jakobsen 1993).

One of the thornier aspects of this subject has been the origin of these background intensities. One obvious source of ionising photons is the integrated emission of quasars which have non-thermal spectra which extend into the far ultraviolet and X-ray wavebands. The general conclusion of a number of studies is that, to account for the inferred background intensity of about 10^{-24} W m^{-2} Hz^{-1} sr^{-1} at $z \sim 2 - 3$, there must be very strong evolution of the quasar population with cosmic epoch. By pushing all the evolutionary parameters of the models to their limits, Bajtlik, Duncan and Ostriker (1988) were able to account for this background intensity. Their computations assumed, however, that Lyman-continuum absorption in both types of absorption system could be neglected and, when this was incorporated into their computations, Jakobsen (1993) found that the predicted background intensity fell somewhat short (Fig. 18.6). It is not at all clear what the sources of the ionising background are at large redshifts. One of the important uncertainties in these calculations is the extent to which dust extinction associated with the Lyman forest has influenced the statistics of large redshift quasars. Maybe a population of faint quasars at large redshifts can supply the deficit. Another possibility is that the ionising radiation from young galaxies, or protogalaxies, may make a contribution to the background. Steidel and Sargent (1989) have pointed out, however, that the predicted far-ultraviolet spectrum of young galaxies is such that the line ratios observed in damped Lyman-α systems would be quite different from those observed – the line ratios agree well with an ionsing continuum with a power law spectrum $I(\nu) \propto \nu^{-0.5}$.

At low redshifts, the background flux density may well be two orders of magnitude less than that at large redshifts and so the problem of accounting for the background is somewhat less severe. The models due to Miralda-Esudé and Ostriker (1990) and Madau (1992) could readily account for the observed background intensity.

18.4 Star Formation Rates from Optical, Ultraviolet and Submillimetre Observations

The analyses of Sect. 18.3 were entirely concerned with the properties of Lyman-α absorption systems observed in the ultraviolet continuum spectra of quasars. As Fall (1997) has noted, in observing these systems, not a single stellar photon was detected. As he expresses it, that discussion concerned the *absorption history* of the formation of stars and heavy elements in the Universe – we now turn to their *emission history*.

18.4.1 The Cowie and Lilly Argument

Lilly and Cowie (1987) first showed how the rate of formation of heavy elements could be inferred from the flat blue continuum spectra of star-forming galaxies in a manner which is independent of the choice of cosmological model. This argument was discussed in a more general context by Cowie (1988). The analysis begins with the observation that a prolonged burst of star formation has a remarkably flat intensity spectrum at wavelengths longer than the Lyman limit at 91.2 nm. This is illustrated by the model star-bursts synthesised by Bruzual using his spectral synthesis codes for predicting the spectra of galaxies at different phases of their evolution (see White 1989). Figure 18.7 shows the spectrum of a starburst galaxy as observed at different ages, assuming that the star formation rate is constant and that the stars are continuously formed with the same Salpeter mass function. The flatness of the spectrum is due to the fact that, although the most luminous blue stars have short lifetimes, these are constantly being replaced by new stars. To a good approximation, it can be assumed that the spectrum of the star-forming galaxy can be described by a power-law $I(\nu) \propto \nu^{-\alpha}$ with $\alpha = 0$ at wavelengths $\lambda > 91.2$ nm, and zero intensity at shorter wavelengths. Furthermore, the intensity of the flat part of the spectrum is directly proportional to the rate of formation of heavy elements in the starburst since, as discussed in Sect. 18.2, the conversion of hydrogen into helium is the essential first stage in the synthesis of the heavy elements in the central regions of massive stars.

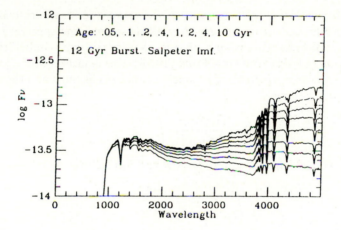

Fig. 18.7. Synthetic spectra for a starburst with constant star-formation rate as observed at the ages indicated. A Salpeter initial stellar mass function $N(M)\,\mathrm{d}M \propto M^{-2.35}\,\mathrm{d}M$ has been assumed with cut-offs at 75 and 0.08 $M_\odot$. The spectra were generated by Gustavo Bruzual using his evolutionary synthesis programmes (White 1989).

From simple physical arguments, the intensity of a star-forming galaxy in the ultraviolet waveband, and hence the rate at which heavy elements are being created, can be determined. The same assumptions made in Sect. 18.2 are adopted, namely, that the total luminosity of the system is $0.007\dot{M}_Z c^2/\alpha_\mathrm{m}$, where $\dot{M}_Z$ is the rate at which heavy elements are synthesised and α_m is the fraction of the synthesised elements returned to the interstellar medium. We now use the result that this radiation has a flat intensity spectrum $\alpha = 0$ up to the Lyman limit ν_{Ly}, so that

$$\int L_\nu \, d\nu = L_\nu \nu_{Ly} = 0.007 \dot{M}_Z c^2 / \alpha_m. \tag{18.14}$$

where L_ν is measured in W Hz^{-1}. This can be rewritten in the form

$$L_\nu = 2 \times 10^{22} \left(\frac{\dot{M}_Z / \alpha_m}{1 M_\odot \text{ year}^{-1}} \right) \quad \text{W Hz}^{-1} \tag{18.15}$$

for all wavelengths longer than the Lyman continuum limit at 91.2 nm. Similar results have been found from detailed modelling of the spectra of star-forming galaxies (see, for example, Meier 1976).

It is now a simple calculation to work out the background intensity of a cosmological distribution of such sources. The integral (17.23) can be used in the form

$$I(\nu_0) = \frac{1}{4\pi} \int_0^\infty \frac{L(\nu_0, z) N_0}{(1+z)^{1+\alpha}} \, dr \tag{18.16}$$

where we have allowed the luminosity of the sources to be an arbitrary function of redshift, and assumed that they all have the same spectral index α. Now $L(\nu_0, z) N_0$ is the *comoving* luminosity density per unit bandwidth due to the formation of heavy elements. From (18.14), this is just $0.007 c^2 \dot{\varrho}_m / \alpha_m \nu_{Ly}$, where $\dot{\varrho}_m$ is the *comoving* rate of formation of heavy elements and the spectral index $\alpha = 0$. Therefore,

$$I(\nu_0) = \frac{0.007 c^2}{4\pi \alpha_m \nu_{Ly}} \int_0^\infty \dot{\varrho}_m \frac{dr}{(1+z)} \tag{18.17}$$

But $dr/(1+z) = c \, dt$ and so the background intensity is

$$I(\nu_0) = \frac{0.007 c^3}{4\pi \alpha_m \nu_{Ly}} \int_0^\infty \dot{\varrho}_m dt = \frac{0.007 c^3}{4\pi \alpha_m \nu_{Ly}} \int_0^\infty \frac{d\varrho_m}{dz} \, dz \tag{18.18}$$

provided the Lyman limit is not redshifted beyond the observing waveband. This is the remarkable result found by Cowie and Lilly – the background intensity due to star-forming galaxies is directly related to the rate at which elements are formed and is *completely independent* of the cosmological model. Cowie (1988) recommended a correction factor of $\alpha_m = 0.67$ to take account of the fraction of metals which is not returned to the general interstellar medium but is locked up in stellar remnants and this results in good agreement with Meier's simulations. Inserting the values of the constants,

$$\Delta I_\nu(z) = 7 \times 10^{-25} \left[\frac{\Delta \varrho_m(z)}{10^{-31} \text{ kg m}^{-3}} \right] \quad \text{W m}^{-2} \text{ Hz}^{-1} \text{ sr}^{-1} \tag{18.19}$$

Notice that the interpretation of (18.19) is that $\Delta I_\nu(z)$ is the observed background intensity of flat spectrum star-forming galaxies in some redshift interval Δz and $\Delta \varrho_m(z)$ is the density of heavy elements observed at the present epoch which were created by these galaxies in that redshift interval. The reference density used in (18.19), 10^{-31} kg m^{-3} of heavy elements, corresponds roughly to $Z = 0.01$ in a Universe in which the density parameter in baryons is $\Omega_B = 0.01$. The beautiful aspect of this result is that the observed intensity of the background radiation originating in a particular redshift interval Δz due to flat spectrum star-forming galaxies determines directly the abundance of heavy elements observed today which were synthesised in that redshift interval.

Cowie, Lilly and their colleagues undertook deep multi-colour surveys to discover flat spectrum star-forming galaxies at large redshifts. Their survey was successful in finding such galaxies, which have roughly equal intensities in the U, B and V wavebands – the background due to such objects amounted to about 10^{-24} W m^{-2} Hz^{-1} sr^{-1}. Cowie et al. (1988) interpreted this result as meaning that a significant fraction of the heavy elements, about 1.5×10^{-31} kg m^{-3}, must have been synthesised at redshifts of about one. This type of analysis has also been used to set limits to star formation rates in the Canada-France Redshift Survey by Lilly et al. (1995) and in the Hubble Deep Field by Madau et al. (1996), which we will discuss later.

18.4.2 The Lyman-break Galaxies

In a remarkable pioneering set of observations, Steidel and his colleagues extended the multicolour technique for finding star-forming galaxies to redshifts $z > 3$. The idea was similar to that developed by Cowie and Lilly, but now the objective was to search for star-forming galaxies in which the Lyman-limit, clearly seen in the models of young galaxies shown in Fig. 18.7, is redshifted into the optical waveband. The technique is illustrated in Fig. 18.8a, in which the spectrum of a star-forming galaxy at redshift $z = 3.15$ is observed through carefully chosen filters in the ultraviolet, blue and red spectral regions. The signature of such a galaxy is that its image should be bright in the two longer wavebands, but should not be present in the ultraviolet waveband. Steidel's original intention was to use this technique to identify the galaxies responsible for the Lyman-limit absorption systems in the spectra of distant quasars and this programme turned out to be remarkably successful (Steidel and Hamilton 1992). It was soon found, however, that the technique was also a remarkably effective means of discovering star-forming galaxies in the general field at $z > 3$ (for a review of recent developments, see Steidel 1997). Observations of the Hubble Deep Field have proved to be ideal for exploiting this approach because, in addition to very precise photometry in four wavebands spanning the wavelength range $300 < \lambda < 900$ nm, high resolution optical images have enabled the morphologies of these galaxies to be studied. HST images in four wavebands of one of the Lyman-break galaxies in the Hubble Deep Field are shown in Fig. 18.8b.

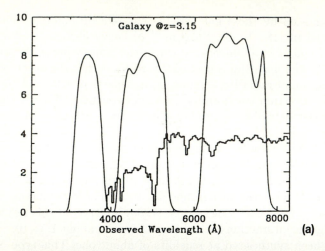

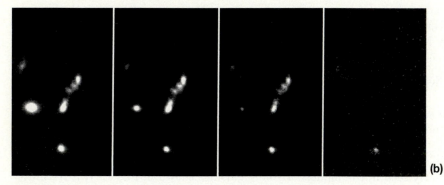

Fig. 18.8. (a) Illustrating how the use of multicolour photometry can be used to isolate star-forming galaxies at redshifts $z > 3$, at which the Lyman-limit is redshifted into the optical region of the spectrum (Steidel 1997). (b) Images of one of the distant star-forming galaxies appearing in the Hubble Deep Field. From left to right, the images were taken at red (I), green (V), blue (B) and ultraviolet (U) wavelengths. Because the Lyman-limit has been redshifted beyond the U waveband, no image of the galaxy appears in the U image (Macchetto and Dickenson 1997).

This technique has been remarkable successful in isolating distant star-forming galaxies. Spectroscopic confirmation of the large redshifts for almost 200 galaxies selected by this technique has been obtained by observations with the Keck Telescope. An example of the redshift distribution found for the galaxies in Steidel's surveys is shown in Fig. 18.9. It can be seen that the redshift distribution more or less follows that expected according to the colour selection criteria. There is a large 'spike' at $z = 3.09$ which is probably associated with the large-scale clustering of galaxies at this redshift. Thus, not only the properties of these star-forming galaxies, but also their three-dimensional spatial distribution can be studied.

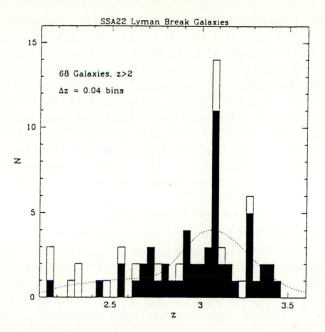

Fig. 18.9. The redshift distribution of Lyman-break galaxies in a single $9' \times 18'$ area of sky, all of which have spectroscopically confirmed redshifts. The differently shaded histograms reflect slightly different selection criteria. The dotted curve shows the expected redshift distribution defined by the colour selection criteria for the complete sample of objects. The 'spike' at $z = 3.09$ is significant at the 99.9% confidence level (Steidel 1997).

These data can be analysed to determine the star formation rates within these galaxies and they have turned out not to be so different from the typical overall star-formation rate in galaxies at the present epoch (see Sect. 18.5 and Fig. 18.13). In addition, it has been possible to learn a great deal about the astrophysical nature of these objects from spectroscopic observations with the Keck Telescope, for example, their internal kinematics, their stellar populations and so on. This project gives some impression of the remarkable astrophysics which will become possible with the coming generation of $8-10$ metre optical-infrared telescopes.

18.4.3 The Hubble Deep Field

The famous picture of the Hubble Deep Field is one of the most startling images in all astronomy. Once the Correction Optics and the second Wide Field-Planetary Camera (WFPC2) were sucessfully installed during the 1993 maintenance and refurbishment mission, the Hubble Space Telescope (HST) recovered its full scientific capability. It is no exaggeration to say that it has exceeded the fondest expectations of even the most optimistic astronomers.

In the light of the remarkable capability of the telescope to obtain deep high-resolution images of distant galaxies, the Director of the Space Telescope Science Institute, Dr. Robert Williams, decided to devote about 2 weeks of Director's discretionary time to the imaging of a single field. In consultation with the astronomical community, an 'empty' region of sky, which could be viewed continuously for 24 hours per day, was selected. The field was observed for over 150 consecutive orbits in four WFPC broad-band filters centred at wavelengths of 300 nm (U), 450 nm (B), 606 nm (R) and 814 nm (I). The image is shown in Fig. 18.10 (Williams *et al.* 1996). Within a month of the field being observed in December 1995, the data were released to the astronomical community and many studies have been made of the objects appearing on the image (Williams 1998).

Some of the more important results can be summarised as follows:

- The Hubble Deep Field is only 2.3 arcmin in angular size. Within this small area of sky, there are at most 9 stars belonging to our own Galaxy. In contrast, there are about 3000 galaxies with surface density 2×10^6 degree^{-2} with apparent magnitudes $V \leq 29$ and these are included in the number counts of galaxies shown in Fig. 2.10. One feature of the image is immediately obvious – there are very large numbers of compact blue galaxies in the field, far in excess of what had been expected. The counts of galaxies in the B, V and I wavebands agree well with those determined from ground-based observations in the magnitude ranges in which they overlap. Specifically, the counts in the Hubble Deep Field extend to about one magnitude fainter than the best ground-based observations in the B waveband, and to about 5 magnitudes fainter in the I waveband, in which the sky noise becomes an increasing problem for the detection of faint objects from the ground. The blue counts show the distinct excess of faint blue galaxies compared with the expectations of uniform world models. At the faintest magnitudes, however, the number counts of faint galaxies converge, about 60% of the integrated background light in the B waveband coming from galaxies with apparent magnitudes less than about B = 24.5 (Madau 1998).

- Unlike samples of bright galaxies, there are large numbers of irregular/interacting systems, comprising about 25% of the objects, compared with less than 1% in samples of galaxies in the local Universe (see Sect. 17.5 and Fig. 17.17).

- Spectroscopic redshifts have now been determined for over 150 objects in the field and these comprise two classes of object. The first consists of complete statistical samples of the brighter galaxies in the field with B < 24.5 and their redshifts extend to just beyond $z = 1$, with a median value of $\langle z \rangle = 0.6$. The spectra of the high surface brightness, compact galaxies in this redshift range have strong emission lines, H & K breaks and blue continua, properties characteristic of star-burst galaxies observed at small redshifts. The second class consists of those objects with blue

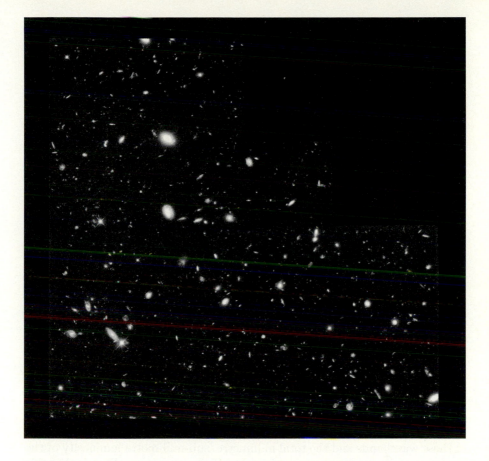

Fig. 18.10 A grey-scale composite image of the Hubble Deep Field, created by combining the exposures in the B (450 nm), V (606 nm) and I (814 nm) wavebands (Williams *et al.* 1996). The image is about 2.7 arcmin in angular size.

colours and U-band drop-outs, which are found to be active star-forming galaxies at large redshifts, exactly the same class of object discovered by Steidel and his colleagues (Sect. 18.4.2). The Lyman continuum drop-out objects are found to possess the most disturbed morphologies and show evidence for interactions.

- Redshifts for the faint galaxies in the field can be estimated from the very precise photometry available from HST observations in the four standard wavebands. Comparison with spectroscopically determined redshifts shows that multicolour photometry can provide excellent redshift estimates. This is particularly important for the faintest galaxies with $B \geq 26$, which are certainly beyond the spectroscopic grasp of even the Keck 10-metre Telescope. Knowledge of the redshifts of these faint galaxies has enabled estimates of heavy element and star-formation rates to be determined using the techniques discussed in Sect. 18.3.

18.4.4 Submillimetre Observations of Star-forming Galaxies

A concern about the interpretation of optical observations of star-forming galaxies is the extent to which their number densities and space distributions have been influenced by the effects of dust extinction. It is well known that galaxies undergoing bursts of star formation are not only sources of intense ultraviolet continuum radiation but are also strong emitters in the far-infrared waveband because of the presence of dust in the star-forming regions. In a study of star-forming galaxies in the Markarian catalogues of ultraviolet-excess galaxies, Mazzarella and Balzano (1986) found that star-forming galaxies are on average stronger emitters in the far-infrared than in the ultraviolet waveband. Similarly, in a sample of star-forming galaxies studied by the IUE, Weedman (1994) found that most of the galaxies emit much more of their luminosity in the far-infrared rather than in the ultraviolet region of the spectrum. These observations have two consequences. First of all, some of the star-forming galaxies may well be obscured by dust and so not be present in optical-ultraviolet multicolour surveys. Secondly, it is quite possible that a significant fraction of the radiation associated with the formation of the heavy elements was not radiated in the ultraviolet-optical region of the spectrum, but at far-infrared wavelengths.

One way of estimating the significance of such obscuration is to make observations of the distant Universe in the millimetre and sub-millimetre wavebands. Dust associated with star-forming regions is an intense emitter in these wavebands and the total millimetre/sub-millimetre luminosity of the galaxy is also a direct measure of the star formation rate. What makes this approach feasible observationally is the fact that the spectrum of dust in these spectral regions is strongly 'inverted'. As discussed in Sect. 17.2.2, the resulting K-corrections are so large and negative that a typical star-forming galaxy is expected to have essentially the same flux density, whatever its redshift in the range $1 < z < 10$ (see Fig. 17.3b and Blain and Longair 1993).

Deep surveys in these wavebands have only recently become feasible as a result of the construction of instruments such as the sub-millimetre bolometer array receiver SCUBA operating on the James Clerk Maxwell Telescope. Smail, Ivison and Blain (1997) have made the first deep submillimetre surveys in the fields of two clusters of galaxies with SCUBA and have discovered a large population of faint background submillimetre sources. Although only six sources were discovered in the two fields, their number density on the sky indicates that the population of these galaxies has evolved strongly with cosmic epoch (Fig. 18.11). The number counts of these sources suggest that there may be more star-forming galaxies at large redshifts than indicated by the numbers of faint blue galaxies in the Hubble Deep Field. These are important new results and open up yet another way of probing the evolution of galaxies over the critical redshift interval $1 < z < 10$.

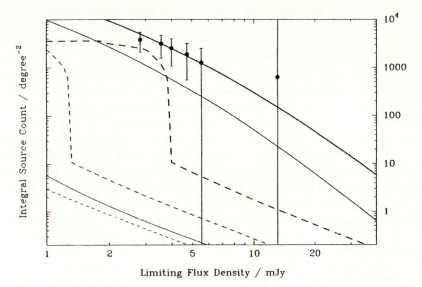

Fig. 18.11. The counts of submillimetre sources from studies of the fields of two rich clusters of galaxies. All the sources are distant background objects and their flux densities have been corrected for the effects of gravitational lensing by the foreground clusters. The solid lines show models for the counts of submillimetre sources based upon the local IRAS luminosity function of far-infrared sources. To account for the observed numbers of sources, the source population must evolve strongly with cosmic epoch. Adopting a model including luminosity evolution, the luminosities would have to evolve as $L \propto (1+z)^3$ out to $z = 2.7$ and then remain constant in the redshift interval $2.7 \leq z \leq 7$. The dashed lines show the expected counts if a population of sources of space density 0.6×10^{-4} Mpc^{-3} at a temperature of 60 K filled the Universe out to $z = 10$ with star-formation rates of 20, 50 and 150 $M_\odot$ yr^{-1} (Smail, Ivison and Blain 1997).

18.5 Putting It All Together – the Equations of Cosmic Chemical Evolution

Fall and Pei (1993) have developed a simple formalism for interpreting the emission and absorption histories of star formation and the build-up of the chemical elements with cosmic epoch. This is based upon what they call the *equations of cosmic chemical evolution.* These are similar to those originally derived by Tinsley (1980) for the chemical evolution of galaxies. Many more details of these types of calculation and of the physics of the chemical evolution of stars and galaxies are given by Pagel (1997).

In the approach adopted by Fall and Pei, all mention of galaxies disappears, all that remains being the global averages of the density parameters for the major constituents of the Universe. At each stage, reasonable simplifications need to be made, but this is in the spirit of the approach, which is to provide a framework for star and metal formation in galaxies, viewed globally. The density parameter in stars Ω_s is zero when star formation begins and builds up to the mean value $\Omega_s(t_0) \approx (4-8) \times 10^{-3}$ by the present epoch. At the same time, the density parameter in gas Ω_g initially comprised 100% of the baryonic matter in galaxies and has decreased to only about 5% of its initial value by the present epoch. The density parameter in heavy elements Ω_m was initially zero and has built up to about 1% of the baryonic mass by the present epoch. Finally, dust was also formed as the abundance of the heavy elements built up. A typical figure would be that about 50% of the heavy elements in the interstellar media of galaxies is in the form of dust. Allowance can also be made for the infall of unprocessed material into galaxies from the intergalactic medium, and also for the expulsion of processed material from galaxies through supernova explosions and galactic winds.

The objective of the calculation is to relate the absorption history, which provides a series of snapshots of the values of the various density parameters at different epochs, to the emission history, which generally provides information about the rates of formation of stars and heavy elements, that is, the derivatives of their density parameters with respect to cosmic time.

The first equation describes the conservation of mass

$$\frac{d}{dt}(\Omega_g + \Omega_s) = \dot{\Omega}_f, \tag{18.20}$$

where $\dot{\Omega}_f$ is the rate of infall into, or of the expulsion of baryonic matter from, galaxies. If this term is zero, (18.20) states that $\dot{\Omega}_g = -\dot{\Omega}_s$, that is, the rate at which gas is depleted is equal to the rate at which mass is condensed into stars.

The second equation describes the rate at which the mass of heavy elements changes with time. The metallicity of the gas at any epoch is defined to be $Z = \Omega_m/\Omega_g$ and so we can write the rate of change of Ω_m as

$$\frac{d\Omega_m}{dt} = \frac{d}{dt}(Z\Omega_g) = y\frac{d\Omega_s}{dt} - Z\frac{d\Omega_s}{dt} + Z_f\dot{\Omega}_f. \tag{18.21}$$

The terms on the right-hand side need some explanation. The first term, $y\,d\Omega_s/dt$, describes the rate of increase of the mass of heavy elements associated with the rate of star formation. The quantity y is called the *yield*. Written in this way, the term involves the assumption of *instantaneous recycling* of processed material to the interstellar medium. In reality, there is bound to be some delay between the formation of a generation of stars, the completion of their evolution and the recycling of processed material to the interstellar gas. The rationale behind the form of this term is that most of the heavy element formation is associated with massive stars, which have lifetimes short compared to the typical time-scales over which the average abundances change significantly (see Sect. 18.2). This term means that, in the time dt, a mass of stars $d\Omega_s$ is formed per unit comoving volume and these eventually return a mass of heavy elements $y\,d\Omega_s$, to the interstellar medium. In the spirit of this analysis, the yield y is assumed to be independent of cosmic epoch.

The second term on the right-hand side of (18.21) describes the loss of heavy elements because of the formation of stars from gas which has already attained a metallicity Z. In the time dt, the loss of heavy elements per unit comoving volume is $-d\Omega_m = -Z\,d\Omega_s$. The third term on the right-hand side represents enhancement of the heavy element abundance by the infall of baryonic material from intergalactic space, assumed to have metallicity Z_f. This term takes account of the fact that the primordial gas might have been enriched by early generations of star-formation.

Fall and Pei (1993) give some simple illustrative examples of the solutions of (18.20) and (18.21).

Closed Box Model. In this example, there is no infall of matter from intergalactic space, nor is gas expelled from the galaxies, $\dot{\Omega}_f = 0$. It is straightforward to show that

$$Z = -y\ln\left[\frac{\Omega_g(z)}{\Omega_g(\infty)}\right] \qquad (18.22)$$

where $\Omega_g(\infty)$ is the initial density parameter in baryons.

Inflow Model. In the simplest picture, the infalling matter consists of unenriched primordial gas, $Z_f = 0$. Some simplifying assumptions need to be made about $\dot{\Omega}_f$. Pei and Fall (1995) adopted a model in which $\dot{\Omega}_f = \nu\dot{\Omega}_s$, that is, the rate of infall of mass is regulated by the rate at which matter is being converted into stars. This is a plausible picture for the discs of spiral galaxies, in which infall of intergalactic material replenishes gas which is continuously being converted into stars. Larsen (1972) noted that, in the case of our own Galaxy, the rate of infall of mass is more or less the same as the rate at which mass is being converted into stars in regions of star formation in the disc of our Galaxy. This picture can account for the fact that the interstellar gas in the disc has not all been converted into stars. Again, there is a simple analytic solution:

$$Z = \frac{y}{\nu}\left\{1 - \left[\frac{\Omega_g}{\Omega_g(\infty)}\right]^{\nu/(1-\nu)}\right\} \qquad (18.23)$$

Outflow Model. The formalism employed in the infall model can be adapted to this case, the assumption being that the rate of loss of mass from a galaxy is proportional to the supernova rate, which, in the instantaneous recycling picture, is proportional to the rate at which stars are formed. This type of outflow picture was proposed by Hartwick (1976) to account for the low abundance of iron in halo stars. The mathematical difference as compared with the inflow model is that, not only is mass lost from galaxies $\dot{\Omega}_f = -\nu\dot{\Omega}_s$, but also heavy elements are lost with $Z_f = Z$. This model also has a simple analytic solution

$$Z = -\frac{y}{(1+\nu)} \ln\left[\frac{\Omega_g}{\Omega_g(\infty)}\right] \qquad (18.24)$$

It should be emphasised that these are simply convenient analytic models for understanding the build-up of the heavy elements in galaxies. All three have the useful feature that the models are completely defined once the evolution of the gaseous content of galaxies as a function of cosmic epoch is prescribed. The type of data required was illustrated in Fig. 18.5a (Storrie-Lombardi *et al.* 1996). The problem is that these numbers refer to the *observed*, rather than the *true*, value of the density parameter of neutral hydrogen in damped Lyman-α systems. The true variation of the density parameter with redshift depends upon correcting for those quasars which have been excluded from the complete samples because of dust obscuration by the damped Lyman-α systems themselves. These corrections have a strong impact upon models of the star formation rate as a function of cosmic epoch, as indicated by the analyses of Fall and Pei (1993) and Pei and Fall (1995).

To illustrate this important point, the results of the calculations of Pei and Fall (1995) for the $\Omega_0 = 1$ model are shown in Fig. 18.12. These models repay some study – for the sake of illustration, let us analyse the case of the closed box model. Pei and Fall fitted the *observed* variation of Ω_{HI} with redshift by the function

$$\Omega_g(z) = \Omega_g(\infty)\, x \quad \text{where} \quad x = \frac{\exp z}{\exp z + 19}, \qquad (18.25)$$

which is shown by the dashed line in Fig. 18.12b. Equations (18.20) and (18.21) can be used to show that the mean metallicity Z and the star formation rate change with redshift as

$$\frac{Z}{Z_0} = \frac{\ln(\exp z + 19) - z}{\ln 20}. \qquad (18.26)$$

$$\dot{\Omega}_s(z) = -\dot{\Omega}_g(z) = -\frac{d\Omega_g}{dt} = -\frac{d\Omega_g}{dz}\frac{dz}{dt}$$
$$= \Omega_g(\infty)\, H_0\, \frac{19\exp z}{(\exp z + 19)^2}(1+z)^{5/2} \qquad (18.27)$$

where Z_0 is the metallicity of the interstellar media in galaxies at the present epoch. These formulae are not particularly elegant, but they describe precisely the dashed curves labelled C in Figs. 18.12c and d. This model is in poor agreement with the observation that, at redshift $z = 2$, the mean metallicity is only 10% of its present value and only about 1% at $z \sim 4$. Furthermore, according to this model, the maximum star-formation rate occurred at a redshift of about 4 and has been declining since that epoch.

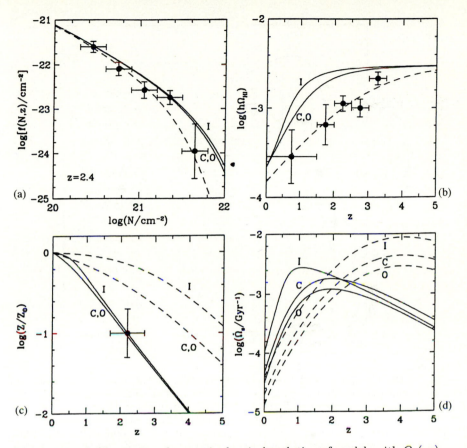

Fig. 18.12a-d. Illustrating the cosmic chemical evolution of models with $\Omega_g(\infty) = 4 \times 10^{-3}h^{-1}$, $\nu = 0.5$ and $\Omega_0 = 1$. (a) The distribution of HI column densities at $z = 2.4$; (b) the comoving density of neutral hydrogen as a function of redshift z; (c) the mean metallicity Z of the interstellar gas relative to its present value Z_0 as a function of redshift; (d) the predicted comoving rate of star formation as a function of redshift. The solid curves represent the true values for the three models discussed in the text: C – closed box model, I – inflow model, O – outflow model. The dashed curves in (a) and (b) represent the corresponding observed quantitites, while the dashed curves in (c) and (d) show the effects of neglecting dust extinction in each of the models. The data points in (a), (b) and (c) are from Lanzetta *et al.* (1991), Lanzetta *et al.* (1995) and Pettini *et al.* (1994), respectively.

The picture changes dramatically when the effects of incompleteness due to dust obscuration of distant quasars is taken into account. The detailed analyses of Fall and Pei (1993) indicate the complications involved in making accurate corrections for dust extinction, which are particularly important in the redshift interval $2 < z < 3$. There is direct observational evidence for reddening in the spectra of distant quasars and so the objective is to find a self-consistent picture for the dust extinction, which is also consistent with the variation of the abundances of the heavy elements with cosmic epoch. The

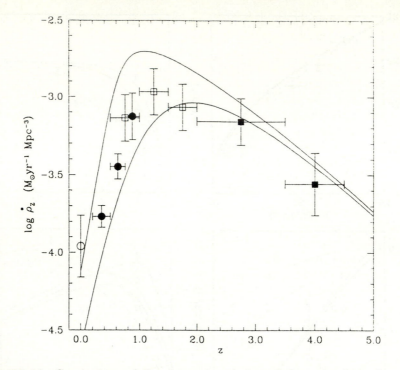

Fig. 18.13. Comparison of the observed variation of the rate of star-formation with cosmic epoch with that predicted on the basis of the absorption history of the interstellar gas in galaxies (Fall 1998). The observed star-formation rates are taken from Gallego *et al.* (1995) at the present epoch, Lilly *et al.* (1995) who used 280 nm continuum emission of star-forming galaxies at $z < 1$, Connally *et al.* (1997) who used the Hα emission of star-forming galaxies at $1 < z < 2$, and Madau *et al.* (1996, 1997), who used the Lyman-limit galaxies and those observed in the Hubble Deep Field to set limits to the star-formation rate at redshifts $z > 3$. Both the observed rates and the model predictions have been corrected in a self-consistent manner for dust extinction.

solid lines in Fig. 18.12a and b show their best estimates of the corrected column density distribution and the variation of the density parameter in neutral hydrogen with redshift. The key point is that the equations of cosmic chemical evolution relate the star-formation rate to the derivative of the density parameter in neutral hydrogen with respect to cosmic time (18.20). The effect of these corrections to the variation of the density parameter in neutral gas with cosmic epoch is to delay the peak star-formation rate to lower redshifts, typically to redshifts $z \sim 1-2$, and to predict a roughly exponential build up of the heavy elements with decreasing redshift (Fig. 18.12c and d). It is left as an exercise to the reader to run the calculation backwards by assuming that the build-up of the heavy elements is exponential with redshift, say, $Z = Z_0 \exp(z/2)$, and to reconstruct from that function the variations of the density parameter in neutral hydrogen and the star-formation rate with cosmic epoch. It can be seen from Fig. 18.12b that the reason for the slower build-up of the heavy elements and the star formation rate with cosmic epoch is the flattening of the relation between the density parameter in neutral hydrogen and redshift in the redshift interval $1.5 < z < 4$. As noted above, the predicted star formation rate depends upon the gradient of this relation.

As Fall (1997) has pointed out, the study of the absorption history of the chemical enrichment of the interstellar gas enables the emission history to be predicted. His latest comparison of the predicted star-formation rate as a function of redshift with a number of independent estimates of the observed star formation rate, derived by the techniques discussed in Sect. 18.3, is shown in Fig. 18.13 (Fall 1998). It is evident that successful self-consistent models can be found, their characteristic being that the greatest rates of star-formation occurred at redshifts $z \sim 1 - 2$.

These important calculations show how a wide range of different observations can now be brought to bear upon the problem of defining the rate at which stars and the heavy elements were built up in galaxies. A number of points should be noted. First of all, the calculations refer to global averages of the history of star-formation and would not necessarily be expected to describe the properties of any particular galaxy or class of galaxies. Second, notice the sensitivity of the inferred star-formation rate to the corrections for incompleteness in the quasar samples. It would not require large changes in the extinction corections to accommodate the larger star-formation rates at redshifts $z \sim 3$ inferred from the submillimetre counts of galaxies (Sect. 18.4.4). Third, the precision of these studies is limited by the available statistics of damped Lyman-α absorption systems. Fourth, the story can be taken further to evaluate the submillimetre emission of the dust responsible for the obscuration of distant quasars. Fall et al. (1996) have shown how their estimates of the amount of dust in these systems is consistent with the present limits to the background emission observed in the submillimetre waveband. These studies will unquestionably be the subject of intensive programmes with the coming generation of 8-metre telescopes.

19 Diffuse Intergalactic Gas

19.1 Introduction

The physical properties of the diffuse intergalactic gas is one of the more tantalising problems of modern cosmology. It is certain that there is diffuse intergalactic gas in rich clusters of galaxies as is indicated by the observation of intense X-ray bremsstrahlung at a temperature of about 10^8 K from clusters such as the Coma and Perseus clusters (Sect. 4.3.2). In such clusters, the mass of intergalactic gas is at least as great as the mass in the visible parts of the cluster galaxies. Observations of the emission lines of highly ionised iron show that the heavy elements are underabundant relative to the cosmic abundances of the elements as a whole. It is inferred that the primordial gas has been enriched by the processes of nucleosynthesis in galaxies.

It is much more difficult to study the diffuse intergalactic gas between clusters of galaxies. These is no definite evidence that there is any diffuse gas at all between the clusters of galaxies, but there are good reasons to suppose that there must be some gas present. For example, there must be some means of confining the hydrogen clouds which make up the Lyman-α forest. It also seems unlikely that the process of galaxy formation was so efficient that all the diffuse intergalactic gas was condensed into galaxies. Furthermore, galactic winds and explosions are capable of expelling interstellar gas well beyond the confines of galaxies.

Let us begin by deriving expressions for the background emission from diffuse intergalactic gas and its optical depth for the absorption and scattering of the emission of background sources.

19.2 The Background Emission of and Absorption by the Intergalactic Gas

In Sect. 17.6.2, it was shown that the background intensity of a uniform distribution of sources can be written

$$I(\nu_0) = \frac{1}{4\pi} \int_0^\infty \frac{L(\nu_1, z)N_0}{(1 + z)} \, \mathrm{d}r, \qquad (19.1)$$

where it is assumed that the comoving number density of sources N_0 is conserved, but the luminosities of the sources have been allowed to change with cosmic epoch. To adapt this expression to the case of the emission of diffuse intergalactic gas, it is simplest to work in terms of the *proper* number density of objects at redshift z, $N(z) = N_0(1+z)^3$. Then, the luminosity per unit proper volume, that is, the *emissivity* of the intergalactic medium, is $\varepsilon(\nu_1) = L(\nu_1, z)N(z)$, and the background intensity is

$$I(\nu_0) = \frac{1}{4\pi} \int_0^\infty \frac{\varepsilon(\nu_1)}{(1+z)^4}\, \mathrm{d}r. \qquad (19.2)$$

Notice that we can obtain the same result if we work in terms of the change of surface brightness per unit frequency with redshift $I(\nu_0) = I(\nu_1)/(1+z)^3$. If we adopt the Friedman models with $\Omega_\Lambda = 0$, we find

$$I(\nu_0) = \frac{c}{4\pi H_0} \int_0^\infty \frac{\varepsilon(\nu_1)}{(\Omega_0 z + 1)^{1/2}(1+z)^5}\, \mathrm{d}z, \qquad (19.3)$$

where $\nu_0 = \nu_1/(1+z)$.

In exactly the same way, we can work out the optical depth of the gas at an observed frequency ν_0 due to absorption by intergalactic matter along the line of sight to redshift z. If $\alpha(\nu_1)$ is the absorption coefficient for radiation at frequency ν_1, then the increment of optical depth for the photons which will eventually be redshifted to frequency ν_0 by the time they reach the Earth is $\mathrm{d}\tau(\nu_0) = \alpha(\nu_1)\,\mathrm{d}l = \alpha(\nu_1)c\,\mathrm{d}t$, where $c\,\mathrm{d}t = \mathrm{d}l$ is the element of proper distance at redshift z. Hence, integrating along the path of the photon, we find

$$\tau(\nu_0) = \int \alpha(\nu_1)\,\mathrm{d}l = \int \alpha(\nu_1)\frac{\mathrm{d}r}{1+z}$$
$$= \frac{c}{H_0} \int_0^z \frac{\alpha(\nu_1)\,\mathrm{d}z}{(\Omega_0 z + 1)^{1/2}(1+z)^2}, \qquad (19.4)$$

where $\nu_0 = \nu_1/(1+z)$. Notice that, in the case of an absorption line, the function $\alpha(\nu_1)$ describes its line profile.

19.3 The Gunn–Peterson Test

One of the most important tests for the presence of intergalactic neutral hydrogen was described independently by Gunn and Peterson (1965) and by Scheuer (1965), soon after the first quasar with redshift greater than 2, 3C9, was discovered. I am particularly fond of 3C9 since I made the optical identification of the quasar as almost my first task for Martin Ryle when I joined the Cambridge Radio Astronomy Group in 1963 (see Ryle and Sandage 1964, Longair 1965). The particular aspect of quasars which is used in the test is the fact that their continuum spectra are non-thermal and extend into the far ultraviolet and X-ray wavebands.

The Gunn–Peterson test makes use of the fact that the cross-section of the Lyman-α transition at 121.6 nm is very large and so, when the ultraviolet continuum of distant quasars is shifted to the redshift at which it has wavelength 121.6 nm, the radiation is absorbed and re-emitted in some random direction many times so that, if there is sufficient neutral hydrogen present at these redshifts, an absorption trough is observed to the short wavelength side of the redshifted Lyman-α line. Only when quasars with redshifts $z > 2$ were discovered was the Lyman-α wavelength redshifted into the observable visible waveband. Let us carry out a simple calculation to illustrate how (19.4) can be used to work out the optical depth for Lyman-α scattering.

The expression for the photo-excitation cross-section of the Lyman-α transition is

$$\sigma(\nu) = \frac{e^2 f}{4\varepsilon_0 m_e c} g(\nu - \nu_{Ly}), \tag{19.5}$$

where ν_{Ly} is the frequency of the Lyman-α transition, f is its oscillator strength, which is 0.416 for this transition, and the function $g(\nu - \nu_{Ly})$ describes the profile of the Lyman-α absorption line. In this form, the function $g(\nu)$ has been normalised so that $\int g(\nu)\,d\nu = 1$. Therefore, inserting $\alpha(\nu_1) = \sigma(\nu_1)N_H(z)$ into (19.4), we find that the optical depth due to Lyman-α scattering is

$$\begin{aligned}
\tau(\nu_0) &= \frac{c}{H_0} \int_0^z \frac{\sigma(\nu_1)N_H(z)\,dz}{(\Omega_0 z + 1)^{1/2}(1+z)^2} \\
&= \frac{e^2 f}{4\varepsilon_0 m_e H_0} \int_{\nu_0}^{\nu_1} \frac{N_H(z)g[\nu_0(1+z) - \nu_{Ly}]}{\nu_0(\Omega_0 z + 1)^{1/2}(1+z)^2}\,d[\nu_0(1+z)], \quad (19.6)
\end{aligned}$$

where $\nu_1 = \nu_0(1 + z_{max})$. Since $g(\nu)$ is very sharply peaked at the wavelength of the Lyman-α line, we can approximate it by a delta function and then

$$\tau(\nu_0) = \frac{e^2 f}{4\varepsilon_0 m_e H_0 \nu_{Ly}} \frac{N_H(z)}{(\Omega_0 z + 1)^{1/2}(1+z)}. \tag{19.7}$$

Inserting the values of the constants, we find

$$\tau(\nu_0) = 4 \times 10^4\, h^{-1} \frac{N_H(z)}{(\Omega_0 z + 1)^{1/2}(1+z)}. \tag{19.8}$$

This continuum absorption feature has been searched for in those quasars which have such large redshifts that the Lyman-α line is redshifted into the observable optical waveband, that is, quasars with redshifts $1 + z \geq (330\,\text{nm})/\lambda_{Ly}$, $z \geq 2$. There is no such redshift restriction for observations with the Hubble Space Telescope. The typical spectrum of a large redshift quasar is shown in Fig. 19.1. It can be seen that there is no evidence for a depression to the short wavelength side of either Lyman-α at 121.6 nm, nor to the short wavelength side of the corresponding line of neutral helium, HeI, which has rest wavelength 58.4 nm. Typically, the upper limit to the optical depth to the short wavelength side of Lyman-α is $\tau(\nu_0) \leq 0.1$. Substituting this value into (19.8), we find that, for a quasar at a redshift of 3, the upper limit to the number density of neutral hydrogen atoms at that redshift is $N_H \leq 10^{-5}$ m^{-3}. This is a very small value indeed compared to typical cosmological baryonic densities, which are about $10\Omega_B h^2 (1+z)^3$ m$^{-3} \approx 6$ m^{-3}, assuming $z = 3$ and $\Omega_B = 0.01$. Thus, if there is a signficant amount of hydrogen in the intergalactic medium, it must be very highly ionised.

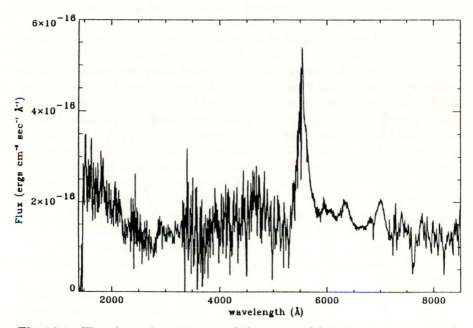

Fig. 19.1. The ultraviolet spectrum of the quasar OQ 172, which has redshift $z = 3.544$, observed by the Faint Object Spectrograph of the Hubble Space Telescope and at the Lick Observatory (Beaver *et al* 1993). Although there are lines present associated with the Lyman-α forest, there is no depression of the continuum intensity to the short wavelength side of either the Lyman-α line at 121.6 nm or the HeI line at 58.4 nm.

Searches have been made for the Lyman-α trough in low redshift quasars which can be observed beyond the redshifted Lyman-α line from space by Davidsen (1993) and from the HST, but there is little evidence for any absorption at all. It is inferred that, even at the present epoch, the intergalactic gas must contain very little diffuse neutral hydrogen.

19.4 The X-ray Thermal Bremsstrahlung of Hot Intergalactic Gas

If the diffuse intergalactic gas is ionised, one possibility is that it is very hot and, for some years, there was the possibility that the isotropic X-ray background emission could be the X-ray bremsstrahlung of this gas. It was pointed out by Marshall *et al.* (1980) that the spectrum of the X-ray background emission in the 1 – 50 keV energy band can be very well described by the spectrum of thermal bremsstrahlung at a temperature of $k_\mathrm{B}T = 40$ keV (Fig. 19.2). If the emission originated from hot diffuse gas at redshift z, the temperature of this gas would be $k_\mathrm{B}T = 40(1 + z)$ keV.

There are, however, a number of problems with this seemingly simple explanation of the origin of the X-ray background emission. First of all, the number density of the hot gas can be evaluated using the formulae for the emissivity of the gas (4.18 and 4.19) and it is found that the corresponding density parameter in baryons would be $\Omega_\mathrm{B}h^2 \geq 0.23$. This value is significantly greater than the upper limit derived from studies of primordial nucleosynthesis, $\Omega_\mathrm{B}h^2 \leq 0.036$ (Sect. 10.6). Second, in this picture, it is natural to attribute the high degree of ionisation of the gas to the same process responsible for heating the gas to a high temperature. Since the heating would have to take place at a large redshift, $z \geq 3$, the energy requirements for heating the gas would be very great. For example, the thermal energy density of the hot gas, if it had $\Omega_\mathrm{B} = 0.23$ and temperature $k_\mathrm{B}T = 160$ keV at a redshift $z = 3$, would be 7.6×10^7 eV m^{-3}, which is the same as the energy density of the Cosmic Microwave Background Radiation at that redshift. To put it in perspective, this energy corresponds to a density parameter $\Omega = 3.3 \times 10^{-3}$ at that redshift.

There is, however, a further serious concern. Such large quantities of hot gas would give rise to distortions of the Planck spectrum of the Cosmic Microwave Background Radiation. As first discussed by Sunyaev and Zeldovich, Compton scattering of the photons of the background radiation leads to a characteristic distortion of the spectrum of the background radiation, by redistributing the photon energies. The perfect Planck spectrum of the background radiation as observed by COBE enabled Mather (1995) to set a powerful constraint to the Compton optical depth of hot diffuse intergalactic gas

$$y = \int \frac{k_\mathrm{B}T_\mathrm{e}(z)}{m_\mathrm{e}c^2} \frac{\sigma_\mathrm{T}N_\mathrm{e}(z)}{(1+z)} \, dr \leq 2.5 \times 10^{-5}. \tag{19.9}$$

The most conservative estimate we can make is to assume that the hot gas fills the intergalactic medium at the present epoch, in which case, we can make the approximation

$$y \approx \frac{c}{H_0} \sigma_\mathrm{T}N_\mathrm{e}(0) \frac{k_\mathrm{B}T_\mathrm{e}(0)}{m_\mathrm{e}c^2}. \tag{19.10}$$

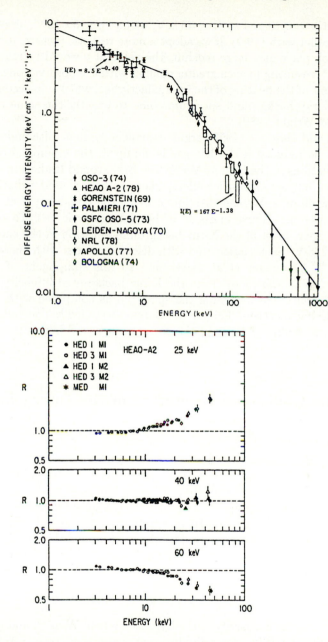

Fig. 19.2. Fits of bremsstrahling spectra to the spectrum of the diffuse X-ray background in the energy range $1 < \varepsilon < 50$ keV. Adopting a temperature $k_B T = 40$ keV results in an excellent fit to the observations (Marshall *et al.* 1980).

Assuming $\Omega_B = 0.23$ and $k_B T_e = 40$ keV, we find $y = 5 \times 10^{-4}$, already in significant conflict with (19.9). If we adopt a more realistic picture, in which the heating took place at a large redshift, the value of y would increase roughly as $(1+z)^{2.5}$, resulting in even greater conflict with the observations. More detailed models of the heating of the intergalactic gas, which can account for the observed X-ray background spectrum come to exactly the same conclusion (Taylor and Wright 1989).

Fabian and Barcons (1992) argue convincingly that, although the value of Ω_B can be reduced if the hot gas is clumped, the clumps also result in large fluctuations in the Cosmic Microwave Background Radiation because of the thermal Sunyaev–Zeldovich effect, when the background radiation passes through the clumps. We can therefore exclude the possibility that the X-ray background is the diffuse emission of hot intergalactic gas. To complicate matters further, most of the X-ray background emission in the energy range $0.5 < \varepsilon < 2$ keV can be associated with discrete sources with energy spectra $I(\varepsilon) \propto \varepsilon^{-0.7}$ (Hasinger et al. 1993), and so this component of the background has to be subtracted from the X-ray background intensity at 1 keV. This has the effect of flattening the spectrum as compared with a thermal bremsstrahlung spectrum. Fabian and Barcons were led to conclude that 'the perfect bremsstrahlung shape of the X-ray background is just a cosmic conspiracy'.

19.5 The Collisional Excitation of the Intergalactic Gas

Although the arguments of Sect. 19.3 rule out a high temperature, high density intergalactic gas, it is entirely possible that collisional excitation by shocks could be responsible for its heating and ionisation. If its temperature lay in the appropriate temperature range, the gas would be expected to be an emitter of collisionally excited HI and HeII Ly-α line emission at $\lambda = 121.6$ nm and $\lambda = 30.4$ nm respectively. The observed intensity of the lines is given by (19.3), in which the emissivity of the Lyman-α line per unit frequency interval can be written

$$\varepsilon(\nu_1) = \varepsilon_{Ly}(z)\, g(\nu_1 - \nu_{Ly}), \tag{19.11}$$

where $\varepsilon_{Ly}(z)$ is the emissivity of the Lyman-α line in W m^{-3}, and $g(\nu_1 - \nu_{Ly})$ describes its line profile, normalised so that $\int g(\nu_1 - \nu_{Ly})\, d\nu_1 = 1$. Therefore, the intensity of the line per unit bandwidth is $\varepsilon_{Ly}(z)\, g(\nu_1 - \nu_{Ly})$. The intensity of background radiation associated with the Lyman-α emission of the intergalactic gas is therefore

$$I(\nu_0) = \frac{c}{4\pi H_0} \int_0^\infty \frac{\varepsilon_{Ly}(z) g(\nu_1 - \nu_{Ly})}{(\Omega_0 z + 1)^{1/2}(1 + z)^5}\, dz. \tag{19.12}$$

As in Sect. 19.2, we can take $g(\nu_1 - \nu_{\mathrm{Ly}})$ to be a δ-function, in which case

$$
\begin{aligned}
I(\nu_0) &= \frac{c}{4\pi H_0} \int_0^\infty \frac{\varepsilon_{\mathrm{Ly}}(z)\delta[\nu_0(1+z)]}{(\Omega_0 z + 1)^{1/2}(1+z)^5}\, \mathrm{d}z \\
&= \frac{c}{4\pi H_0} \frac{\varepsilon_{\mathrm{Ly}}(z)}{\nu_{\mathrm{Ly}}(\Omega_0 z + 1)^{1/2}(1+z)^4},
\end{aligned}
\tag{19.13}
$$

where $\nu_1 = \nu_{\mathrm{Ly}} = \nu_0(1+z)$. Notice that, in this calculation, the intensities are in units of W m^{-2} Hz^{-1} sr^{-1}. In ultraviolet astronomy, it is common to quote intensites in terms of the numbers of photons s^{-1} m^{-2} Hz^{-1} sr^{-1} or of photons s^{-1} m^{-2} nm^{-1} sr^{-1}. To transform (19.13) into these units, we note that $I(\nu_0) = h\nu_0\, I_{\mathrm{ph}}(\nu_0)$ and $\varepsilon_{\mathrm{Ly}}(z) = h\nu_{\mathrm{Ly}}\, N_{\mathrm{Ly}}(z)$ where $I(\nu_0)$ is the observed intensity in photons per unit frequency interval and $N_{\mathrm{Ly}}(z)$ is the emissivity of the intergalactic gas at redshift z in photons m^{-3}. It follows that

$$
I_{\mathrm{ph}}(\nu_0) = \frac{c}{4\pi H_0} \frac{N_{\mathrm{Ly}}(z)}{\nu_{\mathrm{Ly}}(\Omega_0 z + 1)^{1/2}(1+z)^3}.
\tag{19.14}
$$

Alternatively, in terms of photons per unit wavelength interval, we can write $I_{\mathrm{ph}}(\lambda_0) = I_{\mathrm{ph}}(\nu_0)(\mathrm{d}\nu_0/\mathrm{d}\lambda_0)$ and so, recalling that $\lambda_0 = \lambda_{\mathrm{Ly}}(1+z)$, we find

$$
I_{\mathrm{ph}}(\lambda_0) = \frac{c}{4\pi H_0} \frac{N_{\mathrm{Ly}}(z)}{\nu_{\mathrm{Ly}}(\Omega_0 z + 1)^{1/2}(1+z)^5}.
\tag{19.15}
$$

As discussed by Jakobsen (1991), the emissivity of the intergalactic gas can be written in the form

$$
N_{\mathrm{Ly}}(z) = N_0^2 (1+z)^6 \gamma_{\mathrm{Ly}}(T),
\tag{19.16}
$$

where $N_0 = 7.8\Omega_{\mathrm{B}}h^2$ is the number density of neutral hydrogen atoms at the present epoch and Ω_{B} is the density parameter corresponding to the density of the intergalactic gas. γ_{Ly} is a suitably normalised emission coefficient. Fig. 19.3a shows the variation of γ_{Ly} as a function of temperature for a standard mixture of hydrogen and helium in collisional and thermal equilibrium (Jakobsen 1991). The figure shows that collisionally excited HI and HeII Lyman-α lines are particularly intense at what are referred to as the 'thermostat' temperatures of $T \approx 2 \times 10^4$ K and $T \approx 8 \times 10^4$ K, at which the ionisation state of the intergalactic gas changes from HI to HII and from HeII to HeIII respectively. These changes are illustrated by the change in ionisation state as a function of temperature in Fig. 19.3b. It is immediately apparent from Fig. 19.3b that, if the intergalactic gas were to be collisionally ionised, the temperature of the gas would have to be at least 10^5 K in order that $[N_{\mathrm{HI}}/N_{\mathrm{HII}}] \ll 10^{-5}$. These tools were used by a number of authors, including Kurt and Sunyaev (1967) and Weymann (1967), to show how the ultraviolet background radiation could be used to detect a 'luke-warm' intergalactic gas.

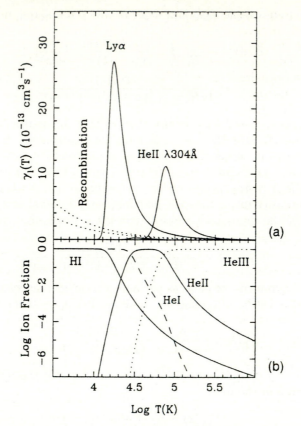

Fig. 19.3. (a) The emissivity per hydrogen atom of collisionally excited HI and HeII Ly-α emission. The dotted lines show the equivalent line emission due to recombination in a fully photoionised gas. (b) The ionisation structure of a cosmological mixture of hydrogen and helium in collisional equilibrium as a function of temperature. Emission from HeI is negligible compared with that of HI and HeII (Jakobsen 1991).

19.6 The Luke-warm Intergalactic Gas

The very low abundances, or absence, of metals in the neutral hydrogen clouds of the Lyman-α forest strongly suggests that they consist of primordial material. Their *column densities* $\int N_H \, dl$ can be determined from the strength of the absorption lines, but it is much more difficult to determine the number density of the neutral hydrogen atoms. Only in a few cases can the number density of neutral hydrogen atoms can be found. A good example is the case of the Lyman-α clouds observed in front of the double quasar 2345+007, in which the same clouds are observed along the lines of sight to each quasar. Knowing the extent of the cloud, a number density of neutral hydrogen atoms of about 3×10^{-3} m^{-3} has been found (Carswell 1988).

Knowing the intergalactic flux of ionising radiation, the number density of the ionised hydrogen atoms can be found by balancing the numbers of photoionisations by the number of the recombinations per second. Typically, the number density of ionised hydrogen amounts to about 10^2 m^{-3} so that the neutral hydrogen fraction is of the order 3×10^{-5}. The total mass of the Lyman-α cloud would be about $10^7 \, M_\odot$. The typical masses of the Lyman-α clouds are thought to lie in the range $10^6 - 10^9 \, M_\odot$.

The connection with the properties of the diffuse intergalactic gas arises from considerations of the means by which the clouds are confined. It should be emphasised that it is not at all certain how the clouds are held together. The binding might be gravitational, or due to the pressure of the intergalactic gas. A popular picture is one in which the Lyman-α clouds are in rough pressure balance with a 'luke-warm' intergalactic gas at a temperature of about 3×10^4 K. The intergalactic flux of ionising radiation at a redshift of $2 - 3$ is so intense that the clouds are almost completely ionised. To attain rough pressure balance, the intergalactic gas at redshifts $z \sim 2 - 3$ would have to be about 10^2 m^{-3}. At low redshifts, this would correspond to a density parameter in the intergalactic gas $\Omega_{\rm IGG} \sim 0.1$ which is probably uncomfortably large.

The full treatment of this problem is quite complicated. A good example of the issues involved in these studies is provided by the papers by Ikeuchi and Ostriker (1983, 1987). A complete study involves consideration of the stability of the clouds, the role of evaporation, photoionisation balance within the clouds and so on. Ostriker (1988) favoured a picture in which the intergalactic gas was shock heated to a high temperature at redshifts $z \geq 5$ and the density parameter of the intergalactic gas corresponded to $\Omega_{\rm IGG} \sim 0.02$.

The interesting question addressed by Jacobsen (1995) is whether or not the redshifted Lyman-α emission of such a luke-warm intergalactic gas would be detectable in the ultraviolet waveband. The easiest case to deal with is that in which it is supposed that the intergalactic gas is kept more or less fully ionised by the diffuse intergalactic ionising background radiation. Jakobsen considers all the possible cases involving the photoionisation and shock heating of the intergalactic plasma. In the case of a pure photoionisation model, it is straightforward to show that the diffuse redshifted Lyman-α background radiation observed at the present epoch must always be significantly less intense that the ionising background itself and so, even if the observed ionising background were responsible for the ionisation of the gas, the redshifted emission could make only a very small contribution to the background radiation. The same conclusion can be drawn for the case in which the intergalactic gas is heated by shocks. The basic reason that these backgrounds are so faint is twofold. First, the number density of neutral hydrogen must be very low because of the null result of the Gunn–Peterson test for neutral hydrogen HI. Second, the constraints of primordial nucleosynthesis indicate that the density of the intergalactic gas is at least one and probably two orders of magnitude less than the critical cosmological density.

An important diagnostic for the ionisation state of the intergalactic gas is the search for He^+ continuum absorption using the equivalent of the Gunn–Peterson test, but now for the redshifted He^+ Lyman-α line, which has a rest wavelength of 30.4 nm. If the intergalactic gas was photoionised by a diffuse intergalactic flux of ionising radiation with spectrum $I_\nu \propto \nu^{-0.5}$, it is expected that the HeII ion would be a order of magnitude more abundant that HI, and so the Gunn–Peterson decrement might be observable in large redshift quasars for which the 30.4 nm absorption line is redshifted into the observable ultraviolet waveband, that is, in quasars with redshifts greater than 3. The problem with this test is that the 30.4 nm line falls below the Lyman-limit absorption edge of neutral hydrogen at 91.2 nm and so is attenuated by absorption in the Lyman-α clouds and in Lyman-limit systems, if these happen to lie along the line of sight to the quasar.

Evidence for HeII absorption troughs due to intergalactic He^+ has been found by Jakobsen *et al.* (1994), Jakobsen (1996) and by Davidsen *et al.* (1996). Jakobsen and his colleagues made a remarkable set of observations with the Faint Object Camera of the Hubble Space Telescope (HST). The camera was operated in a low-resolution spectroscopic mode and, prior to the refurbishment of the HST in 1993, a search was made among 25 large redshift quasars to discover the few examples in which the far-ultraviolet continuum radiation was not absorbed by intervening Lyman-α clouds. The only quasar which presented an unimpeded view of the wavebands shorter than HeII Lyman-α was the quasar Q0302-003 at a redshift $z = 3.286$. In the low resolution spectrum shown in Fig. 19.4a, there is a break in the continuum intensity at precisely the wavelength of the redshifted HeII line at 30.4 nm. At shorter wavelengths, the continuum intensity falls below the noise level due to photon counting statistics. At the 90% confidence limit, the optical depth for HeII absorption is $\tau_{HeII} \geq 1.7$. This limit corresponds to a lower bound to the diffuse intergalactic number density of HeII ions of $N_{HeII} \geq 1.5h \times 10^{-3}$ m^{-3} at redshift $z = 3.29$. A similar result was found for the quasar PKS1935-692 at a redshift $z = 3.185$ which was discovered in a similar spectroscopic survey carried out by Tytler and his colleagues (see Jakobsen 1996). It is remarkable that, between them, these two surveys discovered only two suitable candidates for this test from over 110 bright quasars with $z > 3$ observed by the Faint Object Camera. It is reassuring that these two observations have provided evidence for primordial helium in the intergalactic gas at large redshifts.

In their original interpretation, Jakobsen *et al.* (1994) assumed that the absorption was associated with diffuse intergalactic gas. In the meantime, observations of the Lyman-α forest in the same quasar Q0302-003 with the Keck 10-metre Telescope by Cowie and his colleagues (Songaila *et al.* 1995, Hu *et al.* 1995) indicated that the column density distribution of Lyman-

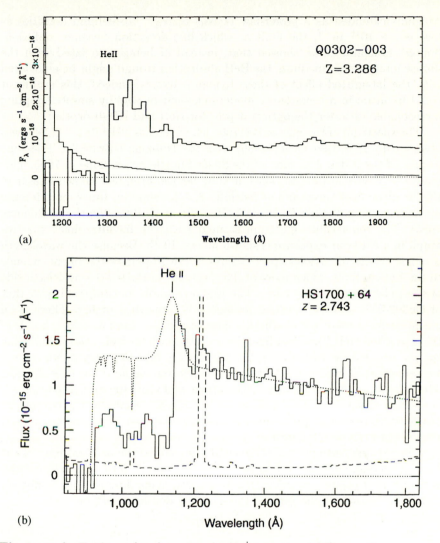

Fig. 19.4a-b. Evidence for absorption by He$^+$ ions at redshifts $z \sim 3$.
(a) The ultraviolet spectrum of the quasar Q0302-003 observed by the Faint Object Camera of the Hubble Space Telescope. The solid thin line shows the 1σ uncertainty per wavelength bin of width $\Delta\lambda = 10\text{Å}$, due to photon statistics. The position of the HeII line at 30.4 nm in the quasar rest-frame is indicated (Jakobsen *et al.* 1994).
(b) The far-ultraviolet spectrum of the quasar HS1700+64 observed by the Hopkins Ultraviolet Telescope as part of the Astro-2 mission of the Space Shuttle. The dotted curve shows an extrapolation of the power-law continuum spectrum beyond the redshifted HeII emission line. The dashed line shows the statistical error at each wavelength (Davidsen *et al.* 1996).

α clouds extended as a power-law $N(N_H) \propto N_H^{-1.5}$ to column densities as low as 2×10^{16} m^{-2}, the limit at which line detection became confusion-limited. Songaila *et al.* showed that, instead of being associated with the diffuse intergalactic medium, the HeII absorption trough might be associated with the integrated effect of these Lyman-α clouds. Indeed, this argument could be made in a remarkably model independent way by simply adopting a reasonable ratio for the optical depths for HeII and HI absorption.

The spectrum of the quasar HS1700+64, which has redshift $z = 2.743$, was observed by Davidsen *et al.* (1996) with the Hopkins Ultraviolet Telescope as part of the Astro-2 mission of the Space Shuttle programme in 1995. This quasar is of special interest, since it is by far the brightest known quasar in the far-ultraviolet waveband at redshifts $z > 2$, being ten times more intense in these wavebands than Q0302-003 (Reimers *et al.* 1989, Vogel and Reimers 1995). The remarkable spectrum secured with the far-ultraviolet spectrograph in a 5.4 hour exposure is shown in Fig. 19.4b. Because the wavelength response of the ultraviolet spectrograph of the Hopkins experiment extends to the Lyman limit, much more of the spectrum to the short wavelength side of the redshifted HeII Lyman-α line was observable as compared with that of Q0302-003, and the spectral resolution is somewhat higher. There is an abrupt depression of the continuum intensity to the short wavelength side of the redshifted HeII Lyman-α line, the optical depth for HeII absorption being $\tau_{HeII} = 1.00 \pm 0.07$. In addition, the flux density falls abruptly to zero at the Lyman limit, at which the interstellar gas in our Galaxy becomes opaque. The signal-to-noise ratio of the observations and their spectral resolution are not sufficient to determine whether or not the variations in the continuum intensity between the HeII emission line and the Galactic Lyman-α cut-off represent the Lyman-α forest of HeII.

Optical observations of HS1700+64 have enabled the optical depth of the Gunn–Peterson effect for HI to be found, $\tau_{HI} = 0.22$. The ratio of opacities is therefore $\tau_{HeII}/\tau_{HI} = 4.5 \pm 0.5$. If the intergalactic gas were uniformly distributed in space, the ratio of these optical depths would be

$$\frac{\tau_{HeII}}{\tau_{HI}} = \frac{\eta}{4}, \qquad (19.17)$$

where $\eta = N(\text{HeII})/N(\text{HI})$ is the ratio of He$^+$ ions to hydrogen atoms. If the gas is clumped, the inferred value of η is somewhat greater. According to Davidsen *et al.* (1996), values of $\eta \sim 80$ are typical of the values predicted for the ionisation of intergalactic gas clouds by the far ultraviolet continuum spectra of quasars.

As discussed by Jakobsen (1996), the present data are not adequate to distinguish between absorption due to diffuse intergalactic HeII and absorption due to HeII belonging to the HeII Lyman-α forest at low column densities. The finite optical depth of $\tau = 1.0 \pm 0.07$ for absorption at $\langle z \rangle = 2.4$ found in HS1700+64 is significantly less than the lower limit of $\tau \geq 1.7$ found in Q0302-003 at $\langle z \rangle = 3.15$. As shown by Jakobsen (1996), such an evolution of the optical depth for HeII Lymanα absorption can be explained in either picture. As we discuss in Sect. 19.8, there may not, in fact, be any difference between these pictures.

19.7 The Lyman Continuum Opacity of the Intergalactic Gas

The low opacities for HI absorption found in the Gunn–Peterson test suggest that the diffuse intergalactic gas is remarkably transparent to far ultraviolet radiation. The cumulative effect of the Lyman-α forest and the Lyman-limit systems, however, produces significant Lyman continuum absorption of the far ultraviolet radiation originating from large redshifts (Bechtold, Weymann, Lin and Malkan 1987, Møller and Jakobsen 1990). The facts that the number densities of both classes of absorption line systems, the range of column densities and their evolution with cosmic epoch are now well defined enable estimates of the opacity of the Universe to far ultraviolet photons to be made. Møller and Jakobsen (1990) have evaluated the opacity of the intergalactic medium in terms of the number density of absorption line systems as a function of redshift $N(z)$ and knowing the average absorption per absorber from the distribution function $N(N_H)$. They first work out the average optical depth of the cloud responsible for the absorption $\langle \tau \rangle = \langle \sigma_H(\nu)N(N_H)\rangle$. Then, the fraction of the background intensity which passes through a typical cloud is $1 - \exp(-\langle \tau \rangle)$. The absorption coefficent σ_H has the form $\sigma_0(\nu_H/\nu)^3$ for all frequencies greater than ν_H where ν_H is the Lyman limit at 91.2 nm; for smaller frequencies, the absorption is taken to be zero. Hence, the total optical depth on passing through a distribution of clouds $N(z)\,\mathrm{d}z$ is

$$\tau_{\text{tot}} = \int_0^{z_e} N(z)[1 - \exp(-\langle \tau \rangle)], \, \mathrm{d}z \qquad (19.18)$$

and the fractional transmitted intensity is $E = \exp(-\tau_{\text{tot}})$.

This expression has been evaluated by Møller and Jakobsen (1990) and their results are displayed in Fig. 19.5. In the figure, z_e is the emission redshift and the ordinate shows the transmitted fraction of the radiation as observed at different ultraviolet wavelengths. The diagram can be interpreted as follows. Suppose we observe at a wavelength of 300 nm. Then, sources at redshifts of 2 and less do not suffer any Lyman continuum absorption because at a redshift of 2, the emission wavelength is 100 nm which is greater than the wavelength of the Lyman limit. By a redshift of 3, however, the emission wavelength is 75 nm and there is strong absorption by the clouds in the redshift interval 3 to $z = (300/91.2) - 1 = 2.29$. It can be seen that the average transmission decreases very rapidly with increasing redshift and this has important consequences for the observability of sources at the very largest

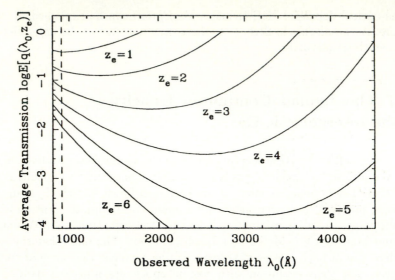

Fig. 19.5. The average transmission of the ultraviolet universe out to large red-shifts as a function of wavelength. The 'Lyman valley' shown in the figure includes the Lyman continuum opacity due to both the Lyman forest and to Lyman-limit systems (Møller and Jakobsen 1990).

redshifts. For example, suppose we wished to observe the Gunn–Peterson decrement for He$^+$ at rest wavelength 30.4 nm or the HeII Lyman-α emission at the same wavelength at an observing wavelength of 160 nm. The radiation would then originate from redshift $z = (160/30.4) - 1 = 4.3$. Inspection of Fig. 19.5 shows that the intensity of such radiation would be attenuated by a factor of about 100, rendering its observation very difficult indeed. Empirically, this is confirmed by the fact that only two good examples out of over 110 quasars with $z > 3$ were found to be suitable for the HeII Lyman-α test. It is apparent that the far-ultraviolet Universe at large redshifts is likely to be heavily obscured. As Jakobsen (1991) expresses, 'Even if the intergalactic medium did go through a phase of intense HeII emission during reheating, the resulting far-ultraviolet radiation will in all likelihood remain forever hidden from our view.'

There is some relief from this conclusion in that the absorption is due to discrete clouds rather than to a diffuse medium and the strongest absorbers are the optically thick Lyman-limit systems. Therefore, along some lines of sight, there may be little absorption, which is what has allowed observations of the HeII absorption troughs in Q0302-003, PKS1935-692 and HS1700+64 to be made successfully. A further consequence is that any large redshift, far-ultraviolet diffuse emission should be patchy.

19.8 Modelling the Evolution of the Intergalactic Medium

While it has been traditional to model the Lyman-α forest in terms of the properties of 'clouds', simulations of the evolution of the primordial intergalactic gas in different dark matter models for structure formation provide a number of important hints about the nature of the structures responsible for Lyman-α absorption systems. Excellent examples of what is now feasible computationally are provided by the simulations of Hernquist *et al.* (1996), Katz *et al.* (1996) and Miranda-Escudé *et al.* (1996). The success of the models of accounting for the large-scale distribution of galaxies, as illustrated by the simulations of Fig. 14.7, has encouraged the numerical cosmologists to tackle the next step of including diffuse baryonic matter into the computations and following the evolution of the gas using the procedures of Single Particle Hydrodynamics and similar computational algorithms.

This much more ambitious programme necessarily involves incorporating many of the physical processes discussed in Chap. 16 into the simulations, in particular, the role of dissipative processes in enabling the baryonic components of galaxies to form the types of structures we observe in the real Universe. Once star formation gets underway, the intense ultraviolet radiation of young massive stars contributes to the ionisation of the diffuse intergalactic gas. In addition, it is essential to incorporate the fact that quasars are formed as an integral part of the galaxy population, and their ionising ultraviolet radiation has an important influence upon the physical state of the gas. Thus, a number of largely empirical rules have to be incorporated into the simulations, those involving the processes of star formation, the form of the initial mass function and the rôle of feedback in limiting the rate of star formation in young galaxies being of special importance. It is nonetheless very impressive how much real physics can be included in the latest supercomputer simulations of the behaviour of the baryonic component.

Of particular interest in this chapter are the simulations of the resulting structure of intergalactic neutral hydrogen clouds, a good example being illustrated in Fig. 19.6, which is taken from the simulations of Katz *et al.* (1996). The underlying cosmology is a standard Cold Dark Matter model with $\Omega_0 = 1$ and $h = 0.5$; the baryonic density parameter has been taken to be $\Omega_B = 0.05$. In this simulation, a background ionising radiation flux is assumed to be present which is consistent with the observations of the proximity effect observed in large redshift quasars (Sect. 18.3.4); it is assumed that this flux originates in quasars. In the even more ambitious models of Cen *et al.* (1996), the ultraviolet continuum radiation is modelled as the emission of quasars which form as part of the process of galaxy formation. An important part of the simulation is that the ionisation of the clouds takes account of their self-shielding by neutral hydrogen in their peripheries (see Sect. 18.3.4).

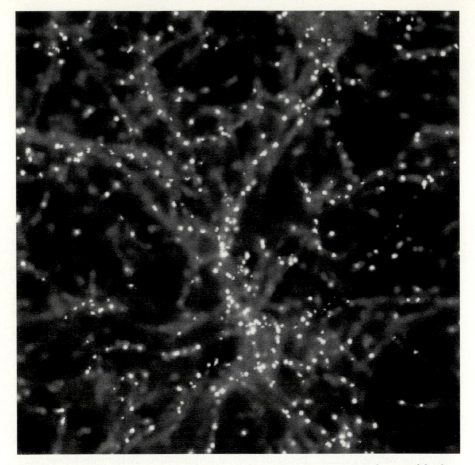

Fig. 19.6 A supercomputer simulation of the expected structure of neutral hydrogen in the intergalactic medium at a redshift $z = 2$ in a standard Cold Dark Matter cosmology with $\Omega_0 = 1$. The size of the box corresponds to a comoving scale of 22.22 Mpc. The simulation includes self-shielding of the neutral hydrogen from the background ultraviolet ionising radiation. The grey-scale is such that the white blobs correspond to column densities $N(HI) \geq 10^{20.5}$ m^{-2}. The faint filamentary structures correspond to column densities $10^{19.5} \geq N(HI) \geq 10^{18.5}$ m^{-2} and the black 'voids' to regions with $N(HI) \leq 10^{18.5}$ m^{-2} (Katz *et al.* 1996, Hernquist *et al.* 1996.)

Fig. 19.6 shows an example of the resulting structure of neutral hydrogen clouds at redshift $z = 2$. The simulations result in a network of filaments with dense knots of neutral hydrogen forming in the vicinity of galaxies. Hernquist *et al.* (1996) then worked out the column density distribution of Lyman-α absorption clouds which would be observed along any line of sight through this distribution to a distant quasar, taking into account the typical spectral resolution of modern spectrographs on large telescopes. Remarkably, the column densities found in these simulations span the range from about 10^{18} to 10^{26} m^{-2}, and their number density distribution follows closely the observed power-law relation $N(N_{\mathrm{H}}) \propto N_{\mathrm{H}}^{-1.5}$ over this range. Similar results have been obtained by Miralda-Escudé *et al.* (1996) for simulations in the $\varLambda$ Cold Dark Matter model discussed in Sect. 14.6.

These are remarkable results and suggest new ways of tackling the problems of the formation of galaxies and the hydrogen clouds associated with them. Although there are many uncertainties in the modelling procedures, it is encouraging that the Lyman-α forest can develop rather naturally in the standard hierarchical picture for the origin of galaxies and larger scale structures. For our present purposes, the important consideration is the nature of the low column density systems. According to the simulations, the high column density knots, shown as white blobs in Fig. 19.6 arise from radiatively cooling gas associated with galaxies which form in high density regions. In constrast, the low density systems are associated with a wide range of different types of structure. To quote Hernquist *et al.* (1996), '... the low column density absorbers are physically diverse: they include filaments of warm gas; caustics in frequency space produced by converging velocity flows; high-density halos of hot collisionally ionised gas; layers of cool gas sandwiched between shocks; and modest local undulations in undistinguished regions of the intergalactic medium. Temperatures of the absorbing gas range from below 10^4 K to above 10^6 K.' Of particular interest is the fact that most of the low column density absorbers are large, flattened structures in which the density constrast $\varDelta_{\mathrm{B}}$ is low. These are present throughout the 'voids' seen in Fig. 19.6. From this perspective, the distinction between the low column density absorbers of the Lyman-α forest and the continuum Gunn–Peterson effect is largely academic.

20 Final Things

In the course of last 19 chapters, we have attempted to tie together many diverse observations and facts about the Universe into a self-consistent picture, within the context of the standard Big Bang scenario. It is, however, incomplete in a number of important ways. A useful comparison may be drawn with the history of our understanding of the origin of the chemical elements, a pleasant analogy drawn to my attention by Dr. Martin Harwit. In the 1930s, the origin of the chemical elements was a mystery. The tools were not available to understand the physical processes by which the synthesis of the chemical elements could have taken place. They might well have been laid down primordially by processes which took place in the inaccessible early Universe. The picture changed dramatically as the role of nuclear processes and the physics of the early Universe became better understood. One of the key events was the discovery of the triple-α resonance by Hoyle in 1953, which showed how the barriers to the synthesis of carbon from three helium nuclei in the central regions of stars could be overcome. Over the succeeding years, Burbidge, Burbidge, Fowler and Hoyle and Cameron elucidated the processes by which the chemical elements could be synthesised in stars. There remained the problem of the synthesis of light elements, such as helium and deuterium, but this was resolved with the realisation that they could be synthesised by non-equilibrium processes in first few minutes of the Big Bang, a topic dealt with in some detail in Chap. 10. It is now univerally accepted that there is no need for the chemical abundances of the elements to be laid down primordially – they can be accounted for by physical processes which take place naturally in the course of primordial and stellar evolution.

The analogy with the problems of contemporary astrophysical cosmology is that it remains to be seen what parts of the most promising scenarios for the origin of galaxies and the large-scale structure of the Universe will be explainable by physical processes which have not yet been established by laboratory experiments, which can be established convincingly by astronomical observations, and which may remain inaccessible to us. To put it another way, what is the minimum set of axioms needed in order to account for the Universe as we know it today? As we will discuss, it might turn out that some parts of the story are simply beyond what can be treated by the physical tools we are likely to possess and then some reasonable set of initial conditions would have to be adopted. On the other hand, optimists, such as the present author, believe that precise astronomical observations and their judicious interpretation are likely to be very fruitful routes to understanding physical processes which cannot be reproduced in the laboratory. Indeed, this is a belief which is held fervently by many of us.

In this final chapter, some of these uncharted areas are surveyed. We begin with topics which are well within current and future observational capababilities, and then deal with deeper and more difficult issues. Specifically, these topics are:

- Can the vast range of observations of galaxies throughout the redshift interval $0 < z < 5$ be reconciled on an empirical basis with some overall scenario for the formation of galaxies and larger-scale structures?
- What is the origin of the rotation of galaxies and their magnetic fields?
- What are the essential features of theories of the very early Universe required by our analysis of the problems of astrophysical cosmology and the formation of galaxies?

These are somewhat diverse topics, but they have been lurking in the background of the astrophysical discussions of much of this book.

20.1 A Synthesis of Observations Related to the Origin and Evolution of Galaxies

It will have become apparent from the considerations of Part 3 that the standard dark matter, gravitational instability picture of the formation of galaxies and large-scale structures is remarkably compelling. Although there are numerous variations on the basic picture, some of which were discussed in Sect. 14.6, a common theme is the formation of galaxies and clusters by hierarchical clustering, as illustrated by the simulations of Fig. 14.3. It is therefore useful to discuss the observations in the context of hierarchical models for the formation of galaxies and clusters and the Press-Schechter formalism, described in Sect. 16.3, provides a convenient framework for that analysis.

The question addressed in this Section is whether or not the vast array of observations and astrophysical analysis described in Chaps. 16 to 19 tells a self-consistent story and whether or not the impressive theoretical apparatus described in Chaps. 11 to 15 can be reconciled with that story. The thoughtful review by Fukugita, Hogan and Peebles (1996) is an important attempt to provide such an observational synthesis on the basis of a careful review of the types of information described in the last four chapters. I will highlight of some of the issues which they raise.

20.1.1 Massive Galaxies

One of the more important results of recent studies has been the fact that the most massive and luminous galaxies, which we can take to be galaxies with $L > L^*$, must have formed the bulk of their stellar populations quite early in the process of galaxy formation. Several pieces of evidence point in this direction:

- The luminosity functions of luminous galaxies in the red and infrared wavebands do not change greatly over the redshift interval $0 < z < 1$, as illustrated by the luminosity functions of Fig. 17.15.

- Qualitatively the same conclusion comes from observations of the massive galaxies associated with strong radio sources. The advantage of this selection procedure is that massive elliptical galaxies can be identified at large redshifts in a relatively straightforward manner. The redshifts of the 3CR radio galaxies extend to $z \approx 1.8$ and their average stellar masses are remarkably constant over this redshift interval (Best $et\ al.$ 1998). As discussed in Sect. 8.4.1, these galaxies are as luminous as the brightest galaxies in clusters at redshifts $z \sim 1$ and have stellar masses $M \approx 3 \times 10^{11}\,M_\odot$ (Fig. 20.1). If account is taken of the dark matter present in these systems, their total masses are likely to be about ten times greater. Direct evidence that the stellar populations of these large redshift radio galaxies are likely to be old was presented by Lilly (1988) for the radio galaxy 0902+34 at a redshift $z = 3.4$ and by Dunlop $et\ al.$ (1996) for the radio galaxy LBDS 53W091 at redshift 1.55. In the latter case, the age of its stellar population is at least 3.5×10^9 years so that these stars must have been formed at large redshifts. A similar result, namely an age of 4.5×10^9 years has been found for the radio galaxy LBDS 53W069 at redshift $z = 1.43$ (Dey 1997). Radio galaxies with qualitatively the same stellar properties have been observed to redshifts greater than 4 (Lacy $et\ al.$ 1994, Spinrad $et\ al.$ 1995).

- The fact that quasars flourished in the redshift interval $2 < z < 3$ can be naturally interpreted as evidence that their host galaxies must have had masses $M \geq 10^{12}\,M_\odot$ (see Sect. 16.3), and so their stellar populations must already have been assembled into galaxies, more or less as we know them today, by a redshift of 3.

- Studies of the Lyman-break galaxies discovered by Steidel and his colleagues at redshifts $z \sim 3$ show that these galaxies are similar in many ways to L^* galaxies, but the star formation activity, which is the signature of these galaxies, seems to be distributed like the light in massive elliptical galaxies observed at the present day (Giavalisco $et\ al.$ 1997). The number density of such galaxies is not so different from that of L^* galaxies observed at the present epoch.

Thus, it seems that a significant fraction of all massive galaxies must have formed their stellar populations at large redshifts, $z \geq 3$. The expected distri-

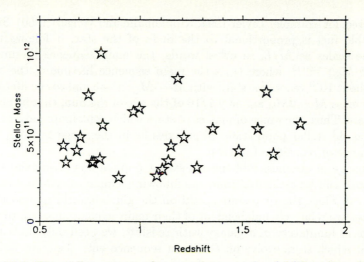

Fig. 20.1 The stellar masses of a complete sample of 3CR radio galaxies as a function of redshift, as inferred from their infrared K luminosities (Best *et al.* 1998).

bution of masses according to the hierarchical clustering picture as described by the Press-Schechter formalism, and illustrated in Fig. 16.5, shows that it is possible to understand how this could have come about, but it is quite a close run thing (Efstathiou 1995). If substantial numbers of massive galaxies were discovered at redshifts $z > 4$, the standard hierarchical picture would probably need to be modified.

The lack of strong evolution of the properties of giant galaxies in the near infrared waveband, as illustrated by Fig. 17.15 and the number counts of galaxies at 2.2 μm (Fig. 2.10), can be understood on the basis of a simple model for the evolution of their stellar populations. At these wavelengths, most of the light originates from stars on the red giant branch and their luminosities and colours are essentially independent of the masses of their main sequence progenitors (Tinsley and Gunn 1976). The lifetimes of red giant stars are much less than the time their progenitors spend on the main sequence, and so the number of red giant stars is given by the product of the rate at which stars evolve off the main sequence onto the giant branch and the time they spend there. In order to understand the likely changes in the luminosity of galaxies in the infrared waveband out to redshifts, say, $z \sim 4$, when the Universe was probably up to about one tenth of its present age, we need only know the rate of evolution of stars off the main sequence for quite a narrow range of masses, as can be understood from the following argument.

The mass-luminosity relation for stars on the main-sequence is assumed to take the form $L/L_\odot = (M/M_\odot)^x$. Then, their main sequence lifetimes are determined by the time it takes the star to burn about 10% of its mass into helium, the well-known *Schönberg-Chandrasekhar limit*. During evolution on the main sequence, the luminosities of stars of a given mass are remarkably constant (see, for example, Tayler 1994, Kippenhahn and Weigert 1990). Since the available fuel is proportional to the mass of the star, it follows that its lifetime scales as M/L, in other words, the main sequence lifetime is $t = t_\odot (M/M_\odot)^{-(x-1)}$, where $t_\odot$ is the main sequence lifetime of the Sun, which is about 10^{10} years. For stars with $M \sim M_\odot$, $x \sim 5$ and so the lifetimes of stars of mass $M = 2M_\odot$ are only $1/16$ of the age of the Sun, that is, about 6×10^8 years. Thus, the range of masses of stars which contribute most of the light of the old stellar populations of galaxies lie in the range $1 < M_\odot < 2$ over the range of redshifts of interest.

The change in luminosity of the old stellar populations of galaxies with cosmic epoch can be estimated from the following simple calculation due to Gunn (1978). Since the time stars spend on the giant branch, t_g, and their luminosities are relatively independent of their main sequence masses, to find the change in luminosity of a galaxy with redshift, we need only determine the rate at which stars evolve off the main sequence onto the giant branch as a function of main sequence mass. In what is known as the *passive evolution model* for the luminosity of the galaxy, it is assumed that all the stars were formed in an initial brief starburst and that the subsequent luminosity evolution of the galaxy is due to the stellar evolution of this population. For illustration, let us assume that the initial mass function of the stars was of Salpeter form, $\mathrm{d}N = N(M)\,\mathrm{d}M \propto M^{-y}\,\mathrm{d}M$, where $y = 2.35$. It is a straightforward calculation to show that the number of stars on the giant branch N_g is

$$N_g = t_g \frac{\mathrm{d}N}{\mathrm{d}t} = t_g \left(\frac{\mathrm{d}N}{\mathrm{d}M} \right) \left(\frac{\mathrm{d}M}{\mathrm{d}t} \right) \tag{20.1}$$

Thus, using the above relation between mass and main sequence lifetime, $t = t_\odot (M/M_\odot)^{-(x-1)}$, we find

$$L(t) = L(t_0)\, t^{-(x-y)/(x-1)} \tag{20.2}$$

Inserting the values $x = 5$ and $y = 2.35$, we find $L \propto t^{-0.66}$. For the case of the critical world model, $t/t_0 = (1+z)^{-3/2}$ and so, to an excellent approximation, $L \propto (1+z)$. Thus, at a redshift of 1, the old stellar populations of galaxies should be about twice as luminous as they are at the present epoch, and at redshift $z = 3$, four times as luminous. Thus, it is not particularly surprising that the changes in the luminosity function of massive galaxies shown in Fig. 17.15 and the deviations of the number counts in the K waveband from the expectations of the uniform world models are quite modest – the galaxies were only about a magnitude more luminous at redshifts $z \sim 1 - 2$, as illustrated by the predicted redshift–K magnitude relations for 3CR radio galaxies in Figs. 8.3 and 8.5, which were derived using stellar population synthesis codes.

20.1.2 Clusters of Galaxies

Dramatic evidence for the evolution of galaxies in rich, regular clusters at relatively small redshifts was first described in the pioneering analyses of Butcher and Oemler (1978). They found that the fraction of blue galaxies in such clusters increased from less than 5% in a nearby sample to percentages as large as 50% at redshift $z \sim 0.4$. The Butcher-Oemler effect has been the subject of a great deal of study and debate, the major observational problems concerning the contamination of the cluster populations by foreground and background galaxies, as well as by bias in the selection criteria according to which the clusters were selected for observation in the first place. These problems are enumerated by Dressler (1984).

Observations with the Hubble Space Telescope have clarified a number of these issues (Dressler and Smail 1997, Dressler *et al.* 1997, Dickinson 1997). First of all, the colours of the spheroidal galaxies in clusters have been determined out to redshifts $z = 0.5$ and these show remarkably little scatter about the relation expected for passively evolving elliptical galaxies (Ellis *et al.* 1996). The inference is that these galaxies formed their stellar populations at large redshifts and that there has been little ongoing star formation, say, associated with mergers, as in some scenarios for the formation of elliptical galaxies. Perhaps the most remarkable result of these studies has been the change in the relative populations of galaxies of different types at $z \sim 0.5$ as compared with a nearby sample. Fig. 20.2a shows the morphology-density relation for 10 clusters in the redshift interval $0.36 < z < 0.57$, in the same format as Fig. 3.3. The surface densities of galaxies are systematically about a factor of three greater than those in Fig. 3.3, probably due to the fact that the clusters are somewhat richer than the nearby sample. There are two remarkable features of the comparison of Figs. 3.3 and 20.2a. First of all, the fraction of spiral galaxies in the large redshift sample is very much greater at high galaxy densities than in the nearby sample. Second, the overall fraction of S0 galaxies is very much less than in the local sample. At the same time, the fraction of spheroidal galaxies is at least as large as in the nearby sample, constituting further evidence that the population of elliptical galaxies was already fully formed at large redshifts. There is a correlation between galaxy type and galaxy density for the four highest concentration, regular clusters in the sample, as illustrated by Fig. 20.2b, but for the lower concentration, irregular clusters, the correlation is not present.

These observations suggest that the elliptical galaxy population was already well formed at large redshifts and that it is unlikely that the bulk of the ellipticals were formed by mergers of spiral galaxies which were members of the cluster. On the other hand, the population of S0 galaxies must have grown considerably over the redshift interval $0 < z < 1$. The likely origin of these population changes is the transformation of a significant fraction of the large abundance of spiral galaxies seen in the large redshift clusters into S0 galaxies. Dynamical interactions between galaxies, ram-pressure stripping,

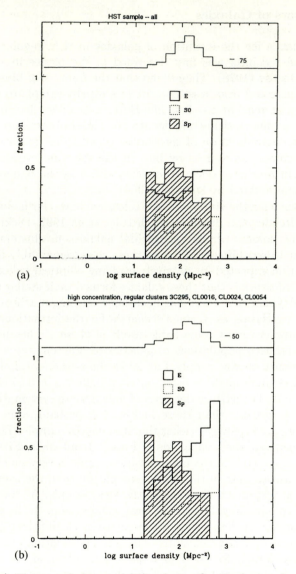

Fig. 20.2 (**a**) The morphology-redshift relation for 10 clusters in the redshift interval $0.36 < z < 0.57$. The diagram shows the relative proportions of galaxies of different morphological classes as a function of surface density. The upper histogram shows the total number of galaxies in each bin. (**b**) The same relation for the 4 high-concentration, regular clusters in the sample, 3C295, Cl0016+16, Cl0024+16 and Cl0054-27 (Dressler and Smail 1997).

accretion of mass and collisions between spiral galaxies are possible ways of enhancing the population of S0 galaxies relative to spiral galaxies. This picture is consistent with the significant number of blue irregular, or merging systems observed in the clusters in the sample observed at $z \sim 0.5$. There is evidence from the Canada-France Redshift Survey that the percentage of blue galaxies increases from a few percent locally to about 25% at $z \sim 0.5$, and to perhaps as much as 80% at $z \sim 0.9$ (Schade 1997).

In contrast to the passive evolution of the elliptical galaxies in rich clusters of galaxies, the redshift–K apparent magnitude relation for the brightest galaxies in clusters shows evidence for evolution of their total luminosities. Aragón-Salamanca et al. (1993) showed that this relation is consistent with the expectations of standard world models in which their luminosities were unchanged with cosmic epoch. This must be the result of a cosmic conspiracy since the stellar populations of the brightest cluster galaxies must have evolved at least passively with cosmic epoch, and so, according to the analysis of Sect. 20.1.1, the galaxies should be about one magnitude fainter at the present epoch as compared with their luminosities at $z \sim 1$. The obvious way of accounting for the constancy of the absolute magnitudes of these galaxies is to appeal to dynamical friction and galactic cannibalism. This process was first described by Hausman and Ostriker (1978) in the context of explaining the small dispersion in the absolute magnitudes of the brightest galaxies in clusters. The expression for the gravitational relaxation time-scale τ_g for the exchange of kinetic energy between the masses M_1 and M_2 in a cluster of point masses has the form

$$\tau_g \propto \frac{v^3}{M_1 M_2}. \tag{20.3}$$

This process is often referred to as *dynamical friction* and leads to the exchange of kinetic energy between the point masses as they attempt to establish thermodynamic, as well as dynamical, equilibrium among the cluster members (for a simple derivation of this formula, see Longair 1997). In the case of a cluster of galaxies, (20.3) shows that the most massive galaxies lose kinetic energy to less massive members most rapidly and so the former drift towards the dynamical centre of the cluster. The expression (20.3) indicates that relaxation is most rapid among the more massive galaxies, leading to the most massive galaxies cannibalising the next most massive galaxies in the clusters. This might be an explanation of the Bautz-Morgan effect, discussed in Sect. 4.2.4. Clearly, the case of galaxies is much more complex than the case of point masses, but it can be appreciated that dynamical friction and cannibalism can naturally account for the increase in mass and luminosity of the brightest galaxies in clusters.

Simulations of the merging of galaxies in clusters have been carried out by Kaufmann and White (1993) within the context of the hierarchical scenario for cluster formation. On the basis of these simulations, they have shown that the luminosity of the brightest galaxies in clusters can increase on average by a factor of between 3 and 5 between redshift $z \sim 1$ and the present epoch, which can account for the cosmic conspiracy which produces an apparently non-evolving redshift-K magnitude relation for the brightest galaxies in clusters.

The general result of these considerations is that it is not so surprising if evolutionary changes are observed in the properties of clusters of galaxies at relatively small redshifts. Within the standard hierarchical clustering scenario, clusters of galaxies form rather late in the Universe, as discussed in Sect. 16.1. At the same time, these considerations provide tests of the standard picture in that it is not expected that there will be many fully formed clusters of galaxies at large redshifts. Dickinson (1997) has given a list of possible associations of galaxies observed at redshifts $z > 2$, many of them being discovered as companion galaxies to active galaxies and quasars, while some of them have been found as clusters of Lyman-α absorbers. In one impressive case, Pascarelle *et al.* (1996) have found an association of galaxies with strong Lyman-α emission at a redshift $z = 2.39$. These are very important observations, and one of the most important future programmes will be to establish precisely how these associations are related to the typical rich, regular clusters observed at the present day.

20.1.3 The Blue Galaxies

One of the more perplexing issues concerns the nature of the excess of faint blue galaxies. This topic has already been discussed in Sects. 17.5 and 18.4.1, in which it was pointed out that the blue galaxies can be convincingly associated with galaxies undergoing bursts of star formation. What is not so clear is the nature of the galaxies involved, the possibilities including giant galaxies, merging or interacting galaxies, and dwarf galaxies. The big problem is that the bulk of the excess of faint blue galaxies occurs at apparent magnitudes $B > 24$, which is beyond the limit at which spectra can be obtained even with 8-m class telescopes. There is evidence, however, that all three types of galaxy contribute to the excess.

The complete galaxy sample of Cowie *et al.* (1995) was selected in the apparent magnitude interval $22.5 < B < 24$ and the redshift distribution was found to span the range $0.2 < z < 1.7$ (Fig. 17.16). The blue galaxies at redshifts $z \geq 0.5$ are luminous galaxies with $L \geq L^*$ undergoing intense bursts of star formation and so some of the blue excess is likely to be associated with star formation in massive galaxies. These objects are likely to be similar to the blue star-forming galaxies observed in the Canada-France Redshift Survey and which are major contributors to the large star-formation rates observed at $z \sim 1$.

Another likely contributor to the blue excess are the merging galaxies, as expected in the hierarchical picture of the formation of galaxies. These events could involve mergers of galaxies of comparable mass, or the coalescence of a dwarf galaxy with a more massive galaxy (Hernquist and Mihos 1995).

Le Fèvre *et al.* (1997) have used HST observations of the galaxies in the Canada-France Redshift Survey and the LDSS Survey described in Sect. 17.5 to confirm that the steep blue number counts are principally associated with irregular/peculiar galaxies. These galaxies are not dwarfs, but are mostly associated with objects with redshifts $0.5 < z < 1$. Le Fèvre and his colleagues have also analysed the number density of strongly merging galaxies as a function of redshift and find that the number of *major mergers* is a strong function of redshift. For example, they find that the percentage of luminous galaxies, brighter than $M_B < -20$, with companions within a projected radial distance of $10h^{-1}$ kpc and apparent magnitudes within 1.5 magnitudes of the brighter galaxy, amounts to 30% of the galaxies at $z \sim 0.8$ compared with a very much smaller percentage at the present epoch (Fig. 20.3). Thus, it seems that mergers and the resulting intense blue starbursts were very much more common at redshifts $z \sim 1$.

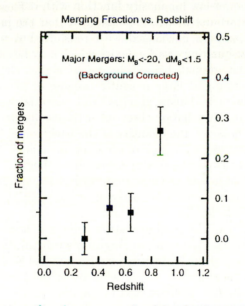

Fig. 20.3 The fraction of major mergers involving luminous galaxies as derived from the Canada-France Redshift Survey and the LDSS Survey (Le Fèvre *et al.* 1997)

There is also evidence that there must be a large number of compact blue star-forming dwarfs among the galaxies fainter than B = 24 in the Hubble Deep Field sample. Many of the faint blue galaxies in the sample have half-light radii $\theta \sim 0.3$ arcsec and, according to Im *et al.* (1995), at $V = 28$, the typical half-light radius is less than $\theta = 0.2$ arcsec, significantly less than the expected angular sizes of non-evolving L^* galaxies. These objects have been identified with compact star-forming dwarf galaxies and bear some resemblance to the low mass HII galaxies observed locally. The ultraviolet luminosities of the compact blue dwarfs observed at $z \sim 1$ would have to decay rapidly, however, because there is little evidence for such a large population of dwarf galaxies in the local Universe.

The problems of interpreting the number counts of galaxies at apparent magnitudes B > 24 has been carefully discussed by Ferguson (1997). For illustrative purposes, he describes in some detail two quite different models which are consistent with the observed redshift distribution of galaxies at $B \sim 24$, but which make quite different predictions about the redshift distribution at fainter magnitudes. In the first example, a pure luminosity evolution model, the luminosity function of galaxies is assumed to change systematically with redshift in such a way as to account for the observed number counts of blue galaxies. To achieve this, the local luminosity function of luminous galaxies is assumed to be of Gaussian form to which is added a population of dwarf galaxies with a power-law luminosity function with $\alpha = -1.3$. In the second model, a dwarf starburst model, it is assumed that the luminosity function of the bright galaxies is unchanged with redshift, but to this is added a population of star-bursting dwarf galaxies which only began forming stars at redshift $z \sim 1$. The rationale behind this model is that low mass galaxies are inhibited from forming at large redshifts because of photoionisation of the pregalactic gas in their shallow gravitational potential wells by the intense ultraviolet background radiation observed at these redshifts (Sect. 18.3.4). By a redshift $z \sim 1$, however, the intensity of the background has decreased to a sufficient degree for recombination of the baryonic matter in the protogalaxy to take place and the process of star formation to begin.

The remarkable aspect of these quite different models is that they can both be tuned to give a reasonable account of the overall blue galaxy number counts and the redshift distribution of galaxies in the apparent magnitude range 22.5 < B < 24. The redshift distributions are, however, very different at fainter magnitudes. According to Ferguson's computations, at I = 28, the redshift distribution of the luminosity evolution model is remarkably uniform over the redshift interval $0 < z < 5$, whereas in the bursting dwarf model, the redshift distribution is dominated by galaxies at $z \sim 1$, which is hardly surprising. Ferguson's thoughtful critique of the problems of accounting for all the existing data on faint galaxy counts is strongly recommended.

My own view is that there is unlikely to be one single explanation of the details of the counts of galaxies, but rather that many different physical processes play a role. This is suggested by the fact that, in addition to the hierarchical clustering and merging of galaxies, the fraction of the baryonic mass condensed into stars in galaxies changes dramatically through the same redshift interval (Fig. 18.5a). The key observations are the redshift distributions of galaxies at faint magnitudes combined with their morphological types. There has been considerable progress in estimating redshifts from their broad-band colours, but the measurement of spectroscopic redshifts for these very faint objects is not feasible at the moment.

20.2 The Origin of the Rotation of Galaxies and their Magnetic Fields

20.2.1 The Origin of Rotation

According to the standard picture of the development of density perturbations in the Friedman world models, rotational velocities are damped as the Universe expands. As shown in Sect. 11.5, rotational velocities decrease in amplitude as $\delta \mathbf{v}_\perp \propto R^{-1}$, this result being no more than the conservation of angular momentum in a uniformly expanding medium. This consideration poses more or less insuperable problems for theories involving primordial turbulence (von Weizsacker 1947), many of which have been discussed in some detail by Jones (1973) and Jones and Peebles (1972). As they point out, if the rotation of spiral galaxies at the present day were the relic of primordial turbulent eddies, these eddies would be highly supersonic at the epoch of recombination and so be dissipated by shocks.

The most natural explanation for the rotation of galaxies is that it is induced by tidal torques acting between perturbations during the linear stages of their development (Strömgren 1934, Peebles 1993). The perturbations associated with the superposition of random Gaussian waves are not spherical, but, in the lowest approximation, are ellipsoidal and so pairs of ellipsoids experience the mutual effect of tidal torques. These torques lead to the build up of angular momentum in each of the bodies, whilst conserving angular momentum overall. This process can continue so long as the perturbations are in the linear regime. Once the perturbations reach amplitude $\Delta\rho/\rho \sim 1$, they rapidly become virialised and the final rotational state of the object is largely determined by the conservation of the angular momentum it acquired by the time $\Delta\rho/\rho \sim 1$.

The angular momentum J of a gravitationally bound object of mass M and gravitational binding energy E can be characterised by the dimensionless *spin parameter* λ, which is defined to be

$$\lambda = \frac{JE^{1/2}}{GM^{5/2}}. \tag{20.4}$$

As pointed out by Peebles (1993), the quantity λ is a measure of the degree of rotational support of the galaxy. The centripetal acceleration acting on unit mass at a typical radial distance r within the system is $a = v_\phi^2/r$ and the gravitational acceleration is $g \sim GM/r^2$. The angular momentum of the system is $J \sim Mrv_\phi$ and its gravitational binding energy GM^2/r. Therefore, eliminating v_ϕ and r from the ratio a/g, we find

$$\frac{a}{g} = \frac{v_\phi^2}{r} \frac{r^2}{GM} \sim \frac{J^2 E}{G^2 M^5} = \lambda^2 \qquad (20.5)$$

Estimates of the typical value of the spin parameter induced by tidal torques during the growth of primordial density perturbations is best found from N-body simulations, the median value found from those of Barnes and Efstathiou (1987) being $\lambda = 0.05$. The implication of (20.5) is that this amount of rotation is quite inadequate to provide rotational support for the system, which must be primarily supported by the random velocities of the stars and clouds within the system. This value of λ would, however, be consistent with the slow rotation of the spheroidal components of elliptical and spiral galaxies. If tidal torques are the origin of the angular momenta of spiral discs, it is essential that they contract in the radial direction whilst conserving their angular momenta – in other words, the discs would have to lose gravitational binding energy at constant angular momentum.

There is a pleasant solution to this problem within the context of models in which the spiral discs of galaxies form by dissipative contraction within dark matter haloes, as pointed out by White and Rees (1978) and Fall and Efstathiou (1980). If we suppose that the dark matter halo follows the density distribution of an isothermal gas sphere at large radii, $\rho \propto r^{-2}$, then, as shown in Sect. 3.4.2, the rotation curve of a centrifugally supported disc within this potential distribution is $v_c = $ constant, consistent with the flat rotation curves of giant spiral galaxies. Let us assume that the material of the disc initially attains the same spin parameter due to tidal torques as the spheroidal component, $\lambda = 0.05$. Then, the initial rotational velocity of the disc v_ϕ can be found from (20.5),

$$v_\phi \sim \lambda v_c. \qquad (20.6)$$

Now, in the dissipative contraction of the disc, the gravitational potential distribution is fixed by the mass distribution of the dark halo and so the disc can contract, conserving its angular momentum $v_\phi r = $ constant until it reaches the radius at which it is in centrifugal equilibrium, that is, when $v_\phi = v_c$. Thus, the final radius of the disc is λr_i. As pointed out by Peebles (1993), this is an attractive scenario to account for the fact that the typical sizes of the discs of spiral galaxies are very much less than the inferred sizes of their dark matter haloes. The typical scale length of the disc of a giant spiral galaxy, such as our own Galaxy, is $r \sim 4$ kpc and, according to this scenario, this disc would have contracted from a halo of dimensions $4/\lambda \sim 100$ kpc, typical of the inferred sizes of the spheroidal components of spiral galaxies. This picture is consistent with the arguments developed by Ostriker and Peebles (1973) to account for the stability of the discs of spiral galaxies.

This line of reasoning has been refined by Fall (1983) who considered the role of dissipation and tidal stripping to account for the rotational properties of the discs and spheroids of spiral and elliptical galaxies. His improved version of the above argument results in essentially the same answer, namely that dissipative contraction of the discs of spiral galaxies in dark matter haloes can account for their relative dimensions. In addition, he showed that, with reasonable assumptions, this picture can account for the constancy of the central surface brightness distributions of spiral galaxies found by Freeman (1970) (Sect. 3.6.1) and the Tully-Fisher relation (Sect. 3.6.2). His conclusion was that this model for the formation of disc galaxies, involving the conservation of mass and angular momentum but the dissipation of the binding energy of the disc, can give a good account of the most important properties of disc galaxies.

There is, however, a problem with the elliptical galaxies. Since angular momentum is conserved in the formation of the disc, the specific angular momentum of the spiral galaxies, that is, the angular momentum per unit mass, should be the same for elliptical galaxies and the discs of spiral galaxies. This expectation is not consistent with his interpretation of the observation of Davies *et al.* (1983). Fall found that the specific angular momenta of spiral discs are about six times those of elliptical galaxies, the comparison being made for galaxies of the same mass. He proposed various possible solutions to this problem. One might be that the tidal torques result in different values of the spin parameter λ in different galactic environments. For example, elliptical galaxies are found preferentially in regions of high galaxy density and so it might be that the value of λ is anticorrelated with the density of the large-scale environment in which the galaxy is formed. Another possibility is that the outer regions of the collapsing galaxy in dense regions are stripped by tidal forces. Since most of the angular momentum is contained in the outer regions, it might be that some stripping of the protospheroid could reduce the specific angular momentum of the galaxies destined to become elliptical galaxies. The problem is to reduce the angular momentum by a sufficient factor without reducing the dimensions of the halo to unacceptably small values.

20.2.2 The Origin of Magnetic Fields

We have had very little to say about magnetic fields but they are omnipresent in astronomy. Excellent surveys of the techniques by which magnetic flux densities can be measured in different astronomical environments and the results of these estimates can be found in the volume *Cosmical Magnetism* (Lynden-Bell 1994) and the review by Vallée (1997) (see also Chap. 17 of Longair 1997). Magnetic fields play a key role in the process of star-formation, in the dynamics of the interstellar medium and in many different aspects of high energy astrophysics. These topics are central to any study of the physics of galaxies and active galactic nuclei. Some impression of the issues involved in understanding the origin of cosmic magnetic fields is provided by the surveys by Rees (1994, 1995), Kulsrud (1997) and Parker (1997).

To summarise briefly, magnetic fields are present in all astronomical object. In stars, the magnetic flux densities range from 10^8 T in the interiors of neutron stars, through values ~ 1 T in sunspots to about 10^{-7} T in protostellar objects. In the case of normal stars, it is probable that the origin and maintenance of their magnetic fields can be attributed to dynamo action in their interiors, associated with the combination of convective motions and reconnection of the lines of force. The key point is that the time-scale for the amplification of the magnetic field inside stars can be quite short and so there can be many e-folding times to create a finite magnetic field strength starting from a tiny seed field.

The situation is somewhat different in the case of the large-scale magnetic fields in the interstellar media in galaxies and in the intracluster medium. In our own Galaxy, the magnetic flux density in the interstellar medium consists of a large scale ordered field of about 2×10^{-10} T superimposed upon which there is a random component of about $1 - 2$ times this value (Taylor and Cordes 1993). In the case of the intracluster medium in clusters of galaxies, evidence for large scale fields is provided by the diffuse synchrotron radio emission observed from a number of clusters, as well as by the observation of depolarisation of the emission of extended extragalactic radio sources by the surrounding intracluster medium. Typical magnetic flux densities are $B \sim 10^{-10}$ T. In the case of the intergalactic medium between clusters of galaxies, there are only upper limits to the strengths of any large-scale magnetic fields from the lack of depolarisation of the emission of distant radio sources, typical limits corresponding to $B \leq 10^{-13}$ T.

In the case of the large-scale magnetic fields in galaxies and clusters, there is a problem in understanding their origin, since the characteristic time-scales over which magnetic dynamos could operate are rather long. For example, in the case of the interstellar field in our own Galaxy, the period of rotation of the interstellar gas about the Galactic Centre is $\sim 2.5 \times 10^8$ years and so there have been at most about 50 complete rotations of the gas about the centre. The differential rotation of the ionised gas in the interstellar medium results in the stretching and amplification of the magnetic field in the disc so that any primordial magnetic field would be tightly wound up. This is not sufficient, however, to produce an ordered uniform field since the winding up of the field lines would result in tightly wound tubes of magnetic flux running in opposite directions. There needs to be some way of reconnecting the lines of force to create a large-scale uniform field. The problems with this picture have been discussed by Parker (1997), who suggested that a solution may lie in the buoyancy of the interstellar cosmic ray gas which can inflate loops of magnetic field which burst out of the plane of the Galactic disc, reconnect and snap off. Even if this process is effective, there is still the question of the origin of the seed field, which must be present to get the process going.

The process by which a very weak seed field can be generated was discovered by Biermann (1950) and is known as the *Biermann battery*. The principles of operation of the primitive battery are clearly described by Kulsrud (1997). Initially, there is no magnetic field in the plasma, but it is supposed

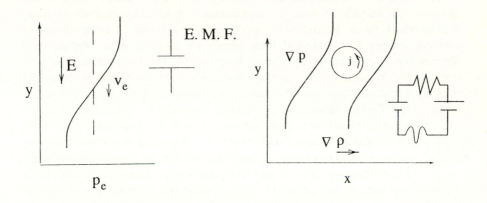

Fig. 20.4a-b. Illustrating the physics of the Biermann battery (from Kulsrud 1997). (a) In this case, there is a variation in electron pressure in the y direction which results in a drift of the electrons which is opposed by the electric field caused by the charge separation of the electrons from the protons. In this case, a static emf is created. (b) Equation (20.7) shows that, if the pressure gradient is fixed, the emf is proportional to N_e^{-1}. If the density of electrons in independent of y, but varies in the x-direction, the emfs at different positions in the x-direction no longer balance and a current can flow, creating a small-scale seed magnetic field.

that there are variations in the pressure of the electrons in the plasma. The electrons flow to lower pressure regions, resulting in a charge imbalance in the plasma, which produces an electic field **E** opposing the flow of electrons.

$$-N_e e\mathbf{E} = \nabla p_e \qquad (20.7)$$

As a result, the flow stops and an emf is created in the plasma. This emf cannot drive a current, however, since the integral round any closed loop in the case of a linear gradient in electron density is zero. If, however, there are variations in the electron density throughout the plasma, different emfs can be induced in different regions and then currents can flow in the plasma, as illustrated schematically in Fig. 20.4 (from Kulsrud 1997). Magnetic fields can then develop and are eventually limited by the self-inductance of the current loop itself. As Kulsrud shows, this process saturates when the magnetic flux density is only about 10^{-25} T, about 10^{15} times less than the strength of the Galactic magnetic field. Thus, the challenge is to find a means of creating a large scale magnetic field, on the scale of galaxies, from these tiny random seed fields, which are generated on the scale of the electron density fluctuations in the plasma. This is not a trivial problem.

In addition to the mechanism described by Parker (1997), Kulsrud has discussed the amplification of the seed magnetic field in a turbulent medium, characterised by a Kolmogorov spectrum of turbulence. In a turbulent medium, energy is transferred from large-scale to small-scale eddies, until the motions are dissipated by viscous forces on the smallest scales. This process has the advantage that, on small scales, the turn-over time of the eddies is short and so effective amplification of the magnetic field can take place on small-scales. The problem is to understand how such a strong chaotic small-scale field could result in a large-scale ordered field. Kulsrud argues that, once equipartition is established between the mean kinetic energy density of the turbulence and the magnetic field energy density, the dominant magnetic fields are those associated with the large-scale eddies, the smaller scale structures being dissipated as energy is transferred to smaller and smaller scales.

An alternative approach to the origin of magnetic fields has been suggested by Rees (1994, 1995). The starting point is the observation that certain high energy astrophysical objects are certainly capable of generating strong ordered magnetic fields over substantial scales. In the case of a supernova remnant such as the Crab Nebula, a convincing story can be told that the magnetic field in the Nebula, which has magnetic flux density about $10^{-(7-8)}$ T over scales of a few pc, originated in the rotating dipole field of the central pulsar which itself was created in the collapse of the progenitor star to a neutron star. The origin of the intense magnetic field of the pulsar can be traced back to the collapse of the magnetic field present in the progenitor star which in turn can be attributed to some form of magnetic dynamo action within the star. The big advantage of this picture is that the time-scales for the amplification of the magnetic field within the progenitor star can be very short and thus there is no problem in principle in generating strong magnetic fields.

The other convincing example of the formation of strong magnetic fields in high energy astrophysical objects concerns the extragalactic radio sources. The extended lobes of these sources can extend to dimensions of 1 Mpc and greater and the fields within them are $B \sim 10^{-10} - 10^{-9}$ T. The standard picture for the origin of these huge magnetic field energies involves the release of energy from the active galactic nucleus in the form of intense jets of high energy particles and magnetic field. Precisely how the fields are amplified is a matter of speculation, but there is no doubt that enormous energy fluxes are generated in active galacti nuclei and that these are sufficient to account for the total magnetic field energy requirements of the radio lobes. Again, it is assumed that, because the jets originate in the compact regions close to the active nucleus, there is no problem, in principle, of accounting for the amplification of the magnetic field because the time-scales in these regions are very short. The presence of a supermassive rotating black hole in the nucleus can result in the generation of strong magnetic fields by electromagnetic

Table 4.1. The Origin of Magnetic Fields in Galaxies (from Rees 1994, 1995)

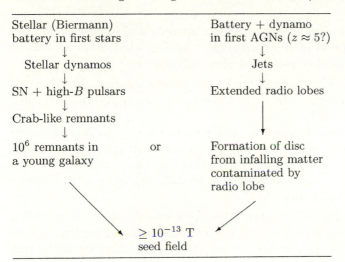

Stellar (Biermann) battery in first stars		Battery + dynamo in first AGNs ($z \approx 5$?)
↓		↓
Stellar dynamos		Jets
↓		↓
SN + high-B pulsars		Extended radio lobes
↓		
Crab-like remnants		
↓		
10^6 remnants in a young galaxy	or	Formation of disc from infalling matter contaminated by radio lobe

$$\geq 10^{-13} \text{ T}$$
seed field

induction if even a tiny magnetic field threads the black hole (Thorne *et al.* 1986).

Rees makes the point that these phenomena indicate how magnetic fields can be generated by high energy astrophysical processes on the scales of supernova remnants, galaxies and clusters of galaxies. To fill the volume of a galaxy with magnetic flux due to supernova explosions, the process would have to be repeated many times in the early history of the galaxy and the net result could well be a seed field which could them be amplified by the processes discussed by Parker and Kulsrud. Notice that a key part of the process involves the smoothing out of the gross inhomogeneities in the field. In the same way, the magnetic field in clusters of galaxies might be the result of repeated radio source events. The magnetic fields created in the radio source events would diffuse through the intracluster plasma and repeated events would add to the total large scale magnetic flux within the cluster. Possible scenarios are indicated schematically in Table 20.1, which is taken from the reviews by Rees (1994, 1995).

There are good reasons why these issues should be taken rather seriously. One of the important means of discovering some of the most distant galaxies we know of is through the identification of extragalactic radio sources, the radio galaxy with the largest redshift being $z = 4.8$. The fact that the galaxy is a strong radio source means that there must already be a strong magnetic field within the galaxy. It seems plausible that magnetic fields were already present in galaxies and clusters at large redshifts. This will certainly have an influence upon many aspects of the formation of subsequent generations of stars and galaxies.

20.3 The Very Early Universe

In the course of this introduction to modern astrophysical cosmology, we have uncovered four aspects of the cosmological models which have no natural explanation in terms of conventional physics. These are:

- The *horizon problem* – why is the Universe isotropic (Sect. 12.1)?
- The *flatness problem* – why is the Universe so close to its critical density, $\Omega_0 = 1$ (Sect. 7.2.6)?
- The *baryon asymmetry problem* – why was the Universe slightly asymmetric with respect to matter and antimatter in its early phases (Sect. 9.6)?
- The *primordial fluctuation problem* – what was the origin of the density fluctuations from which galaxies and large-scale structures in the Universe formed (Sects. 14.2.2 and 14.5)?

In the classical Big Bang picture, these problems are solved by assuming that the Universe was endowed with the appropriate initial conditions in its very early phases. In other words, we have to postulate that the initial conditions from which our Universe has evolved were such that it was isotropic, was set up with density very, very close indeed to the critical density, was slightly matter-antimatter asymmetric, and contained fluctuations with essentially a Harrison-Zeldovich spectrum. To put it crudely, we only get out at the end what we put in at the beginning. The physical processes which may eventually resolve these problems have been surveyed by Kolb and Turner (1990) in their splendid book *The Early Universe*.

I have suggested that there are five possible approaches to solving these problems (Longair 1997).

- That is just how the Universe is — the initial conditions were set up in that way.
- There are only certain classes of Universe in which intelligent life can have evolved. The Universe has to have the appropriate initial conditions and the fundamental constants of nature should not be too different from the values they have today or else there would be no chance of life forming as we know it. This approach is known as the *Anthropic Cosmological Principle* and it asserts that the Universe is as it is because we are here to observe it.
- The inflationary scenario for the early Universe.
- Seek clues from particle physics and extrapolate that understanding beyond what has been confirmed by experiment to the earliest phases of the Universe.
- Something else we have not yet thought of. This would certainly involve new physical concepts.

There is some merit in each of these approaches. For example, even in the first, somewhat defeatist, approach, it might turn out to be just too hard a problem to disentangle the physics responsible for setting up the initial conditions from which our Universe evolved. How can we possibly check that the physics adopted to account for the properties of the very early Universe is correct? Can we do better than boot-strapped self-consistency? Let us look in a little more detail at some of the other possibilities.

20.3.1 The Anthropic Cosmological Principle

In the second approach, there is certainly some truth in the fact that our ability to ask questions about the origin of the Universe must say something about the sort of Universe we live in. The Cosmological Principle asserts that we do not live at any special location in the Universe, and yet we are certainly privileged in that we are able to ask the question at all. There is an intriguing line of reasoning according to which there are only certain types of Universe in which life as we know it could have formed. For example, the stars must live long enough for biological life to form and evolve into sentient beings. Part of the problem stems from the fact that we have only one Universe to study – we cannot go out and investigate other Universes to see if they have evolved in the same way as ours. This line of reasoning is embodied in what is known as the *Anthropic Cosmological Principle* (Carter 1974) and has been dealt with *in extenso* in the books by Barrow and Tipler (1986) and Gribben and Rees (1989). There are a number of versions of the Principle, some of them stronger than others. In extreme interpretations, it leads to statements such as the strong form of the Principle enunciated by Wheeler (1977), 'Observers are necessary to bring the Universe into being.'

Personally, I am not particularly attracted to this line of reasoning, because it suggests that we will never be able to find any physical reason for the initial conditions from which the Universe evolved, or for the relations between the fundamental constants of nature. On the other hand, Weinberg (1997) finds it such a puzzle that the vacuum energy density Ω_Λ, or the cosmological constant Λ, is so very much smaller than the values expected according to current theories of elementary particles (Sect. 7.3.1), that he has invoked a form of anthropic reasoning to account for its smallness. I regard the Anthropic Cosmological Principle as the very last resort if all other physical approaches fail.

20.3.2 The Inflationary Universe and Clues from Particle Physics

One of the more promising ideas for resolving at least some of the fundamental problems has been the idea of the *inflationary universe*, to which we have made a number of references in the text. The ideas first sprang to prominence during the early 1980s, primarily thanks to the pioneering investigations by Guth (1981), although these ideas had been foreshadowed in papers by Linde (1974) and Bludman and Ruderman (1977) (see Guth 1997). I like to think of the inflationary model in a number of stages, in which we add a little more physics at each stage. The most rudimentary of all is *inflation without physics*, in the sense that, if the Universe expanded exponentially by an enormous factor during its very early stages, for whatever reason, the first and third of the basic problems can be eliminated. Let us illustrate this behaviour by some simple calculations.

Let us assume that the scale factor, R, increased exponentially with time as $R \propto e^{t/T}$. Intriguingly, such exponentially expanding solutions were found in some of the earliest applications of Einstein's equations to the Universe, in the guise of empty models driven entirely by what we would now term the vacuum energy density Ω_A (Sect. 7.3.2). The exponential expansion continues for a certain time and then the Universe switches over to the standard radiation-dominated phase of the early Universe. The epochs during which this exponential expansion of the Universe took place is called the *inflationary era*.

Consider a tiny region of the early Universe expanding under the influence of the exponential expansion. Particles within the region are initially very close together and so are in causal communication with each other. Before the inflationary expansion begins, the region has physical scale less than the particle horizon, and so there is time for the small region to attain a uniform, homogeneous state. The region then expands exponentially so that neighbouring parts of the region are driven to such distances that they can no longer communicate with each other by light signals – causally-connected regions are swept beyond their local horizons by the inflationary expansion. At the end of the inflationary epoch, the Universe transforms into the standard radiation-dominated Universe and the inflated region continues to expand as $R \propto t^{1/2}$.

In the most popular version of the inflationary scenario for the very early Universe, the exponential expansion is associated with the breaking of symmetries of Grand Unified Theories of elementary particles at very high energies. According to these theories, at high enough energies, the strong and electroweak forces are unified and it is only at lower energies that they appear as distinct forces. The characteristic energy at which the grand unification phase transition is expected to take place is $E \sim 10^{14}$ GeV, only about 10^{-34} seconds after the Big Bang, which is also the characteristic e-folding time for the exponential expansion. This energy scale is commonly referred to as the GUT scale. In a typical realisation of the inflation picture, the exponential inflationary expansion took place from this time until the Universe was about 100 times older. At the end of this period, there was an enormous release of energy associated with the phase transition and this heats up the Universe to a very high temperature indeed. At the end of the inflationary era, the dynamics become those of the standard radiation-dominated Big Bang.

Let us put some figures into this calculation. Over the interval from 10^{-34} seconds to 10^{-32} seconds, the radius of curvature of the Universe increased exponentially by a factor of about $e^{100} \approx 10^{43}$. The horizon scale at the beginning of this period was only $r \approx ct \approx 3 \times 10^{-26}$ m and this was inflated to a dimension 3×10^{17} m by the end of the period of inflation. This dimension then scaled as $t^{1/2}$, as in the standard radiation-dominated Universe so that the region would have expanded to a size of 3×10^{42} m by the present day – this dimension far exceeds the present dimension of the Universe, which is about 10^{26} m. Thus, our present Universe would have arisen from a tiny region in the very early Universe which was much smaller than the horizon scale at that time. This guarantees that our present Universe would be isotropic on the large scale, resolving the horizon problem. This history of the inflationary Universe is compared with the standard Friedman picture in Fig. 20.5. Notice that this calculation has the same physical content as the estimate of the horizon scale in an exponentially expanding Universe given in Sect. 12.1 (see 12.9).

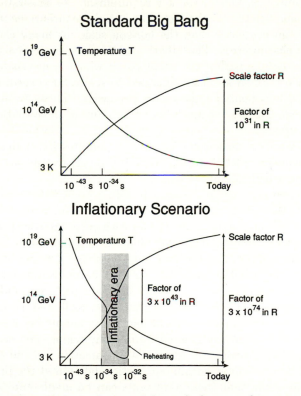

Fig. 20.5 Comparison of the evolution of the scale factor and temperature in the standard Big Bang and Inflationary cosmologies.

A second important consequence of the exponential expansion is that it straightens out the geometry of the early Universe, however complicated it may have been to begin with. Again, suppose the little region of the early Universe had some complex geometry. The effect of the expansion is to increase the radius of curvature of the geometry by the same enormous factor as the exponential expansion. As shown by (5.35), the radius of curvature the geometry scales as $R_c(t) = \Re R(t)$ and so the radius of curvature of the geometry within the tiny region is inflated to dimensions vastly greater than the present size of the Universe – the geometry of the inflated region is driven towards flat Euclidean geometry. This process ends at the end of the inflationary era. Because the geometry of space-time and its energy density are very closely related through (8.8), when the Universe transforms from the exponentially expanding inflationary state into the radiation dominated phase, the geometry is Euclidean $\kappa = 0$ and consequently the Universe must have $\Omega_0 + \Omega_\Lambda = 1$. If $\Omega_\Lambda = 0$, then the Universe should have $\Omega_0 = 1$.

This process has other beneficial features for resolving other aspects of the great problems. One of the key requirements for understanding the origin of cosmic structure is the fact that initial perturbations were set up on scales very much greater than the horizon scale for many classes of object of astrophysical interest. The inflationary hypothesis resolves this problem since perturbations within the small patch which is driven far beyond the local particle horizon are also expanded to scales far exceeding the horizon scale by the exponential inflationary expansion. These eventually re-enter the observable Universe at much later times during the radiation and matter dominated eras.

A further important feature concerns the suppression of an excessive number of monopoles created during the GUT phase transition. Kibble (1976) showed that, when this phase transition took place, a variety of topological defects are expected to be created, including point defects (or monopoles), lines defects (or cosmic strings) and sheet defects (or domain walls). Kibble showed that one monopole is created for each correlation scale at that epoch. Since that scale cannot be greater than the particle horizon at the GUT phase transition, it is expected that huge numbers of monopoles are created. According to the simplest picture of the GUT phase transition, the mass density in these monopoles in the standard Big Bang picture would vastly exceed $\Omega_0 = 1$ at the present epoch (see, for example, Kolb and Turner 1990). The inflationary picture can solve this problem very simply because the whole of our observable Universe originated from a tiny region on a scale very much smaller than the horizon scale at the GUT transition era and so there would be at most one monopole within our particle horizon at the present epoch.

It is interesting that these arguments can be made quite independently of an understanding of the physics of the processes responsible for the inflationary expansion. Physical forces which have qualitatively the appropriate properties have been invoked by particle physicists to explain features of the electroweak force. As explained in Sect. 7.3.1, the Higgs fields were introduced to eliminate the singularities present in electroweak theory and to endow the

$W^\pm$ and Z^0 particles with mass. They have the property of being scalar fields with negative energy equations of state, $p = -\rho c^2$ and, as shown in that section, this is precisely the type of force needed to account for the inflationary exponential expansion of the early Universe.

An enormous amount of theoretical work has been carried out to find a physical realisation of the forces which could cause the inflationary expansion of the early Universe. Although there are many models in the literature, no single preferred picture has emerged. There is general agreement about the properties of the scalar fields during the inflation era. The successful realisations must involve what is called *slow roll-over* so that the inflationary expansion can take place over many e-folding times. At the end of the slow roll-over phase, the field settles down to its true vacuum state and then all the energy stored in the vacuum fields is transferred to particles and radiation fields which are reheated to a high temperature, as illustrated in Fig. 20.5. Indeed, a common procedure is to work backwards and determine the properties of the scalar field from the requirement that an adequate amount of inflation took place, a procedure known as *reconstructing the inflation potential*.

Another key contribution of particle physics concerns the origin of the baryon-asymmetry problem, a subject referred to as *baryogenesis*. In a prescient paper, Sakharov (1967) enunciated the three rules necessary to account for the baryon-antibaryon asymmetry of the Universe. *Sakharov's rules* for the creation of non-zero baryon number from an initially baryon symmetric state are:

- *Baryon number* must be violated;
- C (charge conjugation) and CP (charge conjugation combined with parity) must be violated;
- The asymmetry must be created under *non-equilibrium conditions*.

The reasons for these rules can be readily appreciated from simple arguments (Kolb and Turner 1990). Concerning the first rule, it is evident that, if the baryon asymmetry developed from a symmetric high temperature state, baryon number must be violated at some stage – otherwise, the baryon asymmetry would have to be built in from the very beginning. The second rule is necessary in order to ensure that a net baryon number is created, even in the presence of interactions which violate baryon conservation. The third rule is necessary, because baryons and antibaryons have the same mass and so, thermodynamically, they would have the same abundances in thermodynamic equilibrium, despite the violation of baryon number and C and CP invariance.

Fortunately, there is evidence that all three rules can be satisfied in the early Universe from a combination of theoretical ideas and evidence from particle physics experiments. Concerning the first rule, baryon number violation is a generic feature of Grand Unified Theories which unify the strong and electroweak interactions – the same process is responsible for the predicted instability of the proton. C and CP violation have been observed in the decay of the neutral K^0 and $\bar{K}^0$ mesons. The K^0 meson should decay symmetrically into equal numbers of particles and antiparticles but, in fact, there is a slight preference for matter over antimatter, at the level 10^{-3}, very much greater than the degree of asymmetry necessary for baryogenesis, $\sim 10^{-8}$. The need for departure from thermal equilibrium follows from the same type of reasoning which leads to the primordial synthesis of the light elements (Chap. 10). As in that case, so long as the time-scales of the interactions which maintained the various constituents in thermal equilibrium are less than the expansion time-scale, the number densities of particles and antiparticles of the same mass would be the same. In thermodynamic equilibrium, the number densities of different species do not depend upon the cross-sections for the interactions which maintain the equilibrium. It is only after decoupling, when non-equilibrium abundances are established, that the number densities depend upon the specific values of the cross-sections for the production of different species.

In a typical baryogenesis scenario, the asymmetry is associated with some very massive boson and its antiparticle, $X, \bar{X}$, which are involved in the unification of the strong and electroweak forces and which can decay into final states which have different baryon numbers. Kolb and Turner (1990) provide a clear description of the principles by which the observed baryon asymmetry can be generated at about the epoch of grand unification or soon afterwards, when the very massive bosons can no longer be maintained in equilibrium. Although the principles of the calculations are well defined, the details of these processes are not understood, partly because the energies at which they are likely to be important are not attainable in laboratory experiments, and partly because predicted effects such as the decay of the proton have not been observed. Thus, although there is no definitive evidence that this line of reasoning is secure, there is no question but that there exist well-understood physical processes of the type necessary for the creation of the baryon-antibaryon asymmetry. The importance of these studies go well beyond their immediate significance for astrophysical cosmology. As Kolb and Turner remark, '... in the absence of direct evidence for proton decay, baryogenesis may provide the strongest, albeit indirect, evidence for some kind of unification of the quarks and the leptons'.

The last problem concerns the origin of the fluctuations from which galaxies formed. There are several possible sources of primordial fluctuations in the scenario we have outlined. One possibility is that fluctuations are created during the phase change which was responsible for the inflationary expansion of the very early Universe. Whenever changes of state occur in nature, such as when water freezes or boils, there is the possibility of creating large fluctuations and one version of this scenario involves the formation of various types of topological defect, such as cosmic strings, domain walls and textures. Another possibility is that the fluctuations developed from statistical fluctuations which must have been present during the inflationary era. This topic is discussed by Kolb and Turner (1990) who show how the amplitudes of these fluctuations depend upon the form of the scalar potential. It should be recalled that, during the inflationary expansion, the energy density of the vacuum remains the same because of the work done by the negative energy equation of state (see Sect. 7.3.1). Thus, the fluctuations originate within a medium of constant energy density, despite the fact that the Universe is continually expanding by an enormous factor. These fluctuations are driven beyond the horizon scale by the inflationary expansion and then re-enter the horizon once the Universe has transformed over to the standard radiation-dominated phase long after the end of the inflationary era. For a wide range of theories, a scale-free Harrison-Zeldovich spectrum is predicted.

These ideas look promising but we should bear in mind that there is no evidence for the inflationary picture beyond the need to solve the four great problems listed at the beginning of this Section. There is no question but that enormous progress has been made in understanding the types of physical process necessary to resolve the four great problems but it is not clear how we are to find independent evidence for the physical processes responsible for the inflationary expansion.

A representation of the evolution of the Universe from the Planck era to the present day is shown in Fig. 20.6. The *Planck era* is that time in the very remote past when the energy densities were so great that a quantum theory of gravity is needed. On dimensional grounds, we know that this era must have occurred when the Universe was only about $t_{pl} \sim (hG/c^5)^{1/2} \sim 10^{-43}$ s old. Despite enormous efforts on the part of theorists, there is no quantum theory of gravity and so we can only speculate about the physics of these extraordinary eras.

Fig. 20.6 is drawn on a logarithmic scale and so we are able to encompass the whole of the Universe, from the Planck area at 10^{-43} s to the present age of the Universe which is about 3×10^{17} s or 10^{10} years old. Halfway up the diagram, from the time when the Universe was only about a millisecond old, to the present epoch, we can be reasonably confident that we have the correct picture for the Big Bang despite the four basic problems described above. At times earlier than about 1 millisecond, we very quickly run out of known physics. Indeed, in the model of the very early Universe we have been describing, it is assumed that we can extrapolate across the huge gap from 10^{-3} s to 10^{-43} s using our current understanding of particle physics. Maybe the current thinking will turn out to be correct, but there must be some concern that some fundamentally new physics may emerges at higher and higher energies before we reach the GUT era at $t \sim 10^{-36}$ s and the Planck era at $t \sim 10^{-43}$ s.

The one thing which is certain is that at some stage a quantum theory of gravity is needed. Roger Penrose and Stephen Hawking have developed very powerful singularity theorems which show that, according to classical theories of gravity under very general conditions, there is inevitably a physical

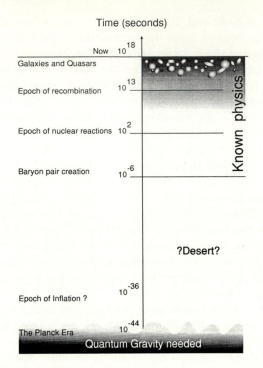

Fig. 20.6 A schematic diagram illustrating the evolution of the Universe from the Planck era to the present time. The shaded area to the right of the diagram indicates the regions of known physics.

singularity at the origin of the Big Bang, that is, as $t \rightarrow 0$, the energy density of the Universe tends to infinity. One of the possible ways of eliminating this singularity is to find a proper quantum theory of gravity. This remains an unsolved problem and we can be certain that our understanding of the very earliest stages of our Universe will remain incomplete until it is solved. Thus, there is no question but that new physics is needed if we are to develop a convincing physical picture of the very early Universe.

References

Chapter 1

Harrison, E.R. (1987). *Darkness at Night: A Riddle of the Universe*. Cambridge, Massachusetts: Harvard University Press.

Herschel, W. (1785) *Phil. Trans. R. Soc.*, **75**, 213.

Hoskin, M. (1976). *J. Hist. Astr.*, **8**, 169.

Hubble, E..P. (1929). *Proc. Natl. Acad. Sciences*, **15**, 168.

Hubble, E. P. and Humason, M. (1934). *Astrophys. J.*, **74**, 43.

Longair, M.S. (1991). *Theoretical Concepts in Physics*. Cambridge: Cambridge University Press.

Longair, M.S. (1995). *Astrophysics and Cosmology* in *Twentieth Century Physics*, (eds L.M. Brown, A. Pais and A.B. Pippard), **3**, 1671. Bristol and Philadelphia, IOP Publishing and New York AIP Press.

Sandage, A.R. (1961). *The Hubble Atlas of Galaxies*. Washington, D.C.: Carnegie Institution of Washington Publication 618.

Smith, R.W. (1982). *The Expanding Universe: Astronomy's 'Great Debate' 1900 – 1931*. Cambridge: Cambridge University Press.

Chapter 2

Bennett, C.L., Banday, A.J., Gorski, K.M., Hinshaw, G., Jackson, P., Koegstra, P., Kogut, A., Smoot, G.F., Wilkinson, D.T. and Wright, E.L. (1996). *ApJ*, **464**, L1.

Dunlop, J.S. and Peacock, J.A. (1990). *MNRAS*, **247**, 19.

Fixsen, D.J., Cheng, E.S., Gales, J.M., Mather, J.C., Shafer, R.A. and Wright, E.L. (1996). *ApJ*, **473**, 576.

Geller, M.J. and Huchra, J.P. (1989). *Science*, **246**, 897.

Gott, J.R. III, Melott, A.L. and Dickenson, M. (1986). *ApJ*, 306, 341.

Gregory, P.C. and Condon, J.J. (1991). *ApJS*, **75**, 1011.

Groth, E.J. and Peebles, P.J.E. (1977). *ApJ*, **217**, 385.

Groth, E.J. and Peebles, P.J.E. (1986). *ApJ*, **310**, 507.

Hubble, E.P. (1929). *Proc. Natl. Acad. Sciences*, **15**, 168.

Hubble, E.P. and Humason, M. (1934). *ApJ.*, **74**, 43.

Kolatt, T., Dekel, A. and Lahav, O. 1995. *MNRAS*, **275**, 797.

Lin, H., Kirshner, R.P., Shectman, S.A., Landy, S.P., Oemler, A., Tucker, D.L. and Schechter, P.L. (1996). *ApJ*, **471**, 617.

Longair, M.S. (1992). *High Energy Astrophysics. Vol. 1*. Cambridge: Cambridge University Press.

Mather, J.C., Cheng, E.S., Eplee, R.E. Jr., Isaacman, R.B., Meyer, S.S., Shafer, R.A., Weiss, R., Wright, E.L., Bennett, C.L., Boggess, N.W., Dwek, E., Gulkis, S., Hauser, M.G., Janssen, M., Kelsall, T., Lubin, P.M., Moseley, S.H. Jr., Murdock, T.L., Silverberg, R.F., Smoot, G.F. and Wilkinson, D.T. (1990). *ApJ.*, **354**, L37.

Maddox, S.J., Efstathiou, G., Sutherland, W.G. and Loveday, J. (1990). *MNRAS*, **242**, 43P.
Melott, A.L., Weinberg, D.H. and Gott, J.R. III (1988). *ApJ*, **306**, 341.
Metcalfe, N., Shanks, T., Campos, A., Fong, R. and Gardner, J.P. (1996). *Nature*, **383**, 236.
Page, L. (1997). In *Critical Dialogues in Cosmology*, (ed. N. Turok), 343. Singapore: World Scientific.
Peebles, P.J.E. (1993). *Principles of Physical Cosmology*. Princeton: Princeton University Press.
Penzias, A.A. and Wilson, R.W. (1965). *ApJ.*, **142**, 419.
Sandage, A.R. (1968). *Observatory*, **88**, 91.
Sunyaev, R.A. and Zeldovich, Ya.B. (1980). *ARA&A*, **18**, 537.
Weiss, R. (1980). *ARA&A*, **18**, 489.
Zeldovich, Ya. B. and Sunyaev, R.A. (1969). *Astrophys. Sp. Science*, **4**, 301.

Chapter 3

Aaronson, M. and Mould, J. (1983). *ApJ*, **265**, 1.
Abell, G.O. (1962). In *Problems of Extragalactic Research*, (ed C.G. McVittie), p. 232. New York: McMillan.
Barnes, J.E. (1992). *ApJ*, **393**, No. 2, Part 1, videotape.
Bertola, F. and Capaccioli, M. (1975). *ApJ*, **200**, 439.
Binggeli, B., Sandage, A.R. and Tammann, A.G. (1988). *ARAA*, **26**, 509.
Binney, J. and Tremaine, S. (1987). *Galactic Dynamics*, Princeton, NJ: Princeton University Press.
Bruzual, G. (1983). *ApJ*, **273**, 105.
Bruzual, G. and Charlot, S. (1993). *ApJ*, **405**, 538.
Davies, R.L., Efstathiou, G., Fall, S.M., Illingworth, G. and Schechter, P.L. (1983). *ApJ.*, **266**, 41.
Disney, M. (1976). *Nature*, **263**, 573.
Djorgovski and Davis (1987). *ApJ*, **313**, 59.
Dressler, A. (1980). *ApJ.*, **236**, 351.
Dressler, A., Lynden-Bell, D., Burstein, D., Davies, R.L., Faber, S.M., Terlevich, R.J. and Wegner, G. (1987). *ApJ*, **313**, 42.
Faber, S.M. and Jackson, (1976). *ApJ*, **204**, 668.
Felten, J.E. 1977. *AJ*, **82**, 869.
Felten, J.E. 1985. *Comments Astrophys.*, **11**, 53.
Fisher, J.R. and Tully, R.B. (1975). *A&A*, **44**, 151.
Freeman, K.C. (1970). *ApJ*, **160**, 811.
Hubble, E.P. (1936). *The Realm of the Nebulae*, 46. New Haven: Yale University Press.
Illingworth, G. (1977). *ApJ Letts*, **218**, L43.
Kennicutt, R.C., Edgar, B.K and Hodge, P.W. (1989). *ApJ*, **337**, 761.
Kormendy, J. (1982). In J. Binney, J. Kormendy and S.D.M. White *Morphology and Dynamics of Galaxies*, (eds L. Martinot and M. Mayor). Sauverny-Versoix, Switzerland: Geneva Observatory Publications.
Kruit, P.C. van der (1989). In G. Gilmore, I. King and P. van der Kruit, *The Milky Way as a Galaxy* (eds R. Buser and I. King). Sauverny-Versoix, Switzerland: Geneva Observatory Publications.
Larson, R.B. and Tinsley, B.M. (1978). *ApJ*, **219**, 46.
Loveday, J., Peterson, B.A., Efstathiou, G. and Maddox, S.J. (1992). *ApJ*, **390**, 338.
Ostriker, J.P. and Peebles, P.J.E. (1973). *ApJ*, **186**, 467.

Roberts, M.S. and Haynes, M.P. (1994). *ARAA*, **26**, 115.

Rocca-Volmerange, B. and Guiderdoni, B. (1988). *A&A Suppl*, **75**, 93.

Salpeter, E.E. (1955). *ApJ*, **121**, 161.

Sandage, A.R. (1975). In *Stars and Stellar Systems – Galaxies and the Universe*, (eds. A.R. Sandage, M. Sandage and J. Kristian), **9**, 1. Chicago: University of Chicago Press.

Schechter, P. (1976). *ApJ*, **203**, 297.

Schwarzschild, M. (1979). *ApJ*, **232**, 236 .

Searle, L., Sargent, W.L.W and Bagnuolo, W.G. (1973). *ApJ*, **179**, 427.

Toomre, A. and Toomre, Yu. (1972). *ApJ*. **178**, 623.

Toomre, A. (1974). In *The Formation and Dynamics of Galaxies*, (ed. J.R. Shakeshaft), 347. Dordrecht: D. Reidel Publishing Company.

Tremaine, S.D. and Richstone, D.O. (1977). *ApJ*, **212**, 311.

Vaucouleurs, G. de (1948). *Ann. d'Astrophys*, **11**, 247.

Vaucouleurs, G. de (1963). *ApJ Suppl.*, **73**, 31.

Vaucouleurs, G. de (1974). In *The Formation and Dynamics of Galaxies*, (ed. J.R. Shakeshaft), 1. Dordrecht: D. Reidel Publishing Company.

Chapter 4

Abell, G.O. (1958). *ApJS*, **3**, 221.

Abell, G.O., Corwin, H.G.and Olowin, R.P (1989). *ApJS*, **70**, 1.

Alcock, C., Allsman, R.A., Axelrod, T.S., Bennett, D.P., Cook, K.H., Park, H.-S., Marshall, S.L., Stubbs, C.W., Griest, K., Perlmetter, S., Sutherland, W., Freeman, K.C., Peterson, B.A., Quinn, P.J. and Rodgers, A.W. (1993a). In *Sky Surveys : Protostars to Protogalaxies*, (ed. T. Soifer), 291. San Francisco: Astron. Soc. Pac. Conf. Ser.

Alcock, C., Akerlof, C.W., Allsman, R.A., Axelrod, T.S., Bennett, D.P., Chan, S., Cook, K.H., Freeman, K.C., Griest, K., Marshall, S.L., Park, H.-S., Perlmutter, S., Peterson, B.A., Pratt, M.R., Quinn, P.J., Rodgers, A.W., Stubbs, C.W. and Sutherland, W. (1993b). *Nature*, **365**, 621.

Alcock, C., Allsman, R.A., Alves, D., Axelrod, T.S., Becker, A.C., Bennett, D.P., Cook, K.H., Freeman, K.C., Griest, K., Guern, J., Marshall, S.L., Peterson, B.A., Pratt, M.R., Quinn, P.J., Rodgers, A.W., Stubbs, C.W., Sutherland, W. and Welch, D.L. (1997). *ApJ*, **486**, 697.

Bahcall, J.N. (1990). *Neutrino Astrophysics*. Cambridge: Cambridge University Press.

Bahcall, N. (1977). *ARAA*, **15**, 505.

Bahcall, N. (1988). *ARAA*, **26**, 631.

Bautz, L. and Morgan, W.W. (1970). *ApJL*, **162**, L149.

Binggeli, B., Tammann, G.A. and Sandage, A.R. (1987). *AJ*, **94**, 251.

Birkinshaw, M. (1990). In *The Cosmic Microwave Background: 25 Years Later*, (eds. N. Mandolesi and N. Vittorio), 77. Dordrecht: Kluwer Academic Publishers.

Blandford, R.D. and Narayan, R. (1992). *ARAA*, **30**, 311.

Böhringer, H. (1994). In *Frontiers of Space and Ground-based Astronomy*, (eds. W. Wamsteker, M.S. Longair and Y. Kondo), 359. Dordrecht: Kluwer Academic Publishers.

Böhringer, H. (1995). *Ann. New York Acad. Sciences*, **759**, 67.

Cavaliere, A. (1980). In *X-ray Astronomy*, (eds. R. Giacconi and G. Setti), 217. Dordrecht: D. Reidel Publishing Co.

Chwolson, O. (1924). *Astr. Nachrichten*, **221**, 329.

Dashevsky, V.M. and Zeldovich, Ya.B. (1964). *Astr. Zh.*, **41**, 1071.

Dressler, A. (1980). *ApJ*, **236**, 351.

Dressler, A. (1984), *ARAA*, **22**, 185.

Dyer, C.C. and Roeder, R.C. (1972). *ApJL*, **174**, L115.

Dyer, C.C. and Roeder, R.C. (1973). *ApJL*, **180**, L31.

Efstathiou, G. (1990). In *Physics of the Early Universe*, (eds. J.A. Peacock, A.F. Heavens and A.T. Davies), 361. Edinburgh: SUSSP Publications.

Einstein, A. (1915). *K. Preuss. Akad. Wiss. (Berlin) Sitzungsber.*, **1**, 142.

Einstein, A. (1936). *Science*, **84**, 506.

Fabricant, D. Lecar, M. and Gorenstein, P. (1980). *ApJ*, **241**, 552.

Forman, W. and Jones, C. (1982). *ARAA*, **20**, 547.

Fort, B. and Mellier, Y. (1994). *A&A Review*, **5**, 239.

Geller, M.J. and Postman, M. (1983). *ApJ*, **274**, 31.

Hawking, S.W. (1975) *Comm. Math. Phys.*, **43**, 199.

Hewitt, J.N., Turner, E.L., Burke, B.F., Lawrence, C.R., Bennett, C.L., Langston, G.I. and Gunn, J.E. (1987). In *Observational Cosmology*, (eds. A. Hewitt, G.R. Burbidge and Fang Li Zhi), 747. Dordrecht: D. Reidel and Co.

Jones, M., Saunders, R., Alexander, P., Birkinshaw, M., Dillon, N., Grainge, K., Hancock, S., Lasenby, A., Lefebvre, D., Pooley, G. Scott, P. Titterington, D and Wilson, D. (1993). *Nature*, **365**, 320.

Kaiser, N. (1992). *ApJ*, **388**, 272.

Kent, S.M. and Gunn, J.E. (1982). *AJ*, **87**, 945.

King, I. (1966). *AJ*, **71**, 64.

Kolb, E.W. and Turner, M.S. (1990). *The Early Universe*. Redwood City, California: Addison–Wesley Publishing Co.

Kneib, J.-P., Ellis, R.S., Smail, I., Couch, W.J. and Sharles, R.M. (1996). *ApJ*, **471**, 643.

Longair, M.S. (1992). *High Energy Astrophysics. Vol. 1.* Cambridge: Cambridge University Press.

Lynds, R. and Petrosian, V. (1986). *Bull. Am. Astr. Soc.*, **18**, 1014.

Merritt, D. (1987). *ApJ*, **313**, 121.

Oemler, A. (1974). *ApJ*, **194**, 1.

Peebles, P.J.E. (1980). *The Large-Scale Structure of the Universe*. Princeton: Princeton University Press.

Perkins, D.H. (1987). *Introduction to High Energy Physics*. Menlo Park, California: Addison–Wesley Publishing Co.

Sandage, A.R.and Hardy, E. (1976). *ApJ*, **183**, 743.

Sandage, A.R.(1988). *ARAA*, **26**, 561.

Schneider, P., Ehlers, J. and Falco, E.E. (1992). *Gravitational Lensing*. Berlin: Springer–Verlag.

Seldner, M. and Peebles, P.J.E. (1977). *ApJ*, **215**, 703.

Soucail, G., Fort, B., Mellier, Y and Picat, J.P. (1987). *A&A*, **172**, L14.

Sunyaev, R.A. and Zeldovich, Ya.B. (1970). *Astrophys. Sp. Sci.*, **7**, 20.

Tremaine, S. and Gunn, J.E. (1979). *Phys. Rev. Letts.*, **42**, 407.

Tremaine, S. and Richstone, D.O. (1977). *ApJ*, **212**, 311.

Turner, E. (1997). In *Critical Dialogues in Cosmology*, (ed. N. Turok), World Scientific (in press).

Zeldovich, Ya.B. (1964). *Astron. Zh.*, **41**, 19. [trans: *Soviet Astronomy – AJ*, **8**, 13.]

Zwicky, F. (1937). *ApJ.*, **86**, 217.

Chapter 5

Berry, M.V. (1989). *Principles of Cosmology and Gravitation*. Bristol: Adam Hilger.
Bondi, H. (1960). *Cosmology* 2nd edition. Cambridge: Cambridge University Press
Friedman, A.A. (1922). *Zeitschrift für Physik*, **10**, 377.
Friedman, A.A. (1924). *Zeitschrift für Physik*, **21**, 326.
Goldhaber, G., Deustra, S., Gabi, S., Groom, D., Hook, I., Kim, A., Kim, M., Lee, J., Pain, R., Pennypacker, C., Perlmutter, S., Small, I., Goobar, A., Ellis, R., McMahon, R., Boyle, B., Bunclark, P., Carter, D., Glazebrook, K., Irwin, M., Newberg, H., Filippenko, A.V., Matheson, T., Dopita, M., Mould, J. and Couch, W. (1997). In *Thermonuclear Supernovae*, (eds. P. Ruiz-Lapuente, R. Canal and J. Isern), 777. Dordrecht: Kluwer Academic Publishers.
Robertson, H.P. (1935). *ApJ.*, **82**, 284.
Walker, A.G. (1936). *Proc. Lond. Math. Soc.*, Ser. 2., **42**, 90.
Weinberg, S. (1972). *Gravitation and Cosmology*. New York: John Wiley and Co.
Weyl, H. (1923). *Phys. Zeitschrifte*, **29**, 230.

Chapter 6

Damour, T. and Taylor, J.H. (1991). *ApJ*, **366**, 501.
d'Inverno, R. (1995). *Introducing Einstein's Relativity*. Oxford: Clarendon Press.
Hellings, R.W., Adams, P.J., Anderson, J.D., Keesey, M.S., Lau, E.L., Standish, E.M., Canuto, V.M. and Goldman, I. (1983). *Phys. Rev. Letts.*, **51**, 1609.
Müller, J., Schneider, M., Soffel, M. and Ruder, H. (1991). *ApJLetts.*, **382**, L101.
Pais, A. (1982). *Subtle is the Lord ...: the Science and Life of Albert Einstein*. Oxford: Oxford University Press.
Penrose, R. (1997). *The Large, the Small and the Human Mind*. Cambridge: Cambridge University Press.
Pound, R.V. and Rebka, G.A.Jr. (1960). *Phys. Rev. Letts.*, **4**, 337.
Pound, R.V. and Snider, J.L. (1965). *Phys. Rev.*, **140**, B788.
Rindler, W. (1977). *Essential Relativity: Special, General and Cosmological*. New York: Springer–Verlag.
Shapiro, I.I. (1990). In *General Relativity and Gravitation, 1989*, (eds) N. Ashby, D.F. Bartlett and W.Wyss, 313. Cambridge: Cambridge University Press.
J. Taylor, J.H. (1992). *Phil. Trans. R. Soc.*, **341**, 117.
Weinberg, S. (1972). *Gravitation and Cosmology*. New York: John Wiley
Will, C.M. (1993). *Theory and Experiment in Gravitational Physics*. Cambridge: Cambridge University Press.

Chapter 7

Bernstein, J. and Feinberg, G. (1986). *Cosmological Constants: Papers in Modern Cosmology*. New York: Columbia University Press.
Blandford, R.D. and Narayan, R. (1992). *ARAA*, **30**, 311.
Bondi, H. (1960). *Cosmology*, 2nd edition. Cambridge: Cambridge University Press.
Carroll, S.M., Press, W.H. and Turner, E.L. (1992). *ARAA*, **30**, 499.
Close, L.M., Hall, P.B., Liu, M.C. and Hege, E.K. (1995). *ApJ*, **452**, L9.
Dashevsky, V.M. and Zeldovich, Ya.B. (1964). *Astr. Zh.*, **41**, 1071.
Dyer, C.C. and Roeder, R.C. (1972). *ApJL*, **174**, L115.
Dyer, C.C. and Roeder, R.C. (1973). *ApJL*, **180**, L31.
Einstein, A. (1917). *Sitzungsberichte Berl. Akad.*, **1**, 142.
Einstein, A. (1922). *Zeitschrift für Physik*, **11**, 326 and (1923). **16**, 228.

Felten, J.E. and Isaacman, R. (1986). *Rev. Mod. Phys.*, **58**, 689.

Friedman, A.A. (1922). *Zeitschrift für Physik*, **10**, 377.

Friedman, A.A. (1924). *Zeitschrift für Physik*, **21**, 326.

Gamow, G. (1970). *My World Line*. New York: Viking Press.

Kaiser, N. (1992). *ApJ*, **388**, 272.

Kolb, E.W. and Turner, M.S. (1990). *The Early Universe*. Redwood City, California: Addison–Wesley Publishing Co.

Longair, M.S. (1994). *Theoretical Concepts in Physics*. Cambridge: Cambridge University Press.

Longair, M.S. and Scheuer, P.A.G. (1970). *MNRAS*, **151**, 45.

Mattig, W. (1959). *Astr. Nach.*, **285**, 1.

Milne, E.A. and McCrea, W.H. (1934). *Q. J. Math.*, **5**, 64 and 73.

Schneider, D.P., Schmidt, M. and Gunn, J.E. (1991). *AJ*, **102**, 837.

Tropp, E.A., Frenkel, V.Ya. and Chernin, A.D. (1993). *Alexander A. Friedmann; the Man who Made the Universe Expand*. Cambridge: Cambridge University Press.

Weinberg, S. (1972) *Gravitation and Cosmology*. New York: John Wiley and Co.

White, S.D.M. (1990). In *Physics of the Early Universe*, (eds. J.A. Peacock, A.F. Heavens and A.T. Davies), 1. Edinburgh: SUSSP Publications.

Zeldovich, Ya.B. (1964). *Astron. Zh.*, **41**, 19. [trans: *Soviet Astronomy – AJ*, **8**, 13.]

Zeldovich, Ya.B. (1968). *Usp. Fiz. Nauk.*, **95**, 209. [trans: *Sov. Phys. – Usp.*, **11**, 381.]

Zeldovich, Ya.B. (1986). *Astrophys. Sp. Phys. Rev.*, **5**, 1.

Chapter 8

Anders, E. (1963). In *The Moon, Meteorites and Comets – The Solar System IV*, (eds. B.M. Middelhurst and G.P. Kuiper), 277. Chicago: University of Chicago Press.

Aragòn-Salamanca, A., Ellis, R.S., Couch, W.J. and Carter, D. (1993). *MNRAS*, **262**, 764.

Best, P.N., Longair, M.S. and Röttgering, H.J.A. (1996) *MNRAS*, **280**, L9.

Best, P.N., Longair, M.S. and Röttgering, H.J.A. (1997) *MNRAS*, **292**, 758.

Best, P.N., Longair, M.S. and Röttgering, H.J.A. (1998) *MNRAS*, **295**, 549.

Blandford, R.D. and Narayan, R. (1992). *ARAA*, **30**, 311.

Bolte, M. (1997). In *Critical Dialogues in Cosmology*, (ed. N. Turok), 156. Singapore: World Scientific.

Branch, D. and Tammann, G.A. (1992). *ARAA*, **30**, 359.

Burke, B.F., Lehàr, J. and Connor, S.R. (1992). In *Gravitational Lenses*, (eds. R. Kayser, T. Schramm and L. Nieser), 237. Berlin: Springer-Verlag.

Carroll, S.M., Press, W.H. and Turner, E.L. (1992). *ARAA*, **30**, 499.

Chaboyer, B. (1998). In *Cosmological Parameters and Evolution of the Universe: IAU Symposium No. 183*, (ed. K. Sato). Dordrecht: Kluwer Academic Publishers (in press).

Chambers, K.C., Miley, G.K. and van Breugel, W.J.M. (1987) *Nature*, **329**, 604.

Cowan, J.J.. Thielemann, F.-K. and Truran, J.W. (1991). *ARAA*, **29**, 447.

Dekel, A. (1995). In *Clustering in the Universe*, (eds. S. Maurogordato, C. Balkowski, C. Tao and J. Trân Thanh Vân), 89. Gif-sur-Yvettes, France: Edition Frontières.

Dekel, A., Burstein, D. and White, S.D.M. (1997). In *Critical Dialogues in Cosmology*, (ed. N. Turok), 175. Singapore: World Scientific.

Eales, S.A., Rawlings, S., Law-Green, D., Cotter, G. and Lacy, M. (1998). *MNRAS*, (in press).

Feast, M.W. and Catchpole, R.M. (1997). *MNRAS*, **286**, L1.

Freedman, W.L. (1997). In *Critical Dialogues in Cosmology*, (ed. N. Turok), 92. Singapore: World Scientific.

Freedman, W.L. (1998). In *Cosmological Parameters and Evolution of the Universe: IAU Symposium No. 183*, (ed. K. Sato). Dordrecht: Kluwer Academic Publishers (in press).

Fukugita, M., Futamase, T., Kasai, M. and Turner, E.L. (1992). *ApJ*, **393**, 3.

Garnavich, P.M., Kirshner, R.P., Challis, P., Tonry, J., Gilliland, R.L., Smith, R.C., Clocchiatti, A., Diercks, A., Filippenko, A.V., Hamuy, M., Hogan, C.J., Leibundgut, B., Phillips, M.M., Reiss, D., Riess, A.G., Schmidt, B.P., Spyromilio, J., Stubbs, C., Suntzeff, N.B. and Wells, L. (1998). *ApJ*, **493**, L53.

Goldhaber, G., Deustra, S., Gabi, S., Groom, D., Hook, I., Kim, A., Kim, M., Lee, J., Pain, R., Pennypacker, C., Perlmutter, S., Small, I., Goobar, A., Ellis, R., McMahon, R., Boyle, B., Bunclark, P., Carter, D., Glazebrook, K., Irwin, M., Newberg, H., Filippenko, A.V., Matheson, T., Dopita, M., Mould, J. and Couch, W. (1997). In *Thermonuclear Supernovae*, (eds. P. Ruiz-Lapuente, R.Canal and J. Isern), 777. Dordrecht: Kluwer Academic Publishers.

Gunn, J.E. (1978) In J.E. Gunn, M.S. Longair and M.J. Rees *Observational Cosmology*, (eds. A. Maeder, L. Martinet and G. Tammann). Geneva: Geneva Observatory Publications.

Hesser, J.E., Harris, W.E., VandenBerg, D.A., Allright, J.W.B., Shott, P. and Stetson, P. (1989). *Publ. Astron. Soc. Pacific*, **99**, 739.

Jaunsen, A.O., Jablonski, M., Pettersen, B.R. and Stabell, R. (1995) *A&A*, **300**, 323.

Jones, M., Saunders, R., Alexander, P., Birkinshaw, M., Dillon, N., Grainge, K., Hancock, S., Lasenby, A., Lefebvre, D., Pooley, G. Scott, P. Titterington, D and Wilson, D. (1993). *Nature*, **365**, 320.

Kapahi, V.K. (1987). In *Observational Cosmology*, (eds. A. Hewett, G. Burbidge and L.Z. Fang), 251. Dordrecht: D. Reidel Publishing Co.

Kaufmann, G. (1995). *MNRAS*, **274**, 153.

Kellermann, K.I. (1993). *Nature*, **361**, 134.

Kochanek, C.S. (1996). *ApJ*, **466**, 638.

Kormendy, J. and Richstone, D.O. (1995). *ARAA*, **33**, 581.

Kundic, T., Turner, E.T., Colley, W., Gott, J.R. III, Rhoads, J.E., Wang, Y., Bergeron, L.E., Gloria, K.A., Long, D.C., Malhorta, S. and Wambsganss, J. (1997). *ApJ*, **482**, 75.

Lilly, S.J. and Longair, M.S. (1984) *MNRAS*, **211**, 833.

Maeder, A. (1994). In *Frontiers of Space and Ground-based Astronomy*, (eds. W. Wamsteker, M.S. Longair and Y. Kondo), 177. Dordrecht: Kluwer Acedemic Publishers.

McCarthy, P.J., van Breugel, W.J.M., Spinrad, H. and Djorgovski, S. (1987). *ApJ*, **321**, L29.

Moaz, D., Bahcall, J.N., Doxsey, R., Schneider, D.P., Bahcall, N.A., Lahav, O. and Yanny, B. (1992). *ApJ*, **402**, 69.

Myers, S.T., Baker, J.E., Readhead, A.C.S, Leitch, E.M. and Herbig, T. (1997). *ApJ*, **485**, 1.

Panagia, N., Gilmozzi, R., Macchetto, F., Adorf, H-H. and Kirschner, R.P. (1991). *ApJ*, **380**, L23.

Perlmutter, S., Boyle, B., Bunclark, P., Carter, D., Couch, W., Deustua, S., Dopita, M., Ellis, R., Fillipenko, A.V., Gabi, S., Glazebrook, K., Goldhaber, G., Goobar, A., Groom, D., Hook, I., Irwin, M., Kim, A., Kim, M., Lee, J., Matheson, T., McMahon, R., Newberg, H., Pain, R., Pennypacker, C. and Smail, I. (1996). *Nuclear Phys. B, Proc. Suppl.* **51B**, 20.

Perlmutter, S., Aldering, G., Della Valle, M., Deustua, S., Ellis, R.S., Fabbro, S., Fruchter, A., Goldhaber, G., Goobar, A., Groom, D.E., Hook, I.M., Kim, A.G., Kim, M.Y., Knop, R., Lidman, C., McMahon, R.G., Nugent, P., Pain, R., Pennypacker, C.R., Panagia, N., Ruiz-Lapuente, P., Schaefer, B. and Walton, N. (1998). *Nature*, **391**, 51.

Rowan-Robinson M (1985). *The Cosmological Distance Ladder*. New York: W H Freeman and Company).

Rowan-Robinson M (1988). *Space Sci. Rev.*, **48** 1.

Sandage, A.R. (1988). *ARAA*, **26**, 561.

Sandage, A.R. (1995) In A.R. Sandage, R. Kron, and M.S. Longair, *The Deep Universe*, (eds. B. Bingelli and R. Buser). Berlin: Springer-Verlag.

Sandage, A.R. and Tammann, G.A. (1997). In *Critical Dialogues in Cosmology*, (ed. N. Turok), 130. Singapore: World Scientific.

Schmidt, B.P., Kirschner, R.P. and Eastman, R.G. (1992). *ApJ*, **395**, 366.

Schramm, D. (1990). In *Astrophysical Ages and Dating Methods*, (eds. E. Vangioni-Flam, M. Casse, J. Audouze and T. Tran Thanh Van, . Gif-sur-Yvettes: Editions Frontieres.

Schramm, D.N. and Wasserburg, G.T. (1970). *ApJ*, **162**, 57.

Surdej, J., Claesens, J.F., Crampton, D., Filippenko, A.V., Hutsemekers, D., Magain, P., Pirenne, B., Vanderreist, C. and Yee, H.K.C. (1993). *AJ*, **105**, 2064.

Tammann, G.A. (1998). In *Cosmological Parameters and Evolution of the Universe: IAU Symposium No. 183*, (ed. K. Sato). Dordrecht: Kluwer Academic Publishers (in press).

Weinberg, S. (1997). In *Critical Dialogues in Cosmology*, (ed. N. Turok), 195. Singapore: World Scientific.

Chapter 9

Cowie, L.L., Gardner, J.P., Hu, E.M., Songaila, A., Hodapp, K.-H. and Wainscoat, P.J. (1994). *Nature*, **371**, 43.

Ge, J., Bechtold, J. and Black, J.H. (1997). *ApJ*, **474**, 67.

Kolb, E.W. and Turner, M.S. (1990). *The Early Universe*. Redwood City, California: Addison–Wesley Publishing Co.

Longair, M.S. 1995. In Sandage, A.R., Kron, R. and Longair, M.S. *The Deep Universe* (eds B. Binggeli and R. Buser), 317. Berlin: Springer–Verlag.

Longair, M.S. and Sunyaev, R.A. (1971). *Uspekhi Fiz. Nauk.*, **105**, 41. [English translation: *Soviet Physics Uspekhi*, **14**, 569.]

Longair, M.S. (1992). *High Energy Astrophysics, Vol. 1*. Cambridge: Cambridge University Press.

Peebles, P.J.E. (1966). *ApJ*, **146**, 542.

Peebles, P.J.E. (1968). *ApJ*, **153**, 1.

Pozdnyakov, L.A., Sobol, I.M. and Sunyaev, R.A. (1983). *Astrophys. Sp. Phys. Rev.*, **2**, 263.

Sunyaev, R.A. and Zeldovich, Ya.B. (1980). *ARA&A*, **18**, 537.

Weymann, R. (1965). *Phys. Fluids*, **8**, 2112.

Chapter 10

Hogan, C.J. (1997). In *Critical Dialogues in Cosmology*, (ed. N. Turok), 50. Singapore: World Scientific.

Kolb, E.W. and Turner, M.S. (1990). *The Early Universe*. Redwood City, California: Addison–Wesley Publishing Co.

Linsky, J.L, Diplas, A, Savage, B. Andrulis, C. and Brown, A. (1994). In *The Frontiers of Space and Ground-based Astronomy*, 27th ESLAB Symposium, (eds. W. Wamsteker, M.S. Longair and Y. Kondo), 301. Dordrecht: Kluwer Scientific Publishers.

Olive, K. and Steigman, G. (1995). *ApJS*, **97**, 49.

Songaila, A., Wampler, E.J. and Cowie, L.L. (1997). *Nature*, **385**, 137.

Spite, M. and Spite, F. (1982). *Nature*, **297**, 483 and *A&A*, **115**, 357.

Steigman, G. (1997). In *Critical Dialogues in Cosmology*, (ed. N. Turok), 63. Singapore: World Scientific.

Tytler, D., Fan, X-M. and Burles, S. (1996). *Nature*, **381**, 207.

Wagoner, R.V. (1973). *ApJ*, **179**, 343.

Wagoner, R.W. Fowler, W.A. and Hoyle, F. (1967). *ApJ*, **148**, 3.

Webb, J.K., Carswell, R.F., Lanzetta, K.M., Ferlet, R., Lemoine, M., Vidal-Madjar, A. and Bowen, D.V. (1997). *Nature*, **388**, 250.

Weinberg, S. (1972) *Gravitation and Cosmology*. New York: John Wiley and Co.

Zeldovich, Ya.B. and Novikov, I.D. (1983). *Relativistic Astrophysics, Vol. 2*, (ed. G. Steigman). Chicago: Chicago University Press.

Chapter 11

Carroll, S.M., Press, W.H. and Turner, E.L. (1992). *ARAA*, **30**, 499.

Coles, P. and Lucchin, F. (1995). *Cosmology – the Origin and Evolution of Cosmic Structure*. Chichester: John Wiley and Sons.

Gunn, J.E. (1978). In *Observational Cosmology* by J.E. Gunn, M.S. Longair and M.J. Rees. Geneva: Geneva Observatory Publications.

d'Inverno, R. (1995). *Introducing Einstein's Relativity*. Oxford: Clarendon Press.

Heath, D.J. (1977). *MNRAS*, **179**, 351.

Jeans, J.H. (1902). *Phil. Trans. Roy. Soc.*, **199**, 1.

Lemâitre, G. (1933). *Comptes Rendus*, **196**, 903.

Lifshitz, E.M. (1946). *J. Phys. USSR Acad. Sci.*, **10**, 116.

Longair, M.S. (1991). *Theoretical Concepts in Physics*. Cambridge: Cambridge University Press.

Padmanabhan, T. (1993). *Structure Formation in the Universe*. Cambridge: Cambridge University Press.

Peebles, P.J.E. (1980). *The Large-Scale Structure of the Universe*. Princeton: Princeton University Press.

Tolman, R.C (1934) *Proc. Nat. Acad. Sciences*, **20**, 169. .

Tonks, L. and Langmuir, I. (1929). *Phys. Rev.*, **33**, 195.

Weinberg, S. (1972). *Gravitation and Cosmology*. New York: John Wiley and Co.

Chapter 12

Coles, P. and Lucchin, F. (1995). *Cosmology – the Origin and Evolution of Cosmic Structure*. Chichester: John Wiley and Sons.

Efstathiou, G. (1990). In *Physics of the Early Universe*, (eds. J.A. Peacock, A.F. Heavens & A.T. Davies), 361. Edinburgh: SUSSP Publications.

Guth, A. (1981). *Phys. Rev.*, **D23**, 347.

Padmanabhan, T. (1993). *Structure Formation in the Universe*. Cambridge: Cambridge University Press.

Peebles, P.J.E. (1980). *The Large-Scale Structure of the Universe*. Princeton: Princeton University Press.

Peebles, P.J.E. (1981). *ApJ*, **248**, 885.

Peebles, P.J.E. (1993). *Principles of Physical Cosmology*. Princeton: Princeton University Press.
Rindler, W. (1956). *MNRAS*, **116**, 662.
Sakharov, A.A. (1965). *ZhETP*, **49**, 345. [English translation: (1966) *Sov. Phys. JETP*, **22**, 241.]
Silk, J. (1968). ApJ, **151**, 459.
Sunyaev, R.A. and Zeldovich, Ya.B. (1970). *Ap. Sp. Sc.*, **7**(1), 1.
Sunyaev, R.A. and Zeldovich, Ya.B. (1972). *A&A*, **20**(2), 189.
Weinberg, S. (1972). *Gravitation and Cosmology*. New York: John Wiley and Co.
Zeldovich, Ya.B. (1970). *A&A*, **5**, 84.
Zeldovich, Ya.B. (1993). In *Selected Works of Yakov Borisevich Zeldovich, Vol. 2. Particles, Nuclei and the Universe* (eds. J.P. Ostriker, G.I. Barenblatt and R.A. Sunyaev). Princeton: Princeton University Press.

Chapter 13

Coles, P. & Lucchin, F. (1995). *Cosmology – the Origin and Evolution of Cosmic Structure*. Chichester: John Wiley & Sons.
Efstathiou, G. (1990). In *Physics of the Early Universe*, (eds. J.A. Peacock, A.F. Heavens & A.T. Davies), 361. Edinburgh: SUSSP Publications.
Kochanek, C.S. (1996). *ApJ*, **466**, 638.
Kolb, E.W. & Turner, M.S. (1990). *The Early Universe*. Redwood City, California: Addison–Wesley Publishing Co.
Lyubimov, V.A., Novikov, E.G., Nozik, V.Z., Tretyakov, E.F. & Kozik, V.S. (1980). *Phys. Lett*, **94B**, 266.
Peebles, P.J.E. (1982). *Astrophys. J.*, **263**, L1.
Peebles, P.J.E. (1993). *Principles of Physical Cosmology*. Princeton: Princeton University Press.
Smith, P.F., Arnison, G.T.J., Horner, G.J., Lewin, J.D., Alner, G.J., Spooner, N.J.C., Quenby, J.J., Sumner, T.J., Bewick, A., Li, J.P., Shaul, D., Ali, T., Jones, W.G., Smith, N.J.T., Davies, G.J., Lally, C.H., Vandersputte, M.J., Barton, J.C. & Blake, P.R. (1996). *Phys. Rev.* **B 379**, 299.
Turner, M.S. (1997). In *Relativistic Astrophysics*, (eds. B.J.T Jones & D. Markovic), 83. Cambridge: Cambridge University Press.
Zeldovich, Ya.B. (1993). In *Selected Works of Yakov Borisevich Zeldovich, Vol. 2. Particles, Nuclei and the Universe*, (eds. J.P. Ostriker, G.I. Barenblatt & R.A. Sunyaev). Princeton: Princeton University Press.

Chapter 14

Bahcall, N. (1988). *ARA&A*, **26**, 631.
Bahcall, N. (1997). In *Critical Dialogues in Cosmology*, (ed. N. Turok), 221. Singapore: World Scientific.
Bardeen, J.M., Bond, J.R., Kaiser, N. and Szalay, A.S. (1986). *ApJ*, **304**, 15.
Bond, J.R. & Szalay, A.S. (1983). *ApJ*, **276**, 443.
Boyle, B., Jones, L.P., Shanks, T., Marano, B., Zitelli, V. and Zamorani, G. (1991). In *The Space Distribution of Quasars*, (ed. D. Crampton), 191. San Francisco: ASP Conference Series.
Coles, P. & Lucchin, F. (1995). *Cosmology – the Origin and Evolution of Cosmic Structure*. Chichester: John Wiley & Sons.
Davis, M. Efstathiou, G., Frenk, C.S. & White, S.D.M. (1985). *ApJ*, **292**, 371.
Davis, M., Efstathiou, G., Frenk, C.S. and White, S.D.M. (1992). *Nat*, **356**, 489.

Dekel, A. (1987). *Comments Astrophys.*, **11**, 235.

Dekel, A. and Rees, M.J. (1987). *Nat*, **326**, 455.

Efstathiou, G. (1990). In *Physics of the Early Universe*, (eds. J.A. Peacock, A.F. Heavens & A.T. Davies), 361. Edinburgh: SUSSP Publications.

Efstathiou and Bond (1986). *MNRAS*, **218**, 103.

Frenk, C.S. (1986). *Phil. Trans. R. Soc. Lond.*, **A 330**, 517.

Hamilton, A.J.S., Kumar, P., Lu, E. and Matthews, A. (1991). *ApJ*, **374**, L1.

Harrison, E.R. (1970). *Phys. Rev.*, **D1**, 2726.

Gott, J.R. (1977). In *Critical Dialogues in Cosmology*, (ed. N. Turok), 519. Singapore: World Scientific.

Iovino, A., Shaver, P.A. and Christiani, S. (1991). In *The Space Distribution of Quasars*, (ed. D. Crampton), 202. San Francisco: ASP Conference Series.

Kaiser, N. (1984). *ApJ*, **284**, L9.

Kaiser, N. (1987).*MNRAS*, **277**, 1.

Kolb, E.W. & Turner, M.S. (1990). *The Early Universe*. Redwood City, California: Addison–Wesley Publishing Co.

Longair, M.S. (1997). *High Energy Astrophysics. Vol. 1.* Cambridge: Cambridge University Press.

Lynden-Bell, D. (1967). *Mon. Not. Roy. Astron. Soc.*, **136**, 101.

Ostriker, J.P. and Cowie, L.L. (1981). *ApJ*, **243**, L127.

Padmanabhan, T. (1997). In *Gravitation and Cosmology: Proc. ICGC-95 Conference, Pune*, (eds. S. Dhurandhar and T. Padmanabhan), 37. Dordrecht: Kluwer Academic Publishers.

Peacock, J.A. and Heavens, A.F. (1985). *MNRAS*, **217**, 805.

Peacock, J.A. and Dodds, S.J. (1994). *MNRAS*, **267**, 1020.

Peebles, P.J.E. (1980). *The Large-Scale Structure of the Universe*. Princeton: Princeton University Press.

Peebles, P.J.E. (1983). *ApJ*, **263**, L1.

Peebles, P.J.E. (1993). *Principles of Physical Cosmology.* Princeton: Princeton University Press.

Shanks, T., Boyle, B.J., Fong, R. and Peterson, B.A. (1987). *MNRAS*, **227**, 739.

Sunyaev, R.A. and Zeldovich, Ya.B. (1970). *Astrophys. Sp. Sci.*, **9**, 368.

Turok, N. (ed.) (1997). *Critical Dialogues in Cosmology.* Singapore: World Scientific.

Zeldovich, Ya.B. (1972). *MNAS*, **160**, 1P.

Zeldovich, Ya.B. (1993). In *Selected Works of Yakov Borisevich Zeldovich, Vol. 2. Particles, Nuclei and the Universe*, (eds. J.P. Ostriker, G.I. Barenblatt & R.A. Sunyaev). Princeton: Princeton University Press.

Zeldovich, Ya.B. and Novikov, I.D. (1983). *Relativistic Astrophysics, Vol. 2.* Chicago: Chicago University Press.

Chapter 15

Bennett, C.L., Banday, A.J., Górski, K.M., Hinshaw, G., Jackson, P., Keegstra, P., Kogut, A., Smoot, G.F., Wilkinson, D.T. and Wright, E.L. (1996). *ApJ*, **464**, L1.

Bersanelli, M., Bouchet, F.R., Efstathiou, G., Griffin, M., Lamarre, J.M., Mandolesi, N., Norgaard-Nielsen, H.U., Pace, O., Polny, J., Puget, J.-L., Tauber, J., Vittorio, N. and Volonté, S. (1996). Phase A study for the *Cobras/Samba* Mission. Paris: European Space Agency D/SCI(96)3.

Blain, A.W, Ivison, R.J. and Smail, I. (1998). *MNRAS*, **296**, L29.

Bond, and Efstathiou, G. (1987). *MNRAS*, **226**, 655.

Bracewell, R.N. (1986). *The Fourier Transform and its Applications*. New York: McGraw–Hill Book Company.

Coles, P. & Lucchin, F. (1995). *Cosmology – the Origin and Evolution of Cosmic Structure*. Chichester: John Wiley & Sons.

Crittenden, R., Bond, R., Davis, R.L., Efstathiou, G. and Steinhardt, P.J. (1993). *Phys. Rev. Letts.*, **71**, 324.

Davis, R.L., (1992). *Phys. Rev. Letts.*, **69**, 1856.

Efstathiou, G. (1988). In *Large-scale Motions in the Universe*, (eds. V.C. Rubin and G.V. Coyne), 299. Princeton: Princeton University Press.

Efstathiou, G. (1990). In *Physics of the Early Universe*, (eds. J.A. Peacock, A.F. Heavens and A.T. Davies), 361. Edinburgh: SUSSP Publications.

Hancock, S., Gutierrez, C.M., Davies, R.D., Lasenby, A.N., Rocha, G., Rebolo, R., Watson, R.A. and Tegmark, M. (1997). *MNRAS*, **289**, 505.

Harrison. E.R. (1970). *Phys. Rev.*, **D1**, 2726.

Hu, W. (1996). In *CMBR etc*, (eds. Martinez-Gonzales, E.), 207. Berlin: Springer-Verlag.

Hu, W. and Sugiyama, N. (1995). *ApJ*, **444**, 489.

Hu, W., Sugiyama, N. and Silk, J. (1997). *Nature*, **386**, 37.

Jones, B.J.T. and Wyse, R.F.G. (1985). *A&A*, **149**, 144.

Kibble, T. (1976). *J. Phys. A*, **9**, 1387.

Kogut, A., Banday, A.J., Bennett, C.L., Górski, K.M., Hinshaw, G., Smoot, G.F. and Wright, E.L. (1996). *ApJ*, **464**, 29.

Kolb, E.W. & Turner, M.S. (1990). *The Early Universe*. Redwood City, California: Addison–Wesley Publishing Co.

Mathews, J. and Walker, R.L. (1973). *Mathematical Methods of Physics*. New York: W.A. Benjamin.

Netterfield, C.B., Devlin, M.J., Jarolik, N., Page, L. and Wollack, E.J. (1997). *ApJ*, **474**, 47.

Padmanabhan, T. (1993). *Structure Formation in the Universe*. Cambridge: Cambridge University Press.

Padmanabhan, T. (1996). *Cosmology and Astrophysics through Problems*, 437-440. Cambridge: Cambridge University Press.

Partridge, R.B. (1998). In *Determination of Cosmological Parameters and the Evolution of the Universe*, IAU Symp. No. 183, (ed. K. Sato *et al.*). Dordrecht: Kluwer Academic Publishers (in press).

Partridge, R.B., Richards, E.A., Fomalont, E.B., Kellermann, K.I. and Windhorst, R.A. (1997). **483**, 38.

Peacock, J.A. and Dodds, S.J. (1994). *MNRAS*, **267**, 1020.

Peebles, P.J.E. (1968). *ApJ*, **153**, 1.

Peebles, P.J.E. (1980). *The Large Scale Structure of the Universe*. Princeton: Princeton University Press.

Peebles, P.J.E. (1993). *Principles of Physical Cosmology*. Princeton: Princeton University Press.

Peebles, P.J.E. and Yu, J.T. (1970). *ApJ*, **162**, 815.

Readhead, A.C.S. and Lawrence, C.R. (1992). *ARA&A*, **30**, 653.

Sachs, R.K. and Wolfe, A.M. (1967). *ApJ*, **147**, 73.

Sakharov, A.D. (1965). *Zh. Exp. Theo. Fiz*, **49**, 345.

Scott, P.F., Saunders, R., Pooley, G., O'Sullivan, C., Lasenby, A.N., Jones, M., Hobson, M.P., Duffett-Smith, P. and Baker, J. (1996). *ApJ*, **461**, 1.

Smail, I., Ivison, R.J. and Blain, A.W. (1997). *ApJ*, **490**, L5.

Starobinsky, A.A. (1985). *Sov. Astr. Letts*, *11*, 133.

Sunyaev, R.A. and Zeldovich, Ya.B. (1980). *MNRAS*, **190**, 413.

Turner, M.S. (1997). In *Relativistic Astrophysics*, (eds. B.J.T Jones & D. Markovic), 83. Cambridge: Cambridge University Press.

Turok, N. (1989). *Phys. Rev. Letts.*, **63**, 2625.

Vishniac, E.T. (1987). *ApJ*, **322**, 597.
White, M., Scott, D. and Silk, J. (1994). *ARA&A*, **32**, 319.
Zeldovich, Ya.B. (1972). *MNRAS*, **160**, 1P.
Zeldovich, Ya.B., Kurt, V.G. and Sunyaev, R.A. (1968). *Zh. Exsp. Teor. Fiz.*, **55**, 278. [English translation: *Sov. Phys. – JETP*, **28**, 146.]

Chapter 16

Bardeen, J.M., Bond, J.R., Kaiser, N. and Szalay, A.S. (1986). *ApJ*, **304**, 15.
Blain, A.W. and Longair, M.S. (1993). *MNRAS*, **265**, 21P.
Blumenthal, G.R., Faber, S.M., Promak, J.R. and Rees, M.J. (1984). *Nature*, **311**, 517.
Coles, P. & Lucchin, F. (1995). *Cosmology – the Origin and Evolution of Cosmic Structure*. Chichester: John Wiley & Sons.
Coles, P., Melott, A.L. and Shandarin, S.F. (1993). *MNRAS*, **260**, 765.
Efstathiou, G. (1990). In *Physics of the Early Universe*, (eds. J.A. Peacock, A.F. Heavens and A.T. Davies), 361. Edinburgh: SUSSP Publications.
Efstathiou, G. (1995). In *Galaxies in the Young Universe*, (eds. H. Hippelein, K. Meisenheimer and H.-J. Röser), 299. Berlin: Springer-Verlag.
Efstathiou, G. and Rees, M.J. (1988). *MNRAS*, **230**, 5P.
Hamilton, A.J.S., Kumar, P., Lu, E. and Matthews, A. (1991). *ApJ*, **374**, L1.
Kauffman and White, S.D. (1993). *MNRAS*, **261**, 921.
Kormendy, J. and Richstone, D. (1995). *ARAA*, **33**, 581.
Lin, C.C., Mestel, L. and Shu, F. (1965). *ApJ*,**142**, 1431.
Lynden-Bell, D. (1967). *Mon. Not. Roy. Astron. Soc.*, *136*, 101.
Ohta, K., Yamada, T., Nakanishi, K., Kohno, K., Akiyama, M. and Kawabe, R. (1996). *Nature*, **382**, 426.
Omont, A. (1996). In *Science with Large Millimetre Arrays*, (ed. P.A. Shaver), 82. Berlin: Springer-Verlag.
Omont, A., Petitjean, P., Guilloteau, S., McMahon, R.G., Solomon, P.M. and Pécontal, E. (1996). *Nature*, **382**, 428.
Peacock, J.A. and Dodds, S.J. (1994). *MNRAS*, **267**, 1020.
Peacock, J.A. and Heavens, A.F. (1985). *MNRAS*, **217**, 805.
Press, W.H. and Schechter, P. (1974). *ApJ*, **187**, 425.
Rees, M.J. and Ostriker, J.E. (1977). *MNRAS*, **179**, 541.
Silk, J. and Wyse, R.F.G. (1993). *Physics Reports*, **231**, 293.
Sutherland, R.S. and Dopita, M.A. (1993). *ApJS*, **88**, 253.
Zeldovich, Ya.B. (1970). *A&A*, **5**, 84.

Chapter 17

Abraham, R.G., Tanvir, N.R., Santiago, B., Ellis, R.S., Glazebrook, K and van den Bergh, S. (1996). *MNRAS*, **279**, L47.
Barcons, X. and Fabian, A.C. (1989). *MNRAS*, **237**, 119.
Blain, A.W. and Longair, M.S. (1993). *MNRAS*, **264**, 509.
Blain, A.W. and Longair, M.S. (1996). *MNRAS*, **279**, 847.
Boyle, B.J., Fong, R., Shanks, T. and Peterson, B.A. (1990). *MNRAS*, **243**, 1.
Boyle, B.J., Jones, L.R., Shanks, T., Marano, B., Zitelli, V. and Zamorani, G. (1991). In *The Space Distribution of Quasars*, (ed. D. Crampton), 191. San Francisco: ASP Conf. Series.
Bracessi, A., Formiggini, L. and Gandolfi, E. (1970). *A&A*, **5**, 204 (erratum: *A&A*, **23**, 159).

Cowie, L.L., Hu, E.M. and Songaila, A. (1995). *Nature*, **377**, 603.

Cowie, L.L., Songaila, A., Hu, E.M. and Cohen, J.D. (1996). *AJ*, **112**, 839.

Dunlop, J.S. (1994). In *The Frontiers of Space and Ground-based Astronomy*, 27th ESLAB Symposium, (eds. W. Wamsteker, M.S. Longair and Y. Kondo), 395. Dordrecht: Kluwer Academic Pubishers.

Dunlop, J.S. (1998). In *Observational Cosmology with the New Radio Surveys*, (eds. M.N. Bremer, N. Jackson and I. Péres-Fournon), 157. Dordrecht: Kluwer Academic Publishers.

Dunlop, J.S. and Peacock, J.A. (1990). *MNRAS*, **247**, 19.

Dunlop, J.S., Peacock, J.A. and Windhorst, R.A. (1995). In *Galaxies in the Young Universe*, (eds. H. Hippelein, K. Meisenheimer and H.-J. Röser), 84. Berlin: Springer-Verlag.

Ekers, R., Fanti, R. and Padrielli, L. (eds.) (1996). *Extragalactic Radio Sources*, IAU Symp. No. 175. Dordrecht: Kluwer Academic Publishers.

Ellis, R.G. (1997). *ARA&A*, **35**, 389.

Ellis, R.G., Colless, M., Broadhurst, T.J., Heyl, J.S. and Glazebrook, K. (1996). *MNRAS*, **280**, 235.

Felten, J.E. (1977). *AJ*, **82**, 869.

Fomalont, E.B., Kellermann, K.I., Anderson, M.C., Weistrop, D., Wall, J.V., Windhorst, R.A. and Kristian, J.A. (1988). *AJ*, **96**, 1187.

Giavalisco, M., Livio, M., Bohlin, R.C., Macchetto, F.D. and Stecher, T.P. (1996). *AJ*, **112**, 369.

Glazebrook, K., Ellis, R.S., Colless, M., Broadhurst, T.J., Allington-Smith, J.R. and Tanvir, N.R. (1995). *MNRAS*, **275**, L19.

Gunn, J.E., Hoessel, J.G., Westphal, J.A., Perryman, M.A.C., and Longair, M.S. (1981). *MNRAS*, **194**, 111.

Harrison, E.R. (1987). *Darkness at Night: A Riddle of the Universe*. Cambridge, Massachusetts: Harvard University Press.

Hasinger, G., Burg, R., Giaconni, R., Hartner, G., Schmidt, M., Trümper, J. and Zamorani, G. (1993). *A&A*, **275**, 1.

Hawkins, M.R.S. (1986). *MNRAS*, **219**, 417.

Helou, G., Soifer, B.T. and Rowan-Robinson, M. (1985). *ApJ*, **298**, L7.

Hewett, P.C, Foltz, C.B. and Chaffee, F. (1993). *ApJ*, **406**, 43.

Hewish, A. (1961). *MNRAS*, **123**, 167.

Hook, I.M., McMahon, R.G., Boyle, B.J. and Irwin, M.J. (1991). In *The Space Distribution of Quasars*, (ed. D. Crampton), 67. San Francisco: ASP Conf. Series.

Hubble E.P. (1936). *The Realm of the Nebulae*. New Haven: Yale University Press.

Irwin, M., McMahon, R.G. and Hazard, C. (1991). In *The Space Distribution of Quasars*, (ed. D. Crampton), 117. San Francisco: ASP Conf. Series.

Kashlinsky, A., Mather, J.C., Odenwald, S. and Hauser, M.G. (1996). *ApJ*, **470**, 681.

Kennefick, J.D., Djorgovski, S.G. and de Carvalo, R.R. (1995). *AJ*, **110**, 2553.

Koo, D.C. and Kron, R.G. (1982). *A&A*, **105**, 107.

Laing, R.A., Riley, J.M. and Longair, M.S. (1983). *MNRAS*, **204**, 1511.

Lilly, S.J., Tresse, L., Hammer, F., Crampton, D. and LeFevre, O. (1995). *ApJ*, **455**, 108.

Longair, M.S. (1966). *MNRAS*, **133**, 421.

Longair, M.S. (1978). In Gunn, J.E., Longair, M.S. and Rees, M.J. *Observational Cosmology*. Geneva: Geneva Observatory Publications.

Longair, M.S. (1992). *High Energy Astrophysics, Vol. 1*. Cambridge, Cambridge University Press.

Longair, M.S. (1995a). In *Twentieth Century Physics*, (eds. L.M. Brown, A. Pais and A.B. Pippard), Volume 3, 1691.

Longair, M.S. (1995b) In A.R. Sandage, R.G. Kron and M.S. Longair, *The Deep Universe*, (eds. B. Binggeli and H. Buser), 317. Berlin: Springer-Verlag.

Longair, M.S. (1997). *A&G*, **38**, No.1, 10.

Longair, M.S. and Scheuer, P.A.G. (1970). *MNRAS*, **151**, 45.

Loveday, J., Peterson, B.A., Efstathiou, G. and Maddox, S.J. (1992). *ApJ*, **390**, 338.

Majewski, S.R., Munn, J.A., Kron, R.G., Bershady, M.A. and Smetanka, J.J. (1991). In *The Space Distribution of Quasars*, (ed. D. Crampton), 55. San Francisco: ASP Conf. Series.

Martin, C. and Bowyer, S. (1989). *ApJ*, **338**, 677.

Metcalfe, N., Shanks, T., Campos, A., Fong, R. and Gardner, J.P. (1996). *Nature*, **383**, 236.

Oliver, S.J., Rowan-Robinson, M. and Saunders, W. (1992). *MNRAS*, **256**, 15P.

Osmer, P.S., (1982). *ApJ*, **253**, 28.

Peacock, J.A. (1985). *MNRAS*, **217**, 601.

Persic, M., de Zotti, G., Boldt, E.A., Marshall, F.E., Danese, L., Francheschini, A. and Palumbo, G.G.C. (1989). *ApJ*, **336**, L47.

Petrosian, V. and Salpeter, E.E. (1968). *ApJ*, **151**, 411.

Rowan-Robinson, M. (1968). *MNRAS*, **141**, 445.

Rowan-Robinson, M., Benn, C.R., Lawrence, A., McMahon, R.G. and Broadhurst, T.J. (1993). *MNRAS*, **263**, 123.

Sandage, A.R. (1965). *ApJ*, **141**, 1560.

Saunders, W., Rowan-Robinson, M., Lawrence, A., Efstathiou, G., Kaiser, N., Ellis, R.S. and Frenk, C.S. (1990). *MNRAS*, **242**, 318.

Schade, D., Lilly, S.J., Crampton, D., Hammer, F., LeFevre, O. and Tresse, L. (1995). *ApJ*, **451**, L1.

Scheuer, P.A.G. (1957). *Proc. Camb. Phil. Soc.*, **53**, 764.

Scheuer, P.A.G. (1974). *MNRAS*, **167**, 329.

Scheuer, P.A.G. (1990). In *Modern Cosmology in Retrospect*, (eds. B. Bertotti, R. Balbinot, S. Bergia and A. Messina), 331. Cambridge: Cambridge University Press.

Schmidt, M. (1968). *ApJ.*, **151**, 393.

Schmidt, M. and Green, R.F. (1983). *ApJ*, **269**, 352.

Schmidt, M., Schneider, D.P. and Gunn, J.E. (1991). In *The Space Distribution of Quasars*, (ed. D. Crampton), 109. San Francisco: ASP Conf. Series.

Schmidt, M., Schneider, D.P. and Gunn, J.E. (1995). *AJ*, **110**, 68.

Shectman, S.A. (1974). *ApJ*, **188**, 233.

Sullivan, W.T. III (1990). In *Modern Cosmology in Retrospect*, (eds. B. Bertotti, R. Balbinot, S. Bergia and A. Messina), 309. Cambridge: Cambridge University Press.

Wall, J.V. (1990). In *The Galactic and Extragalactic Background Radiation*, IAU Symp. No. 139, (eds. S. Bowyer and C. Leinert), 327. Dordrecht: Kluwer Academic Publishers.

Wall, J.V. (1996). In *Extragalactic Radio Sources*, IAU Symposium No. 175, (eds. R. Ekers, C. Fanti and L. Padrielli), 547. Dordrecht: Kluwer Academic Publishers.

Warren, S.J., Hewett, P.C., Irwin, M.J., McMahon, R.G., Kibblewhite, E.J., Bridgeland, M.T. and Bunclark, P.S. (1987). *Nature*, **325**, 131.

Warren, S.J., Hewett, P.C. and Osmer, P.S. (1994). *ApJ*, **421**, 412.

Windhorst, R.A., Dressler, A. and Koo, D.A. (1987). In *Observational Cosmology*, IAU Symp. No. 124, (eds. A. Hewitt, G. Burbidge and L.-Z. Fang), 573. Dordrecht: D. Reidel Publishing Co.

Windhorst, R.A., Fomalont, E.B., Kellermann, K.I., Partridge, R.B., Richards, E., Franklin, B.E., Pascerelle, S.M. and Griffiths, R.E. (1995). *Nature*, **375**, 471.

Woltjer, L. (1990). In R.D. Blandford, H. Netzer and L. Woltjer, *Active Galactic Nuclei*, (eds. T.J.-L. Courviosier and M. Mayor), 1. Berlin: Springer-Verlag.

Chapter 18

Bajtlik, S., Duncan, R.C. and Ostriker, J.P. (1988). *ApJ*, **327**, 570.
Bergeron, J. (1988). In *QSO Absorption Lines: Probing the Universe*, (eds. J.C. Blades, D. Turnshek and C.A. Norman), 127. Cambridge: Cambridge University Press.
Blades, J.C., Turnshek, D. and Norman, C.A. (eds) (1988). *QSO Absorption Lines: Probing the Universe*. Cambridge: Cambridge University Press.
Blain, A.W. and Longair, M.S. (1993). *MNRAS*, **264**, 509.
Bochkarev, N.G. and Sunyaev, R.A. (1977). *Sov. Astr.*, **21**, 542.
Boksenberg, A. (1997). In *The Hubble Space Telescope and the High Redshift Universe*, (eds. N.R. Tanvir, A. Aragón-Salamanca and J.V. Wall), 283. Singapore: World Scientific Publishing Co.
Carswell, R.F. (1988). In *QSO Absorption Lines: Probing the Universe*, (eds. J.C. Blades, D. Turnshek and C.A. Norman), 91. Cambridge: Cambridge University Press.
Corbelli, E. and Salpeter, E.E. (1993). *ApJ*, **419**, 104.
Connally, A.J., Szalay, A.S., Dickenson, M., SubbaRao, M,U, and Brunner, R.J. (1997). *ApJ*, **486**, L11.
Cowie, L.L. (1988). In The Post-Recombination Universe, (eds N. Kaiser and A.N. Lasenby), 1. Dordrecht: Kluwer Academic Publishers.
Cowie, L.L., Lilly, S.J., Gardner, J. and McLean, I.S. (1988). *ApJ*, **332**, L29.
Fall, S.M. (1997). In *The Hubble Space Telescope and the High Redshift Universe*, (eds. N.R. Tanvir, A. Aragón-Salamanca and J.V. Wall), 303. Singapore: World Scientific Publishing Co.
Fall, S.M. (1998). (personal communication).
Fall, S.M., Charlot S. and Pei, Y.C. (1996). *ApJ*, **464**, L43.
Fall, S.M. and Pei, Y.C. (1993). *ApJ*, **402**, 479.
Gallego, J., Zamorano, J., Aragón-Salamanca, A. and Rego, M, (1995). *ApJ*, bf 455, L1.
Hartwick, F.D.A. (1976). *ApJ*, **209**, 418.
Jakobsen, P. (1991). In *The Early Observable Universe from Diffuse Backgrounds – 11th Moriond Conference*, (eds. B. Rocca-Volmerange, J.M. Deharveng and J. Trân Thanh Vân), 115. Gif-sur-Yvettes, France: Edition Frontières.
Kennicutt, R.C. (1989). *ApJ*, **344**, 685.
Kippenhahn, R. and Weigert, A. (1990). *Stellar Structure and Evolution*. Berlin: Springer-Verlag.
Kulkarni, V.P. and Fall, S.M. (1993). *ApJ*, **413**, 62.
Kutyrev, A.S. and Reynolds, R.J. (1989). *ApJ*, **344**, L9.
Lanzetta, K.M., Wolfe, A.M. and Turnshek, D.A. (1995). *ApJ*, **440**, 435.
Lanzetta, K.M., Wolfe, A.M., Turnshek, D.A., Lu, L., McMahon, R.G. and Hazard, C. (1991). *ApJS*, **77**, 1.
Larsen, R.B. (1972). *Nature*, **236**, 21.
Lilly, S.J., Tresse, L., Hammer, F., Crampton, D. and LeFevre, O. (1995). *ApJ*, **455**, 108.
Lilly, S.J. and Cowie, L.L. (1987). In *Infrared Astronomy with Arrays*, (eds. C.G. Wynn-Williams and E.E. Becklin), 473. Honolulu: Institute for Astronomy, University of Hawaii Publications.
Macchetto, F.D. and Dickenson, M. (1997). *Sci. Am.*, **276**, 66.
Madau, P. (1992). *ApJ*, **389**, L1.

Madau, P. (1998). In *The Origins of Galaxies, Stars, Planets, and Life*, (eds. J. M. Shull, C. E. Woodward, & H. A. Thronson). San Francisco: ASP Conference Series, (in press).

Madau, P., Ferguson, H.C., Dickenson, M.E., Giavalisco, M., Steidel, C.C. & Fruchter, A. (1996). *MNRAS*, **283**, 1388.

Madau, P., Pozzetti, L. and Dickenson, M, (1998). *ApJ*, **498**, 106.

Maloney, P. (1993). *ApJ*, **414**, 41.

Mazzarella, J.M. and Balzano, V.A. (1986). *ApJS*, **62**, 751.

Meier, (1976). *ApJ*, **207**, 343.

Meylan, G. (ed.) (1995). *QSO Absorption Lines*. Berlin: Springer-Verlag.

Miralda-Escudé, J. and Ostriker, J.P. (1990). *ApJ*, **350**, 1.

Murdoch, H.S., Hunstead, R.W., Pettini, M. and Blades, J.C. (1986). *ApJ*, **309**, 19.

Pagel, B.E.J. (1997). *Nucleosynthesis and Chemical Evolution of Galaxies*. Cambridge: Cambridge University Press.

Pettini, M., King, D.L., Smith, L.J. and Hunstead, R.W. (1996). *ApJ*, **486**, 665.

Pettini, M., Smith, L.J., Hunstead, R.W. and King, D, (1994). *ApJ*, **426**, 79.

Puget, J.-L., Abergel, A., Bernard, J.-P., Boulanger, F., Burton, W.B., Desert, F.-X. and Hartmann, D. (1996). *A&A*, **308**, L5.

Smail, I., Ivison, R.J. and Blain, A.W. (1997). *ApJ*, **490**, L5.

Songaila, A., Bryant, W. and Cowie, L.L. (1988). *ApJ*, **345**, L71.

Songaila, A., Cowie, L.L. and Lilly, S.J. (1990). *ApJ*, **348**, 759.

Songaila, A., Hu., E.M. and Cowie, L.L. (1995). *Nature*, **375**, 124.

Steidel, C.C. (1998). In *1996 Texas Symposium*, (in press).

Steidel, C.C. and Hamilton, D. (1992). *AJ*, **104**, 941.

Steidel, C.C. and Sargent, W.L.W. (1989). *ApJ*, **343**, L33.

Storrie-Lombardi, L.J., McMahon, R.G. and Irwin, M.J. (1996). *MNRAS*, **283**, L79.

Sunyaev, R.A. (1969). *Astrophys. Lett.*, **3**, 33.

Tanvir, N.R., Aragón-Salamanca, A. and Wall, J.V. (eds.) (1997). *The Hubble Space Telescope and the High Redshift Universe*. Singapore: World Scientific Publishing Co.

Tinsley, B.M. (1980). *Fund. Cosm. Phys.*, **5**, 287.

Toller, (1990). In *The Galactic and Extragalactic Background Radiation*, (eds. S. Bowyer and C. Leinert), 21. Dordrecht: Kluwer Academic Publishers.

Tyson, A. (1990). In *The Galactic and Extragalactic Background Radiation*, (eds. S. Bowyer and C. Leinert), 245. Dordrecht: Kluwer Aademic Publishers.

Weedman, D. (1994). *The First Stromlo Symposium: The Physics of Active Galaxies*, (eds. G.V. Bicknell, M.A. Dopita and P.J. Quinn), 409. San Francisco: ASP Conf. Series.

White, S.D. (1989). In *The Epoch of Galaxy Formation*, (eds. C.S. Frenk, R.S. Ellis, T. Shanks, A.F. Heavens and J.A. Peacock), 1. Dordrecht: Kluwer Academic Publishers.

Williams, R.E., Blacker, B., Dickenson, M., Dixon, W.V.D., Ferguson, H.C., Fruchter, A.S., Giavalisco, M., Gilliland, R.L., Heyer, I., Katsanis, R., Levay, Z., Lucas, R.A., McElroy, D.B., Petro, L., Postman, M., Adorf, H.M. and Hook, R.N. (1996). *AJ*, **112**,1335.

Wolfe, A.M. (1988). In *QSO Absorption Lines: Probing the Universe*, (eds. J.C. Blades, D. Turnshek and C.A. Norman), 297. Cambridge: Cambridge University Press.

Chapter 19

Beaver, E., Burbidge, E.M., Cohen, R., Junkkarinen, V., Lyons, R. and Rosenblatt, E. (1992). In *Science with the Hubble Space Telescope*, (eds. P. Benvenuti and E. Schreier), 53. Garching-bei-München: ESO Publications.

Bechtold, J., Weymann, R.J., Lin, Z. and Malkan, M.A. (1987). *ApJ*, **315**, 180.

Carswell, R.F. (1988). In *QSO Absorption Lines: Probing the Universe*, (eds. J.C. Blades, D. Turnshek and C.A. Norman), 91. Cambridge: Cambridge University Press.

Davidsen, A. (1993). *Science*, **259**, 327.

Davidsen, A.F., Kriss, G.A. and Zheng, W. (1996). *Nature*, **380**, 47.

Fabian, A.C. and Barcons, X. (1992). *ARA&A*, **30**, 429.

Gunn, J.E. and Peterson, B.A. (1965). *ApJ*, **142**, 1633.

Hasinger, G., Burg, R., Giacconi, R., Hartner, G., Schmidt, M., Trümper, J. and Zamorani, G. (1993). *A&A*, **275**, 1.

Hernquist, L., Katz, N., Weinberg, D.H. and Miralda-Escudé, J. (1996). *ApJ*, **457**, L51.

Hu, E.M., Kim, T.-S., Cowie, L.L., Songaila, A. and Rauch, M. (1995). *AJ*, **110**, 1526.

Ikeuchi, S. and Ostriker, J.P. (1987). *ApJ*, **301**, 552.

Jakobsen, P. (1991). In *The Early Observable Universe from Diffuse Backgrounds – 11th Moriond Conference*, (eds. B. Rocca-Volmerange, J.M. Deharveng and J. Trân Thanh Vân), 115. Gif-sur-Yvettes, France: Edition Frontières.

Jakobsen, P. (1995). In *The Extragalactic Background Radiation*, (eds. D. Calzetti, M. Livio and P. Madau), (in press). Cambridge: Cambridge University Press.

Jakobsen, P. (1996). In *Science with the Hubble Space Telescope – II*, (eds. P. Benvenuti, F.D. Macchetto and E.J. Schreier), 153. Paris: European Space Agency.

Jakobsen, P., Boksenberg, A., Deharveng, J.M., Greenfield, P., Jedrzejewski, R. and Paresce, F. (1994). *Nature*, **370**, 35.

Katz, N., Weinberg, D.H., Hernquist, L. and Miranda-Escudé (1996). *ApJ*, **457**, L57.

Kurt, V.G. and Sunyaev, R.A. (1967). *Cosm. Res.*, **5**, 496.

Longair, M.S. (1965). *MNRAS*, **129**, 419.

Marshall, F.E., Boldt, E.A., Holt, S.S., Miller, R.B., Mushotzky, R.F., Rose, L.A., Rothschild, R.E. and Serlemitsos, P.J. (1980). *ApJ*, **235**, 4.

Mather, J. (1995). In *The Extragalactic Background Radiation*, (eds. D. Calzetti, M. Livio and P. Madau), 169. Cambridge: Cambridge University Press.

Miranda-Escudé, J., Cen, R., Ostriker, J.P. and Rauch, M. (1996). *ApJ*, **471**, 582.

Møller, P and Jakobsen, P. (1990). *A&A*, **228**, 299.

Ostriker, J.P. (1988). In *QSO Absorption Lines: Probing the Universe*, (eds. J.C. Blades, D. Turnshek and C.A. Norman), 319. Cambridge: Cambridge University Press.

Ostriker, J.P. and Ikeuchi, S. (1983). *ApJ*, **268**, L63.

Reimers, D., Clavel, J., Groote, D., Engels, D., Hagen, H.J., Naylor, T., Wamsteker, W. and Hopp, U. *A&A*, **218**, 71.

Ryle, M. and Sandage, A.R. (1964). *ApJ*, **139**, 419.

Scheuer, P.A.G. (1965). *Nature*, **207**, 963.

Songaila, A., Hu., E.M. and Cowie, L.L. (1995). *Nature*, **375**, 124.

Taylor, G.B. and Wright, E.L. (1989). *ApJ*, **405**, 125.

Vogel, S. and Reimers, D. (1995). *A&A*, **294**, 377.

Weymann, R. (1967). *ApJ*, **147**, 887.

Chapter 20

Aragón-Salamanca, A., Ellis, R.S., Couch, W.J. and Carter, D. (1993). *MNRAS*, **262**, 764.

Barnes, J. and Efstathiou, G. (1987). *ApJ*, **319**, 575.

Barrow, J.D. and Tipler, F.J. (1986). *The Anthropic Cosmological Principle*. Oxford: Oxford University Press.

Best, P.N., Longair, M.S. and Röttgering, H.J.A. (1998). *MNRAS*, **295**, 549.

Biermann, L. (1950). *Z. Naturforsch.*, **5a**, 65.

Bludman, S.A. and Ruderman, M. (1977). *Phys. Rev. Letts.*, **38**, 255.

Butcher, H. and Oemler, A., Jr. (1978). *ApJ*, **219**, 18.

Carter, B. (1974). In Confrontation of Cosmological Theories with Observational Data, (ed. M.S. Longair), 291. Dordrecht: D. Reidel Publishing Company.

Davies, R.L., Efstathiou, G., Fall, S.M., Illingworth, G. and Schechter, P.L. (1983). *ApJ*, **266**, 41.

Dey, A. (1997). In *The Hubble Space Telescope and the High Redshift Universe*, (eds. N.R. Tanvir, A. Aragón-Salamanca and J.V. Wall), 373. Singapore: World Scientific Publishing Co.

Dickenson, M. (1997). In *The Hubble Space Telescope and the High Redshift Universe*, (eds. N.R. Tanvir, A. Aragón-Salamanca and J.V. Wall), 207. Singapore: World Scientific Publishing Co.

Dressler, A. (1994). *ARAA*, **22**, 185.

Dressler, A., Oemler, A. Jr., Couch, W.J., Smail, I., Ellis, R.G., Barger, A., Butcher, H., Poggianti, B.M. and Sharples, R.M. (1997). *ApJ*, **490**, 577.

Dressler, A. and Smail, I. (1997). In *The Hubble Space Telescope and the High Redshift Universe*, (eds. N.R. Tanvir, A. Aragón-Salamanca and J.V. Wall), 185. Singapore: World Scientific Publishing Co.

Dunlop,J., Peacock, J., Spinrad, H., Dey, A., Jimenez, R., Stern, D. and Windhorst, R. (1996). *Nature*, **381**, 581.

Efstathiou, G. (1995). In *Galaxies in the Young Universe*, (eds. H. Hippelein, K. Meisenheimer and H.-J. Röser), 299. Berlin: Springer-Verlag.

Ellis, R.G., Smail, I., Dressler, A., Couch, W.J., Oemler, A., Butcher, H. and Sharples, R.M. (1997). *ApJ*, **483**, 582.

Fall, S.M. (1983). In *Internal Kinematics and Dynamics of Galaxies*, (ed. E. Athanassoula), 391. Dordrecht: D. Reidel Publishing Co.

Fall, S.M. and Efstathiou, G. (1980). *MNRAS*, **193**, 189.

Ferguson, H.C. (1997). In *The Hubble Space Telescope and the High Redshift Universe*, (eds. N.R. Tanvir, A. Aragón-Salamanca and J.V. Wall), 15. Singapore: World Scientific Publishing Co.

Freeman, K.C. (1970). *ApJ*, **160**, 811.

Fukugita, M., Hogan, C.J. and Peebles, P.J.E. (1996). *Nature*, **381**, 489.

Giavalisco, M., Steidel, C.C., Adelberger, K.L., Pettini, M. and Dickenson, M.E. (1997). In *The Hubble Space Telescope and the High Redshift Universe*, (eds. N.R. Tanvir, A. Aragón-Salamanca and J.V. Wall), 25. Singapore: World Scientific Publishing Co.

Gribben, J. and Rees, M.J. (1989). *Dark Matter, Mankind and Anthropic Cosmology*. New York: Bantam Books.

Gunn, J.E. (1978). In *Observational Cosmology* by J.E. Gunn, M.S. Longair and M.J. Rees, 1. Geneva: Geneva Observatory Publications.

Guth, A. (1981). *Phys. Rev. D*, **23**, 347.

Guth, A. (1997). *The Inflationary Universe. The Quest for a New Theory of Cosmic Origins*. Reading, MA: Addison-Wesley.

Hausman, M.A. and Ostriker, J.P. (1978). *ApJ*, **224**, 320.

Hernquist, L. and Mihos, J.C. (1995). *ApJ*, **448**, 41.

Im, M., Casertano, S., Griffiths, R.E. and Ratnatunga, K.U. (1995). *ApJ*, **441**, 494.

Jones, B.J.T (1973). *ApJ*, **181**, 269.

Jones, B.J.T. and Peebles, P.J.E. (1972). *Comments Ap. Sp.Phys.*, **4**, 121.

Kauffmann, G. and White, S.D.M. (1993). *MNRAS*, **261**, 92.

Kippenhahn, R. and Weigert, A. (1990). *Stellar Structure and Evolution*. Berlin: Springer-Verlag.

Kolb, E.W. and Turner, M.S. (1990). *The Early Universe*. Redwood City, Caifornia: Addison-Wesley Publishing Company.

Kulsrud, R.M. (1997). In *Critical Dialogues in Cosmology*, (ed. N. Turok), 328. Singapore: World Scientific.

Lacy, M., Miley, G., Rawlings, S., Saunders, R., Dickinson, M., Garrington, S., Maddox, S., Pooley, G., Steidel, C., Bremer, M.N., Cotter, G., van Oijk, R., Röttgering, H. and Warner, P. (1994). *MNRAS*, **271**, 504.

Le Fèvre, O., Ellis, R.S., Lilly, S.J., Abraham, R.G., Brinchman, J., Schade, D., Broadhurst, T.J., Colless, M., Crampton, D., Glazebrook, K., Hammer, F. and Tresse, L. (1997). In *The Hubble Space Telescope and the High Redshift Universe*, (eds. N.R. Tanvir, A. Aragón-Salamanca and J.V. Wall), 25. Singapore: World Scientific Publishing Co.

Lilly, S.J. (1988). *ApJ*, **333**, L161.

Lilly, S.J., Tresse, L., Hammer, F., Crampton, D. and Le Fèvre, O. (1995). *ApJ*, **455**, 108.

Linde, A.D. (1974). *Pisma Zh. Eksp. Teor. Fiz*, **5**, 32.

Longair, M.S. (1997). In *Critical Dialogues in Cosmology*, (ed. N. Turok), 285. Singapore: World Scientific.

Longair, M.S. (1997). *High Energy Astrophysics. Vols. 1 & 2* (corrected 2nd editions). Cambridge: Cambridge University Press.

Lynden-Bell, D. (ed.) (1994). *Cosmical Magnetism*. Dordrecht: Kluwer Academic Publishers.

Ostriker, J.P. and Hausman, M.A. (1977). *ApJL*, **217**, 125.

Ostriker, J.P. and Peebles, P.J.E. (1973). *ApJ*, **186**, 467.

Pascarelle, S.M., Windhorst, R.A., Keel, W.C. and Odewahn, S.C. (1996). *Nature*, **383**, 45.

Parker, E.N. (1997). In *Critical Dialogues in Cosmology*, (ed. N. Turok), 309. Singapore: World Scientific.

Peebles, P.J.E. (1993). *Principles of Physical Cosmology*. Princeton: Princeton University Press.

Rees, M.J. (1994). In *Cosmical Magnetism*, (ed. D. Lynden-Bell), 155. Dordrecht: Kluwer Academic Publishers.

Rees, M.J. (1995). *Perspectives in Astrophysical Cosmology*. Cambridge: Cambridge University Press.

Sakharov, A.D. (1967). *JETP Letters*, **5**, 24.

Schade, D. (1997). In *The Hubble Space Telescope and the High Redshift Universe*, (eds. N.R. Tanvir, A. Aragón-Salamanca and J.V. Wall), 199. Singapore: World Scientific Publishing Co.

Spinrad, H., Dey, A. and Graham, J.R. (1995). *ApJ*, **438**, L51.

Strömgren, B. (1934). *ApJ*, **79**, 460.

Tayler, R.J. (1994). *The Stars: their Structure and Evolution*. Cambridge: Cambridge University Press.

Taylor, J. and Cordes, J.M. (1993). *ApJ*, **411**, 674.

Thorne, K.S., Price, R.H. and Macdonald, D.A. (1986). *Black Holes: the Membrane Paradigm*. New Haven: Yale University Press.

Tinsley, B.M. and Gunn, J.E. (1976). *ApJ*, **206**, 525.

Vallée, J.P. (1997). *Fund. Cosm. Phys.*, **19**, 1.

Weinberg, S. (1997). In *Critical Dialogues in Cosmology*, (ed. N. Turok), 195. Singapore: World Scientific.

von Weizsacher, C.F. (1947). *Zeit. Ap*, **24**, 181.

Wheeler, J.A. (1977). In *Foundational Problems in the Special Science*, (eds. R.E. Butts and J. Hintikka), 3. Dordrecht: D. Reidel Publishing Co.

White, S.D.M. and Rees, M.J. (1978). *MNRAS*, **183**, 341.

Index

The principal references to topics and physical processes of special importance in this study have been highlighted in **bold-faced type**.

Printing: Mercedesdruck, Berlin
Binding: Buchbinderei Lüderitz & Bauer, Berlin